MAGNETIC FIELDS AND STAR FORMATION

Theory Versus Observations

Editors:

A.I. GÓMEZ DE CASTRO
Instituto de Astronomía y Geodesia (CSIC-UCM)
Universidad Complutense de Madrid, Spain

M. HEYER
University of Massachussetts, Amherst, MA, U.S.A.

E. VÁZQUEZ-SEMADENI
Centro de Radioastronomía y Astrofísica, UNAM, Morelia, Michoacán 58089, Mexico

R. REBOLO
Instituto de Astrofísica de Canarias, La Laguna, Tenerife, Spain

M. TAGGER
Service d'Astrophysique, CEA-Saclay, Gif-sur-Yvette, France

and

R.E. PUDRITZ
McMaster University, Hamilton, ON, Canada

Reprinted from *Astrophysics and Space Science*
Volume 292, Nos. 1–4, 2004

KLUWER ACADEMIC PUBLISHERS
DORDRECHT / BOSTON / LONDON

Library of Congress Cataloging-in-Publication Data

Magnetic fields and star formation: theory versus observations/edited by Ana I. Gómez de Castro ... [et al.].

p. cm.

ISBN 1-4020-2159-3 (hardback : acid-free paper)

1. Cosmic magnetic fields. 2. Stars–Formation. 3. Cosmic magnetic fields–Observations. I. Gómez de Castro, Ana I.

QB462.8.M36 2004
523.01'88–dc22

2004050738

Published by Kluwer Academic Publishers,
P.O. Box 17, 3300 AA Dordrecht, The Netherlands.

Sold and distributed in the U.S.A. and Canada
by Kluwer Academic Publishers,
101 Philip Drive, Norwell, MA 02061, U.S.A.

In all other countries, sold and distributed
by Kluwer Academic Publishers,
P.O. Box 322, 3300 AH Dordrecht, The Netherlands.

Printed on acid-free paper

Printed in the Netherlands

TABLE OF CONTENTS

INTERACTION DISK-STELLAR MAGNETOSPHERE

PREFACE

Magnetic fields play a key role in the physics of star formation at all scales: from the formation of the large complexes of molecular clouds to the channelling of the accretion flow onto pre-main sequence stars. The plasma physics involved in these problems is non-linear and very complex, requiring the development of large numerical codes as well as analytical studies of a well selected sample of enlightening problems. An additional difficulty is that the detection and study of magnetic fields is not easy from the observational point of view and therefore, theoretical models cannot be easily constrained. In the week from April 21st to 25th of 2003, a meeting was held in the Campus of the Universidad Complutense de Madrid (Spain) to join theoretical and observational efforts to address these issues. The objective was to define a set of relevant problems for the physics of star formation that can be properly addressed with the current or near future instruments. These proceedings summarize the results of this *intensive week of work*. This book is intended to represent a comprehensive review of our current knowledge on the subject, as well as an updated accounting on the ideas and thoughts of the people working in the field of Star Formation.

The contributions are presented in six chapters which correspond to the six fundamental issues (sessions) on which the discussion was focussed during the workshop:

The physics of turbulence in the Interstellar Medium (ISM).
Editor: Mark Heyer
The formation of structure in the ISM.
Editor: Enrique Vázquez-Semadeni
The formation of stars within dense cores of molecular gas.
Editor: Rafael Rebolo
The physics of accretion disks.
Editor: Michel Tagger
The physics of outflows and their interaction with the ISM.
Editor: Ralph E. Pudritz
The interaction between stellar magnetosphere and accretion disk.
Editor: Ana I. Gómez de Castro

Each chapter has an editor who has written a comprehensive introductory summary with the inputs from the contributions and her/his own thoughts about the subject. The editors also acted as *conductors* of the corresponding sessions

Astrophysics and Space Science **292:** 1–3, 2004.

during the workshop. The name *conductor* was chosen to make explicit that we understood the workshop as the performance of a 'world-wide' orchestra with renowned experts, promising young talents and best professionals. The conductor was in charge of selecting the contributed talks for her/his section and making the players: theorists, observers, numerical people and instrumentalist to play together in the best possible harmony.

Conductors also acted as editors of the articles presented in their sessions. They were in charge of the refereeing process, either searching for referees among the specialists in the field or acting as referees themselves. All the articles (from poster papers to review talks) have been refereed. The editors have asked some authors to make modifications in their papers in order to guarantee consistency in formulation and notation throughout the entire book. The objective of the workshop was to identify at least one *very important* problem per session which could be actually addressed in the near future. The task was not easy. Personally, I think that our main achievement was to reach a consensus about the major challenges for the future and be able to spell them out clearly. In this sense, it became apparent the enormous potential of numerical codes not only for the study of complex physical processes but also for designing observing programs suitable to test our understanding of those processes. The best examples were provided in relation to the study of the ISM, e.g., turbulence and fragmentation. Observations provide two-dimensional projections on the plane of the sky (integrated within the telescope beam) of intrinsic three-dimensional distributions of the observables: density, temperature, velocity or magnetic fields. Any meaningful interpretation of the observations requires the recovery of 3D distributions from 2D projections; numerical simulations represent a new and powerful tool for this purpose. Moreover, they allow to determine what physics can be really studied from the observations, e.g., the sensitivity of the observations to the basic parameters of the theory.

Another important aspect is the relevance of the microphysics and, in particular, of dissipative processes associated with the interaction between magnetic fields and matter. Basically, all the chapters (and most of the contributions) address this issue. Unfortunately, our understanding of fundamental processes, as magnetic reconnection, is rather limited.

Currently, our possibilities to actually measure magnetic field strengths are constrained to scales either above some 0.1 pc or associated with the stellar field. Accretion disks, outflows and their engines are not accessible to direct field measurements. The advent of the new generation of radio interferometers, notably ALMA or the VLA-EVLA, may allow direct field measurements either through paramagnetic molecules or through the study of polarized non-thermal radiation. However, these instruments will be unable to probe the wind engine itself since it is located within 1 A.U. (or several milliarcseconds at the distance of the nearby Taurus cloud). Therefore, the role of fields will have to be studied indirectly through the analysis of the kinematics and thermal properties of plasma heated by field dissipation. High-resolution ultraviolet spectroscopy and monitoring programs are instrumental for this purpose.

The workshop was held at the end of April 2003 during turbulent times in international politics. In spite of that, more than 100 scientists from all over the planet met to discuss the relevance of magnetic fields on the formation of stars (obviously some political discussions came in too). This workshop was also an exercise of world-wide human interaction to understand, among many other issues, how the star that shines above all of us and the planet were we all live, were formed.

Finally, I would like to acknowledge the support from the Scientific Organizing Committee and the Local Organizing Committee in the organization of this conference and, very especially, Francisco (Paco) Colomer who has been the responsible of our web page and, therefore, of the interaction among 120 people, during more than 1 year and Julia Coloma who has taken care of the accounting. The workshop was orginized through the collaboration of three Spanish institutions: Universidad Complutense de Madrid, Consejo Superior de Investigaciones Científicas and Observatorio Astronómico Nacional. Partial financial support has been provided by the Universidad Complutense de Madrid, the Ministry of Science and Technology of Spain and the Consejo Superior de Investigaciones Científicas.

Madrid, November 20th 2003
Ana Inés Gómez de Castro

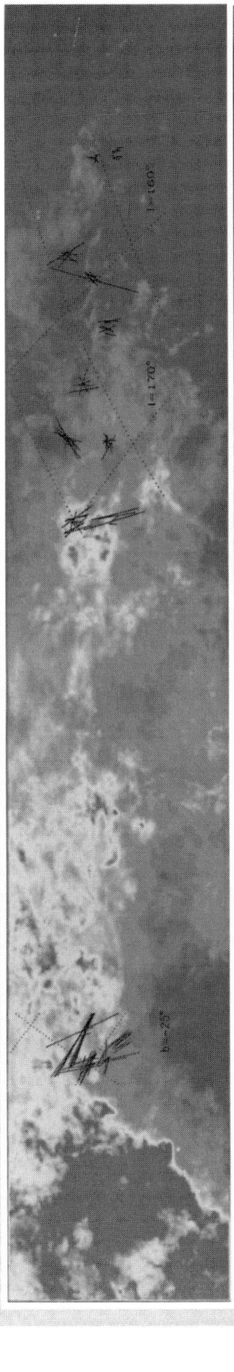

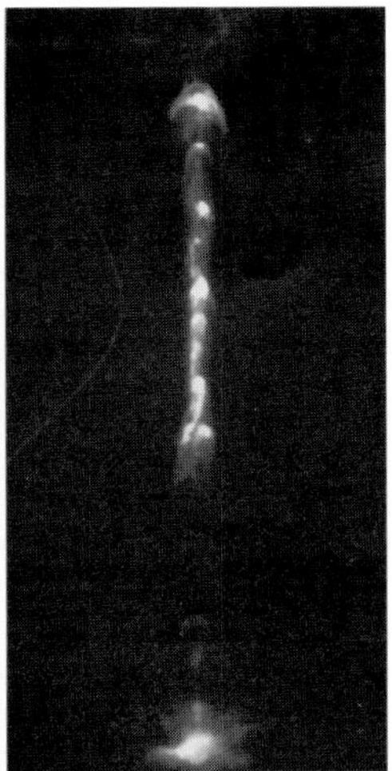

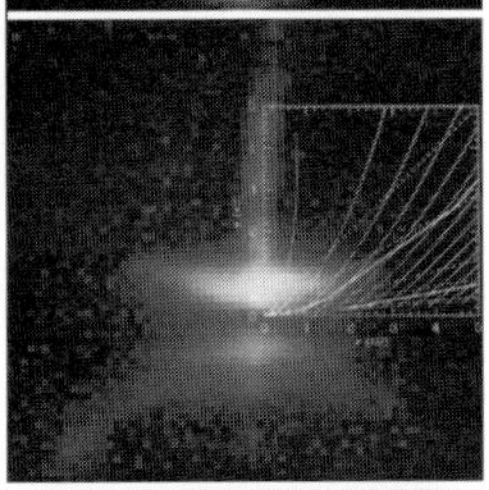

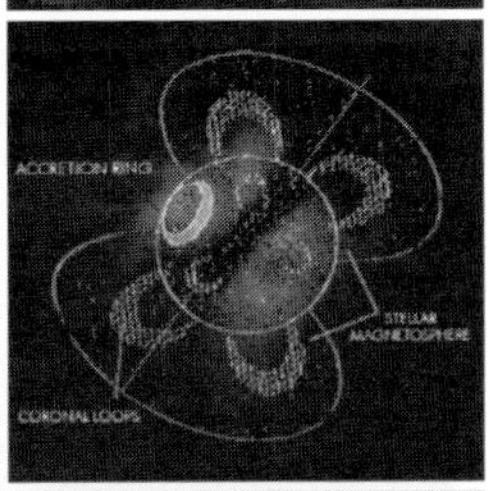

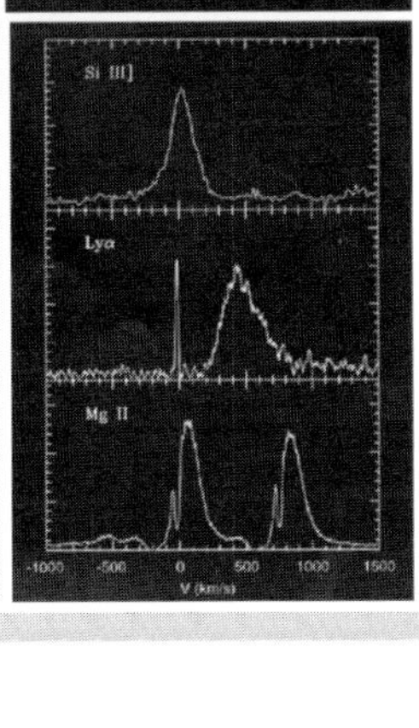

INTERNATIONAL WORKSHOP

ON

MAGNETIC FIELDS

AND

STAR FORMATION

THEORY VERSUS OBSERVATIONS

Madrid (Spain), April 21st-25th 2003

Magnetic fields play a key role in the physics of star formation at al scales: from the formation of the large complexes of molecula clouds to the channelling of the accretion flow onto pre-mai sequence stars. The plasma physics required to solve thes problems is non-linear and very complex requiring the developmen of large numerical codes as well as analytical studies of a wel selected sample of enlightening problems. An additional difficulty i that the detection and study of magnetic fields is not easy from th observational point of view and therefore, theoretical models canno be easily constrained. The aim of this workshop is to join theoretica and observational efforts to define a set of relevant problems for th physics of star formation that can be properly studied with th current or near future instruments.

(URL: www.oan.es/MFSF/)

SOC:
A.I. Gómez de Castro (co-chair),
R.E. Pudritz (co-chair),
R. Bachiller, J. Cernicharo, T. Montmerle,
E. Zweibel, M. Tagger, E. Vázquez-Semadeni,
C.Eiroa, C. Ferro-Fontán,
J. Torrelles, G. Anglada

LOC:
J. Coloma (Secretary), F. Colomer,
A.I. Gómez de Castro, M. Franqueira,
J. Santiago, M. Tafalla, E. Verdugo

CSIC

Conference Hall - Facultad de Informática-Paraninfo- Ciudad Universitaria

c/Juan del Rosal, n. 8 - 28040 Madrid (Spain)

TURBULENCE IN THE ISM

TURBULENCE IN THE STAR-FORMING INTERSTELLAR MEDIUM: STEPS TOWARD CONSTRAINING THEORIES WITH OBSERVATIONS

MARK HEYER[1] and ELLEN ZWEIBEL[2]

[1]*Department of Astronomy, University of Massachusetts, Amherst, U.S.A.*
[2]*Department of Astronomy, University of Wisconsin, Madison, U.S.A.;*
E-mail: heyer@astro.umass.edu

Abstract. Increasingly sophisticated observational tools and techniques are now being developed for probing the nature of interstellar turbulence. At the same time, theoretical advances in understanding the nature of turbulence and its effects on the structure of the ISM and on star formation are occurring at a rapid pace, aided in part by numerical simulations. These increased capabilities on both fronts open new opportunities for strengthening the links between observation and theory, and for meaningful comparisons between the two.

Keywords: interstellar medium, turbulence

1. Introduction

Supernova remnants, expanding HII regions, rotational shear, and spiral arm shocks contribute to the interstellar gaseous maelstrom within galaxies. Even in the high density regime of the interstellar medium, where molecules have condensed and gravity plays an increasingly larger role in the dynamics, the flow of gas is chaotic. In the dense, highly localized cores of giant molecular clouds, self-gravity may overwhelm the countering internal pressure, enabling the generation of newborn stars and stellar clusters. The initial conditions of such protostellar regions are likely set by the overlying turbulent gas. Therefore, understanding the critical process of star formation in galaxies requires more accurate descriptions of interstellar turbulence, especially as it relates to the formation of molecular clouds and within molecular gas itself. Such descriptions demand both insightful theories and relevant observations that confront and constrain physical models.

The study of turbulence within the cold, dense interstellar has greatly benefited from the interplay between theory (analytical and numerical simulations) and observations. Analytical efforts target specific physical processes and typically predict dimensional relationships that may be measured by the observer (Kolmogorov, 1941; Goldreich and Sridhar, 1995; Boldyrev, 2002, papers by Chandran and Boldyrev in these Proceedings). Yet, purely analytical methods do not follow the evolutionary state of a turbulent medium without making overly simplistic assumptions nor do they readily account for complex gas distributions driven by advection.

Astrophysics and Space Science **292:** 9–16, 2004.

The sophistication and dynamic range of the hydrodynamic and magnetohydrodynamic numerical simulations of interstellar turbulence have greatly expanded in recent years owing to ever increasing computational capabilities. The simulations do follow the time evolution of an interstellar volume and provide a three-dimensional spatial view of the gas distribution and kinematics. The primary limitations of current simulations are the low kinetic and magnetic Reynolds numbers relative to expected ISM values (Zweibel et al., 2003), together with the impracticality of using the most highly resolved, and hence highly resource intensive, computations to carry out parameter studies. Nevertheless, the fields of interstellar turbulence and dynamics, and their relationship to star formation, currently are largely driven by the results from these computational efforts [see Mac Low and Klessen (2004) for a recent review].

Observational capabilities have also increased several fold over the last decade. Millimeter wave interferometers can now routinely mosaic primary fields of view to generate high resolution ($<5''$) images of molecular line and continuum emissions from star forming regions. This capability enables investigators to define relationships between compact cores within the field and provides a census of protostellar outflows (Testi and Sargent, 1998; Williams et al., 2003). Sensitive focal plane arrays on single dish, filled aperture telescopes enable the collection of high spatial dynamic range images (see Figure 1). These data reveal both the varying textures of molecular line emissions and environmental connections to the more widely distributed atomic and ionized gas components of the ISM. Moreover, since turbulence couples motions on many spatial scales within the inertial range, the analyses of wide field, spectroscopic images of the molecular gas play an essential role in constraining models of turbulent gas flow.

In this summary of the two sessions on Turbulence in the Interstellar Medium, we will describe important requirements in comparing observations with models of

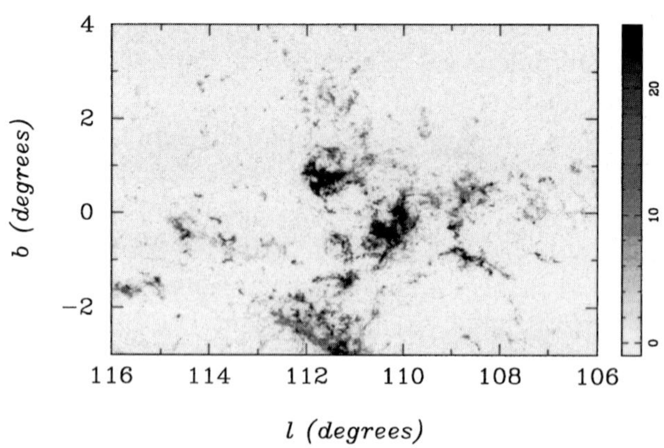

Figure 1. An image of velocity integrated CO J = 1–0 emission from a section of the Perseus spiral arm of the Milky Way (Heyer et al., 1998). The image shows giant molecular clouds, compact globules and flocculent features.

turbulence and the need for future observations and theoretical advances to make progress in this critical field.

2. A Common Platform

A pre-requisite for any useful dialogue between observers and theorists is a common description of the ISM. The ISM and the output from computational studies are best described by three-dimensional scalar (e.g., density, temperature) and vector (e.g., velocity, magnetism) *fields*. However, spectroscopic observations do not directly yield such 3D fields but rather, spectra of a given molecular or atomic transition at different positions in the plane of the sky. The spectrum is a complex integration of scalar and vector field quantities along the line of sight and in principle provides a valuable diagnostic to these fields. In practice, it is very difficult to reliably extract this information. For a turbulent medium, there is a complex re-ordering of the information into the spectrum that precludes an accurate re-construction of the original fields. In addition, spectroscopic observations are subject to selection effects such as line excitation, opacity, and instrumental noise. These biases act to mask an uncertain fraction of the volume from the observer and are not generally considered when parameterizing 3D model fields (e.g., power spectra of the velocity field) or developing predictions. Since it is not possible to uniquely recover the intrinsic fields of density or velocity from the observed spectra, it is necessary to compute synthetic spectra from the model fields with the same observational biases to achieve a common platform to compare with real measurements. Many investigators have recognized this requirement and have developed methods to transform model fields into synthetic spectra taking into account radiative transfer and line excitation (Falgarone et al., 1994; Padoan et al., 1998; Brunt and Heyer, 2002, Padoan et al., 2003).

Diagnostics of turbulence, which are not based on spectroscopy yield valuable complementary information, but are subject to their own uncertainties. Following the original suggestion by Chandrasekhar and Münch (1950) that the statistics of stellar extinction could be used to probe the underlying density distribution of turbulent interstellar gas, Elmegreen (1997) discussed the distribution of column densities in a fractal model, and Padoan and Nordlund (1999) and Ostriker et al. (2001) did the same for numerical models of supersonic MHD turbulence. As the latter set of authors have discussed, spatially and dynamically distinct structures are frequently projected on top of one another, possibly suggesting an intrinsic limitation on the use of such statistics for extracting volumetric quantities.

3. Calibrated Spatial Statistics

Images of molecular line emission from interstellar clouds exhibit a wide range of structural features and textures (see Figure 1). The measured spatial variability arises from the various processes that modify the gas phase (UV fields from massive stars), compress (shocks from outflows, HII regions, or turbulence), redistribute the gas (turbulent eddys), or modulate chemical abundances. While it is tempting to investigate interesting, targeted features, such phenomenological efforts do not often generate quantitative constraints to theoretical descriptions of the ISM. Within a turbulent medium, where structure can be generated by advection of material through the velocity field, localized, high-contrast objects may be short-lived and therefore, do not evolve to higher density configurations that lead to the formation of stars (see Ballesteros-Paredes et al., 1999).

Spatial statistics provide a condensed, unbiased description of observational and model data and are an essential tool for the study of interstellar turbulence. A frequently used example is the power spectrum of the velocity field, which is often further parameterized by the slope of the power law fit. It is essential that any proposed metric derived from the observations reflect the true statistics of interstellar quantities. One cannot assume that there is a direct equivalence given the complexities of the projection described above. Therefore, it is necessary to calibrate the metric against fields with known statistics *prior* to its application on real data. This calibration enables the following results.

- Demonstrate sensitivity to varying true field statistics. A derived metric that is invariant under varying conditions is simply not useful to compare with models.
- Empirically define the algebraic relationship between the true and measured values as this can be distorted in the projection process.
- Evaluate the role of noise, resolution, sample size, and selection effects on the results.

Several semi-calibrated methods have been developed to estimate the scale dependence of the velocity dispersion from spectroscopic data cubes as parameterized by the power law slope of the second-order structure function, or equivalently, the energy spectrum ($E(k) \sim k^{-\beta}$) (Miesch et al., 1999; Lazarian and Pogosyan, 2000; Brunt and Heyer, 2002). However, the accuracy of these methods is not yet sufficient ($\sigma_\beta \sim 0.3$) to even fully distinguish between incompressible Kolmogorov turbulence ($\beta = 5/3$) and shock dominated Burgers turbulence ($\beta = 2$).

In comparison to the use of velocity statistics, diagnostics of turbulence in molecular clouds based on magnetic field measurements are in their infancy. Polarimetric maps of star-forming regions in the far-infrared probe the orientation of the magnetic field in regions of high extinction (Hildebrand et al., 2000). The dispersion of the orientation angle about the mean contains information about the fieldstrength (Chandrasekhar and Fermi, 1953), as well as the fluctuation spectrum, but corrections for line of sight and angular averaging can be substantial (Heitsch et al., 2001; Ostriker et al., 2001). This is demonstrated in Figure 2. Detailed statistical analysis of polarimetric observations of the type emulated in the figure requires more highly resolved spatial sampling than is currently available, but would be quite feasible, for example, with measurements at 100 μm made by SOFIA.

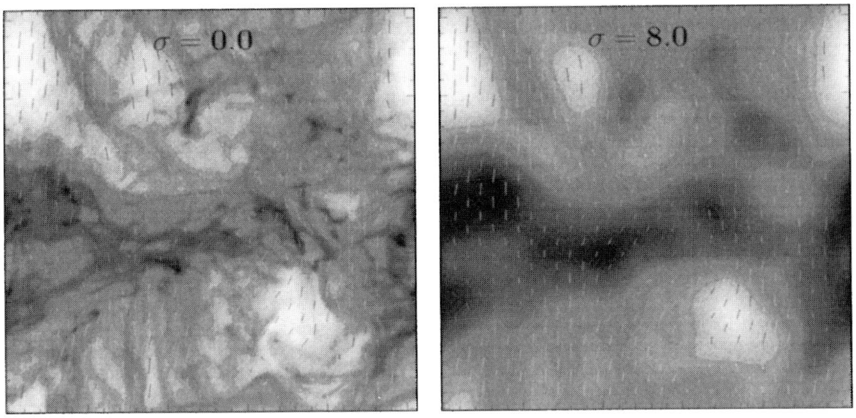

Figure 2. Surface density plots and overlaid polarization vectors for a 256^3 simulation of MHD turbulence driven on global scales. The sonic and Alfvén Mach numbers are 10 and 3.2, respectively. The left panel shows unsmoothed data, and the right panel shows the effect of smoothing with a Gaussian filter with $\sigma = 8$ pixels. After Heitsch et al. (2001).

4. Summary and Future Efforts

The discussions that followed the presentations in this session focussed upon the most essential measurements to conduct in order to constrain models and augment our current understanding of interstellar turbulence.

Mass to Magnetic Flux Ratio—The magnetic field is expected to be frozen to the plasma component of interstellar gas, except on the smallest scales. Unless the Alfvén Mach number of the turbulence is much less than unity, which is unlikely to be the case, one expects a statistical correlation between fieldstrength and gas density. Parameterizing the correlation by a power law exponent $\alpha \equiv d \ln B / d \ln \rho$, one expects $\alpha \sim 2/3$ for very weak fields, and $\alpha \sim 1/2$ for self gravitating clouds supported by moderately strong fields. Published observations of the $B - \rho$ relation suggest that the power law scaling is more of an upper envelope than a tight relationship, but the number of data points is small, and corrections due to line of sight averaging and unresolved angular structure could be substantial. But the absence of a $B - \rho$ relation, if confirmed, is evidence for anomalously strong diffusion in the ISM, possibly turbulent in nature (Zweibel, 2002; Fatuzzo and Adams, 2002; Kim and Diamond, 2002; Heitsch et al., 2001).

The mass to magnetic flux ratio M/Φ is the global counterpart of the density-fieldstrength relation. It measures the degree to which the evolution of a self gravitating mass element of the ISM is regulated by its magnetic field. If M/Φ is above the critical value $(M/\Phi)_c \sim 0.2G^{-1/2}$, the magnetic field cannot prevent collapse.

Measuring the degree of criticality of dense interstellar clouds is necessary in order to determine the dynamical role of magnetic fields in star formation. Observations to date suggest that this ratio, while not far from critical, generally exceeds the critical value (Crutcher, 1999 and these Proceedings; Bourke et al., 2001; Crutcher et al., 2003). Possibly this again is evidence for turbulent diffusion, although these measurements too are subject to small number statistics, line of sight averaging, and resolution effects.

In order to make further progress, it is imperative to obtain direct, reliable observations of the magnetic field strength. In weakly ionized regions this requires measuring Zeeman splitting (difficult), and, in regions with substantial electron density, Faraday rotation. Measurements of the polarization of the far infrared continuum and of radio frequency spectral lines furnish information about the orientation of the field, but not the strength, except indirectly. Reliable measurements of the gas density are required as well; in too many cases one has only column densities, not volume densities. Observing more sources, even if only to obtain upper limits, would improve the statistical reliability of the results. Papers by Crutcher and by Heiles in these Proceedings discuss the situation in molecular and neutral atomic gas, respectively.

It is equally imperative that theoretical models of the B–ρ relation develop observationally testable predictions. For models of turbulent diffusion, this might involve characterizing the turbulent velocity fields, or predicting the structure function for the magnetic fieldstrength or the Stokes parameters. Quantifying the effects of numerical diffusion is a separate but equally important issue in simulations of magnetic field evolution in turbulent media.

Degree of Intermittency—Intermittency describes the filling factor or sparseness of the spatial and temporal dissipation of turbulent energy. Signatures to intermittency include non-gaussian probability density functions (PDFs) of velocity increments (Lis et al., 1996; Miesch et al., 1999; Pety and Falgarone, 2003) and the curvature of the scaling exponent, $\zeta(q)$, with q, where

$$S_q(L) = \langle|\delta v|^q\rangle \propto L^{\zeta(q)}$$

is the generalized structure function and δv refers to the velocity difference between two points in a volume separated by size scale, L. It would be useful to understand the dissipation process as this is a critical step to the formation of stars. At what scales and through which modes is turbulent energy dissipated? Is this energy replenished? If so, what are the primary sources of energy injection to the interstellar cloud (see Miesch and Bally, 1994)? The structure function of velocity fluctuations provides a statistical road-map of energy flow within a volume. Principal Component Analysis (Brunt et al., 2003) and Velocity

Channel Analysis (Lazarian and Pogasyan, 2000) can recover the low-order moments of the structure function. However, intermittent phenomena are more readily measured within the higher-order moments. The complete evaluation of the generalized structure function requires the development of observations and analysis tools that are sensitive to these high-order ($q > 2$) velocity fluctuations.

Density Contrast—Collisions of super-Alfvenic gas streams produce shocks and filamentary density distributions. Such filaments are prominent features within the density fields generated by numerical simulations of compressible interstellar turbulence. The measured angular distribution of molecular line emission is not so readily characterized by filaments. Rather, there are filaments, compact cores, flocculent clouds, and diffuse features. Where are the signatures of hydrodynamic shocks? Are these lost in the projection of the fields or in the observational noise? An important metric to develop is a filamentary index which could describe the fraction of mass contained in filaments or sheets both in three-dimensional density fields and projected images of gas column density.

As stated in the first section of this article, analytical theories of turbulence in the ISM led to a few testable predictions, but do not describe the wealth of detail that observations of the ISM reveal. This made it difficult, in the past, to argue that we understand very basic problems, such as the nature of energy injection, the primary mechanisms for cloud formation, the dynamical role of magnetic fields, and the origin of the observed IMF. Numerical simulations offer a threefold opportunity in this regard. First of all, numerical simulations can be used to study universal problems in data analysis and interpretation, such as recovery of 3D distributions from 2D projections, the role of finite angular resolution, and the role of noise. Studies of this type can be valuable in the interpretation of the observations, and perhaps can lead to new methods of data analysis. Second, it is possible, through parameter studies, to determine the sensitivity of the basic output—the density, velocity, and magnetic fields in a turbulent simulation—to changes in the basic physical input. For example, if turbulence is driven at a particular scale, is that scale imprinted on the output fields? Does it make a difference whether the driving is through addition of energy, or momentum? What is the effect of underlying rotational shear? Finally, the first and second avenues of investigation must be combined to demonstrate the sensitivity, or lack thereof, of observable quantities to physical processes and input parameters. This makes it possible to answer questions such as: What observations are required to determine whether the turbulence in a molecular cloud is driven or decaying? Is there a way to determine, short of direct measurement, whether the magnetic field in a cloud is weak or strong? What is the physical significance of the observed linewidth—size and column density—size relations? We are challenged by these and other questions regarding interstellar turbulence. The continued creative tension between observations and theory provides motivation for future studies to address these critical issues.

Acknowledgements

This work is supported by NSF grant AST 01-00793 to the Five College Radio Astronomy Observatory and NSF grant AST 0098701 to the University of Wisconsin.

References

Ballesteros-Paredes, J., Vazguez-Semadeni, E. and Scalo, J.M.: 1999, *ApJ* **515**, 286.
Boldyrev, S.: 2002, *ApJ* **569**, 841.
Bourke, T.L., Myers, P.C., Robinson, G. and Hyland, A.R.: 2001, *ApJ* **554**, 916.
Brunt, C.M. and Heyer, M.H.: 2002, *ApJ* **556**, 276.
Brunt, C.M., Heyer, M.H., Vazquez-Semadeni, E. and Pichardo, B.: 2003, *ApJ* **595**, 824.
Chandrasekhar, S. and Fermi, E.: 1953, *ApJ* **118**, 113.
Chandrasekhar, S. and Münch, G.: 1950, *ApJ* **112**, 380.
Crutcher, R.M.: 1999, *ApJ* **520**, 706.
Crutcher, R.M., Heiles, C. and Troland, T.H.: 2003, in: E. Falgarone and T. Passot (eds.), *Turbulence & Magnetic Fields in Astrophysics*, Springer, p. 155.
Elmegreen, B.G.: 1997, *ApJ* **477**, 196.
Falgarone, E., Lis, D., Phillips, T.G., Pouquet, A., Porter, D.H. and Woodward, P.R.: 1994, *ApJ* **436**, 72.
Fatuzzo, M. and Adams, F.C.: 2002, *ApJ* **570**, 210.
Goldreich, P. and Sridhar, S.: 1995, *ApJ* **438**, 763.
Heitsch, F., Zweibel, E.G., Mac Low, M.M., Li, P. and Norman, M.L.: 2001, *ApJ* **561**, 800.
Heyer, M.H., Brunt, C., Snell, R.L., Howe, J.E., Schloerb, F.P. and Carpenter, J.M.: 1998, *ApJS* **115**, 241.
Hildebrand, R.H., Davidson, J.A., Dotson, J.L., Dowell, C.D., Novak, G. and Vaillancourt, J.E.: 2000, *PASP* **112**, 1215.
Kim, E.-J. and Diamond, P.H.: 2002, *ApJ* **578**, L113.
Kolmogorov, A.N.: 1941, *Dokl. Akad. Nauk. SSSR* **30**, 301.
Lazarian, A. and Pogasyan, D.: 2000, *ApJ* 537.
Lis, D.C., Pety, J., Phillips, T.G. and Falgarone, E.: 1996, *ApJ* **463**, L623.
Mac Low, M.-M. and Klessen, R.S.: 2004, *Rev. Mod. Phys.* **76**, 125.
Miesch, M.S. and Bally, J.: 1994, *ApJ* **429**, 645.
Miesch, M.S., Scalo, J. and Bally, J.: 1999, *ApJ* **524**, 895.
Ostriker, E.C., Stone, J.M. and Gammie, C.F.: 2001, *ApJ* **546**, 980.
Padoan, P., Goodman, A.A. and Juvela, M.: 2003, *ApJ* **588**, 881.
Padoan, P., Juvela, M., Bally, J. and Nordlund, A.: 1998, *ApJ* **504**, 300.
Padoan, P. and Nordlund, A.: 1999, *ApJ* **526**, 279.
Pety, J. and Falgarone, E.: 2003, *A&A* **412**, 417.
Testi, L. and Sargent, A.A.: 1998, *ApJ* **508**, L91.
Williams, J.P. Plambeck, R.L. and Heyer, M.H.: 2003, *ApJ* **591**, 1025.
Zweibel, E.G.: 2002, *ApJ* **567**, 962.
Zweibel, E.G., Heitsch, F. and Fan, Y.: 2003, in: E. Falgarone and T. Passot (eds.), *Turbulence & Magnetic Fields in Astrophysics*, Springer, p. 101.

A REVIEW OF THE THEORY OF INCOMPRESSIBLE MHD TURBULENCE

BENJAMIN D.G. CHANDRAN
Center for Magnetic Reconnection Studies, Department of Physics & Astronomy, University of Iowa; E-mail: benjamin-chandran@uiowa.edu

Abstract. This brief review provides an introduction to key ideas in the theory of incompressible magnetohydrodynamic (MHD) turbulence.

Keywords: magnetohydrodynamics, MHD turbulence

1. Introduction

Magnetohydrodynamic (MHD) turbulence has been studied in a number of different parameter regimes, with important contributions from many authors, including (Balsara and Pouquet, 1999; Boldyrev et al., 2002; Bhattacharjee and Ng, 2001; Cho and Lazarian, 2003; Cho et al., 2002; Cho and Vishniac, 2000; Dobrowolny et al., 1980; Galtier et al., 2000, 2002; Ghosh and Goldstein, 1997; Goldreich and Sridhar, 1995, 1997; Grappin et al., 1983; Higdon, 1984; Iroshnikov, 1963; Kinney and McWilliams, 1998; Kraichnan, 1965; Lithwick and Goldreich, 2001, 2003; Mac Low, 1999; Maron and Goldreich, 2001; Montgomery and Matthaeus, 1995; Montgomery and Turner, 1981; Milano et al., 2001; Muller and Biskamp, 2000; Ng and Bhattacharjee, 1996, 1997; Ossenkopf and Mac Low, 2002; Oughton et al., 1994; Passot and Vázquez-Semadeni, 2003; Politano et al., 1995; Pouquet et al., 1976; Schekochihin et al., 2002; Shebalin et al., 1983; Sridhar and Goldreich, 1994; Stone and Gammie, 1998; Ting et al., 1986; Vazquez-Semadeni et al., 2000; Vestuto et al., 2003; Winters et al., 2003; Zank and Matthaeus, 1993) This review focuses on incompressible turbulence in which the fluctuating velocities are comparable to or less than the Alfvén speed [defined in Eq. (7)]. It is assumed that there is a mean magnetic field $\mathbf{B}_0$ that is at least as strong as the turbulent magnetic-field fluctuations. Sections 2 and 3 review the MHD equations and MHD waves, and Section 4 describes wave–wave interactions. Section 5 reviews results on weak MHD turbulence, and Section 6 treats strong MHD turbulence. Section 7 describes the role of dynamic alignment in the decay of MHD turbulence.

2. Equations of Incompressible MHD

The governing equations of incompressible MHD are

$$\rho\left(\frac{\partial \mathbf{v}}{\partial t}+\mathbf{v}\cdot\nabla\mathbf{v}\right)=-\nabla\left(p+\frac{B^2}{8\pi}\right)+\frac{1}{4\pi}\mathbf{B}\cdot\nabla\mathbf{B}+\rho\nu\nabla^2\mathbf{v}, \tag{1}$$

Astrophysics and Space Science **292:** 17–28, 2004.

$$\frac{\partial \mathbf{B}}{\partial t} = \nabla \times (\mathbf{v} \times \mathbf{B}) + \eta \nabla^2 \mathbf{B}, \tag{2}$$

$$\nabla \cdot \mathbf{B} = 0, \tag{3}$$

$$\nabla \cdot \mathbf{v} = 0, \tag{4}$$

and

$$\rho = \text{ constant}, \tag{5}$$

where $\mathbf{B}$ is the magnetic field, $\mathbf{v}$ is the velocity, p is the thermal pressure, ρ is the density, and ν is the viscosity. In the limit $\eta \to 0$, Eq. (2) implies that field lines are like threads that are frozen to the plasma and advected at the plasma velocity $\mathbf{v}$.

Upon defining

$$\mathbf{B} = B_0\hat{\mathbf{z}} + \delta\mathbf{B}, \tag{6}$$

where $B_0\hat{\mathbf{z}}$ is the uniform mean field,

$$\mathbf{b} = \frac{\mathbf{B}}{\sqrt{4\pi\rho}} = v_A\hat{\mathbf{z}} + \delta\mathbf{b}, \tag{7}$$

where $v_A = B_0/\sqrt{4\pi\rho}$ is the Alfvén speed, the Elsasser variables

$$\mathbf{a}^{\pm} = \mathbf{v} \pm \delta\mathbf{b}, \tag{8}$$

and the total pressure

$$\Pi = \frac{p}{\rho} + \frac{b^2}{2}, \tag{9}$$

and upon neglecting the dissipation terms ($\nu = \eta = 0$), Eqs. (1) and (2) can be rewritten as

$$\frac{\partial \mathbf{a}^+}{\partial t} - v_A \frac{\partial \mathbf{a}^+}{\partial z} = -\nabla\Pi - \mathbf{a}^- \cdot \nabla\mathbf{a}^+, \tag{10}$$

and

$$\frac{\partial \mathbf{a}^-}{\partial t} + v_A \frac{\partial \mathbf{a}^-}{\partial z} = -\nabla\Pi - \mathbf{a}^+ \cdot \nabla\mathbf{a}^-. \tag{11}$$

In the absence of dissipation, Eqs. (1) through (5) possess three quadratic invariants: the energy $(1/2)\int(v^2 + b^2)\,d^3x$, the cross helicity $(1/2)\int \mathbf{v}\cdot\delta\mathbf{b}\,d^3\mathrm{x}$

$= (1/8) \int (\mathbf{a}^+ \cdot \mathbf{a}^+ - \mathbf{a}^- \cdot \mathbf{a}^-) d^3x$, and the magnetic helicity $\int \mathbf{A} \cdot \mathbf{B}\, d^3x$, where $\mathbf{A}$ is the magnetic vector potential. The magnetic helicity measures linkages of magnetic field lines and is associated with an inverse cascade (Pouquet et al., 1976). The cross helicity describes the excess of one fluctuation type ($\mathbf{a}^+$ or $\mathbf{a}^-$) over the other, and is associated with the phenomenon of dynamic alignment (Section 7).

3. Waves and Wave Packets in Incompressible MHD

If $\mathbf{v} = \delta\mathbf{b}$ (i.e., $\mathbf{a}^- = 0$), the solution to equation (10) for $\nabla \times \mathbf{a}^+$ is a wave that travels at speed v_A in the $-z$ direction. Similarly, if $\mathbf{v} = -\delta\mathbf{b}$ (i.e., $\mathbf{a}^+ = 0$), the solution to Eq. (11) for $\nabla \times \mathbf{a}^-$ is a wave that travels at speed v_A in the $+z$ direction. These wave solutions are valid for arbitrarily large wave amplitude (when either $\mathbf{a}^+ = 0$ or $\mathbf{a}^- = 0$).

There are two types of waves in incompressible MHD, the Alfvén wave and the slow wave, with polarizations illustrated in Figure 1. Both waves propagate along (or anti-parallel to) $\mathbf{B}_0$ at speed v_A. A 1D $\mathbf{a}^-$-Alfvén-wave-packet (with $\mathbf{v} = -\delta\mathbf{b}$) is illustrated in Figure 2. A cubical portion of a 3D wave packet is illustrated in Figure 3. In general, a wave packet is associated with a displacement $\Delta\mathbf{r}$ of a field line, as in Figures 2 and 3. For a small-amplitude wave packet that perturbs the field lines only slightly,

$$\Delta\mathbf{r} = \frac{1}{B_0} \int_{-\infty}^{\infty} \mathbf{B}_\perp(z)\, dz = \frac{\tilde{\mathbf{B}}_\perp(k_z = 0)}{B_0}, \tag{12}$$

where $\mathbf{B}_\perp$ is the component of $\delta\mathbf{B}$ in the xy-plane, and $\tilde{\mathbf{B}}_\perp$ is the Fourier transform of $\mathbf{B}_\perp$ in z. Thus, the field-line displacement is given by the $k_z = 0$ component of a wave packet. (Ng and Bhattacharjee, 1996) If an Alfvén wave packet's length

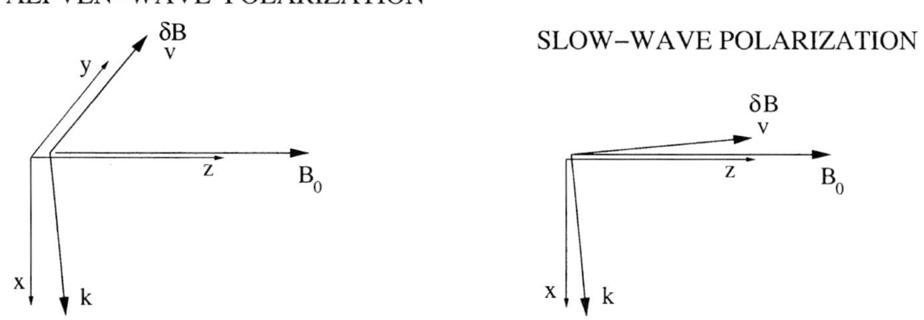

Figure 1. For an Alfvén wave, $\delta\mathbf{v}$ and $\delta\mathbf{B}$ are $\perp$ to both $\mathbf{B}_0$ and the wave vector $\mathbf{k}$. For a slow wave, $\delta\mathbf{v}$ and $\delta\mathbf{B}$ are $\perp$ to $\mathbf{k}$, but in the plane of $\mathbf{k}$ and $\mathbf{B}_0$. If $\mathbf{k}$ is nearly $\perp$ to $\mathbf{B}_0$, then $\delta\mathbf{v}$ and $\delta\mathbf{b}$ are nearly along $\mathbf{B}_0$ for a slow wave.

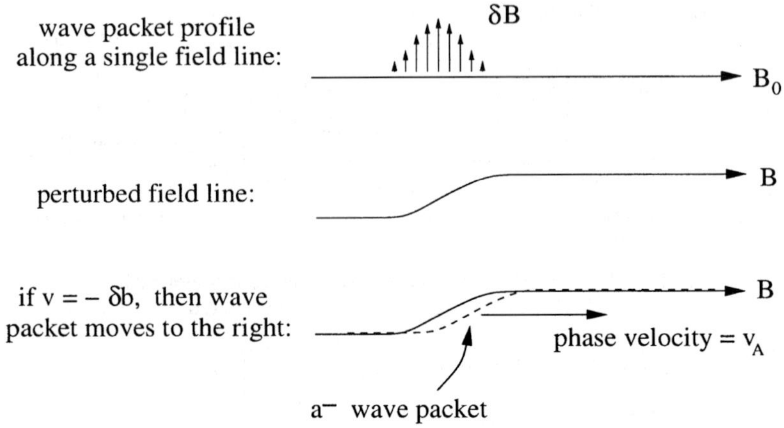

Figure 2. The profile of a 1D Alfvén wave packet.

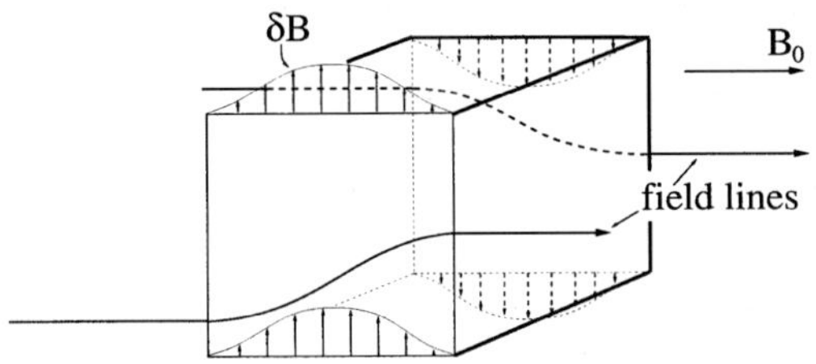

Figure 3. An example of a cubical portion of a 3D wave packet.

along the field is $\lambda_\parallel$, its width across the field is $\lambda_\perp$, and its amplitude is $B_{\lambda_\perp}$, then Eq. (12) implies that

$$\Delta r \sim \frac{\lambda_\parallel B_{\lambda_\perp}}{B_0} \sim \frac{\lambda_\parallel a_{\lambda_\perp}}{v_{\mathrm{A}}}, \tag{13}$$

where $a_{\lambda_\perp}$ is the wave-packet amplitude in Elsasser variables.

4. Wave–Wave Interactions in Incompressible MHD

For the nonlinear $\mathbf{a}^\pm \cdot \nabla \mathbf{a}^\mp$ terms in Eqs. (10) and (11) to be nonzero, both $\mathbf{a}^+$ and $\mathbf{a}^-$ fluctuations must be present at the same point in space. Wave–Wave interactions can thus be thought of as collisions between oppositely directed wave packets (Kraichnan, 1965). When $\mathbf{a}^+$ and $\mathbf{a}^-$ wave packets collide, to lowest order in wave amplitude each wave packet follows the field lines associated with the other wave packet (Sridhar and Goldreich, 1994; Maron and Goldreich, 2001). For example, before a collision the value of $\mathbf{a}^-$ remains constant at a point that moves along $\mathbf{B}_0$

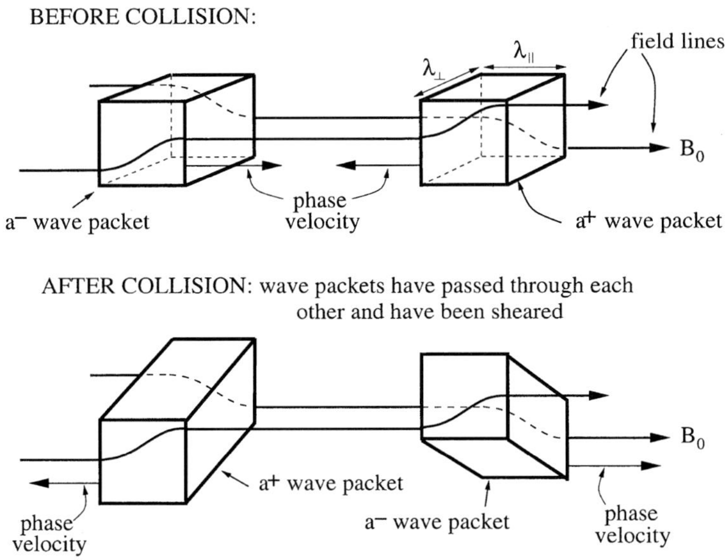

Figure 4. During a wave-packet collision, the $\mathbf{a}^-$ ($\mathbf{a}^+$) packet follows the field lines of the $\mathbf{a}^+$ ($\mathbf{a}^-$) packet.

at speed v_A. During the collision, the value of $\mathbf{a}^-$ remains approximately constant at a point that moves at speed $\sim v_A$ along the field lines formed from the sum of $\mathbf{B}_0$ and the field fluctuations of the $\mathbf{a}^+$ wave packet, with a pressure-induced correction to ensure $\nabla \cdot \mathbf{a}^- = 0$ (Maron and Goldreich, 2001). This is illustrated in Figure 4 for cubical portions of two colliding wave packets. In both cubes, $\delta\mathbf{B}$ is upwards in the near-face of the cube, and downward in the rear-face of the cube, as in Figure 3.

The different displacements of different parts of a wave packet during a collision shear the wave packet in the plane perpendicular to $\mathbf{B}_0$. For colliding wave packets of length $\lambda_\parallel$ along $\mathbf{B}_0$ and width $\lambda_\perp$ across $\mathbf{B}_0$, the change in the value of $\mathbf{a}^\pm$ induced by the collision is roughly the duration of a collision, $\lambda_\parallel/v_A$, times the magnitude of the nonlinear $\mathbf{a}^\mp \cdot \nabla \mathbf{a}^\pm$ term. If $|\mathbf{a}^+| \sim |\mathbf{a}^-| \sim a_{\lambda_\perp}$, and if $\lambda_\perp \lesssim \lambda_\parallel$, then the fractional change χ in $\mathbf{a}^\pm$ induced by the collision is given by (Goldreich and Sridhar, 1995)

$$\chi \sim \frac{a_{\lambda_\perp}\lambda_\parallel}{v_A \lambda_\perp}, \tag{14}$$

or, from Eq. (13),

$$\chi \sim \frac{\Delta r}{\lambda_\perp}. \tag{15}$$

When $\chi \ll 1$, the right-hand side of, for example, the $\mathbf{a}^-$ wave packet in Figure 4 is displaced in almost the same way as the left-hand side, since both sides encounter essentially the same $\mathbf{a}^+$ wave packet, since the $\mathbf{a}^+$ packet is only slightly altered during the collision. Changes to the profile of a wave packet along the magnetic field are thus weaker than changes in the profile of a wave packet in the plane $\perp$ to $\mathbf{B}_0$ (Shebalin et al., 1983; Ng and Bhattacharjee, 1997; Goldreich and Sridhar, 1997). As a result, the cascade of energy to small $\lambda_\parallel$ is much less efficient than the cascade of energy to small $\lambda_\perp$ in weak incompressible MHD turbulence, as discussed in the following section.

5. Weak Incompressible MHD Turbulence

In weak incompressible MHD turbulence, $\chi \ll 1$ and nonlinear wave–wave interactions are a small perturbation to a wave's linear oscillatory behavior. In this limit, the wave kinetic equation for the power spectra can be derived using perturbation theory (Galtier et al., 2000). When only Alfvén waves are excited, the wave kinetic equation is of the form (Galtier et al., 2000)

$$\frac{\partial E^{\pm}(\mathbf{k})}{\partial t} = \int d^3k_1 d^3k_2 \big[N_{\mathbf{k},\mathbf{k}_1,\mathbf{k}_2} E^{\pm}(\mathbf{k}) - P_{\mathbf{k},\mathbf{k}_1,\mathbf{k}_2} E^{\pm}(\mathbf{k}_1)\big] E^{\mp}(\mathbf{k}_2) \times \delta\big(\omega^{\pm}_{\mathbf{k}} - \omega^{\pm}_{\mathbf{k}_1} - \omega^{\mp}_{\mathbf{k}_2}\big)\delta(\mathbf{k} - \mathbf{k}_1 - \mathbf{k}_2)\big], \tag{16}$$

where $\mathbf{k}$ is the Fourier-space wave number,

$$E^{\pm}(\mathbf{k})\delta(\mathbf{k} + \mathbf{k}_1) = \langle \tilde{\mathbf{a}}^{\pm}(\mathbf{k}) \cdot \tilde{\mathbf{a}}^{\pm}(\mathbf{k}_1)\rangle, \tag{17}$$

$\tilde{\mathbf{a}}^{\pm}$ is the Fourier transform of $\mathbf{a}^{\pm}$, $\langle \ldots \rangle$ indicates an ensemble average, the kernels $N_{\mathbf{k},\mathbf{k}_1,\mathbf{k}_2}$ and $P_{\mathbf{k},\mathbf{k}_1,\mathbf{k}_2}$ are independent of $E^{\pm}$ and given by (Galtier et al., 2000), and

$$\omega^{\pm}_{\mathbf{k}} = \mp k_z v_{\mathrm{A}} \tag{18}$$

is the frequency of small-amplitude $\mathbf{a}^{\pm}$-waves, from Eqs. (10) and (11). Eq. (16) contains only the leading-order terms in the weak-turbulence expansion, called the three-wave-interaction terms. The delta functions in the integrand give the three-wave resonance conditions,

$$\mathbf{k} = \mathbf{k}_1 + \mathbf{k}_2, \tag{19}$$

and

$$\omega^{\pm}_{\mathbf{k}} = \omega^{\pm}_{\mathbf{k}_1} + \omega^{\mp}_{\mathbf{k}_2}. \tag{20}$$

The integrand in Eq. (16) contains the product $E^{\pm}E^{\mp}$ and no terms of the form $E^{\pm}E^{\pm}$, consistent with the fact that only oppositely directed wave packets interact.

Upon dividing Eq. (20) by v_A, one finds that

$$-k_z = -k_{1z} + k_{2z} \tag{21}$$

Combining Eq. (21) with the z-component of Eq. (19) yields

$$k_{2z} = 0, \quad \text{and} \quad k_z = k_{1z}. \tag{22}$$

If turbulence is not excited initially at some value of k_z, three-wave interactions will not excite turbulence at that value of k_z. Indeed, three-wave interactions do not transfer energy to larger values of k_z, only to larger values of $k_\perp = \sqrt{k_x^2 + k_y^2}$ (Shebalin et al., 1983; Galtier et al., 2000). Also, the rate of energy transfer for $E^\pm$ is in essence proportional to the $k_z = 0$ part of $E^\mp$. These conclusions remain valid when slow waves are also excited. (Galtier et al., 2000)

What physical meaning can be attributed to the $k_z = 0$ part of the power spectrum? It is simpler to answer this question by working with a 1D spectrum (M, defined below) obtained by integrating E^+ over k_x and k_y. Assuming homogeneous turbulence, let

$$\langle \mathbf{B}^+(x, y, 0) \cdot \mathbf{B}^+(x, y, z) \rangle = g(z), \tag{23}$$

where $\mathbf{B}^+$ is the fluctuating magnetic field associated with all $\mathbf{a}^+$ wave packets ($\mathbf{B}^+ = \mathbf{a}^+\sqrt{\pi\rho}$), and let

$$M(k_z) = \int_{-\infty}^{\infty} e^{-ik_z z} g(z) dz. \tag{24}$$

Eqs. (17) and (24) imply that $M(k_z) = [\rho/32\pi^4] \int dk_x dk_y E^+(\mathbf{k})$. Assuming that all wave packets have coherence length $\lambda_\parallel$ along the magnetic field, coherence length $\lambda_\perp$ across the field, and rms fluctuating field $B_{\lambda_\perp}$, one finds that

$$M(0) = \left\langle \int_{-\infty}^{\infty} \mathbf{B}^+(x, y, 0) \cdot \mathbf{B}^+(x, y, z) dz \right\rangle \sim B_{\lambda_\perp}^2 \lambda_\parallel. \tag{25}$$

If the wave packets have the Alfvén-wave polarization, $\delta\mathbf{B} \perp \mathbf{B}_0$, then from Eq. (13),

$$M(0) \sim B_{\lambda_\perp} B_0 \Delta r, \tag{26}$$

where $\Delta r \sim B_{\lambda_\perp} \lambda_\parallel / B_0$ is the field-line displacement associated with a single wave packet. Thus, for Alfvén waves, $M(0)$ [and similarly $E^\pm(k_z = 0)$] is a measure of the field-line displacements caused by wave packets. The dependence of the cascade rate on $E^\mp(k_z = 0)$ indicates that energy cascade arises from the turbulent wandering of field lines, as described in Section 4. (Ng and Bhattacharjee, 1996;

Goldreich and Sridhar, 1997; Lithwick and Goldreich, 2001; Lithwick and Goldreich, 2003) As illustrated by Eq. (26), the $k_z = 0$ part of the spectrum should not be equated with wave packets that have infinite coherence length along the magnetic field (Galtier et al., 2002; Lithwick and Goldreich, 2003); structures with finite $\lambda_\|$ lead to nonzero Δr, $M(0)$, and $E(k_z = 0)$.

The power spectrum of weak incompressible MHD turbulence can be estimated as follows (Ng and Bhattacharjee, 1997; Goldreich and Sridhar, 1997; Galtier et al., 2000; Bhattacharjee and Ng, 2001). The fractional change χ in a wave packet of width $\lambda_\perp$ and length $\lambda_\|$ during one collision is $\ll 1$. Since the changes induced by successive collisions add randomly, χ^{-2} collisions are required for an order-unity fractional change. The time required for χ^{-2} collisions is $\chi^{-2}\lambda_\|/v_{\rm A}$, or $\sim v_{\rm A}\lambda_\perp^2/(a_{\lambda_\perp}^2\lambda_\|)$. This is roughly the time $\tau_{\rm cascade}$ required for eddies of width $\lambda_\perp$ to pass their energy to eddies of width $\sim\lambda_\perp/2$. The cascade power ϵ is roughly the energy per unit mass in wave packets of width $\lambda_\perp$, i.e., $a_{\lambda_\perp}^2/2$, divided by $\tau_{\rm cascade}$:

$$\epsilon \sim \frac{a_{\lambda_\perp}^4\lambda_\|}{v_{\rm A}\lambda_\perp^2}. \tag{27}$$

Assuming that local interactions in wave-number space dominate the cascade, ϵ is independent of $\lambda_\perp$ when $\lambda_\perp \ll l$ and $\lambda_\perp \gg d$, where l is the outer scale (stirring scale) and d is the dissipation scale. Since $\lambda_\|$ is effectively constant,

$$a_{\lambda_\perp} \propto \lambda_\perp^{1/2} \tag{28}$$

(Ng and Bhattacharjee, 1997; Goldreich and Sridhar, 1997; Galtier et al., 2000). If the one-dimensional power spectrum $E_{1D}(k_\perp)$ is defined so that the energy of modes with $k_\perp$ in the interval $(k_1, 2k_1)$, i.e., $a_{k_1^{-1}}^2$, is given by $\int_{k_1}^{2k_1} E_{1D}(k_\perp)dk_\perp \sim k_1 E_{1D}(k_1)$, then

$$E_{1D}(k_\perp) \propto k_\perp^{-2} \tag{29}$$

(Ng and Bhattacharjee, 1997; Goldreich and Sridhar, 1997; Galtier et al., 2000).

Inserting Eq. (28) into Eq. (14) yields (Goldreich and Sridhar, 1997)

$$\chi \propto \lambda_\perp^{-1/2}. \tag{30}$$

Thus, if the dissipation scale d is sufficiently small, then at sufficiently small $\lambda_\perp$ the value of χ increases to 1, and the turbulence becomes strong.

6. Strong Incompressible MHD Turbulence

In strong MHD turbulence, for which $\chi \gtrsim 1$, fluctuations are not waves, since they are significantly distorted by nonlinear interactions during a single wave period. Nor

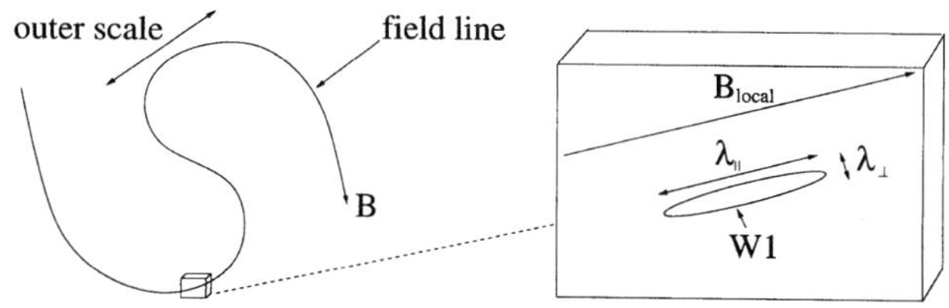

Figure 5. In strong MHD turbulence, eddies are elongated along the local magnetic field.

can propagation of fluctuations along the magnetic field be ignored, since, for example, for $\chi \sim 1$ propagation is in some sense equally important as wave–wave interactions. The combination of propagation and nonlinear interaction determines the length of eddies in strong incompressible MHD turbulence as follows. (Goldreich and Sridhar, 1995; Maron and Goldreich, 2001; Lithwick and Goldreich, 2001)

Consider a volume of width $\lambda_{\perp}$ across the magnetic field and length $\lambda_{\parallel}$ along the field, with $\lambda_{\parallel} \gg \lambda_{\perp}$. The $\mathbf{a}^-$ fluctuations within this volume will be loosely described as a "wave packet" with the label W1. It is useful to consider the "large-scale" magnetic field that is obtained by filtering out all magnetic fluctuations on length scales comparable to or smaller than W1. If one assumes that the dominant nonlinear interactions are between structures of similar size (local in k-space), one can neglect the shearing of W1 by the large-scale magnetic field. Within a box that is large compared to W1 but small compared to the outer scale l, one can thus approximate the large-scale magnetic field as a uniform mean field denoted $\mathbf{B}_{\text{local}}$, as in Figure 5. From Eq. (11), W1 propagates along $\mathbf{B}_{\text{local}}$ at speed $B_{\text{local}}/\sqrt{4\pi\rho}$, while at the same time undergoing nonlinear interactions. Let W2 be a "wave packet" of $\mathbf{a}^+$ fluctuations, also of width $\lambda_{\perp}$ and length $\lambda_{\parallel}$, propagating anti-parallel to $\mathbf{B}_{\text{local}}$ while undergoing nonlinear interactions. Let $a_{\lambda_{\perp}}$ be the amplitude of $\mathbf{a}^-$ within W1 and also of $\mathbf{a}^+$ within W2. Let $\lambda_{\parallel,\text{crit}}$ be the value of $\lambda_{\parallel}$ such that $\chi = 1$ for W1 and W2: $\lambda_{\parallel,\text{crit}} = \lambda_{\perp} v_{\text{A}}/a_{\lambda_{\perp}}$.

Let us suppose that W1 and W2 collide, and that $\lambda_{\parallel} \gg \lambda_{\parallel,\text{crit}}$ (i.e., $\chi \gg 1$), so that initially the $\mathbf{a}^-$ and $\mathbf{a}^+$ fluctuations are coherent over a distance $\gg \lambda_{\parallel,\text{crit}}$ along $\mathbf{B}_{\text{local}}$. After W1 and W2 have propagated into each other a distance $\lambda_{\parallel,\text{crit}}$, as depicted in the bottom part of Figure 6, the $\mathbf{a}^-$ ($\mathbf{a}^+$) fluctuations at the leading edge of W1 (W2) above point Q (P) have been altered by a factor of order unity by nonlinear interactions. As time proceeds, the $\mathbf{a}^-$ fluctuations above points P and Q will evolve in different ways, because they are interacting with significantly different $\mathbf{a}^+$ fluctuations: the $\mathbf{a}^-$ fluctuations above point Q interact with the pristine, undistorted version of W2; the $\mathbf{a}^-$ fluctuations above point P interact with $\mathbf{a}^+$ fluctuations in W2 that have already been changed by the collision. As a result, the collision introduces structure with parallel coherence length $\lambda_{\parallel,\text{crit}}$ into the two wave packets.

According to these heuristic arguments, if χ is initially $\gg 1$ (but finite), the combination of nonlinear interaction and propagation along field lines reduces χ

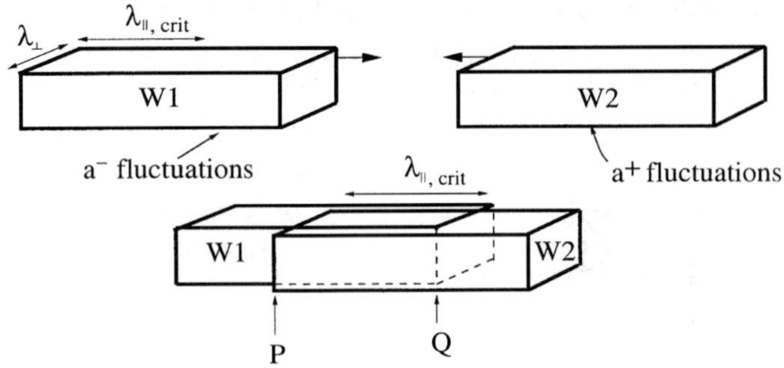

Figure 6. Wave packet collision in strong MHD turbulence.

until $\chi \sim 1$. On the other hand, weak turbulence theory shows shown that if χ is initially $\ll 1$, nonlinear interactions act to increase χ until $\chi \sim 1$, as discussed in Section 5. Thus, 1 is in some sense the "stable equilibrium value" of χ in incompressible MHD turbulence. (Goldreich and Sridhar, 1995).

When $\chi \sim 1$, the turbulence is strong. A theory for the spectrum and anisotropy of strong incompressible MHD turbulence was developed by Goldreich and Sridhar (1995). The time $\tau_{\rm cascade}$ for eddies of width $\lambda_\perp$ and length $\lambda_\parallel$ to pass their energy to smaller eddies is $\sim \lambda_\parallel / v_{\rm A} \sim \lambda_\perp / a_{\lambda_\perp}$. Assuming local interactions in k-space, the cascade power ϵ is given by $a^2_{\lambda_\perp}/\tau_{\rm cascade} \sim a^3_{\lambda_\perp}/\lambda_\perp$. Assuming that ϵ is independent of $\lambda_\perp$,

$$a_{\lambda_\perp} \propto \lambda_\perp^{1/3}, \tag{31}$$

which corresponds to

$$E_{1D}(k_\perp) \propto k_\perp^{-5/3}. \tag{32}$$

The condition $\chi \sim 1$ combined with Eq. (31) yields (Goldreich and Sridhar, 1995)

$$\lambda_\parallel \propto \lambda_\perp^{2/3}. \tag{33}$$

If the turbulence is stirred at scale l, and if $a_l \simeq v_{\rm A}$, then (Goldreich and Sridhar, 1995)

$$\lambda_\parallel = l^{1/3} \lambda_\perp^{2/3}. \tag{34}$$

Eqs. (31) through (34) apply to both Alfvén-mode eddies and slow-mode eddies (Goldreich and Sridhar, 1995). However, Alfvén-mode fluctuations dominate the shearing of both Alfvén-mode eddies and slow-mode eddies (Goldreich and Sridhar, 1995). The reason is that the nonlinear $\mathbf{a}^\pm \cdot \nabla \mathbf{a}^\mp$ terms are much larger for Alfvén modes than for slow modes, since fluctuations vary most rapidly across the magnetic field, and $\mathbf{a}^{\pm}$ is perpendicular to the local magnetic field for Alfvén modes, but nearly along the magnetic field for slow modes when $\lambda_{\perp} \ll \lambda_{\parallel}$.

Eqs. (31) through (34) are consistent with the numerical results of (Cho and Vishniac, 2000; Cho et al., 2002). The numerical simulations of (Müller et al., 2000) agree with Eq. (32), but do not address local anisotropy. On the other hand, the numerical simulations of (Maron and Goldreich, 2001) are consistent with Eq. (34), but find $E_{1D}(k_{\perp}) \propto k_{\perp}^{-3/2}$. Thus, although Eqs. (31) through (34) are supported by physical arguments and some direct numerical simulations, areas of uncertainty remain.

7. Decay and Dynamic Alignment

If only one type of fluctuation, either $\mathbf{a}^{+}$ or $\mathbf{a}^{-}$, is present, nonlinear interactions vanish. Such a state is called "maximally aligned," since either $\mathbf{v} = \mathbf{b}$ or $\mathbf{v} = -\mathbf{b}$. Decaying "turbulence" that is maximally aligned decays only on a viscous or resistive time scale, and can thus be long-lived.

If decaying turbulence initially contains both $\mathbf{a}^{+}$ and $\mathbf{a}^{-}$ fluctuations, but an excess of one over the other, then it decays to a maximally aligned state, in which only the initially predominating fluctuation type remains. (Dobrowolny et al., 1980; Grappin et al., 1983; TiNg and Bhattacharjee, 1986; Galtier et al., 2000; Maron and Goldreich, 2001; Cho et al., 2002; Lithwick and Goldreich, 2003). The reason for this is the following. Whether the turbulence is weak or strong, isotropic or anisotropic, the energy-cascade time of $\mathbf{a}^{\pm}$ fluctuations at a perpendicular scale $\lambda_{\perp}$ is proportional to some power of the energy in $\mathbf{a}^{\mp}$ fluctuations at scale $\lambda_{\perp}$. If turbulence is excited with an excess of $\mathbf{a}^{+}$ waves, the decay time for the $\mathbf{a}^{-}$ waves is short compared to the decay time of the $\mathbf{a}^{+}$ waves. The $\mathbf{a}^{-}$ waves thus decay faster than the $\mathbf{a}^{+}$ fluctuations, with a discrepancy in decay rates that increases in time. It has been suggested by Cho and Lazarian (2003) that this mechanism also operates in compressible MHD turbulence, in which case it provides a potential mechanism for long-lived turbulence in astrophysical environments.

Acknowledgements

The author acknowledges the support of NSF grant AST-0098086 and DOE grants DE-FG02-01ER54658 and DE-FC02-01ER54651 at the University of lowa.

References

Balsara, D. and Pouquet, A.: 1999, *Phys. Plasmas* **6**, 89.
Boldyrev, S., Nordlund, A. and Padoan, P.: 2002, *ApJ* **573**, 678.

Bhattacharjee, A. and Ng, N.: 2001, *ApJ* **548**, 318.
Cho, J. and Lazarian, A.: 2003, astro-ph/0301062.
Cho, J., Lazarian, A. and Vishniac, E.: 2002, *ApJ* **564**, 291.
Cho, J. and Vishniac, E.: 2000, *ApJ* **539**, 273.
Dobrowolny, M., Mangeney, A. and Veltri, P.: 1980, *Phys. Rev. Letters* **45**, 144.
Galtier, S., Nazarenko, S.V., Newell, A.C. and Pouquet, A.: 2000, *J. Plasma Phys.* **63**, 447.
Galtier, S., Nazarenko, S., Newell, A. and Pouquet, A.: 2002, *ApJ* **564**, L49.
Ghosh, S. and Goldstein, M.: 1997, *J. Plasma Phys.* **57**, 129.
Goldreich, P. and Sridhar, S.: 1995, *ApJ* **438**, 763.
Goldreich, P. and Sridhar, S.: 1997, *ApJ* **485**, 680.
Grappin, R., Pouquet, A. and Léorat, J.: 1983, *Astron. Astrophys.* **126**, 51.
Higdon, J.C.: 1984, *ApJ* **285**, 109.
Iroshnikov, P.: 1963, *Astron. Zh.* **40**, 742.
Kinney, R.M. and McWilliams, J.C.: 1998, *Phys. Rev. E* **57**, 7111.
Kraichnan, R.H.: 1965, *Phys. Fluids* **8**, 1385.
Lithwick, Y. and Goldreich, P.: 2001, *ApJ* **562**, 279.
Lithwick, Y. and Goldreich, P.: 2003, *ApJ* **582**, 1220.
Mac Low, M.M.: 1999, *ApJ* **524**, 169.
Maron, J. and Goldreich, P.: 2001, *ApJ* **554**, 1175.
Montgomery, D. and Matthaeus, W.: 1995, *ApJ* **447**, 706.
Montgomery, D. and Turner, L.: 1981, *Phys. Fluids* **24**, 825.
Milano, L., Matthaeus, W., Dmitruk, P. and Montgomery, D.C.: 2001, *Phys. Plasmas* **8**, 2673.
Müller, W. and Biskamp, D.: 2000, *Phys. Rev. Lett.* **84**, 475.
Ng, C.S. and Bhattacharjee, A.: 1996, *ApJ* **465**, 845.
Ng, C.S. and Bhattacharjee, A.: 1997, *Phys. Plasmas* **4**, 605.
Ossenkopf, V. and Mac Low, M.: 2002, *A & A* **390**, 3070.
Oughton, S., Priest, E. and Matthaeus, W.: 1994, *J. Fluid Mech.* **280**, 95.
Passot, T. and Vazquez-Semadeni, E.: 2003, *A&A* **398**, 845.
Politano, H., Pouquet, A. and Sulem, P.-L.: 1995, *Phys. Plasmas* **2**, 2931.
Pouquet, A., Frisch, U. and Léorat, J.: 1976, *J. Fluid Mech.* **77**, 321.
Schekochihin, A., Maron, J. and Cowley, S.: 2002, *ApJ* **576**, 2002.
Shebalin, J.V., Matthaeus, W. and Montgomery, D.: 1983, *J. Plasma Phys.* **29**, 525.
Sridhar, S. and Goldreich, P.: 1994, *ApJ* **432**, 612.
Stone, J., Ostriker, E. and Gammie, C.: 1998, *ApJ* **508**, L99.
Ting, A., Matthaeus, W. and Montgomery, D.: 1986, *Phys. Fluids* **29**, 3261.
Vazquez-Semadeni, E., Ostriker, E., Passot, T., Gammie, C. and Stone, J.: 2000
Vestuto, J., Ostriker, E. and Stone, J.: 2003, *ApJ* **590**, 858.
Winters, W., Balbus, S. and Hawley, J.: 2003, *MNRAS* **340**, 519.
Zank, G.P. and Matthaeus, W.H.: 1993, *Phys. Fluids* **A5**, 257.

BASIC PROPERTIES OF COMPRESSIBLE MHD TURBULENCE: IMPLICATIONS FOR MOLECULAR CLOUDS

A. LAZARIAN and J. CHO
University of Wisconsin-Madison, Dept. of Astronomy, U.S.A.; E-mail: lazarian@astro.wisc.edu

Abstract. Recent advances in understanding of the basic properties of compressible Magnetohydrodynamic (MHD) turbulence call for revisions of some of the generally accepted concepts. First, the MHD turbulence is not so messy as it is usually believed. In fact, the notion of strong non-linear coupling of compressible and incompressible motions is not tenable. Alfven, slow and fast modes of MHD turbulence follow their own cascades and exhibit degrees of anisotropy consistent with theoretical expectations. Second, the fast decay of turbulence is not related to the compressibility of fluid. Rates of decay of compressible and incompressible motions are very similar. Third, the viscosity by neutrals does not suppress MHD turbulence in a partially ionized gas. Instead, MHD turbulence develops magnetic cascade at scales below the scale at which neutrals damp ordinary hydrodynamic motions. The implications of those changes of MHD turbulence paradigm for molecular clouds require further studies. Those studies can benefit from testing of theoretical predictions using new statistical techniques that utilize spectroscopic data. We briefly discuss advances in development of tools using which the statistics of turbulent velocity can be recovered from observations.

Keywords: turbulence, molecular clouds, MHD

1. What is Compressible MHD Turbulence?

It is well known that molecular clouds are magnetized with compressible magnetic turbulence determining most of their properties (see Elmegreen and Falgarone, 1996; Stutzki, 2001). Star formation (see McKee and Tan, 2002; Elmegreen, 2002; Pudritz, 2001), cloud chemistry (see Falgarone, 1999 and references therein), shattering and coagulation of dust (see Lazarian and Yan, 2003 and references therein) are examples of processes for which knowledge of turbulence is absolutely essential. There are many excellent reviews that deal the theory of turbulent molecular clouds and numerical simulations (see Falgarone, 1999; Vazquez-Semadeni et al., 2000; Mac Low and Klessen, 2003). This *short* review is focused on the recently uncovered basic properties of MHD turbulence and their implications for understanding of molecular clouds[1]. We also briefly deal with the recovering

[1]It is not possible to cite all the important papers in the area of MHD turbulence and turbulent molecular clouds. An incomplete list of the references in a recent review on the statistics of MHD turbulence by Cho et al. (2003a; henceforth CLV03a) includes about two hundred entries. The list of references in the molecular cloud review by Mac Low and Klessen (2003) is even longer.

Astrophysics and Space Science **292:** 29–43, 2004.

of 3D statistics of turbulent velocity from observations, which is a theoretical problem in itself. A more detailed discussion of MHD turbulence can be found in the companion review by Chandran, whereas observational aspects of turbulence in molecular clouds are discussed in the review by Falgarone (this volume).

Why do we expect molecular clouds to be turbulent? A fluid of viscosity ν becomes turbulent when the rate of viscous dissipation, which is $\sim\nu/L^2$ at the energy injection scale L, is much smaller than the energy transfer rate $\sim V_L/L$, where V_L is the velocity dispersion at the scale L. The ratio of the two rates is the Reynolds number $Re = V_L L/\nu$. In general, when Re is larger than 10–100, the system becomes turbulent. Chaotic structures develop gradually as Re increases, and those with $Re \sim 10^3$ are appreciably less chaotic than those with $Re\sim 10^8$. Observed features, such as star-forming clouds are very chaotic for $Re > 10^8$. This makes it difficult to simulate realistic turbulence. The currently available 3D simulations containing 512 grid cells along each side can have Re up to $\sim O(10^3)$ and are limited by their grid sizes. Therefore, it is essential to find "*scaling laws*" to extrapolate numerical calculations [$Re \sim O(10^3)$] to real astrophysical fluids ($Re > 10^8$). We show below that even with its limited resolution, numerics is a great tool for *testing* scaling laws.

Kolmogorov theory provides a scaling law for *incompressible non*magnetized hydrodynamic turbulence (Kolmogorov, 1941). This law provides a statistical relation between the relative velocity v_l of fluid elements and their separation l, namely, $v_l \sim l^{1/3}$. An equivalent description is to express spectrum $E(k)$ as a function of wave number k ($\sim 1/l$). The two descriptions are related by $kE(k)\sim v_l^2$. The famous Kolmogorov spectrum is $E(k) \sim k^{-5/3}$. The applications of Kolmogorov theory range from engineering research to meteorology (see Monin and Yaglom, 1975), but its astrophysical applications are poorly justified and the application of the Kolmogorov theory can lead to erroneous conclusions [see reviews by Lazarian et al. (2003) and Lazarian and Yan (2003)]

Let us consider *incompressible* MHD turbulence first. There has long been understanding that the MHD turbulence is anisotropic (e.g., Shebalin et al., 1983). Substantial progress has been achieved recently by Goldreich and Sridhar (1995; hereafter GS95), who made an ingenious prediction regarding relative motions parallel and perpendicular to magnetic field **B** for incompressible MHD turbulence. An important observation that leads to understanding of the GS95 scaling is that magnetic field cannot prevent mixing motions of magnetic field lines if the motions are perpendicular to the magnetic field. Those motions will cause, however, waves that will propagate along magnetic field lines. If that is the case, the time scale of the wave-like motions along the field, i.e., $\sim l_\parallel/V_A$, ($l_\parallel$ is the characteristic size of the perturbation along the magnetic field and $V_A = B/\sqrt{4\pi\rho}$ is the local Alfven speed) will be equal to the hydrodynamic time-scale, $l_\perp/v_l$, where $l_\perp$ is the characteristic size of the perturbation perpendicular to the magnetic field. The mixing motions are hydrodynamic-like[2]. They obey Kolmogorov scaling, $v_l \propto l_\perp^{1/3}$, because incompressible turbulence is assumed. Combining the two relations above we can get the GS95 anisotropy, $l_\parallel \propto l_\perp^{2/3}$ (or $k_\parallel \propto k_\perp^{2/3}$ in terms of wave-numbers). If we interpret $l_\parallel$ as the eddy size in the direction of the local magnetic field[3] and $l_\perp$ as that in the perpendicular directions, the relation implies that smaller eddies are more elongated.

GS95 predictions have been confirmed numerically (Cho and Vishniac, 2000; Maron and Goldreich, 2001; Cho et al., 2002a, hereafter CLV02a; see also CLV03a); they are in good agreement with observed and inferred astrophysical spectra (see CLV03a). However, the GS95 model considered incompressible MHD, but the media in molecular clouds is *highly compressible*. Does any part of GS95 model survives? Literature on the properties of compressible MHD is very rich (see reviews by Pouquet, 1999; Cho and Lazarian, 2003 and references therein). Higdon (1984) theoretically studied density fluctuations in the interstellar MHD turbulence. Matthaeus and Brown (1988) studied nearly incompressible MHD at low Mach number and Zank and Matthaeus (1993) extended it. In an important paper Matthaeus et al. (1996) numerically explored anisotropy of compressible MHD turbulence. However, those papers do not provide universal scalings of the GS95 type.

The complexity of the compressible magnetized turbulence with magnetic field made some researchers believe that the phenomenon is too complex to expect any universal scalings for molecular cloud research. High coupling of compressible and incompressible motions is often quoted to justify this point of view.

Below, we shall provide arguments that are suggestive that the fundamentals of compressible MHD can be understood and successfully applied to molecular clouds.

In molecular clouds the regular magnetic field is comparable with the fluctuating one. Therefore, for most part of our discussion, we shall discuss results obtained for $\delta V \sim \delta B/\sqrt{4\pi\rho} \sim B_0/\sqrt{4\pi\rho}$, where δB is the r.m.s. strength of the random magnetic field.

2. Does the Decay of MHD Turbulence Depend on Compressibility?

Turbulent support of molecular clouds (see review by McKee, 1999) critically depends on the rate of turbulence decay. For a long time magnetic fields were thought to be the means of decreasing dissipation of turbulence. Numerical calculations by Mac Low et al. (1998) and Stone et al. (1998) indicated that compressible MHD turbulence decays as fast as the hydrodynamic turbulence. This gives rise to a

[2]Recent simulations (Cho et al., 2003) suggest that perpendicular mixing is indeed efficient for mean magnetic fields of up to the equipartition value.
[3]Incidentally, if we identify the size of molecular cloud with the longer dimension of an eddy, we shall get the Larson relations $v_l \sim l^{1/2}$ (Larson, 1981) for velocity and cloud size.

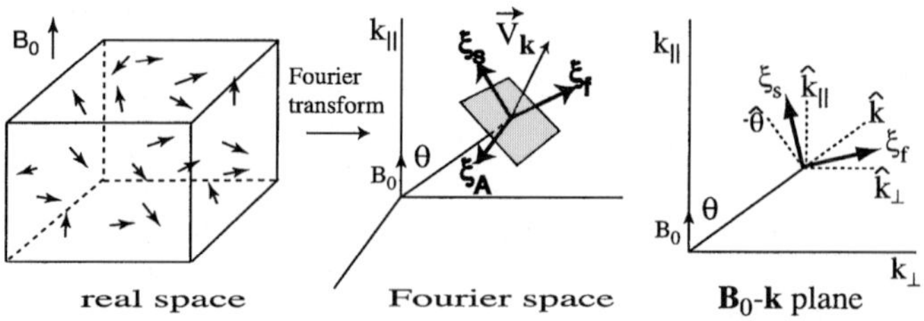

Figure 1. Separation method. We separate Alfven, slow, and fast modes in Fourier space by projecting the velocity Fourier component $\mathbf{v}_\mathbf{k}$ onto bases ξ_A, ξ_s, and ξ_f, respectively. Note that $\xi_A = -\hat{\varphi}$. Slow basis ξ_s and fast basis ξ_f lie in the plane defined by $\mathbf{B}_0$ and $\mathbf{k}$. Slow basis ξ_s lies between $-\hat{\theta}$ and $\hat{\mathbf{k}}_{\|}$. Fast basis ξ_f lies between $\hat{\mathbf{k}}$ and $\hat{\mathbf{k}}_{\perp}$.

belief that it is the compressibility that is responsible for the rapid decay of MHD turbulence.

This point of view has been recently subjected to scrutiny in Cho and Lazarian (2002a, 2003a; henceforth CL02 and CL03, respectively). In these papers a technique of separating different MHD modes was developed and used (see Figure 1). This allowed us to follow how the energy was redistributed between these modes.

How is this idealized incompressible model related to the actual turbulence in molecular clouds? Compressible MHD turbulence is a highly nonlinear phenomenon and it has been thought that different types of perturbations or modes (Alfven, slow and fast) in compressible media are strongly coupled. Nevertheless, one may question whether this is true. A remarkable feature of the GS95 model is that Alfven perturbations cascade to small scales over just one wave period, while the other nonlinear interactions require more time. Therefore one might expect that the nonlinear interactions with other types of waves should affect Alfvenic cascade only marginally. Moreover, since the Alfven waves are incompressible, the properties of the corresponding cascade may not depend on the sonic Mach number.

The generation of compressible motions (i.e., *radial* components in Fourier space) from Alfvenic turbulence is a measure of mode coupling. How much energy in compressible motions is drained from Alfvenic cascade? According to closure calculations (Bertoglio et al., 2001; see also Zank and Matthaeus, 1993), the energy in compressible modes in *hydrodynamic* turbulence scales as $\sim M_s^2$ if $M_s < 1$. CL03 conjectured that this relation can be extended to MHD turbulence if, instead of M_s^2, we use $\sim(\delta V)_A^2/(a^2+V_A^2)$. (Hereinafter, we define $V_A \equiv B_0/\sqrt{4\pi\rho}$, where B_0 is the mean magnetic field strength.) However, since the Alfven modes are anisotropic, this formula may require an additional factor. The compressible modes are generated inside the so-called Goldreich–Sridhar cone, which takes up $\sim(\delta V)_A/V_A$ of the wave vector space. The ratio of compressible to Alfvenic energy inside this cone is the ratio given above. If the generated fast modes become isotropic (see below), the diffusion or, "isotropization" of the fast wave energy in the wave vector space

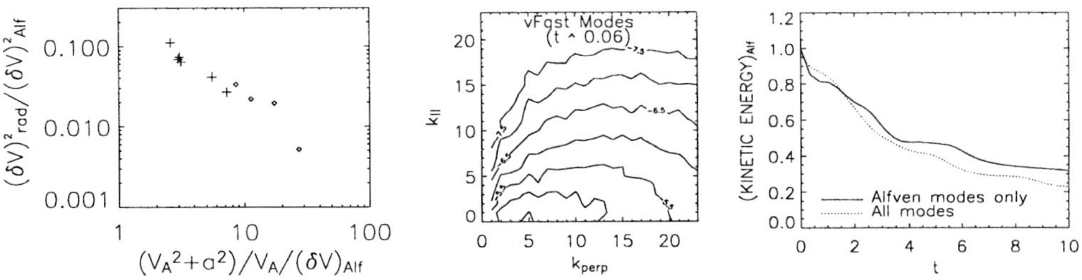

(a) (b) (c)

Figure 2. Mode coupling studies. (a) *left:* Square of the r.m.s. velocity of the compressible modes. We use 144^3 grid points. Only Alfven modes are allowed as the initial condition. "Pluses" are for low β cases ($0.02 \leq \beta \leq 0.4$). "Diamonds" are for high β cases ($1 \leq \beta \leq 20$). (b) *middle:* Generation of fast modes. Snapshot is taken at $t = 0.06$ from a simulation (with 144^3 grid points) that started off with Alfven modes only. Initially, β (ratio of gas to magnetic pressure, P_g/P_{mag}) $= 0.2$ and M_s (sonic Mach number) ~ 1.6. (c) *right:* Comparison of decay rates. Decay of Alfven modes is not much affected by other (slow and fast) modes. We use 216^3 grid points. Initially, $\beta = 0.02$ and $M_s \sim 4.5$ for the solid line and $M_s \sim 7$ for the dotted line. Note that initial data are, in some sense, identical for the solid and the dotted lines. The sonic Mach number for the solid line is smaller because we removed fast and slow modes from the initial data before the decay simulation. For the dotted line, we did *not* remove any modes from the initial data. From CL03.

increase their energy by a factor of $\sim V_A/(\delta V)_A$. This results in

$$\frac{(\delta V)^2_{rad}}{(\delta V)^2_A} \sim \left[\frac{V_A^2 + a^2}{(\delta V)^2_A} \frac{(\delta V)_A}{V_A}\right]^{-1}, \tag{1}$$

where $(\delta V)^2_{rad}$ and $(\delta V)^2_A$ are energy of compressible and Alfven modes, respectively. Eq. (1) suggests that the drain of energy from Alfvenic cascade is marginal[4], when the amplitudes of perturbations are weak, i.e., $(\delta V)_A \ll V_A$. Results of calculations shown in Figure 2 support the theoretical predictions.

We may summarize this issue in the following way. For the incompressible motions to decay fast, there is no requirement of coupling with compressible motions[5]. The marginal coupling of the compressible and incompressible modes allows to study modes separately.

[4]The marginal generation of compressible modes is in agreement with earlier studies by Boldyrev et al. (2002b) and Porter et al. (2002), where the velocity was decomposed into a potential component and a solenoidal component. A recent study by Vestuto et al. (2003) is also consistent with this conclusion.

[5]The reported (see Mac Low et al., 1998) decay of the *total* energy of turbulent motions E_{tot} follows t^{-1} which can be understood if we account for the fact that the energy is being injected at the scale smaller than the scale of the system. Therefore some energy originally diffuses to larger scales through the inverse cascade. Our calculations (Cho and Lazarian, unpublished), stimulated by illuminating discussions with Chris McKee, show that if this energy transfer is artificially prevented by injecting the energy on the scale of the computational box, the scaling of E_{tot} becomes closer to t^{-2}.

3. Is Turbulence in Molecular Clouds Really Messy?

Is it feasible to obtain scaling relations for the compressible MHD turbulence? Some hints about effects of compressibility can be inferred from the GS95 seminal paper. More discussion was presented in Lithwick and Goldreich (2001), which deals with electron density fluctuations in the gas pressure dominated plasma, i.e., in high β regime ($\beta \equiv P_{\mathrm{gas}}/P_{\mathrm{mag}} \gg 1$). The incompressible regime corresponds to $\beta \to \infty$, so it is natural to expect that for $\beta \gg 1$ the GS95 picture would persist. Lithwick and Goldreich (2001) also speculated that for low β plasmas the GS95 scaling of slow modes may be applicable. A detailed study of compressible mode scalings is given in CL02 and CL03.

Our considerations above about the mode coupling can guide us to predict the mode scaling. Indeed, if Alfven cascade evolves on its own, it is natural to assume that slow modes exhibit the GS95 scaling. Indeed, slow modes in gas pressure dominated environment (high β plasmas) are similar to the pseudo-Alfven modes in incompressible regime (see GS95; Lithwick and Goldreich, 2001). The latter modes do follow the GS95 scaling. In magnetic pressure dominated environments (low β plasmas), slow modes are density perturbations propagating with the sound speed a parallel to the mean magnetic field. Those perturbations are essentially static for $a \ll V_A$. Therefore, Alfvenic turbulence is expected to mix density perturbations as if they were passive scalar. This also induces the GS95 spectrum.

The fast waves in low β regime propagate at V_A irrespectively of the magnetic field direction. In high β regime, the properties of fast modes are similar, but the propagation speed is the sound speed a. Thus the mixing motions induced by Alfven waves should marginally affect the fast wave cascade. It is expected to be analogous to the acoustic wave cascade and hence be isotropic.

Results of numerical calculations from Cho and Lazarian (CL03) for magnetically dominated media similar to that in molecular clouds are shown in Figure 3. They support theoretical considerations above.

Why would we care about those scalings? How wrong is it to use Kolmogorov scalings instead? Dynamics, chemistry and physics of molecular clouds (see Falgarone, 1999) presents a complex of problems for which the exact scalings may be required to a different degree. If we talk about dynamics of interstellar dust or propagation of cosmic rays, one *must* account for the actual scalings and couplings of different modes (see reviews by Lazarian and Yan, 2003; Lazarian et al., 2003). There are other problems, e.g., turbulent heat transport where the exact scaling of modes seems to be less important (Cho et al., 2003).

4. What is the Effect of Partial Ionization?

Our considerations above assume that the gas is fully ionized. Molecular clouds are partially ionized. What is the consequence of this? An obvious effect of neutrals is that they do not follow magnetic field lines and thus produce viscosity.

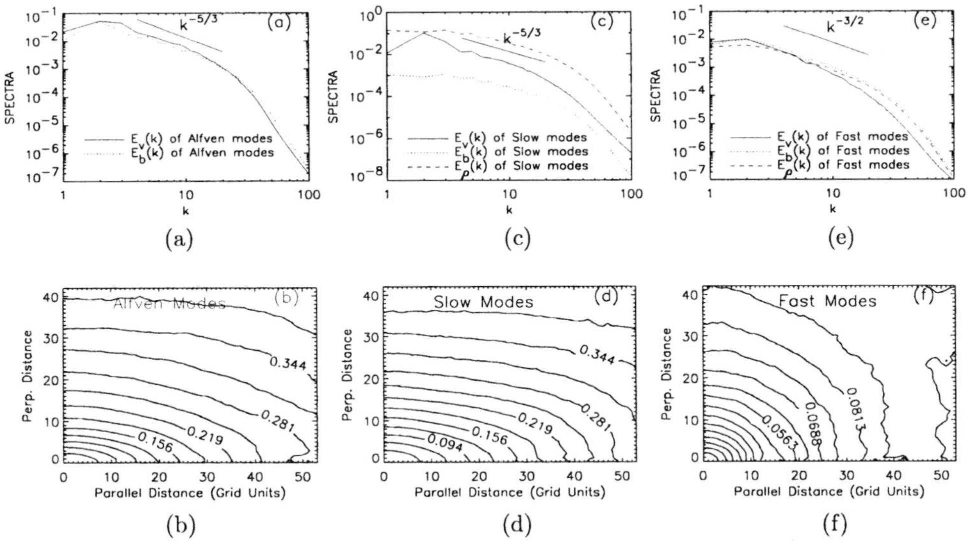

Figure 3. $M_s \sim 2.2$, $M_A \sim 0.7$, $\beta \sim 0.2$, and 216^3 grid points. (a) Spectra of Alfvén modes follow a Kolmogorov-like power law. (b) Eddy shapes (contours of same second-order structure function, SF_2) for velocity of Alfvén modes shows anisotropy similar to the GS95 ($r_\parallel \propto r_\perp^{2/3}$ or $k_\parallel \propto k_\perp^{2/3}$). The structure functions are measured in directions perpendicular or parallel to the local mean magnetic field in real space. We obtain real-space velocity and magnetic fields by inverse Fourier transform of the projected fields. (c) Spectra of slow modes also follow a Kolmogorov-like power law. (d) Slow mode velocity shows anisotropy similar to the GS95. We obtain contours of equal SF_2 directly in real space without going through the projection method, assuming slow mode velocity is nearly parallel to local mean magnetic field in low β plasmas. (e) Spectra of fast modes are compatible with the IK spectrum. (f) The magnetic SF_2 of fast modes shows isotropy. From CL02

In hydrodynamic turbulence viscosity sets a minimal scale for motion, with an exponential suppression of motion on smaller scales. Below the viscous cutoff, the kinetic energy contained in a wavenumber band is dissipated at that scale, instead of being transferred to smaller scales. This means the end of the hydrodynamic cascade, but in MHD turbulence this is not the end of magnetic structure evolution. For viscosity much larger than resistivity, there will be a broad range of scales, where viscosity is important but resistivity is not. On these scales magnetic field structures will be created by the shear from nondamped turbulent motions, which amounts essentially to the shear from the smallest undamped scales. Indeed, this new regime of turbulence has been discovered (see Figure 4)!

A theoretical model for this new regime can be found in Lazarian, Vishniac and Cho (2003; hereafter LVC03). It explains the spectrum $E(k) \sim k^{-1}$ as a cascade of magnetic energy to small scales under the influence of shear at the marginally damped scales. The spectrum is similar to that of the viscous-convective range of passive scalar in hydrodynamic turbulence (see, for example, Batchelor, 1959; Lesieur, 1990), although the study in Lazarian et al. (LVC03) suggests that the

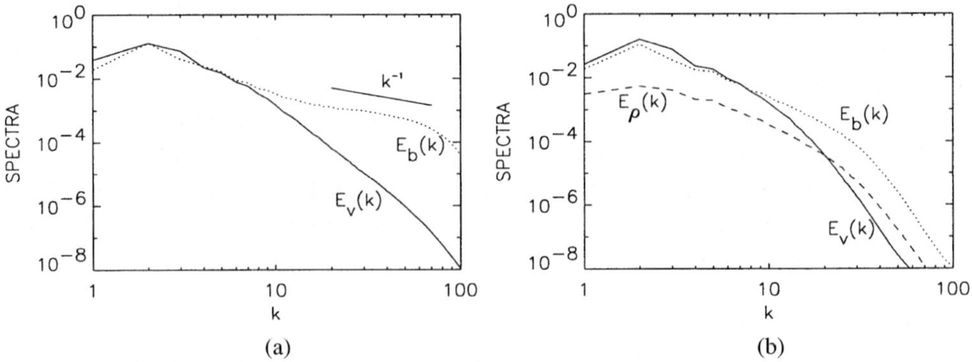

Figure 4. Viscous damped regime (viscosity > magnetic diffusivity). Due to large viscosity, velocity damps after $k \sim 10$. (a) *Left:* Incompressible case with 384^3 grid points. Magnetic spectra show a shallower slope ($E_b(k) \propto k^{-1}$) below the velocity damping scale. We achieve a very small magnetic diffusivity through the use hyper-diffusion. From Cho et al. (2002b). (b) *Right:* Compressible case with 216^3 grid points. Magnetic and density spectra show structures below the velocity damping scale at $k \sim 10$. The structures are less obvious than the incompressible case because it is relatively hard to achieve very small magnetic diffusivity in the compressible run. From CL03.

physical origin of it is different. A study confirming that the k^{-1} spectrum is not a bottleneck effect is presented in Cho et al. (2003b). The mechanism is based on the solenoidal motions and therefore the compressibility should not alter the physics of this regime of turbulence.

For compressible simulations (see Figure 4) the inertial range is much smaller due to numerical reasons, but it is clear that the viscosity-damped regime of MHD turbulence persists. The magnetic fluctuations, however, compress the gas and thus cause fluctuations in density. These density fluctuations may have important and yet unexplored consequences for the small-scale structure of the molecular clouds. There are indications that small-scale structure is indeed present in molecular gas (Marscher et al., 1993) as well as in other partially ionized phases of the ISM (see Heiles, 1997). Those structures can be produced by turbulence with a spectrum substantially more shallow than the Kolmogorov one, e.g., with the spectrum $E(k) \sim k^{-1}$ (see Deshpande, 2000). Below the viscous scale the fluctuations of magnetic field obey the damped regime shown in Figure 4b and produce density fluctuations. For typical Cold Neutral Medium gas (see Draine and Lazarian, 1998), these fluctuations can be as large as ~70 AU and less for denser gas.

An important prediction in LVC03 is that the intermittency of magnetic structures increases with the decrease of the scale. This prediction was confirmed by numerical simulations in Cho et al. (2003), which showed that the filling factor of magnetic field was decreasing with the increase of the wavenumber.

Does the effect of neutrals amounts only to emergence of the viscosity-damped regime of MHD turbulence? The answer to this question is negative. Partial ionization provides the whole range of new interesting effects. Some of them are studied in LVC03. Here we mention those.

First of all, it is clear that whether ions and neutrals act as one fluid in molecular clouds depends on whether the eddy turnover rate $t_{eddy}^{-1} \sim kv_k$ is longer or shorter than the rate t_{ni}^{-1} of ion-neutral collisions. If $t_{eddy}^{-1} > t_{ni}^{-1}$, neutrals decouple from ions and develop *hydrodynamic* Kolmogorov-type cascade. Indeed, the damping rate for those hydrodynamic motions t_{ni}^{-1} and below the decoupling scale the hydrodynamic motions evolve without much hindrance from magnetic field. Magnetic fields with entrained ions develop the viscosity-damped MHD cascade until ion-neutral collisional rate gets longer than the dynamical rate of the intermittent magnetic structures. After that the turbulence reverts to its normal MHD cascade which involves only ions.

If $t_{eddy}^{-1} < t_{ni}^{-1}$ up to the scale at which neutral viscosity damps turbulent motions, the viscosity-damped regime emerges at the scale where kinetic energy associated with turbulent eddies is dissipated. Similarly to the earlier case when the when ion-neutral collisions get insufficient to preserve pressure confinement of the small scale magnetic filaments, outbursts of ordinary ionic MHD turbulence will take place. The turbulence will be intermittent both in time and space because of the disparity of time scales at which turbulence evolves in the viscosity-damped and free ionic MHD regimes. We plan to test those predictions with a two fluid MHD code.

5. How Can One Study Turbulence in Molecular Clouds?

Substantial advances in understanding of MHD turbulence in ordinary and viscosity-damped regime make it most important to test the correspondence of theoretical expectations and observational reality (see review by Ostriker, 2003). Important advances in obtaining high resolution spectral data (see Falgarone et al., 1988) make the testing feasible.

The turbulence spectrum characterizes the distribution of energy over spatial scales and reflects the processes of dissipation and energy injection. The GS95 model would also gives the Kolmogorov spectrum of Alfvenic motions if averaging over $\mathbf{k}$ directions is performed. This stems from the fact that for the inertial range $k_{\perp} \gg k_{\|}$ and therefore, $k \sim k_{\perp}$.

It is known that studies of stochastic density provide only indirect insight into turbulence. A stationary (not evolving) density distribution is indistinguishable from the result of active turbulence. However, Doppler shifted spectral lines carry information about turbulent velocity. The problem is that both velocity and density fluctuations contribute to the observed fluctuations. Therefore, numerous attempts to study turbulence in diffuse interstellar medium and molecular clouds using emission lines (see Munch, 1958; O'Dell, 1986; O'Dell and Castaneda, 1987; Miesch and Bally, 1994; reviews by Scalo, 1987; Falgarone, 1999; Lazarian, 1999) were facing uncertainty of what quantity was actually measured.

Studies of the velocity field have been attempted at different times with velocity centroids (e.g., Munch, 1958; Miesch et al., 1999; Ossenkopf and Mac Low,

2002; Miville-Deschenes et al., 2003). However, it has long been realized that the centroids are affected, in general, by both velocity and density fluctuations (see Stenholm, 1989). A criterion of when centroids indeed reflect the velocity statistics was obtained in Lazarian and Esquivel (2003; henceforth LE03). Without this testing, it is not clear *a priori* what is actually being measured[6].

An important recent development is a quantitative description of spectral data from turbulent media obtained in Lazarian and Pogosyan (2000, 2003; henceforth LP00 and LP03, respectively), where the fluctuations of intensity in channel maps were related to the statistics of velocity and density. LP00 introduced a new technique that was termed Velocity Channel Analysis (henceforth VCA). Within VCA, the separation of velocity and density contributions are obtained by changing the thickness of the analyzed slice of the Position-Position-Velocity (PPV) data cube. The VCA was successfully tested numerically in Lazarian et al. (2001) and Esquivel et al. (2003) and applied to the Small Magellanic Cloud (Stanimirovic and Lazarian, 2002) (see Figure 5).

LP03 accounts for the absorption in the turbulent media. Thus it substantially extends the range of spectral lines that can be used for turbulence studies. Note that the importance of LP00 and LP03 studies is not limited to the development of particular toolk-it of how to interpret channel maps. These works provide theoretical description that can be used within different techniques. For instance, LP03 establishes limitations on turbulence spectra that can be recovered by another promising technique, namely, Spectral Correlation Function (see Rosolowsky et al., 1999).

Other important tools have also been studied recently. Principal Component Analysis (PCA) of the emission data (Heyer and Schloerb, 1997; Brunt and Heyer, 2002a,b) provide a new promissing way to characterise observations. Recent testing showed that it provides statistics different from power spectra (Brunt et al., 2003). Wavelet analysis (see Gill and Henriksen, 1990), spectral correlation functions (see Rosolowsky et al., 1999; Padoan et al., 2001), genus analysis (see Lazarian et al., 2002) (see Figure 5) are other statistical tools. They can provide statistics and insight complementary to power spectra. Synergy of different techniques should provide the necessary insight into turbulence and enable the comparison of observational statistics with theoretical expectations.

6. What is the Future of the Field?

There are whole classes of processes for which we are not sure even about the sign of the effect, for instance, whether turbulence supports or compresses molecular clouds. Or consider imbalanced turbulence, i.e., the turbulence where the flow of

[6]A more optimistic claim about utility of centroids was obtained in Miville-Deschenes et al. (2003) on the basis of experiments with Brownian noise. However, to make the density in their experiments positively defined the authors added additional large mean density. According to the criterion in LE03 this made their centroids bound to be dominated by velocity.

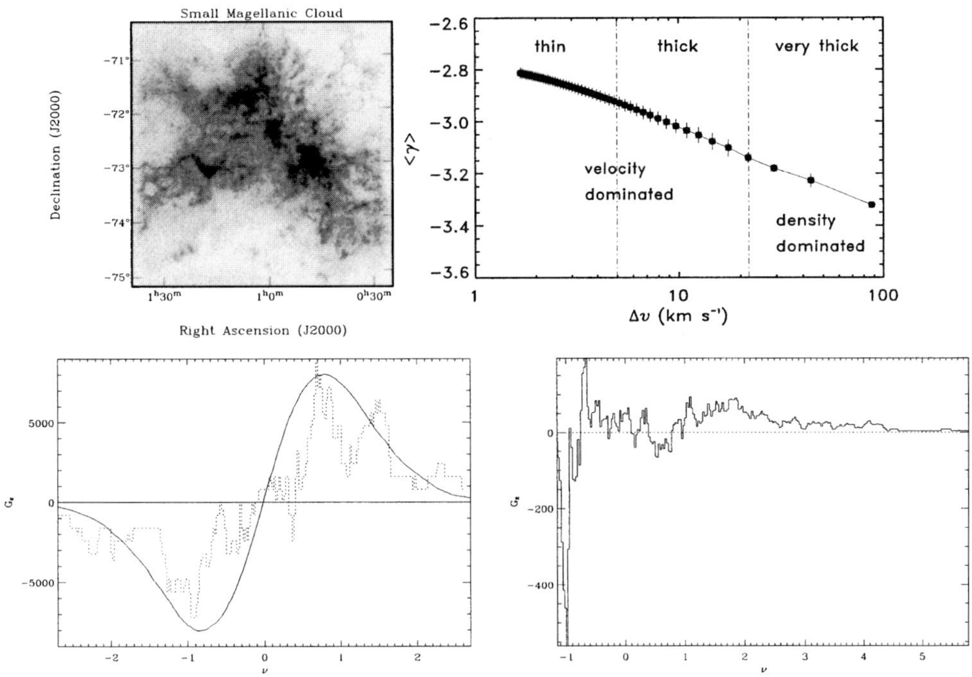

Figure 5. Dealing with Observations: Testing and Applying New Statistical Tools. Upper Left: the 21 cm image of SMC, exhibiting strong density structure. According to our analysis this density has a spectrum close to that expected from MHD turbulence. *Upper Right:* variations of 2-D 21 cm spectral slope with the velocity slice thickness (from Stanimirovic and Lazarian, 2001). The LP00 study predicts that the thick slice reflects the density statistics, while the thin slice is influenced by the velocity. *Lower Left:* the 2D genus (which counts the number of holes versus the number of islands as the threshold contrast changes) of the Gaussian distribution (smooth curve) against the genus for the isothermal compressible MHD simulations with Mach number 2.5 (dotted curve). Interstellar turbulence shows much more intermittency, with consequences for interstellar physics and chemistry. *Lower Right:* the genus of HI distribution in SMC *that has the same power spectrum as the dotted curve in the left panel* (Lazarian et al., 2002). The genus (topology) of the distributions are very different!

energy in one direction is larger than the flow of energy in the opposite direction. This situation is typical for interstellar medium with its localized sources of energy. CLV02a speculated that imbalanced turbulence can propagate over larger distances and feed energy to clouds without star formation. To what degree does this process get modified in the presence of compressibility? Alfvenic modes in imbalanced turbulence live longer and the interaction between density fluctuations and the Alfven mode becomes more important.

In the field of observational studies, the situation looks remarkably promising. With high resolution surveys available (see Falgarone et al., 2000), studies of turbulence statistics should at last become a mainstream research. The prospects of studies of turbulent velocity are most encouraging. Indeed, for the first time, one has understanding of how to get statistics of velocity and density from channel maps (LP00, LP03), and when the velocity centroids reflect the statistics of velocity (LE03). Moreover, a quantitative description of the statistics of the spectral line data cubes (see LP00, LP03) allows us to devise new techniques of turbulence studies. Apart from testing of the particular scaling laws, this research should identify sources and injection scales of the turbulence (see the companion review by Falgarone et al., 1998). Is turbulence in molecular clouds a part of a large scale ISM cascade (see Armstrong et al., 1995)? How does the share of the energy within compressible versus incompressible motions vary within the Galactic disk? There are examples of questions that can be answered in future.

While this review deals mostly with power spectra, higher order statistics will be widely used in the future. Recent numerical research that employed higher order statistics (Muller and Biskamp, 2000; CLV02a, Boldyrev et al., 2002; Cho et al., 2003b) showed it to be a promising tool. For instance, the distinction between the old Iroshnikov-Kraichnan and the GS95 model is difficult to catch using power spectra with a limited inertial range, but is quite apparent for fourth-order statistics. The difference in physical consequences of whether the turbulence dissipates in shocks or in intermittent vortices may be very substantial. Our discussion in §2 suggests that incompressible motions do dissipate via vortices. The scaling of their intermittency with the Reynolds number is still to be established. In the meantime, obtaining higher order statistics with spectral line observations is a challenging problem. Higher order statistics obtained from observational data were reported for observed velocity in Falgarone et al. (1994) and for density in Padoan et al. (2003). According to Falgarone and Puget (1995) and Falgarone et al. (1995) (see also review by Falgarone), the intermittency in vorticity distribution can result in the outbursts of localized dissipation that make tiny regions within cold diffuse clouds chemically active. The testing of those ideas is done in Pety and Falgarone (2000), where synthetic maps obtained using *hydrodynamic* simulations were analyzed. Since then, more results indicating the hydrodynamic simulations may to some degree reflect the physics of MHD turbulence have emerged. First of all, CLV02a found out that the fluid motions perpendicular to **B** are *identical* to hydrodynamic motions. Moreover, for low ionization the turbulence in neutrals and ions decouple with turbulence in neutrals forming a hydrodynamic cascade (see Section 5).

Studies of the statistics of magnetic field in molecular clouds is the next challenging problem (see Crutcher et al., 2003; Ostriker, 2003). Better understanding of grain alignment (see review by Lazarian 2003 and references therein) allows to better identify variations of polarization with the variations of magnetic field. However, this important research has not gained sufficient momentum yet.

7. Summary

1. Understanding of molecular clouds requires understanding of the basics of MHD turbulence. MHD turbulence is not a mess. Scaling relations for its modes have been established recently.
2. Fast decay of MHD turbulence is not due to strong coupling of compressible and incompressible motions. The transfer of energy from Alfven to compressible modes is small. The Alfven mode develops on its own and decays fast.
3. Doppler shifts imprinted in spectra lines provide an excellent way of testing theoretical expectations. Advances in understanding how this information can be extracted from spectroscopic data allow this.

Acknowledgments

A. Lazarian acknowledges the support of NSF Grant AST-0307869.

References

Armstrong, J.W., Rickett, B.J. and Spangler, S.R.: 1995, *ApJ* **443**, 209.
Batchelor, G.K.: 1959, *J. Fluid Mech.* **5**, 113.
Bertoglio, J.-P., Bataille, F. and Marion, J.-D.: 2001, *Phys. Fluids* **13**, 290.
Boldyrev, S., Nordlund, Å. and Padoan P.: 2002a, *ApJ* **573**, 678.
Boldyrev, S., Nordlund, Å. and Padoan P.: 2002b, *Phys. Rev. Lett.* **89**, 031102.
Brunt, C.M. and Heyer, M.H.: 2002a, *ApJ* **566**, 276.
Brunt C.M. and Heyer, M.H.: 2002b, *ApJ* **566**, 289.
Brunt, C.M., Heyer, M.H., Vazquez-Semadeni, E. and Pichardo, B.: 2003, in press, *ApJ*.
Cho, J. and Lazarian, A.: 2002a, *Phy. Rev. Lett.* **88**, 245001 (CL02).
Cho, J. and Lazarian, A.: 2002b, *ApJ* **575**, L63.
Cho, J. and Lazarian, A.: 2003a, in press, *MNRAS*, **345**, 325 (astro-ph/0301062) (CL03).
Cho, J. and Lazarian, A.: 2003b, preprint (astro-ph/0301462).
Cho, J., Lazarian, A., Honein, A., Knaepen, B., Kassinos, S. and Moin, P.: 2003, *ApJ* **589**, L77.
Cho, J., Lazarian, A. and Vishniac, E.: 2002a, *ApJ* **564**, 291 (CLV02a).
Cho, J., Lazarian, A. and Vishniac, E.: 2002b, *ApJ* **566**, L49 (CLV02b).
Cho, J., Lazarian, A. and Vishniac, E.: 2003a, in: E. Falgarone and T. Passot (eds.), *Turbulence and Magnetic Fields in Astrophysics*, Springer LNP, p. 56 (astro-ph/0205286) (CLV03a).
Cho, J., Lazarian, A. and Vishniac, E.: 2003b (astro-ph/0305212) *ApJ* **595**, 812.
Cho, J. and Vishniac, E.: 2000, *ApJ* **539**, 273.
Crutcher, R., Heiles, C. and Troland, T. 2003 in: E. Falgarone and T. Passot (eds.), *Turbulence and Magnetic Fields in Astrophysics*, Springer LNP, 155.
Del Zanna, L., Velli, M. and Londrillo, P.: 2001, *A&A* **367**, 705.
Deshpande, A.A.: 2000, *MNRAS* **317**, 199.
Draine, B.T. and Lazarian, A.: 1998, *ApJ* **508**, 157.
Elmegreen, B.: 2002, *ApJ* **577**, 206.
Elmegreen, B. and Falgarone, E.: 1996, *ApJ* **471**, 816.
Esquivel A., Lazarian A., Pogosyan D. and Cho J.: 2003, *MNRAS* **342**, 325.

Falgarone, E.: 1999, in: J. Franco and A. Carraminana (eds.), *Interstellar Turbulence*, CUP, (henceforth *Interstellar Turbulence*) p. 132.
Falgarone, E., Lis, D.C., Phillips, T.G., Pouquet, A., Porter, D.H. and Woodward, P.R.: 1994, *ApJ* **436**, 728.
Falgarone, E., Panis, J.-F., Heithausen, A., Perault, M., Stutzki, J., Puget, J.-L. and Bensch, F.: 1998, *A&A* **331**, 669.
Falgarone, E., Pineau des Forets, G. and Roueff, E.: 1995, *A&A* **300**, 870.
Falgarone, E and Puget, J.-L.: 1995, *A&A* **293**, 840.
Gill, A.G. and Henriksen, R.N.: 1990, *ApJ* **365**, L27.
Goldreich, P. and Sridhar, S.: 1995, *ApJ* **438**, 763.
Heiles, C.: 1997, *ApJ* **481**, 193.
Heyer, M.H. and Schloerb. F.P: 1997, *APJ* **475**, 173.
Higdon, J.C.: 1984, *ApJ* **285**, 109.
Kolmogorov, A.: 1941, *Dokl. Akad. Nauk SSSR* **31**, 538.
Larson, R.: 1981, *MNRAS* **194**, 809.
Lazarian, A.: 1999, in: *Interstellar Turbulence*, p. 95 (astro-ph/9804024).
Lazarian, A. and Esquivel, E.: 2003, *ApJ* **592**, L37 (LE03).
Lazarian, A., Petrosian, V., Yan, H. and Cho, J.: 2003, in: Rachid Ouyed (ed), *Beaming and Jets in Gamma Ray Bursts*, p. 45, in press (astro-ph/0301181).
Lazarian, A. and Pogosyan, D.: 2000, *ApJ* **537**, 720 (LP00).
Lazarian, A. and Pogosyan, D.: 2003, *ApJ*, in press (LP03).
Lazarian, A. and Pogosyan, D. and Esquivel, A.: 2002, in: A.R. Taylor et al. (eds), *Seeing Through the Dust*, *ASP Conf. Proc.* vol. 276, San Francisco, p. 182.
Lazarian, A. and Pogosyan, D., Vazquez-Semadeni, E., Pichardo, B., Landecker, T.L. and Willis, A.G.: 2001, *ApJ* **555**, 130.
Lazarian, A., Vishniac, E.T. and Cho, J.: 2004, *ApJ* **603**, 180 (LVC03).
Lazarian, A. and Yan, H.: 2003, in: A. Witt and B. Draine (eds.), *Astrophysical Dust*, in press, *APS* (astro-ph/0311370).
Lesieur, M.: 1990, *Turbulence In Fluids*, Dordrecht: Kluwer.
Lithwick, Y. and Goldreich, P.: 2001, *ApJ* **562**, 279.
Mac Low, M.-M. and Klessen, R.: 2003, preprint (astro-ph/0301093).
Mac Low, M.-M., Klessen, R., Burkert, A. and Smith, M.: 1998, *Phys. Rev. Lett.* **80**, 2754.
Maron, J. and Goldreich, P.: 2001, *ApJ* **554**, 1175.
Marscher, A.P., Moore, E.M. and Bania, T.M.: 1993, *ApJ* **419**, L101.
Matthaeus, W.H. and Brown, M.R.: 1988, *Phys. Fluids* **31**(12), 3634.
Matthaeus, W.H., Ghosh, S., Oughton, S. and Roberts, D.A.: 1996, *J. Geophysical Res.* **101**(A4), 7619.
McKee C.: 1999, in: J.L. Charles and D.N. Kylafis. (eds.), *The Origin of Stars and Planetary Systems*, Kluwer, p.29.
McKee, C.F. and Tan, J.C.: 2002, *Nature* **416**, 59.
Miesch, M.S. and Bally, J.: 1994, *ApJ* **429**, 645.
Miesch, M.S., Scalo, J. and Bally, J.: 1999, *ApJ* **429**, 645.
Miville-Deschenes, M.A., Levrier, F. and Falgarone, E.: 2003, *ApJ* in press.
Miville-Deschenes, M.A., Joncas, G., Falgarone, E. and Boulanger, F.: 2003, *A&A*, in press, astro-ph/0306570.
Monin, A.S. and Yaglom, A.M.: 1975, *Statistical Fluid Mechanics: Mechanics of Turbulence*, vol. 2, The MIT Press.
Müller, W.-C. and Biskamp, D.: 2000, *Phys. Rev. Lett.* **84**(3), 475.
Munch, G.: 1958, *Rev. Mod. Phys.* **30**, 1035.
O'Dell, C.R.: 1986, *ApJ* **304**, 767.

O'Dell, C.R. and Castaneda, H.O.: 1987, *ApJ* **317**, 686.
Ostriker, E.: 2003a, in: E. Falgarone and T. Passot (eds.), *Turbulence and Magnetic Fields in Astrophysics*, Springer LNP, p. 252.
Ossenkopf, V. and Mac Low, M.-M.: 2002, *A&A* **390**, 307.
Padoan, P., Boldyrev, S., Langer, W. and Nordlund, A.: 2003, *ApJ* **583**, 308.
Padoan, P. and Nordlund, A.: 1999, *ApJ* **526**, 279.
Padoan, P., Rosolowsky, E. and Goodman, A.: 2001, *ApJ* **547**, 862.
Porter, D., Pouquet, A. and Woodward, P.: 2002, *Phys. Rev. E* **66**, 026301.
Pouquet, A.: 1999, in: J. Franco and A. Carraminana (eds.), *Interstellar Turbulence*, p. 87.
Pudritz, R.E.: 2001, in: T. Montmerle and P. Andre (eds.), *From Darkness to Light: Origin and Evolution of Young Stellar Clusters*, ASP, Vol. 243, San Francisco, p.3
Rosolowsky E., Goodman, A., Wilner, D. and Williams, J.: 1999, *ApJ* **524**, 887.
Scalo, J.M.: 1987, in: D.F. Hollenback and H.A. Thronson (eds.), *Theoretical Approaches to Interstellar Turbulence*, *Interstellar Processes*, Reidel, Dordrecht, p. 225.
Stutzki, J.: 2001, *ApJS* **277**, 39.
Shebalin, J.V., Matthaeus, W.H. and Montgomery, D.C.: 1983, *J. Plasma Phys.* **29**, 525.
Stanimirovic, S. and Lazarian, A.: 2001, *ApJ*, **551**, L53.
Stone, J., Ostriker, E. and Gammie, C.: 1998, *ApJ* **508**, L99.
Vestuto, J.G., Ostriker, E.C. and Stone, J.M.: 2003, *ApJ* **590**, 858.
Zank, G.P. and Matthaeus, W.H.: 1993, *Phys. Fluids A* **5**(1), 257.

MAGNETIC FLUX TRANSPORT IN THE ISM THROUGH TURBULENT AMBIPOLAR DIFFUSION

FABIAN HEITSCH[1], ELLEN G. ZWEIBEL[1], ADRIANNE, D. SLYZ[2] and JULIEN E.G. DEVRIENDT[2]

[1]*University of Wisconsin-Madison, U.S.A.*

[2]*University of Oxford, U.K.*

Abstract. Under ideal MHD conditions the magnetic field strength should be correlated with density in the interstellar medium (ISM). However, observations indicate that this correlation is weaker than expected. Ambipolar diffusion can decrease the flux-to-mass ratio in weakly ionized media; however, it is generally thought to be too slow to play a significant role in the ISM except in the densest molecular clouds. Turbulence is often invoked in other astrophysical problems to increase transport rates above the (very slow) diffusive values. Building on analytical studies, we test with numerical models whether turbulence can enhance the ambipolar diffusion rate sufficiently to explain the observed weak correlations. The numerical method is based on a gas-kinetic scheme with very low numerical diffusivity, thus allowing us to separate numerical and physical diffusion effects.

Keywords: ISM, MHD, ambipolar diffusion, turbulence

1. Motivation

Under ideal MHD conditions, the magnetic field strength should be correlated with density in the interstellar medium (ISM). Isotropic compression of a sphere of gas permeated by a uniform magnetic field results in $B \propto n^{2/3}$, whereas gravitationally controlled regions should theoretically scale as $B \propto n^{1/2}$ (e.g., Fiedler and Mouschovias, 1993). On the other hand, there still is substantial debate about the observational evidence on the $B(n)$ relation. While Zeeman detections (not including upper limits) in the dense molecular gas (Crutcher, 1999; Bourke et al., 2001) indicate a relation consistent with $B \propto n^{1/2}$, B does not seem to correlate with density in the diffuse medium at all (Troland and Heiles, 1986; Heiles, 2003).

Under flux-freezing conditions typical for the ISM, it is difficult to keep the magnetic field strength weak. Of course one could imagine dense regions assembling through flows predominantly along the field lines, in which case we would not expect any correlation between B and n. However, Mestel (1985) pointed out that in this case gas would have to be gathered over distances of about 1 kpc in order to reach the required masses. Moreover, it is not clear whether magnetic fields are strong enough to collimate the flows over such large distances (see however, Ballesteros-Paredes et al., 1999).

Astrophysics and Space Science **292:** 45–51, 2004.

Two mechanisms can break the frozen-flux condition in the ISM: reconnection and ion-neutral drift (a.k.a. ambipolar diffusion, hereafter AD). Invoking reconnection to break the frozen-flux conditions requires an efficient mechanism to raise the reconnection rate far above its intrinsically slow rate in the ISM (e.g., Petschek, 1964; Lazarian and Vishniac, 1999). While we do not rule this out [especially in combination with AD, (Heitsch and Zweibel, 2003)], in this contribution, we investigate numerically the rôle of turbulence in a weakly ionized medium.

For parameters typical of the diffuse to dense ISM, the ambipolar diffusion time scale

$$\tau_{AD} \equiv \frac{L^2}{\lambda_{AD}} = 9.4 \times 10^{-3}\, L^2 \left(\frac{n_n}{B}\right)^2 \frac{\mu_i\, \mu_n}{\mu_i + \mu_n}\, x_i \ \text{[yrs]} \tag{1}$$

is approximately a factor 10 to 100 larger than the dynamical timescales (see e.g., Zweibel, 2002). In Eq. (1), the characteristic length scale L is given in parsecs, the neutral density in cm^{-3}, x_i is the ionization fraction, and $\mu_{i,n}$ are the ion and neutral molecular weights, respectively. Thus, in order to keep the magnetic field weak by ambipolar diffusion, we would have to decrease τ_{AD} by at least one order of magnitude. Zweibel (2002) analytically investigated the effect of an ensemble of stagnation point flows on the ambipolar decay rate and found that such an idealized setup can indeed accelerate the diffusion by a combination of transport and stretching. Kim and Diamond (2002), for mean field theory and Fatuzzo and Adams (2002) emphasizing the importance of field perturbations came to similar results.

This paper is a first attempt at generalizing the analytical results via numerical models. It is based on a presentation at the MFSF Conference in Madrid (2003) to which this volume is devoted. More details and further results will be given in Heitsch et al. (2003).

2. The Models

A numerical study of diffusion processes requires us to separate all relevant diffusion scales, from the global ambipolar diffusion scale down to the numerical diffusion scale, equivalent to the grid scale. We require a numerical method with very low intrinsic numerical diffusion. We used a modified version of the first order gas-kinetic flux-splitting method (Xu, 1999; Tang and Xu, 2000), including ambipolar diffusion in the two-fluid description (see e.g., Mac Low and Smith, 1997) and an Ohmic resistivity, which allows us to control the numerical dissipation of the magnetic field.

For simplicity, we assume a two-dimensional domain, with the field component perpendicular to the domain, corresponding to a strongfield-case in 3D. The Lorentz-force reduces to the magnetic pressure gradient. By restricting the problem to 2D and treating the field (nearly) as an advected quantity, we evade ambiguities arising from field line stretching, which complicate the determination of a diffusion rate. The recombination time scale is by far the shortest of all relevant time scales in the regime of interest (Heitsch and Zweibel, 2003). Assuming ionization equilibrium, the ion density is then slaved to the neutral density. To further simplify, the neutral flow is set up to be incompressible. Although this is certainly not true for large parts of the ISM, it will allow us to isolate the turbulent diffusion effects more clearly. We use an isothermal equation of state, as otherwise the ion-neutral friction would heat the gas, an effect we would have to counter with appropriate cooling descriptions.

With all of the above assumptions, the ion momentum equation reads

$$\rho_i \left(\frac{\partial}{\partial t} \mathbf{u}_i + (\mathbf{u}_i \cdot \nabla)\mathbf{u}_i \right) = -\frac{1}{8\pi} \nabla B^2 - \rho_i \nu_{\text{in}} (\mathbf{u}_i - \mathbf{u}_n), \tag{2}$$

where $\mathbf{u}_i$ and $\mathbf{u}_n$ are the ion and neutral velocities, ρ_i is the (constant) ion density, B the z-component of the magnetic field and ν_{in} is the ion-neutral collision frequency. The induction equation reduces in our case to

$$\frac{\partial}{\partial t} B = -\frac{\partial}{\partial x} u_{i,x} B - \frac{\partial}{\partial y} u_{i,y} B + \lambda_\Omega \nabla^2 B, \tag{3}$$

where λ_Ω is the resistivity which we require to dominate the numerical diffusion.

A major simplification results from treating the neutral velocity as known. From a physical point of view, this corresponds to choosing a spatial scale so small that the neutrals are not coupled to the magnetic field.[1] We choose the two-dimensional version of the "circularly polarized" flow of Galloway and Proctor (1992), (hereafter GP-flow), which we write as

$$\begin{aligned} u_{n,x} &= \sqrt{\frac{3}{2}} v_f \cos(2\pi(k_f y + \epsilon \sin(2\pi \omega k_f v_f t))) \\ u_{n,y} &= \sqrt{\frac{3}{2}} v_f \sin(2\pi(k_f x + \epsilon \cos(2\pi \omega k_f v_f t))), \end{aligned} \tag{4}$$

where $u_{n,x}$ and $u_{n,y}$ are the x and y components of the neutral velocity $\mathbf{u}_n$, respectively. The GP flow velocity is v_f, the wave number is $k_f = \kappa_f / L$, where κ_f is the number of GP cells per box length L. We set the number of GP-cells $\kappa_f = 5$, resulting in 25 eddy structures in the domain.

The GP flow is by itself a single-scale flow. For $\epsilon = 0$, the flow has a regular cellular structure with stationary stagnation points, with an eddy turnover time of

[1]Because of the asymmetry in the ion-neutral and neutral-ion collision frequency, this does not necessarily mean that the ions do not feel the neutral flow.

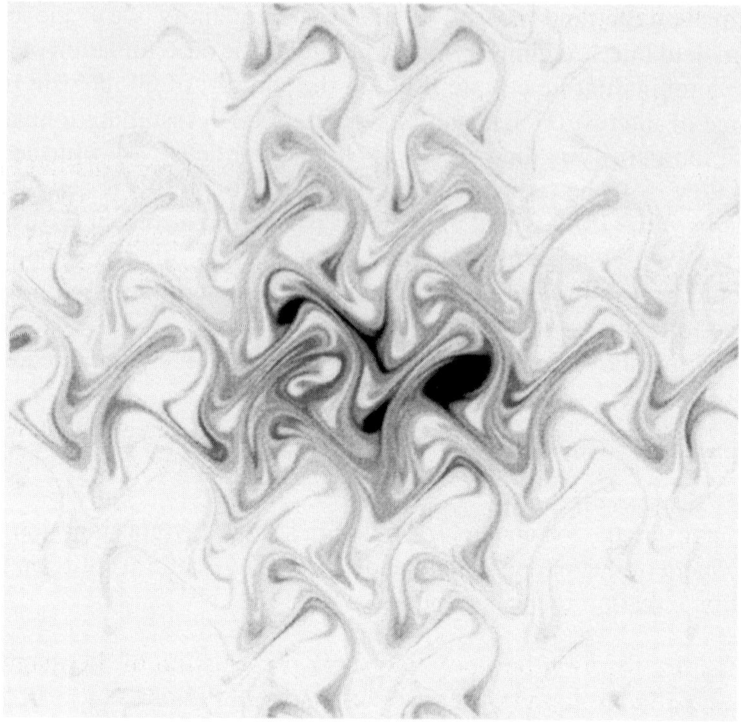

Figure 1. Pattern caused by a tracer field initially placed within a circle with diameter $L/4$ at the center of the domain, after ≈4 GP-eddy turnover times. Black is highest density. Resolution: $N = 1601^2$ grid points.

$\tau_f = \frac{L}{\kappa_f v_f}$. Increasing ϵ increases the complexity of the flow patterns, for small ϵ at the cell boundaries, for larger ϵ over growing fractions of the domain. The flow pattern for $\epsilon > 0$ can be visualized as a group of evenly spaced eddies travelling in snake-like patterns across the domain. The eddies themselves act as a kind of sink to any advected quantities, so that turbulent transport for $\epsilon = 0$ should be at a minimum. Here, the term "sink" means that once the advected quantity enters an eddy, it will circle around the eddy's center indefinitely, and only diffuse in the (stationary) stagnation points. For $\epsilon > 0$, the eddies break up and re-form, so that advected quantities are not bound to the eddies any more, but can travel over several eddy diameters, leading to turbulent transport. Thus, the flow combines the required transport and stretching properties. Figure 1 gives an example of the transport of a tracer quantity initially placed within a circle with diameter $L/4$, after ≈4 GP-eddy turnover times.

Note that the ion (and neutral) density in the domain remains constant. However, the ion flow generally will *not* be incompressible, due to the Lorentz force and the nonlinear Reynolds stress terms in the momentum equation. The initial conditions for the magnetic field are given by

$$B(x, y, 0) = B_0 + \frac{1}{2} B_1 (1 + \cos(kx)\cos(ky)), \tag{5}$$

where we choose $k = 1$, with $x, y \in [0, 2\pi]$. The offset prevents field reversals, which are known to cause instabilities (Brandenburg and Zweibel, 1994). We report here on results with $\kappa_f = 5$, or 25 eddy structures in the domain. Validity conditions and acceptable parameter ranges for the model will be discussed elsewhere.

3. Results

Our goal is to define an average diffusion constant, in order to establish whether we can treat turbulent ambipolar diffusion as an actual diffusion process. We compute the average diffusion constant D as

$$D = \frac{\partial}{\partial t} \int_A \langle B_{k=1} \rangle \, da \Big/ \oint_C \mathbf{n} \cdot \nabla \langle B_{k=1} \rangle \, dl \tag{6}$$

for a square area covering the initial ($k = 1$) perturbation of the magnetic field. $\nabla \langle B_{k=1} \rangle$ refers to a time average over one (local) GP period.

From Figure 2 we conclude,

1. The quiescent diffusion rates (open symbols) agree with the theoretical predictions from the dispersion relation of a magnetosonic wave including AD. For $\nu_{\text{in}} = 23.0$, the numerical rate is slightly larger, since we reach a regime where the Ohmic resistivity gets important.
2. Numerical results for turbulent diffusion rates at resolutions of $N = 801^2$ (filled symbols) and $N = 1601^2$ (thick lines) are indistinguishable, given the plot's resolution: The models are numerically resolved.
3. For an increasing collision frequency ν_{in}, the quiescent decay rate drops linearly as expected, whereas the turbulent decay rate is nearly independent of ν_{in}. Thus, the "amplification" of turbulent over quiescent decay increases with increasing collision frequency, reaching a factor of more than 30 for $\nu_{\text{in}} = 23.0$. Turbulent diffusion by definition should be independent of the microscopic diffusion mechanism. For the lowest collision frequency (filled and open diamonds), the quiescent decay rate starts to overwhelm the turbulent decay.

4. Summary

We presented first results of a numerical study on turbulent ambipolar diffusion, based on previous analytical work by Zweibel (2002). Within the idealized models presented, turbulent transport and stretching can enhance the diffusion rate well

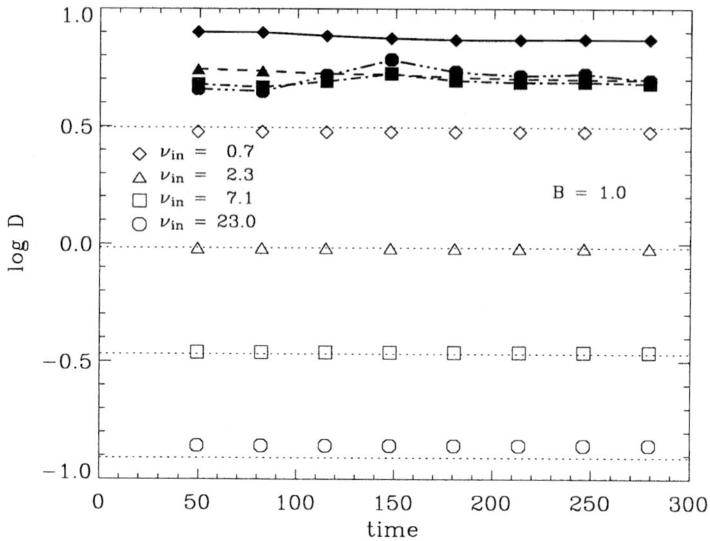

Figure 2. Average diffusion rate D as given in Eq. (6) against time. Open symbols denote results for the quiescent case, filled symbols ($N = 801^2$) and lines ($N = 1601^2$) for the turbulent case. The quiescent rates agree with the theoretical predictions (dotted lines) from the dispersion relation for a magnetosonic wave including AD.

above the quiescent rate. The models display a turbulent diffusion rate independent of the microscopic diffusion mechanism, a canonical feature of turbulent diffusion.

Limitations of the current study include an idealized domain geometry and flow pattern, a non-fully dynamic problem set-up (e.g., the backreaction of the neutrals is not included) and a restricted parameter space. These issues will be addressed in subsequent work.

Acknowledgements

F. Heitsch is grateful for support by a Feodor-Lynen fellowship of the Alexander-von-Humboldt Foundation. This work was supported by NSF grants AST-0098701 and AST-0328821, and the UW Madison Graduate School. A. Slyz acknowledges the support of a Fellowship from the UK Astrophysical Fluids Facility (UKAFF). The research of J. E. G. D. at Oxford is funded by the Leverhulme Trust. Computations presented here were in parts performed on the SGI Origin 2000 machines of the National Center for Supercomputing Applications at Urbana-Champain/IL.

References

Ballesteros-Paredes, J., Vázquez-Semadeni, E. and Scalo, J.: 1999, *ApJ* **515**, 286.
Bourke, T.L., Myers, P.C., Robinson, G. and Hyland, A.R.: 2001, *ApJ* **554**, 916.

Brandenburg, A. and Zweibel, E.G.: 1994, *ApJL* **427**, 91.
Crutcher, R.M.: 1999, *ApJ* **520**, 706.
Fatuzzo, M. and Adams, F.C.: 2002, *ApJ* **570**, 210.
Fiedler, R.A. and Mouschovias, T.Ch.: 1993, *ApJ* **415**, 680.
Galloway, D.J. and Proctor, M.R.E.: 1992, *Nature* **356**, 691.
Heiles, C.: 2003, *Observations of Turbulence in the Diffuse ISM, Proceedings of the Workshop: Magnetic Fields and Star Formation: Theory Versus Observations,* Kluwer.
Heitsch, F. and Zweibel, E.G.: 2003, *ApJ* **583**, 229.
Heitsch, F., Zweibel, E.G., Slyz, A.D. and Devriendt, J.E.G.: 2003, in preparation.
Kim, E.-J. and Diamond, P.H.: 2002, *ApJ* **578**, 113.
Lazarian, A. and Vishniac, E.T.: 1999, *ApJ* **517**, 700.
Mac Low, M.-M. and Smith, M.D.: 1997, *ApJ* **491**, 596.
Mestel, L.: 1985, in: D.C. Black and M.S. (eds.), *Protostars and Planes II*, Matthews Tucson: Univ. Arizona Press, 320.
Petschek, A.G.: 1969, in: W.N. Hess (ed.), *AAS-NASA Symp. on the Physics of Solar Flares*, NASA SP-50, Greenbelt: NASA, p. 425.
Tang, H.Z. and Xu, K.: 2000, *J. Comp. Phys.* **165**, 69.
Troland, T.H. and Heiles, C.: 1986, *ApJ* **339**, 345.
Xu, K.: 1999, *J. Comp. Phys.* **153**, 334.
Zweibel, E.G.: 2002, *ApJ* **567**, 962.

HIGH-RESOLUTION SIMULATIONS OF NONHELICAL MHD TURBULENCE

N.E.L. HAUGEN[1], A. BRANDENBURG[2] and W. DOBLER[3]

[1]*Department of Physics, The Norwegian University of Science and Technology, Høyskoleringen 5, N-7034 Trondheim, Norway; E-mail: nils.haugen@phys.ntnu.no*

[2]*NORDITA, Blegdamsvej 17, DK-2100 Copenhagen Ø, Denmark*

[3]*Kiepenheuer-Institut für Sonnenphysik, Schöneckstraße 6, D-79104 Freiburg, Germany*

Abstract. According to the kinematic theory of nonhelical dynamo action, the magnetic energy spectrum increases with wavenumber and peaks at the resistive cutoff wavenumber. It has previously been argued that even in the dynamical case, the magnetic energy peaks at the resistive scale. Using high resolution simulations (up to 1024^3 meshpoints) with no large-scale imposed field, we show that the magnetic energy peaks at a wavenumber that is independent of the magnetic Reynolds number and about five times larger than the forcing wavenumber. Throughout the inertial range, the spectral magnetic energy exceeds the kinetic energy by a factor of two to three. Both spectra are approximately parallel. The total energy spectrum seems to be close to $k^{-3/2}$, but there is a strong bottleneck effect and we suggest that the asymptotic spectrum is instead $k^{-5/3}$. This is supported by the value of the second-order structure function exponent that is found to be $\zeta_2 = 0.70$, suggesting a $k^{-1.70}$ spectrum. The third-order structure function scaling exponent is very close to unity,—in agreement with Goldreich–Sridhar theory.

Adding an imposed field tends to suppress the small-scale magnetic field. We find that at large scales the magnetic energy spectrum then follows a k^{-1} slope. When the strength of the imposed field is of the same order as the dynamo generated field, we find almost equipartition between the magnetic and kinetic energy spectra.

Keywords: interstellar medium, turbulence

1. Introduction

Magnetic fields may play an important role during star formation. Stars are generally formed in strongly magnetized regions, and the magnetic pressure that builds up in shocks and the initial collapse is likely to determine the detailed evolution.

Early simulations of hydromagnetic turbulence have always suggested that the magnetic field is more intermittent than the velocity field, if the field is generated by dynamo action (Meneguzzi et al., 1981; Kida et al., 1991). Furthermore, linear theory (Kazantsev, 1968) suggests that the magnetic spectrum should peak at the resistive scale, and it has been argued that this may hold even in the nonlinear regime (Maron and Blackman, 2002).

On the other hand, if there is an imposed large scale field, there is no doubt that most of the magnetic energy resides at large scales e.g., Cho and Vishniac (2000). The obvious question is therefore, are the cases of dynamo-generated and imposed fields really drastically different?

Astrophysics and Space Science **292:** 53–60, 2004.

The purpose of this paper is to compare the case of an imposed field with that of a dynamo-generated one. We begin by briefly reviewing the main results of our recent paper (Haugen et al., 2003), where we show that in the dynamo case the magnetic energy does *not* peak at the resistive scale.

2. Equations

Our approach is the same as in Haugen et al. (2003), which is similar to that of Brandenburg (2001), except that the flow is now forced without helicity. We adopt an isothermal equation of state with a constant sound speed $c_{\rm s}$, so the pressure p is related to the density ρ by $p = \rho c_{\rm s}^2$. The equation of motion is written in the form,

$$\frac{\mathrm{D}\boldsymbol{u}}{\mathrm{D}t} = -c_{\rm s}^2 \boldsymbol{\nabla} \ln \rho + \frac{\boldsymbol{J} \times \boldsymbol{B}}{\rho} + \boldsymbol{F}_{\rm visc} + \boldsymbol{f}, \tag{1}$$

where $\mathrm{D}/\mathrm{D}t = \partial/\partial t + \boldsymbol{u} \cdot \boldsymbol{\nabla}$ is the advective derivative, $\boldsymbol{J} = \boldsymbol{\nabla} \times \boldsymbol{B}/\mu_0$ is the current density, μ_0 is the vacuum permeability, $\boldsymbol{F}_{\rm visc}$ is the viscous force, and $\boldsymbol{f}$ is a random forcing function that consists of nonhelical plane waves. The continuity equation is written in terms of the logarithmic density,

$$\frac{\mathrm{D} \ln \rho}{\mathrm{D}t} = -\boldsymbol{\nabla} \cdot \boldsymbol{u}, \tag{2}$$

and the induction equation is solved in terms of the magnetic vector potential $\boldsymbol{A}$, where $\boldsymbol{B} = \boldsymbol{\nabla} \times \boldsymbol{A}$, so

$$\frac{\partial \boldsymbol{A}}{\partial t} = \boldsymbol{u} \times \boldsymbol{B} + \eta \boldsymbol{\nabla}^2 \boldsymbol{A}, \tag{3}$$

where $\eta = \text{const}$ is the magnetic diffusivity. We use periodic boundary conditions in all three directions for all variables.

The solutions are characterized by the kinetic and magnetic Reynolds numbers that are based on the forcing wavenumber $k_{\rm f}$ and defined as

$$\mathrm{Re} = u_{\rm rms}/(\nu k_{\rm f}), \quad R_{\rm m} = u_{\rm rms}/(\eta k_{\rm f}), \tag{4}$$

respectively. The ratio of the two is the magnetic Prandtl number,

$$P_{\rm m} = \nu/\eta = R_{\rm m}/\mathrm{Re}. \tag{5}$$

We use the Pencil Code[1], which is a cache and memory efficient high-order finite-difference code (sixth order in space and third order in time) for solving the compressible MHD equations.

[1]http://www.nordita.dk/data/brandenb/pencil-code.

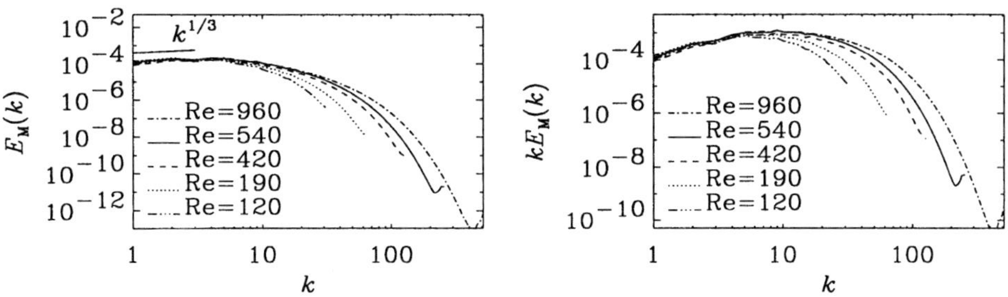

Figure 1. Left: Magnetic energy spectra for all our runs. We see that the peak of the magnetic energy spectrum is at $k = 5$ for all values of Re. *Right*: The peak of $kE_M(k)$ is found at $k \approx 9$ for large values of Re.

3. Results

3.1. The peak of the magnetic energy spectrum

We have run simulations with up to 1024^3 meshpoints to show that the magnetic energy spectrum does not peak at the resistive scale, as has previously been claimed (Maron and Blackman, 2002). In the left panel of Figure 1 we see that the peak of the spectrum is around $k = 5$ for all our runs, i.e., it is independent of Re. We also note that we find a $k^{1/3}$ slope for small values of k (Batchelor, 1950).

A more stringent measure is to look at the magnetic energy per unit logarithmic wavenumber interval, $kE_M(k)$, which would be flat if the contribution from small and large wavenumbers was equal. This is shown in the right hand panel of Figure 1. We see that the peak of $kE_M(k)$ is shifted toward smaller scales compared to $E_M(k)$, but it is still not at the resistive scale. We do indeed see that for the largest runs it seems to settle at $k \approx 10$, which is well within the inertial range.

3.2. The inertial range

In Figure 2 we see that there seems to be a clear inertial range for $7 \lesssim k \lesssim 25$, where $E_M(k)$ and $E_K(k)$ are parallel and have a slope of $k^{-3/2}$. The $k^{-3/2}$ slope is suggestive of the Iroshnikov (1963) and Kraichnan (1965) (IK) theory, and may seem incompatible with the Goldreich and Sridhar (1995) (GS) theory. We also note that in the inertial range, the relative fractions of magnetic and kinetic energy seem to be saturated at $E_M(k)/E_K(k) \approx 2.3$.

Knowing that IK theory predicts that the fourth-order structure function scales linearly ($\zeta_4 = 1$), while GS theory predicts linear scaling for the third-order structure function ($\zeta_3 = 1$), we now calculate the logarithmic derivatives for these structure functions; see Figure 3. From these plots we see that the IK theory cannot be correct since the fourth-order structure function is clearly steeper than linear (i.e., $\zeta_4 = 1.3 \neq 1$). The third-order structure function on the other hand scales linearly.

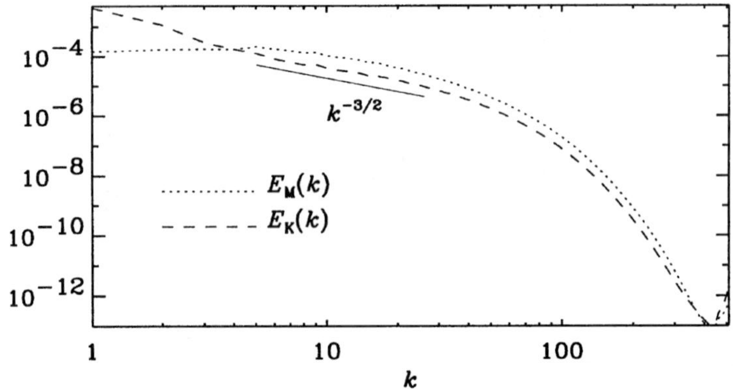

Figure 2. Kinetic and magnetic energy spectra of our largest run; 1024^3 mesh points.

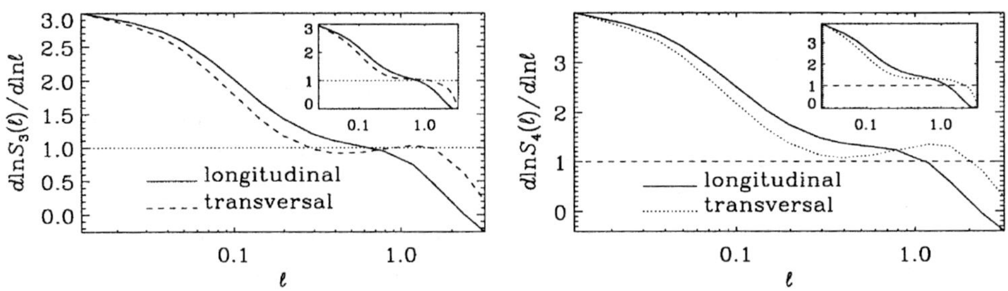

Figure 3. Left: logarithmic derivative of third-order structure function. The inset is for a run with 256^3 mesh points, while the large plot is for a run with 512^3 mesh points. The result is consistent with $\zeta_3 = 1$ *Right:* logarithmic derivative of fourth-order structure function. We see that this is clearly *not* compatible with IK theory, which would predict $\zeta_4 = 1$.

The second-order structure function scaling exponent of the Elsasser variable, $z^{\pm} = \boldsymbol{u} \pm \boldsymbol{B}/\sqrt{\rho\mu}$, is also indicative of GS theory being applicable, since we find $\zeta_2 = 0.7$, which implies $E_T(k) = E_M(k) + E_K(k) \propto k^{-(1+\zeta_2)} = k^{-1.7}$.

As we have argued earlier (Haugen et al., 2003), the reason that Figure 2 shows a $k^{-3/2}$ inertial range is that there is a strong bottleneck effect (Falkovich, 1994). It turns out that this bottleneck is much stronger in three-dimensional than in one-dimensional energy spectra (Dobler et al., 2003). We therefore plot the one-dimensional counterpart of Figure 2 in Figure 4. Here we see that the inertial range indeed has a slope close to $-5/3$, as suggested by our previous findings from the structure functions. From this we conclude that also the three-dimensional energy spectra will show a $k^{-5/3}$ inertial range far enough away from the diffusive subrange.

3.3. Imposed Magnetic Field

Until now, we have only been looking at the case with no externally imposed field. We, therefore, want to see what effect such imposed fields of varying strengths

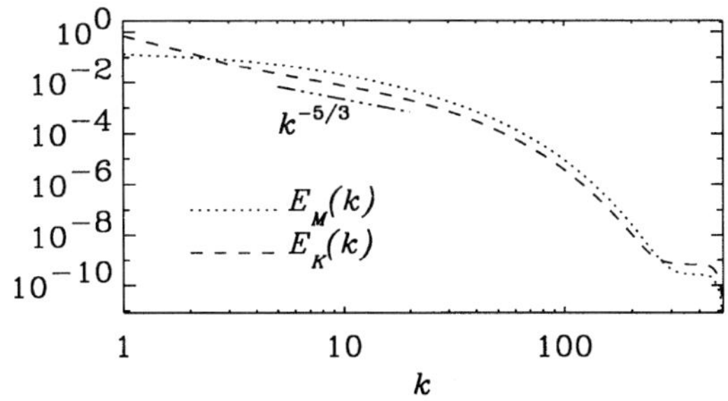

Figure 4. One-dimensional energy spectra of our largest run. We see here that the inertial range is consistent with a $k^{-5/3}$ slope.

have on the dynamo. In Figure 5, we plot energy spectra for simulations with 128^3 meshpoints (left panel) and 256^3 meshpoints (right panel) for different imposed fields. From the figure we see that having an imposed field seems to increase the large-scale magnetic field and at the same time *decreases* the small-scale field. The stronger the external field, the stronger the suppression of the small-scale fields. The same effect is clearly seen in Figure 7. We also note that with an imposed field, the peak of the magnetic energy is found at the largest scale ($k = 1$), not at $k = 5$ as in the case without an imposed field. Looking at Figure 6 we see that for $1 < k < 4$ there seems to be a k^{-1} slope for $E_M(k)$. Such a slope for the large-scale magnetic field has been suggested previously (Ruzmaikin and Shukurov, 1982; Kleeorin and Rogachevskii, 1994; Brandenburg et al., 1996; Matthaeus and Goldstein, 1986). From $k \approx 6$ there seems to be a short inertial range with a $k^{-5/3}$ slope, as expected from GS theory.

When the strength of the imposed field is comparable to the dynamo generated field ($B_0 = 0.06$ for 128^3 meshpoints, and $B_0 = 0.3$ for 256^3 meshpoints in Figure 5), we see that there is almost equipartition between magnetic and kinetic energy spectra, at least for the smaller scales. On the other hand, Cho and Vishniac (2000) find almost perfect equipartition for all scales. The difference is small, and could perhaps be explained by a difference in the forcing function. We did check, however, that changing from a delta-correlated forcing function to one with a renewal time comparable to the turnover time does not resolve this relatively minor discrepancy.

4. Conclusions

Our simulations suggest that both magnetic and kinetic energy spectra show power-law behavior and are proportional to $k^{-5/3}$. The peaks of these spectra are therefore located at relatively small wavenumbers; $k = 5$ and $k = k_f = 1$, respectively. When applied to the interstellar medium, where the forcing wavenumber corresponds to a scale of up to 100 pc, the scale of the magnetic peak would be at a wavenumber

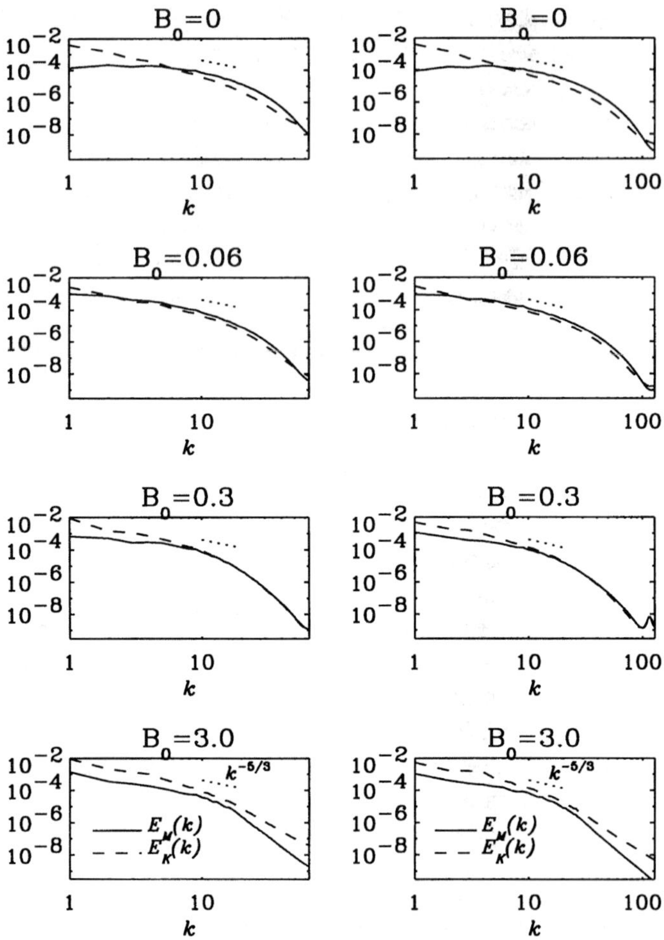

Figure 5. Energy spectra for simulations with imposed fields of different strengths. Solid lines show magnetic spectra; dashed lines show kinetic spectra. *Left:* resolution 128^3 meshpoints; Reynolds numbers vary between 200 and 400. *Right:* 256^3 meshpoints; Reynolds numbers vary between 320 and 500. In both cases the large Reynolds numbers correspond to the cases with $B_0 = 0.3$ and $B_0 = 3.0$, where the kinetic motions are least suppressed by the magnetic fields.

corresponding to a scale of about 10 pc. Thus, the peak is not anywhere near the resistive scale, as has previously been suggested. There is of course still energy at scales as small as the resistive scale ($\sim 10^8$ cm), as evidenced by interstellar scintillation measurements, but this energy can only be a small fraction of the total magnetic energy.

When a magnetic field is imposed, several features change, the magnetic spectrum peaks at $k = 1$, but at higher wavenumbers the magnetic energy is suppressed compared to the case without an imposed field. We believe this is because of a suppression of dynamo action. This highlights once again that imposed fields are

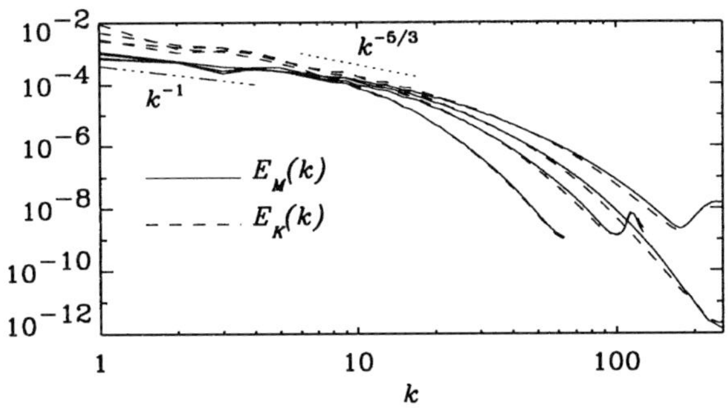

Figure 6. Convergence of energy spectra for runs with increasing Re (=280, 500, 650, and 1170), and with $B_0 = 0.3$. We see a k^{-1} for $1 < k < 4$, and we also start to see the appearance of a short $k^{-5/3}$ inertial range beginning at $k \approx 8$.

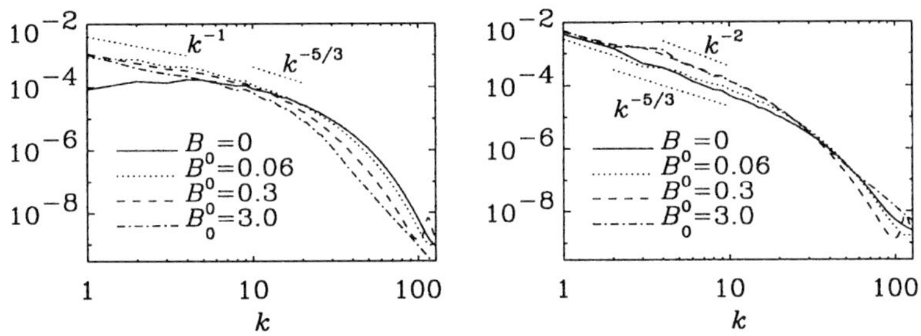

Figure 7. Power spectra for runs with different imposed fields; Re = 320–500. *Left:* magnetic energy spectra; we see that the the larger the external field the more the magnetic field at small scales is suppressed. *Right:* kinetic energy spectra.

somewhat artificial and do not represent the case of a self-generated large-scale magnetic field in a small subvolume of a large turbulent dynamo system.

Acknowledgements

Use of the parallel computers in Trondheim (Gridur), Odense (Horseshoe), Leicester (Ukaff), and Bergen (Fire) is acknowledged.

References

Batchelor, G.K.: 1950, *Proc. Roy. Soc. Lond.* **A201**, 405.
Brandenburg, A.: 2001, *ApJ* **550**, 824.
Brandenburg, A., Jennings, R.L., Nordlund, Å., Rieutord, M., Stein, R.F. and Tuominen, I.: 1996, *J. Fluid Mech.* **306**, 325.
Cho, J. and Vishniac, E.: 2000, *ApJ* **539**, 273.

Dobler, W., Haugen, N.E.L., Yousef, T.A. and Brandenburg, A.: (2003), *Phys. Rev. E* **68**, 026304.
Falkovich, G.: 1994, *Phys. Fluids* **6**, 1411.
Goldreich, P. and Sridhar, S.: 1995, *ApJ* **438**, 763.
Haugen, N.E.L., Brandenburg, A. and Dobler, W.: (2003), *ApJ* **597**, L141.
Iroshnikov, R.S.: (1963), *Sov. Astron.* **7**, 566.
Kazantsev, A.P.: (1968), *Sov. Phys. JETP* **26**, 1031.
Kida, S., Yanase, S. and Mizushima, J.: 1991, *Phys. Fluids* **A3**, 457.
Kleeorin, N. and Rogachevskii, I.: (1994), *Phys. Rev.* **E 50**, 2716.
Kraichnan, R.H.: 1965, *Phys. Fluids* **8**, 1385.
Maron, J. and Blackman, E.G.: (2002), *ApJ* **566**, L41.
Matthaeus, W.H. and Goldstein, M.L.: (1986), *Phys. Rev. Lett.* **57**, 495.
Meneguzzi, M., Frisch, U. and Pouquet, A.: (1981), *Phys. Rev. Lett.* **47**, 1060.
Ruzmaikin, A.A. and Shukurov, A.M.: (1982), *Astrophys. Spa. Sci.* **82**, 397.

SCALING RELATIONS OF SUPERSONIC TURBULENCE IN MOLECULAR CLOUDS

S. BOLDYREV[1], P. PADOAN[2], R. JIMENEZ[3] and Å. NORDLUND[4]

[1]*Department of Asronomy and Astrophysics, University of Chicago, Chicago, IL 60637*

[2]*Department of Physics, University of California at San Diego, San Diego, CA 92093*

[3]*Department of Physics and Astronomy, University of Pennsylvania, Philadelphia, PA 19104*

[4]*Copenhagen Astron. Observatory and Theor. Astrophys. Center, DK-2100 Copenhagen, Denmark*

Abstract. We discuss a model for driven supersonic, super-Alfvénic MHD turbulence that is believed to govern the structure of molecular clouds. Such turbulence is highly intermittent; we describe its statistical properties by obtaining scaling of velocity-difference structure functions. This scaling was analytically predicted in Boldyrev (2002), confirmed in numerical simulations by Boldyrev et al. (2002), and discovered in observations by Padoan et al. (2003).

Keywords: interstellar medium, molecular clouds, MHD turbulence

Introduction

Turbulent gas motion is consistently observed in interstellar molecular clouds. Although the nature of the driving force may vary, observations establish that the turbulence is highly supersonic. The gas that is cooled down to ~10K can have bulk velocities of the order 1–10 km/sec, thus having Mach numbers of the order 30 at scales ~100pc (Larson, 1979, 1981; Falgarone and Philips, 1990; Falgarone et al., 1992; Myers and Gammie, 1999; Padoan and Nordlund, 1999; Ossenkopf and Mac Low, 2000; Ostriker et al., 2001). The density contrast induced by such motion can also be high, $n > 10^4$, and the cloud shapes can be irregular, with fractal boundaries. In other words, it is impossible to speak about a 'typical' molecular cloud. Rather, clouds sizes and shapes, magnetic fields, density and velocity profiles should be described statistically, in terms of probability distribution functions or correlators.

Such first attempts were undertaken by Weizsäcker (1951) , and later by Larson (1981), where the spectrum of the gas velocity was inferred from observations. In both cases it was found that the spectrum was steeper than the Kolmogorov one, $E_k \sim k^2|u_k|^2 \sim k^{-5/3}$, and in both cases it was realized that compressibility of the turbulence was responsible for the steepening of the spectrum.

Let us recall that the Kolmogorov theory of incompressible hydrodynamic turbulence has the exact result

$$\langle (u(\mathbf{x}+\mathbf{L}) - u(\mathbf{x}))^3 \rangle = -\frac{4}{5}\epsilon_L L, \tag{1}$$

where u is the component of the velocity field along the vector $\mathbf{L}$, angular brackets denote ensemble average, and ϵ_L is the density of energy dissipation, which is a

Astrophysics and Space Science **292:** 61–68, 2004.

constant in the Kolmogorov theory. In this formula, L is much smaller than the scale where the turbulence is forced, and much larger than the scale where the turbulence is dissipated. Analogously, one can consider structure functions of different orders, $S_p = \langle(u(\mathbf{x}+\mathbf{L}) - u(\mathbf{x}))^p\rangle \sim L^{\zeta_p}$, however, their scalings, ζ_p, are not known for $p \neq 3$. If the velocity field was not strongly fluctuating, one could conclude from Eq. (1) that $u_L \sim L^{1/3}$, where $u_L \equiv u(\mathbf{x}+\mathbf{L}) - u(\mathbf{x})$, and, therefore, $S_p \sim L^{p/3}$. The non-intermittent scaling, $\zeta_p = p/3$, is also called the Kolmogorov scaling. Such a scaling would be observed if the velocity difference, u_L, had a Gaussian distribution, and such a theory is sometimes referred to as the 'mean-field theory'. For low-order moments, the mean-field theory works rather well, for example, it predicts $\zeta_2 = 2/3$, and the velocity spectrum is found as $E_k \sim k^{-1-\zeta_2} = k^{-5/3}$. However, experimental and numerical investigations of structure functions revealed departure from the linear Kolmogorov scaling, and this departure is attributed to (or may serve as the definition of) intermittency (Benzi et al., 1993; Gotoh, 2002).

Intermittency is rather strong in the compressible turbulence, which makes such turbulence rather difficult even for a qualitative understanding. Indeed, even if we succeed in developing a model for the spectrum of such turbulence, i.e., for the second-order structure function $S_2(L)$, this may give us little insight about the physical processes governing the high-order moments. This is understood, since very different distributions can share the same second moment. The numerical simulations presented in Boldyrev et al. (2002) were devoted to supersonic, super-Alfvénic turbulence ($M \sim 10$, $M_A \sim 3$), driven on the largest scales by a solenoidal random force. It was observed that the velocity field in such turbulence is mostly solenoidal (shear-dominated), with the intensity of its potential part being about 15% of the intensity of its solenoidal part. The other important result is that the most intense dissipative structures are two-dimensional shocks. In the works of (Pety and Falgarone, 2000; Porter et al., 2002) it was also observed that shocks started to play a role already at $M \sim 1$.

These two numerical findings motivated the development of a new analytical model for supersonic turbulence in Boldyrev (2002). It should be noted that early attempts to describe supersonic turbulence in the frameworks of the standard Kolmogorov or Burgers models were not successful. (The Burgers model describes shock turbulence and would give $\zeta_p = 1$ for any $p > 1$, see e.g., Boldyrev, 1997, 1998). The reason is simple: Kolmogorov theory is valid for incompressible velocity field, where pressure effects cannot be ignored, while the Burgers model assumes that the velocity field, is potential. In the supersonic turbulence both these assumptions break down.

Model for Supersonic Turbulence

In 1994 She and Lévêque suggested a model for incompressible turbulence, which turned out to be in a good agreement with experimental and numerical data (She and Lévêque, 1994; She and Waymire, 1995). The model included certain assumptions about the statistics of the turbulent cascade. In general, the She and Lévêque model has three input parameters. Two of them are the exponents of the naive (i.e., non-intermittent) scalings of the velocity fluctuation, $u_l \sim l^{\Theta}$, and of the characteristic time of the energy transfer at this scale, $t_l \sim l/u_l \sim l^{\Delta}$. The first parameter, Θ, enters the theory through the so-called Kolmogorov refined similarity hypothesis that relates the scaling of the fluctuating velocity field to the scaling of the fluctuating energy dissipation (Frisch, 1995). The second parameter, Δ, enters the theory since it is assumed that the fluctuations having the smallest scales available in the inertial range, and occurring near the strongest dissipative singularities, are not intermittent. In our case these parameters are related, $\Delta = 1-\Theta$, i.e., the energy transfer time scales as the eddy turn-over time, $t \sim l/u_l$. In general this does not need to be so, as e.g., in the Iroshnikov–Kraichnan model of incompressible MHD turbulence (Grauer et al., 1994; Politano and Pouquet, 1995; Müller et al., 2003). The other parameter is the dimension of the most intense dissipative structure, D. The She–Lévêque formula, as generalized by Dubrulle (1994), then reads

$$\zeta(p)/\zeta(3) = \Theta\,(1-\Delta)\,p + (3-D)\left(1 - \Sigma^{\Theta p}\right), \tag{2}$$

where $\Sigma = 1-\Delta/(3-D)$. Since in our case the velocity field in the inertial interval is mostly incompressible, one can imagine that the non-intermittent scaling is the Kolmogorov one, i.e., $\Theta = 1/3$, $\Delta = 1-\Theta = 2/3$, while the dissipative structures are shocks that are distributed in sheets, not filaments, so $D = 2$ (Boldyrev, 2002). With such input parameters, the formula gives

$$\zeta(p)/\zeta(3) = \frac{p}{9} + 1 - \left(\frac{1}{3}\right)^{p/3}. \tag{3}$$

In the next section we demonstrate excellent agreement of this formula with direct numerical simulations of supersonic turbulence.

To conclude this section we demonstrate that the She–Lévêque model corresponds to the Log–Poisson statistics of velocity differences. This explanation is due to (She and Waymire, 1995). Let us assume that an eddy of size L has density of energy dissipation ϵ_L. During turbulent fragmentation, the eddy transfers the energy to smaller ones, with size $l < L$, such that ϵ_l is the density of energy dissipation in the smaller eddy. The process of energy transfer is random, so $\epsilon_l = W_{lL}\epsilon_L$, where W_{lL} is a random transfer function. She and Lévêque assumed that this function has a Log–Poisson distribution. Namely, $W_{lL} = (L/l)^{\gamma}\exp(\beta N)$, where N is an integer variable that has a Poisson distribution, $P(N) = \lambda_{lL}^{N}\exp(-\lambda_{lL})/N$. In this formula, $\lambda_{lL} = \alpha\log(L/l)$, therefore, the process of energy transfer depends on the three parameters, α, β, and γ. One now needs to relate the statistics of the energy dissipation, ϵ_l, to those of the velocity field difference, u_l. She and Lévêque did this using the Kolmogorov refined similarity hypothesis stating that u_l scales as $(l\epsilon_l)^{1/3}$, see e.g. (Frisch, 1995). One can now check that from the presented Log–Poisson distribution of ϵ_l one can derive the She–Lévêque formula in Eq. (2), where three parameters α, β, and γ of the Log–Poisson distribution correspond to the three parameters Θ, Δ, and Σ in Eq. (2). The reader may wish to do this as an exercise. We should also note that these statistics are to a large extent phenomenological, they have not been derived from the first principles yet.

Numerical Results

The numerical simulations were performed with 250^3 and 500^3 resolutions for MHD turbulence with an isothermal equation of state, using the same method and program as in Padoan et al. (1997), Padoan and Nordlund (1999), Padoan et al. (2000), Padoan et al. (2001). The turbulence is driven on large scales by a random external force that is generated in the Fourier space and has a normal distribution, with power only in the range of wave numbers $1 \leq kL/2\pi \leq 2$, where $k = 1$ corresponds to the size of the periodic box. A Helmholtz decomposition is performed on both the initial velocity and the random force, and only the solenoidal components are used. In order to produce a force that varies continuously in time, we specify randomly distributed Fourier components of the time derivative of the acceleration, and compute the acceleration as a time integral. The time derivatives of the Fourier components of the acceleration are regenerated at time intervals of about one dynamical time, i.e., the turn-over time on the scale of the box. The solenoidal character of the forcing is not crucial for the velocity scaling in the inertial region—this was checked by comparing with runs with mixed compressional and solenoidal driving. The solenoidal driving was chosen to provide a better ‘boundary condition’ for the inertial interval in k space, since it turns out that the compressional to solenoidal ratio tends to become small in the inertial range.

The external force sustains the supersonic gas motion ($M = 10$) in the simulations. The motions would otherwise decay on a time scale of the order of a crossing time and become sub-sonic due to the dissipation in shocks (Stone et al., 1998; Padoan and Nordlund, 1999). The real forcing is probably due to a turbulent cascade from large scales, driven by supernovae and superbubbles (Korpi et al., 1999; de Avillez, 2000), although other possible driving mechanisms, such as thermal instabilities, have been considered (Kritsuk and Norman, 2002a, 2002b). The initial average Alfvenic Mach number was chosen to be $M_a \sim 3$, since we wanted to ensure an effective dynamic gas compression not only along magnetic field lines but also in the transverse direction. However, the turbulent magnetic field exhibited a rather intermittent behaviour, being dynamically important in the regions of strong turbulent compression (downstream regions of shocks).

To quantitatively characterize the statistical properties of the turbulence, we have measured the structure functions of the velocity field, which are defined using the absolute values of the velocity differences,

$$S_p(L) = \langle |u(x+L) - u(x)|^p \rangle \sim L^{\zeta(p)}, \tag{4}$$

where u is the component of the velocity field perpendicular or parallel to the vector **L**. According to the chosen component, the structure functions are called transversal or longitudinal, respectively. In the inertial interval, the structure functions obey scaling laws, and the exponents $\zeta(p)$ may be determined; it is usually expected that in the homogeneous isotropic case both transversal and longitudinal functions have the same scaling. We performed measurements of the structure functions up to $p = 10$. Since the Reynolds number in our simulations was not large enough to observe good scaling behaviour of the structure functions, the method of Extended Self-Similarity (ESS) was applied. Namely, instead of plotting the structure functions themselves, we plotted them versus the third-order structure function. As was discovered in Benzi et al. (1993) and Camussi and Benzi (1996), such ratios exhibit rather good (and correct) scaling behaviour, even in systems with moderate Reynolds numbers. Then, if one knows from the theory that the third-order structure function scales with $\zeta(3) = 1$, or if any other scaling exponent $\zeta(p)$ is known from numerics or observations with good precision, one can obtain the scalings of all the other structure functions. We chose to deal with the transversal rather than longitudinal functions since the flow is *shear-dominated*, and the transversal functions are therefore found to have a better scaling behaviour. As illustrated by Figure 1, the relative scaling of the structure functions is well established in a rather large interval. As has been suggested in (Dubrulle, 1994), the ratios of scaling exponents may be universal, while the scaling exponents themselves may be not.

From Figure 1 we conclude that the observed properties of the supersonic turbulence can hardly be predicted in the framework of any known model of

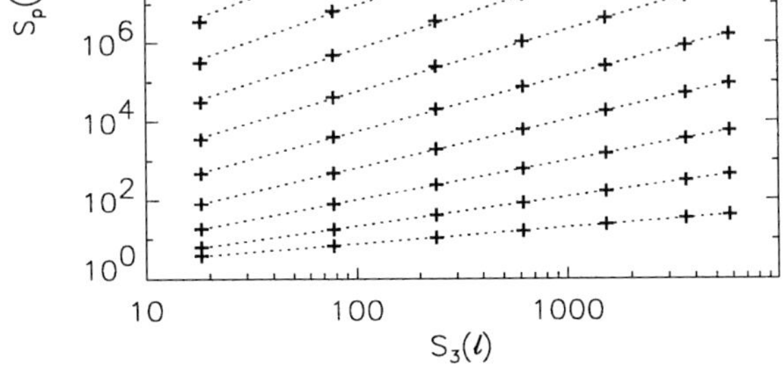

Figure 1. Transversal structure functions computed for $p = 1, \ldots, 10$ (correspondingly, from bottom to top). Log–Log plots of the structure functions are shown. The plots exhibit excellent scalings, in agreement with the Extended Self-Similarity hypothesis. Their slopes are well described by formula in Eq. (3). Note the strong difference of the scalings of the structure functions from the scalings given by the Kolmogorov and Burgers models.

strong turbulence. Indeed, the scalings of the high-order structure functions are clearly far from the scalings predicted either by the Kolmogorov or by the Burgers model. For the first 10 structure functions our model gives: $\zeta(1)/\zeta(3) = 0.42$, $\zeta(2)/\zeta(3) = 0.74$, $\zeta(4)/\zeta(3) = 1.20$, $\zeta(5)/\zeta(3) = 1.40$, $\zeta(6)/\zeta(3) = 1.55$, $\zeta(7)/\zeta(3) = 1.70$, $\zeta(8)/\zeta(3) = 1.84$, $\zeta(9)/\zeta(3) = 1.96$, and $\zeta(10)/\zeta(3) = 2.09$ in excellent agreement with the numerical results presented in Figure 1. The plotted numerical slopes are found to be, correspondingly, 0.42, 0.74, 1.21, 1.38, 1.52, 1.65, 1.76, 1.89, 1.99, which is clearly in favour of our Eq. (3) rather than the Kolmogorov model $\zeta(p) = p/3$, the incompressible She–Lévêque model, $\zeta(p) = p/9 + 2 - 2(2/3)^{p/3}$, or the Burgers model, $\zeta(p) = 1$. The detailed presentation of our numerical results can be found in (Boldyrev et al., 2002a, 2002b; Padoan et al., 2003).

Application to Molecular Clouds

Observational relation of molecular cloud mass to the cloud size is close to $M \sim L^D$, where D is estimated to be greater than two (see, e.g., Elmegreen and Falgarone, 1996), where the estimate $D = 2.3 \pm 0.3$ was presented. If the density were concentrated in elongated shocks, the scaling power would be two. A plausible resolution was suggested in Boldyrev (2002). It was proposed that over a range of scales (not available in numerical simulations due to limited resolution) dense structures are folded to form a complex distribution having *fractal* dimension $D \geq 2$; a complicated fractal (even multifractal) density distribution is indeed inferred from observations (Chappell and Scalo, 2001). Substitution of the fiducial value $D = 2.3$ in our formula in Eq. (2) leads to agreement of the theory with some recent observations (see e.g., Brunt and Heyer, 2002).

It is interesting to note that in Eq. (2) can provide the *upper boundary* for the possible dimension, D, of the most intense dissipative structures. The parameter Σ introduced in Eq. (2) measures the degree of intermittency of turbulence. For example, for $\Sigma \to 1$ one recovers the Kolmogorov, non-intermittent, scaling $\zeta(p) = p/3$. By its definition, $0 \leq \Sigma \leq 1$, and we therefore obtain the upper boundary, $D \leq 3 - \Delta = 2\frac{1}{3}$; for $D > 2\frac{1}{3}$, the theory would be inconsistent. It is interesting that the observed value, $D = 2.3$, is rather close to the upper boundary.

The conjecture about non-integer, fractal, dimensions of the most intense dissipative structures received recently a confirmation in numerical simulations (Padoan et al., 2003). It was observed that compressible turbulence with different Mach numbers can be very well described by Eq. (2), where only one parameter, D, needs to be changed to reproduce the scalings of the structure functions for the investigated range of Mach numbers, $0 < M < 10$. The case $D = 1$ corresponds to incompressible turbulence ($M = 0$, dissipation in filaments), while the other limit, $D = 2$, represents the supersonic turbulence ($M = 10$, dissipation in shocks). For the intermediate values of the Mach number, D changes gradually from $D = 1$ to $D = 2$ as a function of the Mach number. Recently, non-integer dimensions of the most intense dissipative structures were also discussed in the case of incompresible MHD turbulence (Haugen et al., 2003).

The observational evidense for the Log–Poisson distributions of velocity difference in molecular clouds was obtained in (Padoan et al., 2003). Interestingly enough, the same statistics seem to be valid for the line-of-sight integrated density fields (Padoan et al., 2002).

Acknowledgements

The work of S. Boldyrev was supported by the ASCI Flash Center at the University of Chicago, under DoE subcontract B523820. S. Boldyrev and Å. Nordlund would like to thank the Aspen Center for Physics, where a part of this work was done. A part of the computational work was performed at the Danish Center for Scientific Computing.

References

de Avillez, M.A.: 2000, *MNRAS* **315**, 479.
Benzi, R., Ciliberto, S., Tripiccione, R., Baudet, C., Massaioli, F. and Succi, S.: 1993, *Phys. Rev. E* **48**, R29.
Boldyrev, S.: 2002, *ApJ* **569**, 841.
Boldyrev, S., Nordlund, Å and Padoan, P.: 2002a, *ApJ* **573**, 678.
Boldyrev, S, Nordlund, Å and Padoan, P.: 2002b, *Phys. Rev. Lett.* **89**, 031102.
Boldyrev, S.: 1997, *Phys. Rev. E* **55**, 6907.
Boldyrev, S.: 1998, *Phys. Plasmas* **5**, 1681.
Brunt, C.M. and Heyer, M.H.: 2002, *ApJ* **566**, 289.
Camussi, R. and Benzi, R.: 1996, *Phys. Fluids Letters* **9**, 257.
Chappell, D. and Scalo, J.: 2001, *ApJ* **551**, 712.
Dubrulle, B.: 1994, *Phys. Rev. Lett.* **73**, 959.
Elmegreen, B.G.: 2001, in: Montmerle, T. and Andre, P. (eds.), *From Darkness to Light*, ASP Conference Series, in press; astro-ph/0010582.
Elmegreen, B.G. and Falgarone, E.: 1996, *ApJ* **471**, 816.
Falgarone, E. and Phillips, T.G.: 1990, *ApJ* **359**, 344.
Falgarone, E., Puget, J.-L. and Perault, M.: 1992, *A&A* **257**, 715.
Frisch, U.: 1995, *Turbulence*, Cambridge.
Gotoh, T.: 2002, *Comp. Phys. Comm.* **147**, 530.
Grauer, R., Krug, J. and Marliani, C.: 1994, *Phys. Lett.* **A195**, 335.
Haugen, N.E.L., Brangenburg, A. and Dobler, W.: 2003, astro-ph/0303372.
Korpi, M.J., Brandenburg, A., Shukurov, A., Tuominen, I. and Nordlund, Å.: 1999, *ApJL* **514**, L99.
Kritsuk, A. and Norman, M.L.: 2002a, *ApJ* **569**, L127.
Kritsuk, A. and Norman, M.L.: 2002b, *ApJ* **580**, L51.
Larson, R.B.: 1979, *MNRAS* **186**, 479.
Larson, R.B.: 1981, *MNRAS* **194**, 809.
Müller, W.-C., Biskamp, D. and Grappin, R.: 2003, physics/0306045.
Myers, P.C. and Gammie, C.F.: 1999, *ApJ* **522**, L141.
Ossenkopf, V. and Mac Low, M.-M.: 2000, astro-ph/0012247.
Ostriker, E.C., Stone, J.M. and Gammie, C.F.: 2001, *ApJ* **546**, 980; astro-ph/0008454.

Padoan, P., Boldyrev, S., Langer, W. and Nordlund, Å.: 2003, *ApJ* **583**, 308.
Padoan, P., Cambressy, L. and Langer, W.: 2002, *ApJ* **580**, L57.
Padoan, P., Jimenez, R., Nordlund, Å. and Boldyrev, S.: 2003, astro-ph/0301026.
Padoan, P., Juvela, M., Goodman, A.A. and Nordlund, Å.: 2001, *ApJ* **553**, 227.
Padoan, P. and Nordlund, Å.: 1999, *ApJ* **526**, 279.
Padoan, P., Nordlund, A. and Jones, B.J.T.: 1997, *MNRAS* **288**, 145.
Padoan, P., Nordlund, Å., Rögnvaldsson, Ö.E. and Goodman, A.: 2000, astro-ph/0011229.
Pety, J. and Falgarone, É.: 2000, *Astron. Astrophys.* **356**, 279.
Politano, H. and Pouquet, A.: 1995, *Phys. Rev. E* **52**, 636.
Porter, D., Pouquet, A. and Woodwad, P.: 2002, *Phys. Rev. E* **66**, 026301.
She, Z.-S. and Lévêque, E.: 1994, *Phys. Rev. Lett.* **72**, 336.
She, Z.-S. and Waymire, E.C.: 1995, *Phys. Rev. Lett.* **74**, 262.
Stone, J.M., Ostriker, E.C. and Gammie, C.F.: 1998, *ApJ* **508**, L99.
von Weizsäcker, C.F.: 1951, *Astron. J.* **114**, 165.

HYDRODYNAMICAL SIMULATIONS OF MOLECULAR DYNAMICS IN SUPERSONIC TURBULENT FLOW

GEORGI PAVLOVSKI[1], MICHAEL D. SMITH[1], MORDECAI–MARK MAC LOW[2] and ALEXANDER ROSEN[1]

[1]*Armagh Observatory, College Hill, Armagh, BT61 9DG, N.Ireland, U.K.; E-mail:gbp@star.arm.ac.uk*

[2]*Department of Astrophysics, American Museum of Natural History, Central Park West at 79th Street, New York, NY 10024-5192, U.S.A.*

Abstract. Here we present results from simulations of turbulence in star forming environments obtained by coupling three-dimensional hydrodynamical models with appropriate chemical processes. We investigate regimes of decaying high-speed molecular turbulence. Here we analyse PDFs of density for the volume, mass, molecular mass and the energy distribution over the range of scales. We compare our results to those previously obtained for isothermal turbulence and suggest possible explanations.

Keywords: hydrodynamics, turbulence, molecular processes, clouds, ISM

1. Motivations

It is generally accepted now that supersonic turbulence is of fundamental importance to many processes related to the formation of stars (Vázquez-Semadeni et al., 2000; Padoan et al., 2001; Mac Low and Klessen, 2003). Turbulent motions redistribute energy inside molecular clouds, giving rise to their hierarchical structure and determining cloud fragmentation (Padoan and Nordlund, 2002). In this article we report results of our ongoing study of molecular turbulence, i.e., turbulence in the fluid with active chemistry and cooling appropriate for the star-forming environments.

Molecular chemistry and cooling is critical to cloud formation and evolution (Langer et al., 2000; Lim, 2001; Lim et al., 1999). Molecular hydrogen forms most efficiently, where the gas is warm but the grains are cool (H_2 forms mainly when atoms combine after colliding and sticking to dust grains). Simple molecules like OH, CO and H_2O form in the gas phase with H_2 as the reactive agent. These molecules are not only important coolants, but associated emission lines provide a means of measuring the cloud properties. Molecules are dissociated as a consequence of fast shocks, UV radiation, X-rays and cosmic rays (Herbst, 2000). We thus need to study molecular turbulence to determine the distribution and abundances of molecular species.

1.1. Methods

Numerically, we solve the time-dependent flow equations as follows:

$$\frac{\partial \rho}{\partial t} + \nabla \cdot (\rho \boldsymbol{v}) = 0, \tag{1a}$$

$$\frac{\partial (\rho \boldsymbol{v})}{\partial t} + (\boldsymbol{v} \cdot \nabla)\, \boldsymbol{v} = -\frac{1}{\rho} \nabla p, \tag{1b}$$

$$\frac{\partial e}{\partial t} + \boldsymbol{v} \cdot \nabla e = -p \nabla \cdot \boldsymbol{v} + \Lambda\, (T, n, f)\, , \tag{1c}$$

$$\frac{\partial (f n)}{\partial t} + \nabla \cdot (f n \boldsymbol{v}) = R\, (T, n, f) - D\, (T, n, f)\, , \tag{1d}$$

where n is the hydrogen nuclei density, e is the internal energy density and f is the molecular hydrogen abundance (i.e., $n(H_2) = fn$). We consider the gas as a mixture of atomic and molecular hydrogen with 10% of helium (i.e., $n(\mathrm{He}) = 0.1\, n$), therefore the total particle density is $n_{tot} = (1.1 - f)n$ and the temperature is $T = p\, /\, (k\, n_{tot})$. Λ is internal energy loss through radiation and chemistry per unit volume, the function consists of 13 separate parts (some of which heat the gas), their detailed description can be found in Smith and Rosen, Pavlovski et al. (2003, 2002). R and D are reformation and dissociation rates of molecular hydrogen respectively (see Appendix A and B in Smith et al., 2002).

As a basis, we employ the ZEUS-3D code (Stone and Norman, 1992). This is a second order in space and first order in time, grid-based code, which is using van Leer type advection. We study here compressible hydrodynamics without external gravity, self-gravity or thermal conduction. A small amount of linear physical viscosity is modelled, and von Neumann type of artificial viscosity used to capture shocks and determine the dissipation in the shock front.

For computation of cooling we have employed the simultaneous implicit method discussed by Suttner et al. (1997), in which the time step is adjusted so as to limit the change in internal energy in any zone to 30%. This limit implies much shorter time–steps in comparison to any dynamical timescale, and it is one of the most restraining factors in our simulations.

The cooling is appropriate for dense cloud material of any atomic–molecular hydrogen mixture. We include H_2 ro–vibrational and dissociative cooling, CO and H_2O ro–vibrational cooling, gas–grain, thermal bremsstrahlung and a steady–state approximation to atomic cooling (see Appendix A of Smith et al., 2002 and Pavlovski et al., 2002).

We take a very basic network of chemical reactions. Time-dependent hydrogen chemistry is included [Eqs. (2a), (2b), and (2c)], but C and O chemistry is limited to the reactions with H and H_2 which generate OH, CO and H_2O [Eqs. (3a), (3b), and (3c)]. Equilibrium abundances are calculated, which are accurate for our purposes within the shocks, where molecules are rapidly formed and destroyed.

$$H + H + (\text{grain}) \longrightarrow H_2 + (\text{grain}) \qquad (2a)$$

$$H_2 + H \longrightarrow 3H \qquad (2b)$$

$$H_2 + H_2 \longrightarrow 2H + H_2 \qquad (2c)$$

$$O + H_2 \longleftrightarrow OH + H \qquad (3a)$$

$$OH + C \longleftrightarrow CO + C \qquad (3b)$$

$$OH + H_2 \longleftrightarrow H_2O + H \qquad (3c)$$

2. Simulations

We have run a set of simulations (with resolution ranging from 32^3 to 256^3) of *decaying* turbulence in different velocity regimes: initial r.m.s. velocity of 15 km s^{-1}, 30 km $s^{-1}$and 60 km s^{-1}. The number density has been taken to be $n = 10^6$ cm^{-3}, physical box size $L = 10^{16}$ cm and the initial temperature has a homogeneous distribution of $T_0 = 100$ K. The hydrogen was fixed to be *fully molecular* at the beginning: fraction $f = 0.5$. The high number density was selected to create a sufficiently high average column (10^{22} cm^2), which ensures that the simulated region is optically thick. Initial stress was introduce by perturbations applied to model velocities with a spectrum extending over a narrow range of wave numbers $3 \leq |\boldsymbol{k}| \leq 4$ (see Figure 1)

2.1. Results

We find that the dynamical behaviour of the molecular turbulence is not dramatically different from behaviour of the turbulence with the isothermal equation of state.

The decay law of kinetic energy in the simulations are similar to the decay laws of the high Mach number isothermal simulations uncovered in Smith et al. (2000). However, in our simulations, thermal energy is not constant, and decays only a

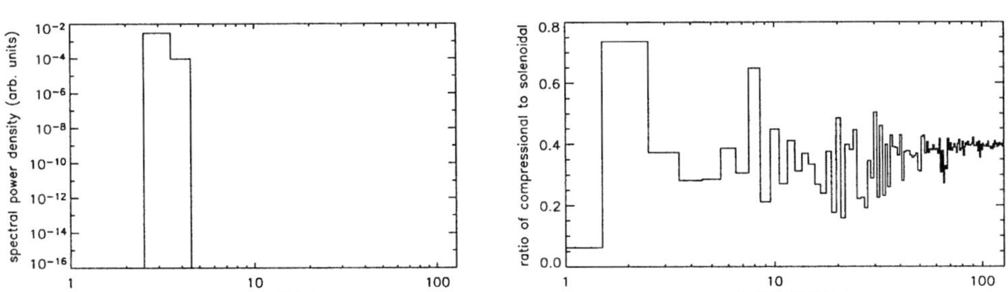

Figure 1. Initial energy distribution over different scales. L.H.S. plot – total (solenoidal plus compressional) energy distribution; R.H.S. plot – ratio of compressional to solenoidal energy.

little bit slower than kinetic energy (for full details see Pavlovski et al., 2002). This fact can result in sustaining Mach number >1 for a longer, compared to isothermal regime, time.

Statistical analysis of the turbulence showed that probability density functions (PDFs) of fractional volume per unit density ($dV/d\rho$), fractional mass per unit density ($dM/d\rho$) and fractional molecular mass per unit density ($d\mathfrak{M}/d\rho$) can be approximated by log-normal distributions as in the isothermal case (Padoan et al., 1997; Passot and Vázquez-Semadeni, 1998; Ostriker et al., 2001),

$$p_{V,M,\mathfrak{M}}(\log(x)) = \frac{1}{\sigma_{V,M,\mathfrak{M}}\sqrt{2\pi}} \exp\left\{ -(\log(x) \pm |\mu_{V,M,\mathfrak{M}}|)^2 / \left(2\sigma^2_{V,M,\mathfrak{M}}\right)\right\}, \tag{4}$$

where $x = \rho/\langle\rho\rangle$, $\langle\rho\rangle$ – volume mean of density. We find, that modules of means for the distribution have close values, but generally the following is true:

$$|\mu_V| \lesssim |\mu_M| \lesssim |\mu_{\mathfrak{M}}|, \quad \sigma_V > \sigma_M \gtrsim \sigma_{\mathfrak{M}}. \tag{5}$$

Typical distributions of these values are presented in Figures 2 and 3. The values of the parameters of the distributions are: in the isothermal case, $\mu_V = -0.43$, $\mu_M = 0.47, \sigma_V = 0.74, \sigma = 0.51$; in the molecular case, $\mu_V = -0.42, \mu_M = 0.46$, $\mu_{\mathfrak{M}} = 0.55$, $\sigma_V = 0.71$, $\sigma_M = 0.52$, $\sigma_{\mathfrak{M}} = 0.50$. This is in agreement with the results discussed in Ostriker et al. (2001).

2.2. Molecular Hydrogen Evolution

The molecular fraction 'f' is displayed in Figure 4. The three initial states correspond to three distinct physical regimes. With an r.m.s. velocity of 15 km s^{-1}, dissociative shocks are not present but some localised dissociation still occurs. With

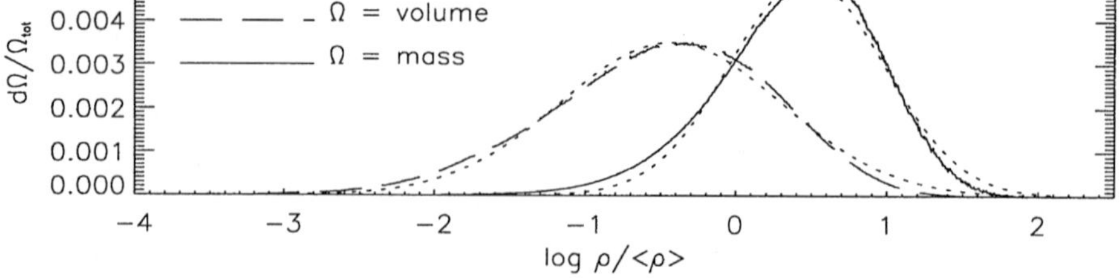

Figure 2. PDFs of volume and mass distributions with density in the initially high speed isothermal turbulence. Data taken from 256^3 simulations with initial r.m.s. velocity of 60 km s^{-1}after 130 yr of evolution (average Mach number: $\mathcal{M} \approx 9$).

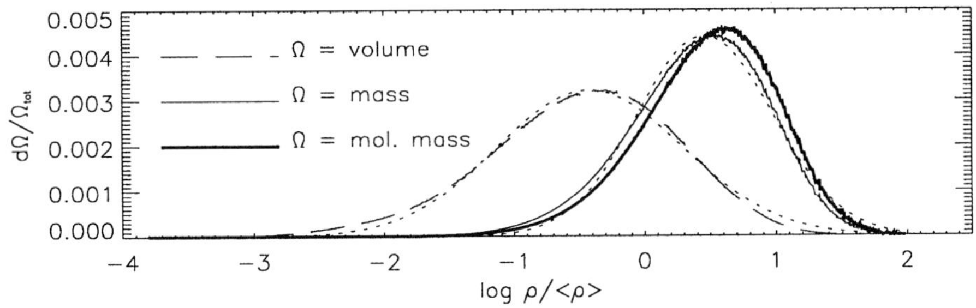

Figure 3. PDFs of volume, mass and molecular mass distributions with density in the decaying molecular turbulence. Data taken from 256^3 simulations with initial r.m.s. velocity of 60 km s^{-1} after 130 yr of evolution (average Mach number: $\mathcal{M} \approx 10$).

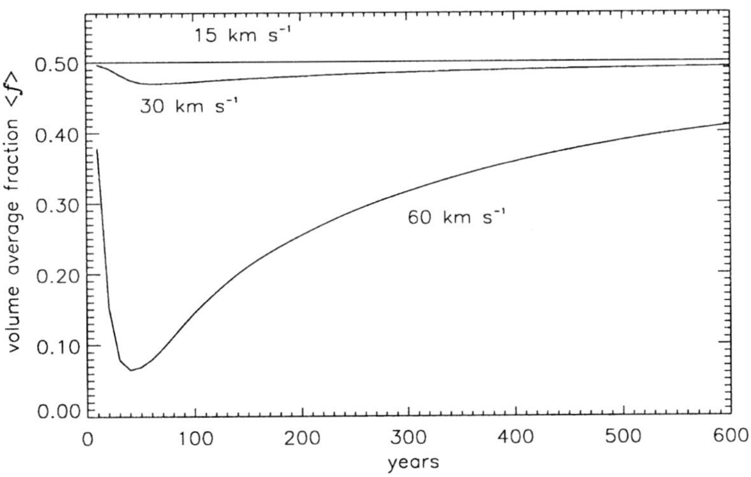

Figure 4. Evolution of the average molecular fraction with time in the runs with different initial r.m.s. velocity. Data from the 256^3 runs.

30 km s^{-1}, a few per cent of the molecules are dissociated, whereas at 60 km s^{-1}, the gas becomes over 80% atomic.

Reformation of molecular hydrogen is unexpectedly rapid. The expected H_2 reformation time at 20 K and 10^6 cm^{-3} is $t_R = 3200\,yr$ (see details in Pavlovski et al., 2002) a factor of 5, larger than the simulation time. At 100 yr, the temperature is $\sim$ 80 K, predicting a reformation time of $t_R = 2000$ yr. Yet, reformation is occurring over $\sim$400 yr. This speed up is caused by the turbulence itself: the molecules preferentially reform in the denser and cooler locations. As weak shocks propagate through the gas, different regions are compressed and expanded. Hence, the reformation time is not only controlled by the 'average' reformation time, but also by the strength of the turbulence. Given a turbulent dynamical timescale shorter than the average reformation time-scale, then we can expect reformation to be accelerated.

We have checked the degree of convergence of both the dissociation and reformation processes. High resolution is paramount to correctly follow the degree of dissociation. This is critical to the 30 km s^{-1}turbulence, since there are numerous intermediate speed shocks, partially dissociating the gas. For the low-and high-speed examples, however, the dissociation is basically zero and complete, respectively. Exhaustive resolution studies of one-dimensional shocks with this code are presented elsewhere (Smith and Rosen, 2003; Rosen and Smith, 2003).

3. Summary

We have presented the properties of a specific model for molecular turbulence. We carried out three– dimensional hydrodynamical simulations of decaying supersonic turbulence in molecular gas. We included a detailed cooling function, molecular hydrogen chemistry and equilibrium C and O chemistry. We studied three cases in which the applied velocity field straddles the value for which wholesale dissociation of molecules occurs. The parameters chosen ensure that for the high-speed turbulence, the molecules are initially destroyed in shocks and gradually reform in a distinct phase.

We find the following:

- An initial phase of slow dissipation and shock formation.
- An extended phase of power-law kinetic energy decay, as in the isothermal case.
- The thermal energy, initially raised by the introduction of turbulence, decays only a little slower than the kinetic energy.
- The reformation of hydrogen molecules, as the fast turbulence decays, is several times faster than expected from the average density.
- The molecular fraction increases quite uniformly, so that density and molecular density are almost identically distributed at any one time.

We mainly wish here to emphasize the insight these simulations provide into how molecular chemistry and supersonic dynamics combine. We have found that

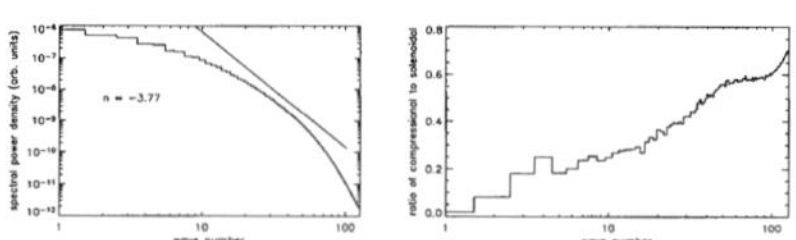

Figure 5. Distribution of energy over different scales at the end of the simulation [600 yr: Mach number $\approx$3; initial r.m.s. velocity 60 km s^{-1}(corresponds to initial Mach number $\approx$55), 256^3]. L.H.S. plot – total (solenoidal plus compressional) energy distribution, straight line is Kolmogorov ($n = -11/3$) 3D spectrum. The total energy distribution does not exhibit clear power law behaviour in the inertial range (i.e., intermediate range of wave numbers). Linear fitting in the wide range of wave numbers (as designated by the reference ($-11/3$) power law) gives steeper than Kolmogorov's ($n = -3.77$) energy cascade as indication of shock dissipation (see Boldyrev et al. in this volume). R.H.S. plot – ratio of compressional to solenoidal energy.

isothermal simulations are indeed very useful, not only for the rate of energy decay but also to trace the molecules.

A simple reason for the fast decay is that a sufficient number of strong shocks survive. As shown by Smith et al. (2000), the rate of energy decay in *decaying* turbulence is dominated by the vast number of weak shocks. These shocks are less efficient at energy dissipation. Energy distribution across different scales is displayed in Figure 5.

References

Herbst, E.: 2000, in: F. Combes and G. Pineau des Forêts (eds.), *Molecular Hydrogen in Space.*, Cambridge University Press, pp. 85–88.
Langer, W.D., van Dishoeck, E.F., Bergin, E.A., Blake, G.A., Telens, A.G.G.M. and Whittet, D.C.B.: 2000, in: V. Mannings, A.P. Boss and S.S. Russel (eds.), *Protostars and Planets IV*, The University of Arisona Space Science Series. The University of Arizona Press, Tuscon.
Lim, A.J., Rawlings, J.M.C. and Williams, D.A.: 1999, *MNRAS* **308**, 1126–1132.
Mac Low, M.-M. and Klessen, R.: 2003, astro-ph/0301093.
Ostriker, E.C., Stone, J.M. and Gammie, C.F.: 2001, *ApJ* **546**, 980–1005.
Padoan, P., Jones, B.J.T. and Nordlund, A.P.: 1997, *ApJ* **474**, 730.
Padoan, P., Juvela, M., Goodman, A.A. and Nordlund, Å.: 2001, *ApJ* **553**, 227–234.
Padoan, P. and Nordlund, Å.: 2002, *ApJ* **576**, 870–879.
Passot, T. and Vázquez-Semadeni, E.: 1998, *Phys. Rev. E* **58**, 4501–4510.
Pavlovski, G., Smith, M.D., Mac Low, M.-M. and Rosen, A.: 2002, *MNRAS* **337**, 477–487.
Smith, M.D., Mac Low, M.-M. and Zuev, J.M.: 2000, *A&A* **356**, 287–300.
Smith, M.D., Pavlovski, G., Mac Low, M.-M., Khanzadyan, T., Gredel, R. and Stanke, T.: 2002, in: de Avillez, M. and Breitschwerdt, D. (eds.), *From Observations to Self-Consistent Modelling of the ISM in Galaxies. JENAM 2002*. Astrophysics and Space Science Series (Kluwer), **289**.
Smith, M.D. and Rosen A.: 2003, *MNRAS* **339**, 133–147.
Rosen, A. and Smith, M.D.: 2003, *MNRAS* **343**, 181–191.
Stone, J.M. and Norman, M.L.: 1992, *ApJS* **80**, 753–790.
Suttner, G., Smith, M.D., Yorke, H.W. and Zinnecker, H.: 1997, *A&A* **318**, 595–607.
Vázquez-Semadeni, E., Ostriker, E.C., Passot, T. and Stone, C.F.: 2000, in: Mannings, V. Boss, A.P. and Russel, S.S. (eds.), *Protostars and Planets IV*, The University of Arisona Space Science Series. The University of Arizona Press, Tuscon.

OBSERVATIONAL MAGNETOGASDYNAMICS: 21 YEARS OF HI ZEEMAN SPLITTING MEASUREMENTS. . . AND MORE*

CARL HEILES
Astronomy Department, University of California, Berkeley, CA, U.S.A.;
E-mail: cheiles@astron.berkeley.edu

Abstract. We review observations of the physical properties of the diffuse ISM HI components, namely the Cold and Warm Neutral Media (CNM and WNM). There is somewhat more WNM than CNM, and at least half of the WNM is not thermally stable. The CNM has typical turbulent Mach number 3. Magnetic fields in the CNM are not as large as expected from the classical flux-freezing argument; neither are magnetic fields always strong enough for the Alfven velocity to equal the turbulent velocity. Nevertheless, they are usually strong enough to put CNM clouds in the magnetically subcritical regime. We identify a probable new source of turbulence for the diffuse ISM. We discuss one very cold cloud that has considerable internal turbulence and, because of its extreme thinness $\sim$0.05 pc, a turbulent crossing time of only $\sim 5 \times 10^4$ yr.

Keywords: interstellar medium, magnetic fields

1. Introduction

This meeting is about magnetic fields and star formation. Most people consider star formation to be the province of molecular clouds, and this is true unless one takes the broader view: molecular clouds themselves are formed somehow, probably from diffuse atomic gas by some coalescence mechanism. Flux freezing applies almost rigorously in the diffuse gas, even in the HI, and the transition from diffuse to molecular gas must occur under the constraints imposed by the field. Part of this story is turbulence, because it is generally recognized that linewidths exceed thermal by considerable factors except in some dense molecular cores. So in addition to the fairly well-understood forces of gas thermal pressure and gravity, we have two difficult physical phenomena at work: magnetic fields and turbulence. Of course, cosmic rays are important too, but they are even harder to measure than magnetic fields so we ignore them for the present.

The difference between diffuse and star-forming gas is gravity. Gravity is usually not important in diffuse HI gas. So the diffuse gas affords another regime in which to study magnetogasdynamics although it is related only indirectly to dense clouds and star formation.

*I dedicate this paper to those worldwide, and especially those in the U.S., U.K., and Spain, who made clear their revulsion at the invasion of Iraq.

Astrophysics and Space Science **292:** 77–88, 2004.

However, let me state an important caveat to the above paragraph. Most molecular clouds are surrounded by huge HI halos. This is very evident in the Taurus/Perseus, Orion, and also the Ophiuchus regions. These HI halos have mass that is comparable to the embedded molecular clouds. The totality of this mass puts the HI halos in rough virial equilibrium (Heiles and Kulkarni, 1986; Section 1.1). For example, the HI mass that envelops the Orion molecular cloud is about 10^5 $M_\odot$, almost as much as the molecular mass. Thus the molecular clouds, which lie toward their centers, are subject to the nonzero pressure of the hydrostatic-equilibrium HI halos. This defines the pressure to be nonzero as the boundary condition for the molecular clouds.

We begin in Section 2 by discussing the diffuse ISM as a nongravitational turbulent medium. We introduce new results from the Arecibo Millennium 21-cm Absorption Line Survey (some are currently published, some not; see Heiles and Troland, 2003a, b). This survey studies individual, randomly-selected components of the HI Cold Neutral Medium (CNM). We first discuss the *nonmagnetic* aspects of this gas in Section 2.1 and then the *magnetic* aspects in Section 2.2. We then explore the angular variations of line-of-sight magnetic field $B_{||}$ in the CNM (Section 2.3.1) and the WNM (Section 2.3.2) to show that the turbulence is not sufficiently strong to completely tangle the magnetic field (i.e., randomize $B_{||}$), which is hardly surprising. In Section 2.4 we use Leiden-Dwingeloo Survey (LDS: Hartmann and Burton, 1997), to introduce a new mode of diffuse medium turbulence—probably, in fact, a new mode of *turbulent energy injection*. The diffuse ISM sometimes exhibits distinct surprises in its manifestations of turbulence; we discuss a spectacular example in Section 3.

2. The ISM as a Turbulent Medium

2.1. The Non-Magnetic Millennium Arecibo Survey

This non-magnetic portion of the survey concentrates on the thermal properties of the diffuse HI. We split both the emission and absorption lines profiles into Gaussian components, for which we derive the basic physical parameters. There are two types of gas, the Warm and Cold Neutral Media (WNM and CNM). The WNM is observationally distinguished by its absence of detectable opacity, but the WNM and CNM are also physically distinct, with highly different temperature distribution functions. For the WNM, we measure column density and linewidth. The linewidth provides an upper limit on kinetic temperature, called T_{kmax}. For the CNM we measure these and also the opacity, from which we derive the kinetic ("spin") temperature T_{k}.

The median column densities for both the WNM and CNM are much smaller than what we think of as a standard cloud, particularly the Spitzer (1978) standard cloud which has $N(HI)_{20} \approx 4.0$. This difference might possibly reflect biases in the selection and definition of a cloud: Spitzer used reddening statistics in the Galactic plane, while we use velocity as a third distinguishing dimension. (The subscript 20 means the units are 10^{20} cm $^{-2}$). A line of sight typically has several components, which suggests (but not necessarily strongly) that the individual components can be considered as turbulence elements.

2.1.1. *The WNM*

The median column density of WNM components is $N(HI)_{20} \sim 1.3$. We find that about 60% of the HI is in the warm phase (the WNM). This is in reasonable agreement with modern theoretical predictions (Wolfire et al., 2003; WMHT), although classical predictions (McKee and Ostriker, 1978) gave much smaller ratios. More surprising is the result that at least half of the WNM is not in stable thermal equilibrium: instead of having $T \sim 8000$ K, which is the thermally stable equilibrium value, it has $T_{\rm kmax} \lesssim 5000$ K. Because $T_{\rm kmax}$ is an upper limit to $T_{\rm k}$, more than half of the WNM is not in thermal equilibrium. Modern theory (WMHT) explains this because heating is a series of impulsive events (shocks).

Finally we estimate the volume filling factor of the WNM. For $z = 0$, i.e. the Galactic plane, our result is 50%, but this very, very rough. The WNM should become even more important with increasing $|z|$.

2.1.2. *The CNM*

The median column density of CNM components is $N(HI)_{20} \sim 0.5$, about 40% of the median for the WNM components. The median kinetic temperature $T_{\rm k}$ of the components is 48 K; the median temperature weighted by column density is higher, 70 K. The difference between these numbers reveals a correlation between $N(HI)$ and T_k.

2.2. Magnetic fields in the CNM

It is easier to measure $B_{||}$ in the CNM than in the WNM because CNM measurements are the (*ON–OFF*) type and are much less subject to instrumental effects (Heiles and Troland, 2003b). This is important because we find that CNM field strengths are surprisingly weak. We have only 14 detections that exceed 2.5σ out of a total of 51 measurements whose uncertainties are low enough to make them interesting. So we must discuss our results statistically. Our statistical analysis is in the very early stages so the present discussion will display the data and draw only the obvious conclusions, which should be considered preliminary.

If flux freezing were to apply, then we would have obtained many clear detections of $B_{||}$. We expect higher field strengths in the CNM than in other diffuse gas phases because the ISM should exhibit approximate thermal pressure equality among the phases. The CNM volume density is higher than in other phases because $T_{\rm k}$ is small, so flux freezing should lead to higher field strengths in the CNM.

However, the CNM fields are no stronger than WIM fields (obtained from pulsars) or the volume-average fields obtained from synchrotron emissivity (e.g. Beck, 2001). This absence of field strength increase for small $n(HI)$ is well known from past studies (e.g. Crutcher, et al., 2003, Section 2), so this is hardly news; nevertheless, we tend to forget these things and, moreover, from an observer's standpoint the paucity of detectable fields is disappointing.

Figure 1 shows the variation of $|B_{||}|$ with gas column density in the diffuse ISM. The solid circles exhibit the observed values of $|B_{||}|$ and $N(HI)_{20}$. The large crosses are reasonably certain detections, those with $|B_{||}| > 2.5\sigma$; obviously, these detections statistically favor higher-than-average measured $|B_{||}|$. The dashed line, from Nakano and Nakamura (1978), separates magnetically subcritical (above the line) from supercritical. Most of the observed points are subcritical. These clouds are far from equilibrium, but if they were to coalesce into a gravitationally-important cloud then magnetic forces would prevent gravitational collapse—unless the field were destroyed in the process, e.g. by the annihilation of oppositely-directed fields during coalescence of individual clouds.

On a more speculative level, the points in Figure 1 look random, and perhaps they are. On the other hand, at higher column densities $[N(HI)_{20} \gtrsim 5]$ most of the points fall below the line. This indicates a possible trend for smaller fields with higher $N(HI)$. It is tempting to suggest a physical reason: high $N(HI)$ can only occur if the field is weak enough to allow gas to come together into larger (but still nongravitational) clouds.

There are no obvious correlations with linewidth or T_k, so we refrain from showing those figures. However, for the entire set of CNM components we can derive the turbulent Mach numbers M_{turb}, and for the restricted set having interesting magnetic measurements we can compare the turbulent and Alfven velocities. Before doing this, however, we make some definitions. These definitions are important because they convert measured quantities to physically meaningful ones.

For the CNM we independently measure the kinetic temperature T_k and the linewidth ΔV, so we can unambiguously derive the turbulent velocity dispersion ΔV_{turb}. We define the turbulent Mach number M_{turb} as

$$M_{turb}^2 = \frac{\Delta V_{turb}^2}{C_s^2} = \frac{3\Delta V_{turb,1D}^2}{C_s^2} \tag{1}$$

With this factor of 3, we assume that the turbulence is isotropic and correct the measured 1D turbulent linewidth to 3D. In this paper, all linewidths ΔV are dispersions: for example, the one-dimensional thermal linewidth—the one we measure for HI with a radio telescope—is $\Delta V_{th,1D}^2 = kT_k/m_H$. For the sound speed C_s we use the isothermal one and include the contribution of He to the mean atomic weight.

In comparing the Alfven velocity with the turbulent velocity, we adopt the same pressure-based approach as is normally done for a thermal magnetic plasma, for

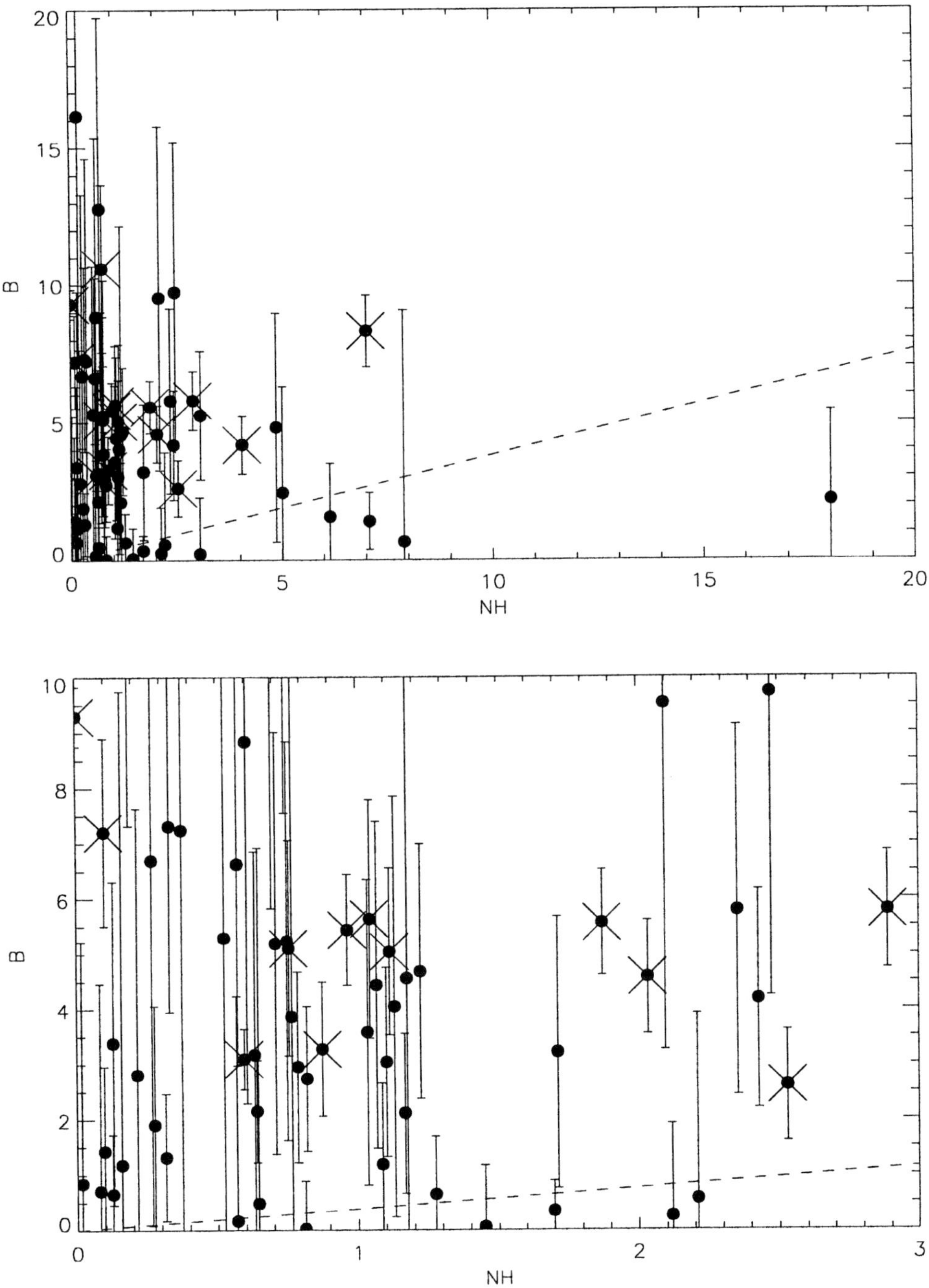

Figure 1. $|B_{||}|$ versus $N(HI)_{20}$, with the lower panel having expanded scales. Filled circles include all datapoints; crosses are those having $|B_{||}| > 2.5\sigma$. The line is from Nakano and Nakamura (1978); points above the line are magnetically subcritical.

which the conventional parameter is the thermal to magnetic pressure ratio, i.e.

$$\beta_{th} = \frac{P_{th}}{P_{mag}} = \frac{2\Delta V^2_{th,1D}}{V^2_A} f_{He,th} \tag{2}$$

The factor $f_{He,th}$ accounts for the differing nuclear masses of H and He and is about 1.3. We can regard the factor 2 (in $2\Delta V^2_{th,1D}$) as arising because the field exerts pressure in two dimensions. Analogously, for a turbulent magnetized plasma we can define

$$\beta_{turb} = \frac{2\Delta V^2_{turb,1D}}{V^2_A} f_{He,turb} \tag{3}$$

Here the factor $f_{He,turb}$ depends on M_{turb}; for $M_{turb} \gg 1$, $f_{He,turb} = 1$, and we assume this case. In our plots, we use this as the relevant ratio. Specifically, we multiply the measured 1D turbulent velocity variance by 2. Similarly, in calculating V_A we make the statistically correct correction and use $B^2 = 3B^2_{||}$. In V_A we obtain the mass density ρ by using our measured T_k and assuming $\frac{P_{CNM}}{k} = 3000$ cm^{-3} K (Jenkins and Tripp, 2001; WMHT).

Figure 2 exhibits the statistics for the Turbulent Mach Number M_{turb}. The top panel plots M_{turb} versus T_k. For points with relatively low errorbars, there seems to be a tendency for M_{turb} to rise with T_k. This trend seems to be violated by points at low T_k with high errorbars. We haven't yet analyzed these trends carefully. The bottom plot, which shows the histogram of M_{turb}, bypasses the uncertainties and we conclude that most CNM components are highly supersonic, with M_{turb} typically $\sim$3.

This supersonic turbulence is not affected by the magnetic field in an obvious way. Figure 3 exhibits V_A versus V_{turb}. The majority of points lie below the line. Even with these large errorbars, we can conclude that many CNM clouds have turbulent velocities that significantly exceed the Alfven velocity.

2.3. Variability in the Signs of $B_{||}$

We investigate the variability in the sign of $B_{||}$ in two ways. One uses Zeeman splittings measured in the Millennium Arecibo survey (HI absorption); the other uses measurements in HI emission.

2.3.1. *Variation of $B_{||}$ in Absorption*

In this comparison we exclude Cas A and Tau A, because they are in the Galactic plane where lines of sight are long and components can be physically separate even if their velocities are comparable. At higher latitudes we have three sources with absorption lines containing more than a single Zeeman splitting result exceeding

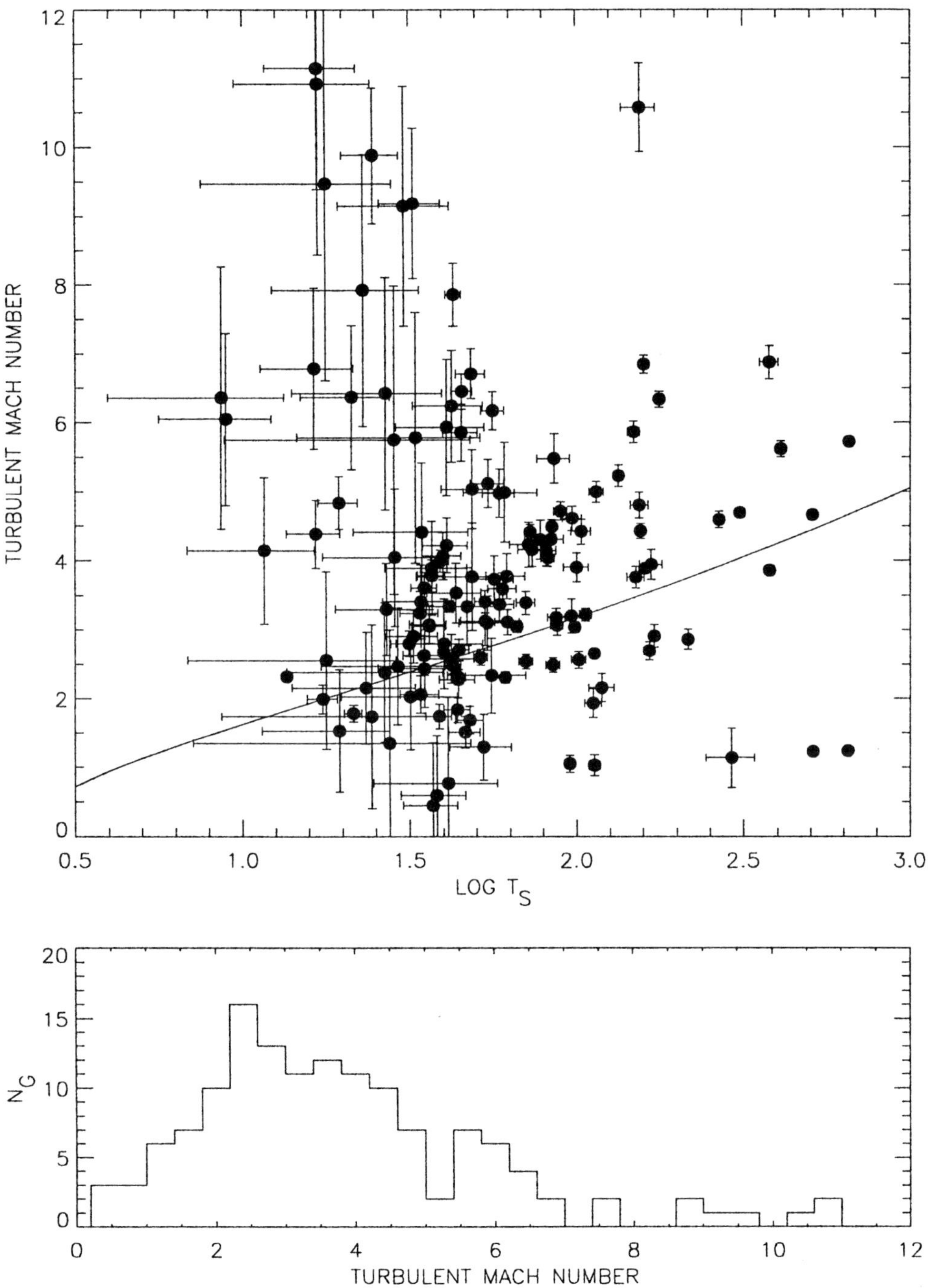

Figure 2. Turbulent Mach Number M_{turb} versus T_k (top), and the histogram of M_{turb} (bottom).

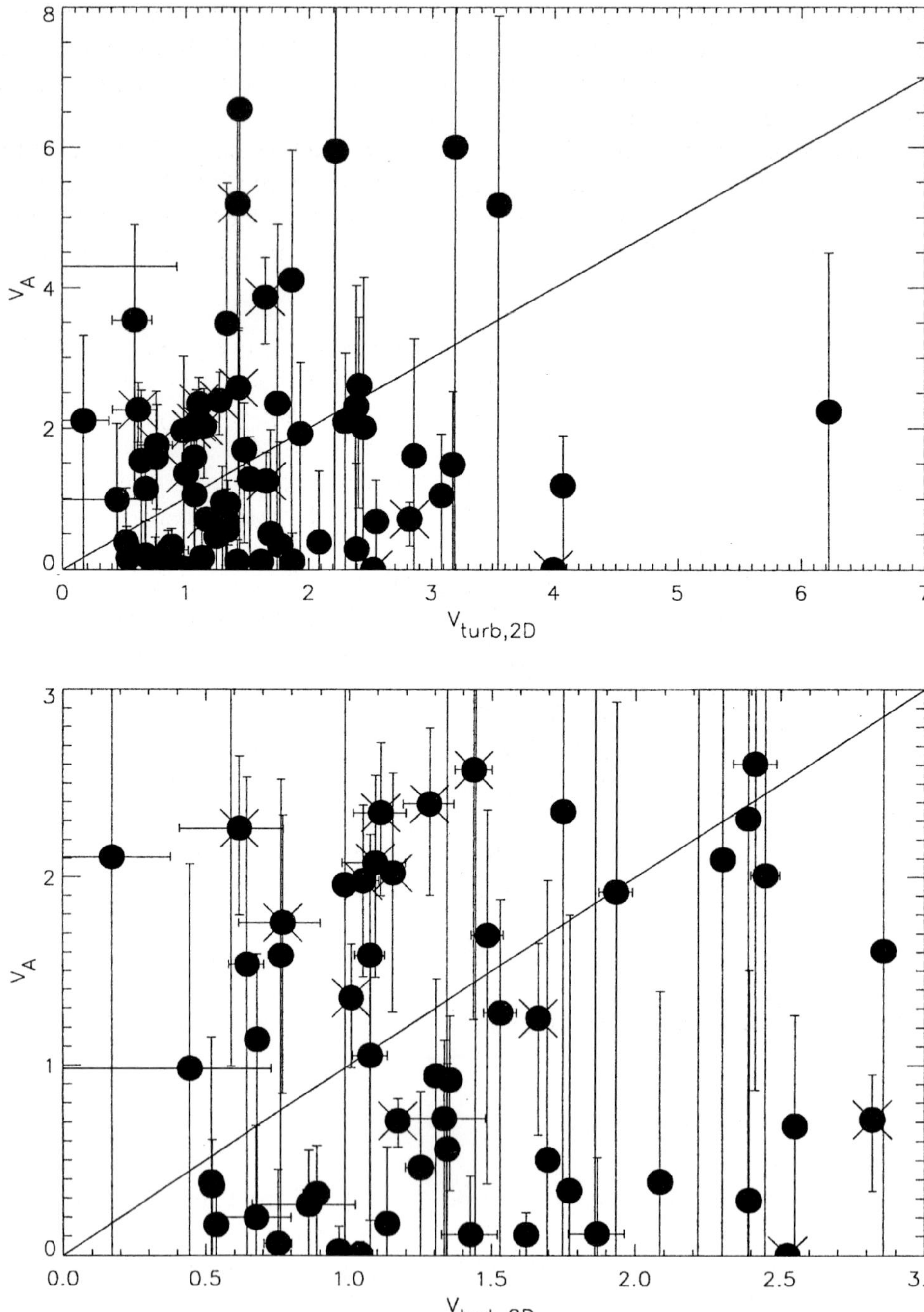

Figure 3. Alfven velocity V_A versus 2D turbulent velocity $V_{turb,2D}$, with the lower panel having expanded scales. Filled circles include all datapoints; crosses are those having $|B_{||}| > 2.5\sigma$. Field strengths and velocity widths have been corrected for the 3D universe. The line shows equality.

2.5σ. These are 3C138, 3C353, and 3C409. The latter two each have two detections; for each, the two detections have the same sign of $B_{||}$. For 3C138 there are three detections, two of one sign and one with a discrepant sign. The one with a discrepant sign is a marginal detection—it is just *barely* 2.5σ.

If we discount the marginal detection in 3C138, then we conclude that along a given line of sight away from the Galactic plane, the magnetic field does not change in sign. Statistically speaking, this means only a little because there are only three sightlines. But it is an indication, one which is bolstered by the emission results.

2.3.2. *Variation of $B_{||}$ in Emission*

The Hat Creek 85 foot telescope was devoted almost exclusively to Zeeman splitting during the years before its catastrophic demise (Heiles, 1993). It made many Zeeman splitting detections in HI emission. These were severely criticized by Verschuur (1995), who stated that all published HI emission Zeeman-splitting detections should be considered as invalid because of beam polarization effects. However, Heiles (1996) discussed these effects from both the theoretical and empirical standpoints and showed the typical 1σ uncertainty with the Hat Creek telescope to be $\sim 1.4\ \mu$G. Moreover, his Hat Creek results for the North Celestial Pole are confirmed with independent measurements with the Green Bank Telescope (not yet published). All this means that Hat Creek reliably measured strong fields in HI emission, but not weak fields.

The Hat Creek telescope mapped $B_{||}$ in a number of morphologically obvious regions. These included several supernova or superbubble shells such as Eridanus, the North Polar Spur, and the North Celestial Pole Loop (Heiles, 1989; Myers et al., 1995) and the HI in the vicinity of Orion (Heiles, 1997). In every morphologically obvious structure except Orion, the fields were strong ($\gtrsim 5\ \mu$G) and the field retained the same sign over the feature. In Orion the region is separated into two subregions, each with strong but oppositely directed fields.

We conclude that turbulence in the diffuse ISM is not strong enough to tangle the mean magnetic field. One would not expect the turbulence to be so dominant, so this result is consistent with our understanding of magnetoturbulence.

2.4. An Heretofore Unrecognized Possible Source of Turbulence in the Atomic Gas

Figure 4 exhibits a gray-scale velocity-longitude plot from the LDS. We have cranked up the gain to show weak line wings. This image reveals a very common phenomenon: there are small regions where the line wings stick out relative to the surroundings. On this image we have chosen a fairly typical line wing to illustrate the parameters. We show its profile, together with the surrounding profiles and also the difference. There is excess HI in the line wing, centered at about 15 km s^{-1}; it is perhaps matched by a deficiency in the lower-velocity gas in the main peak. If at 100 pc distance, this peak contains about $0.4 M_{\odot}$ and about 10^{45} erg.

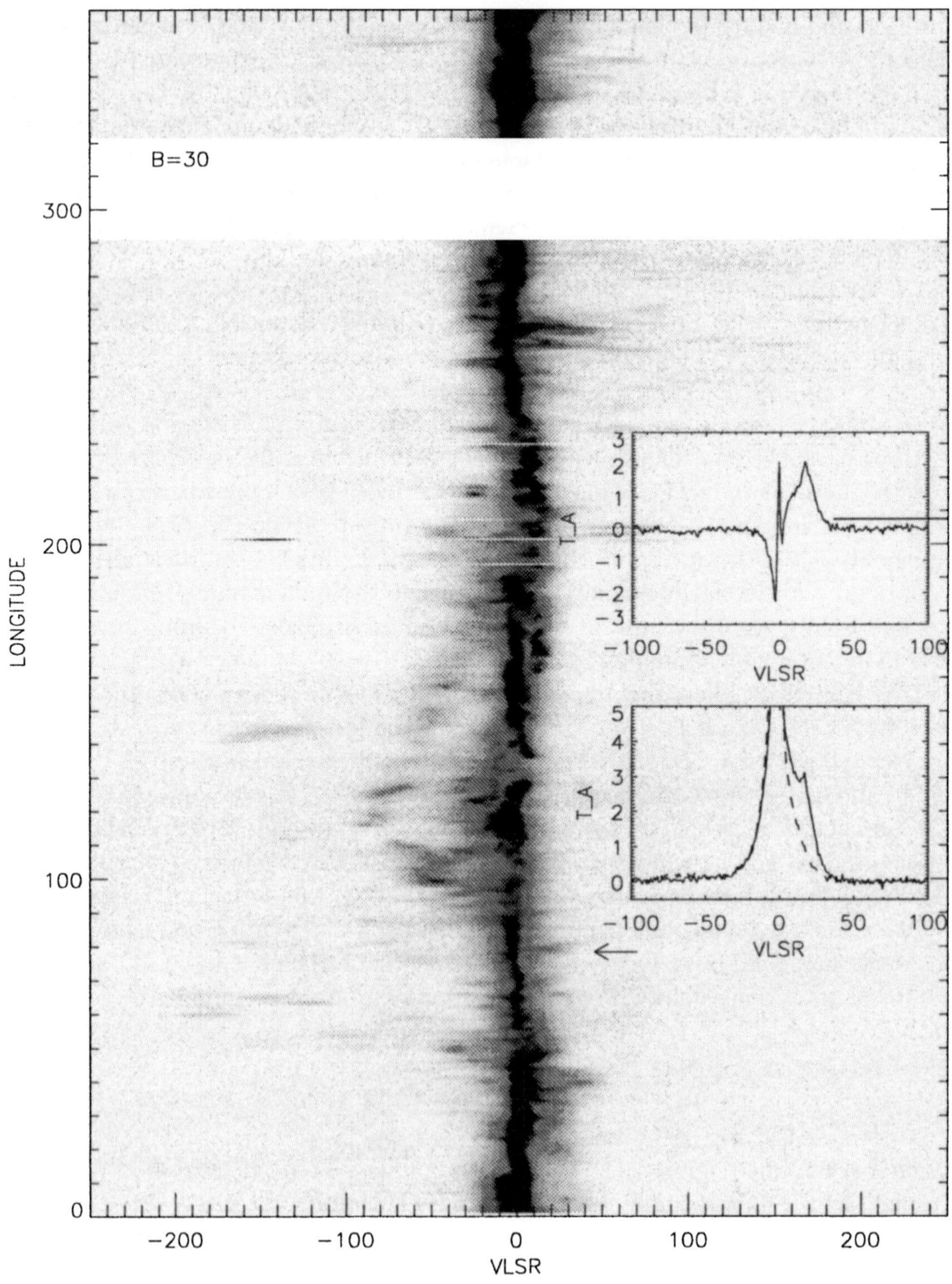

Figure 4. Velocity-longitude image from the LDS survey, for $b = 30^\circ$. The arrow marks an extended line wing. The plots show the HI profiles on and off the wing (bottom), and the difference (top).

We have not yet done a careful analysis, but it is clear from visual inspection that these line wing excesses are quite common. We suspect that they are places where energy is injected into the diffuse ISM, perhaps by stars; for example, evolved AGB stars lose mass prodigiously (Knapp, 1991). These wings reveal sources of direct mechanical energy for the diffuse ISM, different from the familiar point sources related to massive hot stars (supernovae, stellar winds, and HII regions), and different from large-scale sources such as Galactic spiral density wave shocks. Interstellar astronomers are perpetually searching for sources energy for interstellar turbulence, and this may be one; if so, it has been overlooked.

3. A Puzzling Manifestation of Turbulence

In the Millennium Arecibo survey, Heiles and Troland (2003a) rediscovered the cold cloud of Knapp and Verschuur (1972). The column density $N(HI)_{20} \sim 0.2$. The kinetic temperature $T_k \sim 20$ K and ranges down to 17 K. The linewidth exceeds thermal, with $\Delta V_{\mathrm{turb,1D}} \sim 1$ km s^{-1}.

This is a remarkable cloud because of the following combination of reasons.

Firstly, consider the line-of-sight distance $L_{||}$ occupied by this CNM component. We calculate it from the ratio of column to volume density. We infer the volume density $n(HI)$ from the typical CNM pressure $\frac{P}{k} = 3000$ cm^{-3} K (WMHT); at $T_{\mathrm{k}} = 20$ K, $n(HI) = 150$ cm^{-3}. All this gives $L_{||} = 0.05$ pc. That, in itself, is remarkable. But there is more!

The component is visible in emission and can be easily followed in the LDS survey, mainly because the emission from other unrelated gas is so weak. It forms three clouds lying along a ribbon of width $\sim 2^\circ$ and length $\gtrsim 20^\circ$. Overall, it occupies an angular area ~ 66 deg^2. For discussion purposes, we adopt the characteristic angular extent to be $\sim 6^\circ$. If the distance is 100 pc, which is not unreasonable for a cloud at its Galactic latitude $b \sim 44^\circ$, then the linear extent on the plane of the sky $L_\perp \sim 10$ pc.

The ratio $\frac{L_\perp}{L_{||}} \sim 200$. This is a huge aspect ratio, far exceeding the diameter-to-thickness ratio of an old-fashioned long-playing (LP) record. Sheets in the ISM are nothing new (Heiles, 1967), and this argument applied to most CNM clouds produces large aspect ratios. But an aspect ratio this large is extraordinary!

Secondly, consider the effect of turbulence. With the line-of-sight component of turbulence velocity $\Delta V_{\mathrm{turb,1D}} \sim 1$ km s^{-1}, the line-of-sight crossing time $\sim \frac{L_{||}}{\Delta V_{\mathrm{turb,1D}}} \sim 5 \times 10^4$ yr. This is very short and implies the possibility of interstellar diffuse gas kinematical evolution over human history! Moreover, we would expect the turbulence to produce observable structure on the sky over scales of $L_{||}$; if the distance is 100 pc, this corresponds to an angle ~ 1.5 arcmin. Preliminary observations with the Arecibo 3.3 arcmin beam show little structure (but we emphasize that these are preliminary).

Similarly, it seems reasonable to expect that this object would be torn apart by the slightest *external* perturbation. Turbulence in the adjacent medium would perturb the boundary. Such an influence would likely generate recognizable damage to the morphology, such as the rifts seen in two thin sheets by Heiles (1967). Again, the timescale for such perturbations should be comparable to the 5×10^4 years mentioned above.

If this object is spatially, i.e. a singular entity both along and across the line of sight as it appears to be, the timescale for disruption should be $\sim 5 \times 10^4$ years, which is very short. This suggests that the object is transitory, a manifestation of turbulence. Do numerical simulations of turbulence reveal the formation of such morphologically extreme overdensity perturbations?

Acknowledgements

It is a real pleasure to acknowledge the pleasurable Zeeman-splitting collaborations over the years with Tom Troland (especially!), Dick Crutcher, and Alyssa Goodman. I thank Paul Demorest for help with the cold cloud. This work was supported in part by NSF grant AST-0097417.

References

Crutcher, R.M., Heiles, C. and Troland, T.H.: 2003, in: E. Falgarone and T. Passot (eds.), *Turbulence and Magnetic Fields in Astrophysics*, Springer, p. 155.
Hartmann, D. and Burton, W.B.: 1997, *Atlas of Galactic Neutral Hydrogen*, Cambridge University Press (LDS).
Heiles, C.: 1967, *ApJS* **15**, 97.
Heiles, C.: 1989, *ApJ* **336**, 808.
Heiles, C.: 1993, *BAAS* **25**, 829.
Heiles, C.: 1997, *ApJS* **111**, 245.
Heiles, C. and Kulkarni, S.: 1986, in: G.E. Morfill and M. Scholer (eds.), *Physical Processes in Interstellar Clouds*, Reidel, Dordrecht, The Netherlands, p. 13.
Heiles, C. and Troland, T.: 2003a, *ApJ* **586**, 1067.
Heiles, C. and Troland, T.: 2003b, *ApJS* **145**, 329.
Jenkins, E.B. and Tripp, T.M.: 2001, *ApJS* **137**, 297.
Knapp, G.R.: 1991, in: *Frontiers of Stellar Evolution, ASP Conf Ser,* Vol. 20, p. 229.
Knapp, G.R. and Verschuur, G.L.: 1972, *AJ* **17**, 717.
Myers, P.C., Goodman, A.A., Gusten, R. and Heiles, C.: 1995, *ApJ* **442**, 177.
Nakano, T. and Nakamura, T.: 1978, *PASJ* **30**, 67.
McKee, C.F. and Ostriker, J.P.: 1978, *ApJ* **218**, 148.
Spitzer, L.: 1978, *Physical Processes in the Interstellar Medium*, Wiley, p. 143.
Verschuur, G.L.: 1995, *ApJ* **451**, 645.
Wolfire, M.G., McKee, C.F., Hollenbach, D. and Tielens, A.G.G.M., 2003, *ApJ* **587**, 278 (WMHT).

STRUCTURE OF MOLECULAR CLOUDS

EDITH FALGARONE, PIERRE HILY-BLANT and FRANÇOIS LEVRIER
LERMA/LRA—Ecole Normale Supérieure & Observatoire de Paris, France;
E-mail: edith@lra.ens.fr

Abstract. This review, far from exhaustive, focuses on the specific properties of molecular clouds in their least condensed phase. Emphasis is placed on their connection with the diffuse medium, their multi-fractal characteristics and the diversity of their small-scale structure, where cold and dense gas may coexist with dilute and warm gas. Stuctures of essentially pure velocity origin are uncovered and seen as a possible signature of the intermittency of turbulence dissipation.

Keywords: molecular clouds, turbulence, fractals, intermittency, magnetic fields

1. Introduction

Paradoxically, our understanding of the evolution of molecular clouds and star formation in galaxies progresses more slowly than that of the early universe. The main reason is that the relevant physics is highly non-linear and the entities under study are not isolated systems. The range of coupled scales, in space and time, by far exceeds those accessible to numerical simulations: star formation occurring at the AU scale is nonetheless linked to events occurring at galactic and intergalactic scales. The interplay of gravity and turbulence, the importance of dissipative processes and microphysics (chemistry on dust grain surfaces, radiative transfer, coupling of matter to the magnetic field) are a few among the most challenging facets of this physics.

This review is intended to complement those written recently by Williams et al. (2000) and Mac Low and Klessen (2003) and puts emphasis on a few points of relevance to further confrontation between observations and numerical simulations.

There is a growing consensus that molecular clouds are ephemeral structures, i.e. their lifetime is much smaller than their so-called crossing time. The argument, developed in Hartmann et al. (2001), is manifold. There is a lack of old stars in the vicinity of significant amounts of molecular gas, turbulence is fed at large scales and magnetic fields are so weak that the observed supersonic motions are super-Alfvénic. Molecular clouds are therefore described as shock-dominated structures, formed at the convergence of large-scale flows in the atomic medium. This new paradigm naturally explains coeval star formation within large-scale shocks but has a number of problems among which the low average star formation rate observed in our Galaxy and the lack of large-scale shock signatures in molecular gas

Astrophysics and Space Science **292:** 89–101, 2004.

(Brunt, 2003). Yet, it has the merit of proposing a novel view of molecular clouds as out-of-equilibrium structures which we think is closer to reality than the earlier descriptions.

2. A Broad Overview

2.1. Large pressure fluctuations

Molecular clouds extend into the diffuse medium. Almost half of the hydrogen in the cold neutral medium (CNM) is molecular (Savage et al., 1977) and unexpectedly large abundances of molecules are found in the CNM. The formation of CH^+, HCO^+, OH (see references in Lucas and Liszt, 2002) and water (Neufeld et al., 2002, 2003) in the CNM is not well understood because it requires energy inputs apparently exceeding those available in the medium. Interestingly, Jenkins and Tripp (2001) have established that the thermal pressure in the diffuse medium exhibits large excursions up to $P/k \sim 10^5$ K cm^{-3}, 50 times larger than the median value. They ascribe the existence of wisps of high-pressure material in the CNM to either small-scale density enhancements or warm turbulent boundary layers. This result and the recent claim by Heiles (2001) that a large fraction of the diffuse medium is in the thermally unstable range separating the CNM from the warm neutral medium (WNM) have promoted the picture of a rapidly evolving diffuse medium, often driven out of thermal and chemical equilibrium (Hennebelle and Pérault, 2000; Joulain et al., 1998).

A similar perspective may be relevant to molecular clouds. In a recent ^{13}CO(J = 1–0) survey of giant molecular clouds (GMCs) in the molecular ring of the Galaxy, Simon et al. (2001) identified structures in position–velocity space. They found that a large fraction of these stuctures are not bound by self-gravity but by an external pressure force as large as $P/k \sim 10^6$ K cm^{-3}. The weight of the atomic gas layer in the Galaxy, $P_{\rm HI}/k \sim 2 \times 10^4$ K cm^{-3} (Boulares and Cox, 1990), and the gravitational pressure of the parent GMC, $P_{\rm G}/k = 5540 A_{\rm v}^2$ K cm^{-3} (Williams et al., 2000) contribute to this external pressure but the non-thermal pressure of the environment, including magnetic stresses, has to be considered.

2.2. Characteristic sizescales

This highly dynamical perspective contrasts with the existence of characteristic sizescales that are preserved along the evolution of molecular clouds towards increasing gravitational binding. This is true for GMCs which keep similar sizes while their gas mass and gravitational binding differ by orders of magnitude (see the review of Falgarone, 1998), but it is also true for much smaller structures, dense cores and filaments. A few conspicuous and opaque filaments in nearby GMCs

TABLE I

Properties of the filamentary structures in a few GMCs from $^{13}CO(J = 1\text{–}0)$ data of Bally et al. (1987), Loren (1989), Abergel et al. (1994) and Hily-Blant et al. (this volume)

	$W(^{13}CO)$ ($K\,km\,s^{-1}$)	Mass/length ($M_\odot pc^{-1}$)	Width (pc)	Length (pc)
OriA	54	390	0.5	15
ρ Oph	6–12	30–50	0.4–0.8	~6
Taurus	8	40	0.6	18
L1512	3	10	0.16	~1

(Table I) have comparable widths while their masses per unit length, derived from the same tracer, $^{13}CO(J = 1\text{–}0)$, differ by a factor of 40.

2.3. The Dual Nature of Molecular Clouds

Molecular clouds comprise large amounts of diffuse molecular gas of low column density ($A_V < 2$ mag) whose boundaries are fractal (Falgarone et al., 1991). This component, traced by the ^{12}CO ($J = 1\text{–}0$) line emission, is well described as a fractal-Brownian-motion (fBm) structure (i.e. a superposition of Gaussian fluctuations with amplitudes following a power law, power spectrum and random phases) (Stutzki et al., 1998). Diffuse molecular gas screens structures of much larger column densities (A_V up to 300 mag in the inner Galaxy), best traced by mid-IR absorption (Abergel et al., 1996; Pérault et al., 1996; Bacmann et al., 2000; Price et al., 2002; Omont et al., 2003) and submillimetric dust thermal emission (Motte et al., 1998; Johnstone et al., 2000, 2001; Dupac et al., 2001; see also André, this volume). These opaque, often small scale, structures do not have fractal boundaries. They are connected to true filaments (i.e. not sheet-like structures seen in projection) with volume density much larger than that of their environment (see Table I; Lada et al., 1994; Alves et al., 1999). Filaments are likely threaded by helical magnetic fields as suggested by the decrease of polarisation of the dust continuum emission (Fiege and Pudritz, 2000c; Matthews et al., 2001) and/or the kinematic signatures predicted by Fiege and Pudritz (2000a, b) and uncovered by Falgarone et al. (2001). Helical fields and therefore currents have also been inferred in molecular clouds from a morphological analysis of patterns seen in extinction by Carlqvist et al. (1998).

In summary of this brief overview, molecular clouds in our Galaxy appear as dual structures. Their diffuse component (i.e. the bulk of the mass in clouds of the solar neighborhood) is fractal and highly dynamical., It screens more opaque, condensed and coherent (as opposed to fractal) structures, the dense cores and filaments. The rest of this review focuses on the diffuse component. Dense cores are reviewed by André (this volume).

3. Statistical Analysis of the Clouds Density and Velocity Structure.

3.1. Scaling Laws

Molecular clouds in our Galaxy, as traced by CO, have a remarkable self-similar structure (Larson, 1981). Figure 1 is an illustration of the mass–size and size–linewidth scalings of connex structures identified in the position–velocity space of ^{12}CO and ^{13}CO(J = 1–0) maps (see references in Falgarone, 1998). This method is disputable on the justified basis that similar projected velocities do not imply actual spatial connexity of the structures identified in *(l,b,v)* space (Ostriker et al., 2001). However, a similar size–linewidth scaling law has been obtained more recently between 0.05 pc and 20 pc with a different method (Ossenkopf and Mac Low, 2002).

Three points have to be stressed here: (i) the size–linewidth relation has quite a large dispersion, (ii) one single fractal dimension γ, defined by $M(L) \propto L^{\gamma}$, cannot characterize the mass–size scaling law and values between $\gamma = 2$ (solid line) at large scales and $\gamma = 2.3$ (dashed line) at scales smaller than $\sim$1 pc may be more relevant, (iii) at scales smaller than about 0.5 pc, the mass in a given sizescale varies by orders of magnitude between self-gravitating dense cores and much less massive structures of the same size. The size–density scaling $\rho \propto L^{-1}$ is thus meaningless, especially at those scales, as already noted by Falgarone et al. (1992).

Similar methods provide mass spectra, $dN/dM \propto M^{-\alpha}$. Different samples provide $\alpha = 1.83$ between $10^3\ M_\odot < M < 2\,10^6\ M_\odot$ in the early CO survey of the inner galactic plane (Solomon et al., 1987), $\alpha = 1.80$ between $500\ M_\odot < M < 10^6\ M_\odot$ from the CO survey of the outer galaxy (Heyer et al., 2001) and $\alpha = 1.84$ over $10^{-3}\ M_\odot < M < 50\ M_\odot$ in a high-latitude cloud (Heithausen et al., 1998). The slopes of the mass spectra are therefore found to be the same over 9 orders of magnitude of masses, in the inner and outer galaxy, in GMCs and

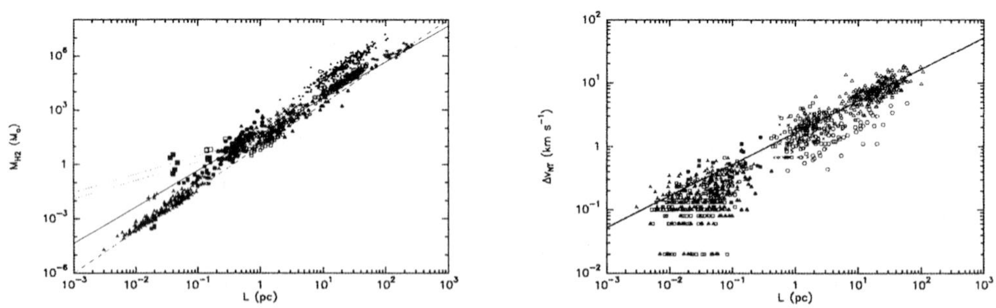

Figure 1. (Left) Gas mass derived from ^{12}CO(J = 1–0) and ^{13}CO(J = 1–0) lines versus sizescale. Different fractal dimensions γ = 2 (solid line), γ = 2.3 (dashed line) defined by $M(L) \propto L^{\gamma}$ are shown. The scaling of isothermal dense cores with size (γ = 1) is also shown for two different gas temperatures (dotted lines). (Right) Non-thermal velocity dispersion versus sizescale. The solid line shows $\Delta v_{\rm NT} \propto L^{0.5}$. From Falgarone (1998).

cirrus clouds, in star forming and inactive regions. Similar values, with larger error bars, have been found in a number of smaller samples covering the mass range 10 to $10^3\ M_\odot$.

The mass spectra of the dense cores found in the maps of the dust submillimeter thermal emission of molecular clouds have significantly different slopes: $\alpha = 2.1$ and 2.5 (Motte et al., 1998; Testi and Sargent 1998). The comparison of the different mass spectra is difficult, however, because the observing techniques and the methods used to derive the mass spectra differ.

Several other interesting statistical approaches have been developed for the confrontation of observations with models of molecular clouds such as the Principal Component Analysis (Brunt and Heyer, 2002), the Spectral Correlation Function method (Rosolowsky et al., 1999; Padoan et al., 2003), high-order moments statistics (see Boldyrev, this volume) and others discussed in the review of Lazarian and Cho (this volume). They cannot be discussed in these few pages.

3.2. The Molecular Cloud Hierarchy and the Diffuse Medium

The Δ-variance method, introduced by Stutzki et al. (1998), has been used on maps of line integrated areas (i.e. close to column density maps) to determine the power spectra $P(k) \propto k^{-\beta}$ of the molecular cloud emission (Bensch et al., 2001; Simon and Bensch, 2002). The power spectral index β is found to increase steadily when measured over decreasing sizescales suggesting smoother structures at small scales. Bensch et al. (2001) note that the origin of this increase is still unclear, possibly related to optical depth effects in the CO lines.

Figure 2 displays the power spectral indices β of a variety of maps versus the range of sizescales over which the power law $P(k) \propto k^{-\beta}$ is observed. Values derived from HI observations in emission and absorption and from 100 μm dust thermal emission are included for comparison with the ^{12}CO and ^{13}CO maps of line integrated areas. In the LMC, a discontinuity of the slope of the power spectrum is observed, and two values of β (connected by a dotted line) are displayed over each relevant range of scales. This change of slope is ascribed to the transition from 2- to 3-dimensional turbulence at about 100 pc, the estimated thickness of the HI layer in that galaxy, according to the criterion of Lazarian and Pogosyan (2000).

Figure 2 provides quantitative insights on the link between turbulence in the diffuse medium and that in molecular clouds. First, the values of β obtained from HI in emission (thus probing the CNM and the WNM) and in absorption (CNM only) are significantly different. Second, the spectral indices of molecular clouds compare well with those of cold HI only and FIR dust emission, at the same scales. The origin of the increase of β towards small scales in molecular clouds cannot therefore be ascribed to optical depth effects in the CO lines.

Most likely, this variation is a manifestation of the multi-fractal facet of molecular clouds. As shown by Stutzki et al. (1998), there is a relation $\beta = \gamma(3 - \alpha)$

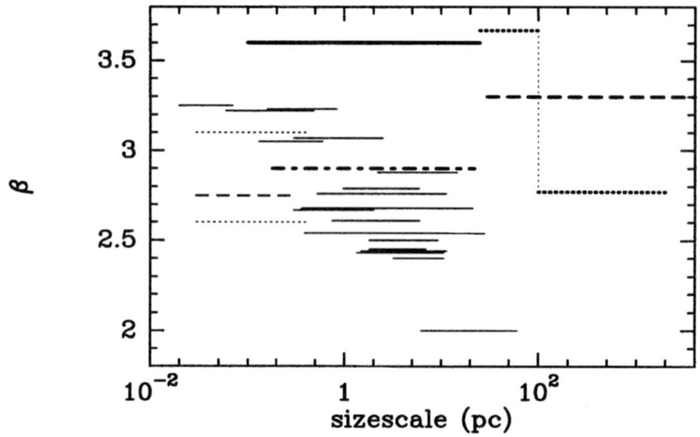

Figure 2. Spectral indices β of brightness maps versus the range of sizescales where the power law $P(k) \propto k^{-\beta}$ is observed. ^{12}CO and ^{13}CO line integrated area maps (thin solid lines) of high-latitude clouds, quiescent and star–forming clouds from Bensch et al. (2001), Plume et al. (2000), Simon and Bensch (2002) and Hily-Blant et al. (in preparation, thin dotted lines). IRAS 100 μm maps of high-galactic latitude sky (thick dot-dashed line) (Gautier et al., 1992). Extragalactic HI in emission: LMC (thick dotted lines) (Elmegreen et al., 2001), SMC (thick dashed line) (Stanimirovic and Lazarian, 2001). Galactic HI in emission: North Polar Spur (thick solid line) (Miville-Deschênes et al., 2003a). Galactic HI in absorption: VLBI maps of HI optical depth against Cas A (thinner dashed lines) (Deshpande et al., 2000). The error bars (not shown for clarity) on all the values of β range between 0.05 and 0.3.

between the power spectrum β of a fBm structure, its fractal dimension γ and the slope of the mass spectrum α. Since α does not seem to vary (see Section 3.1), the increase of β from 2 to 3.3 (Figure 2) is nothing more than the steady increase of the fractal dimension of the cold interstellar medium as the size scale decreases, or in other words a decrease of lacunarity in the maps of the cold medium, be it mostly atomic or molecular. The numerical values, though, are interesting. For $\alpha = 1.8$, β increases from 2.4 to 3.6 as the fractal dimension increases from $\gamma = 2$ to 3. Spectral indices of 3.6 are thus expected for molecular clouds at scales small enough that $\gamma = 3$, i.e. at scales where isotropy is recovered. It is also remarkable that the threshold value for this fractal hierarchy of molecular clouds, as observed, is $\beta = 3.6$, the value found in the HI emission maps.

The fact that the cold HI (CNM only) and molecular clouds have shallower spectra than the HI seen in emission, and that their power spectrum steepens at small scales, up to that of the HI seen in emission, suggests that they are a multi-fractal subset of large-scale HI turbulence. By extrapolation of the broad trend drawn by existing data towards small scales in Figure 2, one would predict that the scale at which isotropy is recovered in molecular gas is smaller than 10^{-2} pc, possibly smaller than the ambipolar diffusion scale (see Mac Low and Klessen, 2003). From this, one might identify the whole hierarchy of cold gas (atomic and

TABLE II

Characteristics of the turbulence observed in various components of the interstellar medium (solar neighborhood). ϵ is expressed in $L_\odot/M_\odot$ for comparison with the energy provided by stellar radiation, $\frac{1}{2}\overline{\rho} v_l^3/l$ in erg cm^{-3} s^{-1} and $P_{\rm turb}$ in erg cm^{-3}

	CNM + WNM	Molecular clouds	Dense cores
$\overline{n}$ (cm^{-3})	30	200	10^4
l (pc)	10	3	0.1
σ_l (km s^{-1})	$\approx$3.5	1	0.1
B (μG)	10	20	100
v_A (km s^{-1})	3.4	2	1.4
$\epsilon = \frac{1}{2} v_l^3/l$	10^{-3}	10^{-4}	10^{-6}
$\frac{1}{2}\overline{\rho} v_l^3/l$	2×10^{-25}	1.7×10^{-25}	2.5×10^{-25}
$P_{\rm turb} = \frac{1}{3}\overline{\rho} v_l^2$	3×10^{-11}	2×10^{-11}	10^{-11}

molecular) with the subset of space where the dissipation of large-scale turbulence in the diffuse medium is concentrated. This subset is known to be fractal in incompressible turbulence (Frisch et al., 1978). In this perspective, molecular clouds cannot be considered as isolated entities with turbulent interiors fed from inside only.

The characteristics of interstellar turbulence as observed in the diffuse medium (HI in emission), molecular clouds and dense cores (see Table II) are such that the two possible invariants of the cascade, the turbulent pressure $P_{\rm turb}$ and the energy transfer rate $\frac{1}{2}\overline{\rho} v_l^3/l$, encompass the warm neutral medium and the coldest and densest structures. The small average values of ϵ, the transfer rate of specific energy in the cascade, show that turbulence dissipation, if homogeneously distributed, cannot be a dominant source of heating for the diffuse medium, in comparison with that provided by starlight. If the dissipation is concentrated on a subset of space with a volume filling factor smaller than $\sim 10^{-3}$, it may become *locally* the dominant heating source. In this case, the local value of ϵ may become larger than $1 L_\odot/M_\odot$ (see Table II).

Incidentally, the spectral index $\beta = 3.6$ of the power spectrum of the HI integrated emission (i.e. the HI column density) in the North Polar Spur is the same as that found for the line velocity centroid maps. Miville-Deschênes et al. (2003b) have shown, by using fBm simulations, that in the case of an optically thin medium, the former is the spectral index of the 3-dimensional density field and the latter, that of the velocity field. The value $\beta = 3.6$ being consistent with the Kolmogorov scaling of incompressible turbulence, the galactic HI value in emission (also found in other samples) provides one more example of Kolmogorov scaling found in the compressible interstellar turbulence (see the review of Lazarian and Cho, this volume; Falgarone et al., 1994).

4. Small Scale Structure in Low-Brightness Molecular Clouds

4.1. High Density and Low Temperature or Low Density and High Temperature?

Regions of low brightness in molecular clouds have long been described as uniform low-density material. Unexpectedly, it has been found to be structured down to scales smaller than 10^{-2} pc. Large densities and scales smaller than ~200AU were first inferred from the smoothness of line profiles and uniformity of the excitation temperature (Tauber et al., 1991; Falgarone et al., 1998; Ingalls et al., 2000). High J CO observations (up to $J = 4$–3) of high-latitude clouds (Ingalls et al., 2000) confirm the existence of high-density $n_{H_2} = 3 \times 10^4$ cm^{-3} and low temperature $T_k \approx 8$K gas (HDLT solution). The sizescale of the HDLT structures, set by their column density, is less than 0.01 pc. Falgarone and Phillips (1996) and Falgarone et al. (2001) use the same set of CO lines but limit their analysis to the line wings, to avoid the optically thick portions of the lines. Their range of solutions is broader, extending from the HDLT solution up to a low density, high-temperature (LDHT) solution, $T_k = 250$K and $n_{H_2} = 180$ cm^{-3}.

These determinations rely on the large observed ^{12}CO(4–3)/^{12}CO(1–0) ratio which rules out the usual solution of moderate density and temperature derived from low J lines in the Large Velocity Gradient (LVG) approximation which may introduce large biases in the results. Hegmann and Kegel (2000) have shown that correlation lengths in the density and velocity distributions of molecular clouds affect the radiative transfer of CO line photons. The cloud physical properties derived from CO lines are thus sensitive to the dynamical structure of clouds, most critically at scales close to the mean free path of CO photons, $\lambda \approx 10$ AU. These observations teach us, however, that the gas emitting in the line wings and line cores may not be at the same density and temperature, and that a small fraction may have densities and temperatures (and possibly pressures) significantly different from those of the bulk of the gas, a result to be compared to that of Jenkins and Tripp (2001) mentioned earlier.

4.2. ^{12}CO ($J = 1$–0) Milliparsec-Scale Structure

Milliparsec-scale structure has been found in ^{12}CO ($J = 1$–0) maps of the Owens Valley Radio Observatory (OVRO; Ingalls et al., in preparation) and Plateau de Bure (PdBI; Falgarone and Pety, in preparation) interferometers. The synthesized beams are 5 and 3.7 arcsec, respectively, corresponding to about 3 milliparsec or 500 AU. In the two datasets, the power seen by the interferometer is found exclusively at velocities belonging to the line wings of the single dish spectrum. No signal is detected by the interferometers at the velocities of the line cores. This effect is ascribed to the smaller surface filling factor of the emitting structures in the tails of the velocity distribution.

In the PdBI maps, many structures appear as resolved elongated and straight filaments of thickness $\sim$1000 AU. Two of them, separated by 3.5 km s^{-1}, are almost parallel to each other and separated in projection by less than 10" or 1500 AU. If this projected separation is a valid size scale to compute a velocity derivative, the inferred shear [1] is as large as $\sim$500 km s^{-1} pc^{-1}, still larger than the value $\sim$40 km s^{-1} pc^{-1} observed over 0.05 pc in high-latitude clouds by Sakamoto (2002). The associated timescale is only $\approx 2 \times 10^3$ yr. Velocity shears, occasionally much larger than the average value observed in molecular clouds $\sim$1 km s^{-1} pc^{-1}, seem therefore to be found at small-scales in the diffuse parts of molecular clouds. Such large local shears have also been found in HI maps (see Heiles, this volume). The density of these small-scale structures cannot be determined because one line only is available.

4.3. Small-scale Structure of Essentially Pure Velocity Origin

The rare and large small-scale velocity shears discussed may be the kinematic signatures of intermittent dissipation of interstellar turbulence (Falgarone and Phillips, 1990). They may trace small-scale shocks or coherent vortices. Lis et al. (1996) have shown that the extrema of the small-scale increments of line centroids (i.e. extrema of observed velocity shears) trace large non-Gaussian excursions of vorticity in turbulence. Shocks bear the same statistical signatures, though, probably because they generate vorticity. Pety and Falgarone (2000) have used the numerical simulations of Porter et al. (1994) to compute synthesized line spectra and the statistical signatures in the maps of the most intense shocks (3% positions of largest negative divergence $-\nabla.\mathbf{v}$) and vortices (3% positions of largest $\mid \nabla \times \mathbf{v} \mid$). Both subsets are at the origin of the non-Gaussian wings in the probability distribution functions (PDF) of line centroid increments at small-scale, and have a similar morphology of narrow filaments with very heterogeneous surface filling factor (Figure 3).

Interstellar turbulence being magnetized, with neutrals only partially coupled to the magnetic fields, the dominant dissipation is due to ion–neutral friction, either in MHD shocks (Flower and Pineau des Forêts, 1998), Alfvén waves (Zweibel and Josafatsson, 1983) or coherent magnetized vortices (Joulain et al., 1998). Reconnection in thin current layers promoted by ion–neutral drift may also operate (Zweibel and Brandenburg, 1997; Heitsch and Zweibel, 2003a, b). In all cases, this dissipation is intermittent in space and time and has signatures most visible at most scales.

Are such intense shocks or vortices observed in molecular clouds? They may be identified by the enhancement of their non-Gaussian behavior at small-scale. They have been searched for in the distributions of line centroid increments (i.e. the shear of the first moment of the velocity in molecular clouds, a substitute of vorticity). One

[1] Velocity shear is the only velocity derivative accessible to interstellar studies, i.e. variations in the plane of the sky of the line of sight velocity.

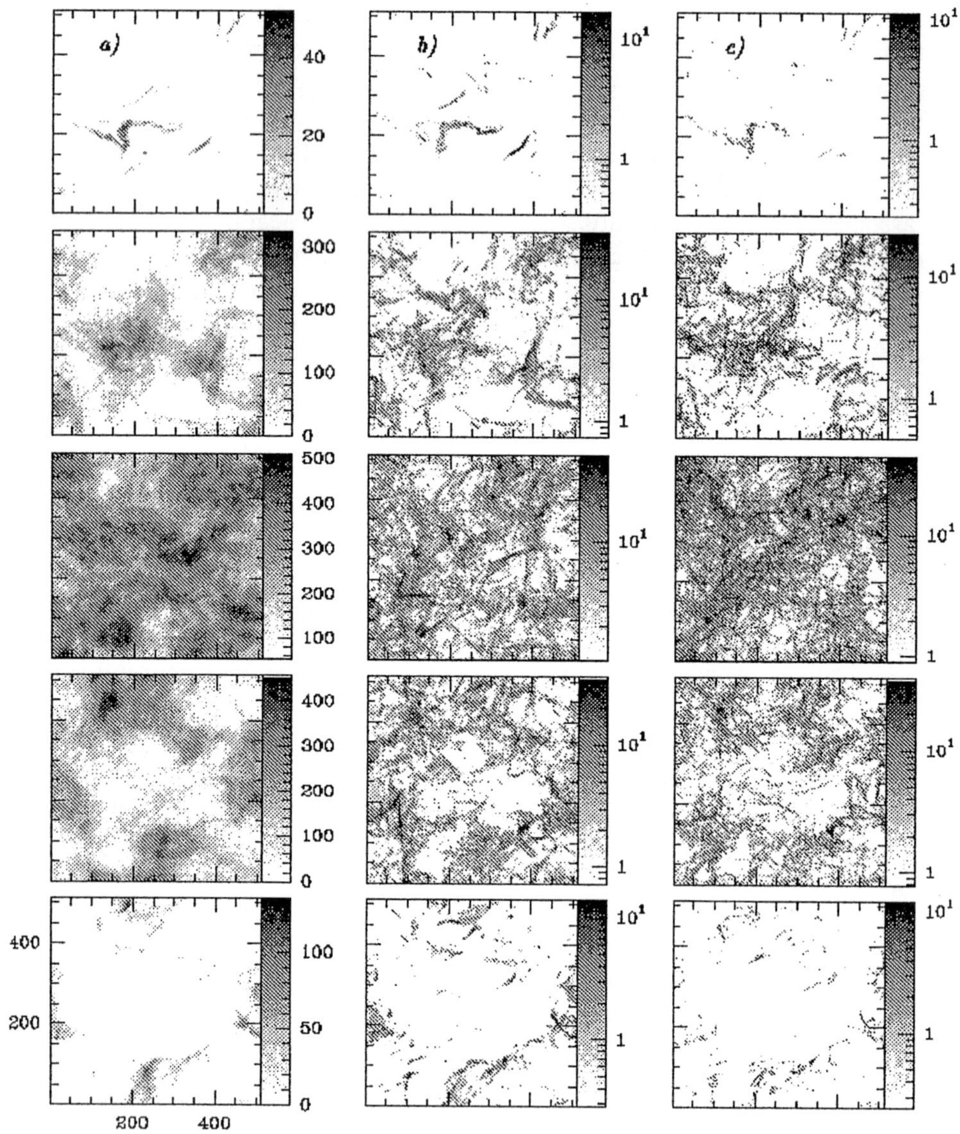

Figure 3. Maps of gas surface filling factor at different projected velocities, across a cube of midly supersonic turbulence (Porter et al., 1994). The velocity tails are in the top and bottom panels. The samples shown are: (a) the complete velocity cube, (b) the subset of the most intense vortices (3% of the data points), and (c) the subset of the largest negative divergence values. From Pety and Falgarone (2000).

example is shown in Figure 4 (Pety and Falgarone, 2003). It displays the locations in the maps where the line centroid increments exceed a threshold value, close to the width of the Gaussian core of the PDF. Similar structures have been found in other fields.

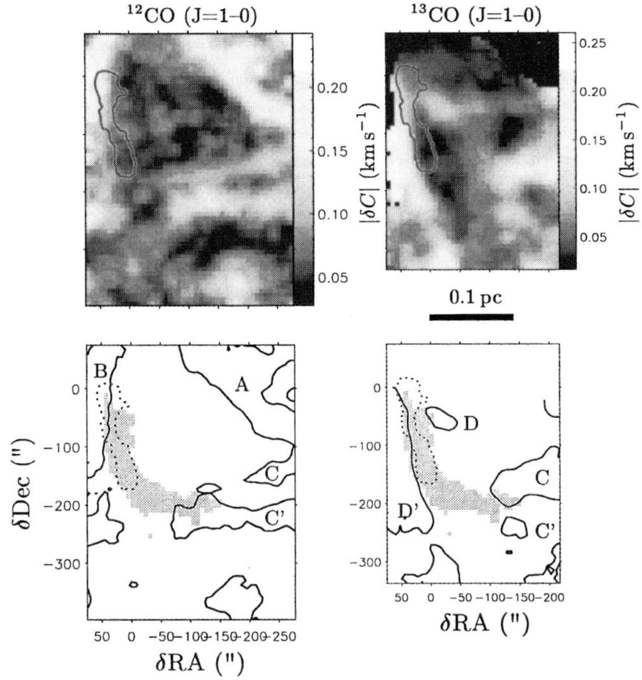

Figure 4. (Top panels) Maps of ^{12}CO and ^{13}CO line centroid increments (or velocity shear) in a high-latitude cloud. The dense core in the field is shown as a solid contour. (Bottom panels). Sketches of the regions of large shear (black contours) relative to the dense core (dotted contour) and the ^{13}CO peak (grey area). The linear scale is shown. From Pety and Falgarone (2003).

These are a novel kind of small-scale structures in molecular clouds because they are essentially pure velocity structures (i.e. not coinciding with any small-scale column density enhancement). Most are elongated and straight, with variable thickness ranging from 0.03 pc down to less than 0.01 pc for those unresolved. Further observations at higher angular resolution are required to determine their smaller scale morphology and infer their nature and link to magnetic fields.

5. Concluding Remarks

This review, meant to promote doubts and questions, is focused on purpose on the characteristics of molecular clouds which may be connected to the out-of-equilibrium aspects of their physics. It discloses the difficulties met at interpreting the observations, at reaching statistical significance in datasets, at recognizing even individual physical processes (e.g. shocks). It uncovers the variety of density and temperature conditions susceptible to coexist at small-scale, even in the apparently most simple parts of molecular clouds, away from protostellar cores. It shows that diffuse molecular clouds are structured down to at least $\sim$500 AU and provides elements in support of a link between turbulence in the diffuse medium and the multi-fractal geometry of molecular clouds. Including all these aspects in numerical simulations is presently out of reach. Yet, we need permanent and thorough confrontation between theory, numerical simulations and observations to progress in that field.

Acknowledgements

E.F. thanks J. Ingalls for providing her with data ahead of publication and for stimulating discussions.

References

Abergel, A., Boulanger, F., Mizuno, A. and Fukui, Y.: 1994, *ApJL* **423**, L59.
Abergel, A., Bernard, J.P., Boulanger, F. et al.: 1996, *A&A* **315**, L329.
Alves, J., Lada, C.J. and Lada, E.A.: 1999, *ApJ* **515**, 265.
Bacmann, A., André, P., Puget, J.-L., Abergel, A., Bontemps, S. and Ward-Thompson, D.: 2000, *A&A* **361**, 555.
Bally, J., Langer, W.D., Stark, A.A. and Wilson, R.: 1987, *ApJL* **312**, 45.
Bensch, F., Stutzki, J., and Ossenkopf, V.: 2001, *A&A* **366**, 636.
Boulares, A. and Cox, D.P.: 1990, *ApJ* **365**, 544.
Brunt, C.M.: 2003, *ApJ* **583**, 280.
Brunt, C.M. and Heyer, M.: 2002, *ApJ* **566**, 289.
Carlqvist, P., Kristen, H. and Gahm, G.F.: 1998, *A&A* **332**, L5.
Deshpande, A.A., Dwarakanath, K.S. and Goss, W.M.: 2000, *ApJ* **543**, 227.
Dupac, X., Giard, M. Bernard, J.-P. et al.: 2001, *ApJ* **553**, 604.
Elmegreen, B.G., Kim, S. and Staveley-Smith.: 2001, *ApJ* **548**, 749.
Falgarone, E. and Phillips, T.G.: 1990, *ApJ* **359**, 344.
Falgarone, E., Phillips, T.G. and Walker, C.K.: 1991, *ApJ* **378**, 186.
Falgarone, E., Puget, J.-L. and Pérault, M.: 1992, *A&A* **257**, 715.
Falgarone, E., Lis, D.C., Phillips, T.G., Pouquet, A., Porter, D.H., and Woodward, P.R.: 1994, *ApJ* **436**, 728.
Falgarone, E. and Phillips, T.G.: 1996, *ApJ* **472**, 191.
Falgarone, E., Panis, J.-F., Heithausen, A., Pérault, M., Stutzki, J., Puget, J.-L. and Bensch, F.: 1998, *A&A* **331**, 669.
Falgarone, E.: 1998, in: B. Guiderdoni and A. Kembhavi (eds.), *Starbursts: Triggers, Nature, and Evolution*, Les Houches School, ESP Sciences.
Falgarone, E., Pety, J. and Phillips, T.G.: 2001, *ApJ* **555**, 178.
Fiege, J.D. and Pudritz, R.E.: 2000a, *MNRAS* **311**, 85.
Fiege, J.D. and Pudritz, R.E.: 2000b, *MNRAS* **311**, 105.
Fiege, J.D. and Pudritz, R.E.: 2000c, *ApJ* **544**, 830.
Flower, D.R. and Pineau des Forêts, G.: 1998, *MNRAS* **297**, 1182.
Frisch, U., Sulem, P.-L. and Nelkin, M.: 1978, *JFM* **87**, 719.
Gautier, T.N., Boulanger, F., Pérault, M. and Puget, J.-L.: 1992, *AJ* **103**, 1313.
Hartmann, L., Ballesteros-Paredes, J. and Bergin, E.A.: 2001, **562**, 852 .
Hegmann, M. and Kegel, W.H.: 2000, *A&A* **369**, 405 .
Heiles, C.: 2001, *ApJ* **551**, L105.
Heithausen, A., Bensch, F., Stutzki, J., Falgarone, E. and Panis, J.-F.: 1998, *A&A* **331**, L65.
Heitsch, F. and Zweibel, E.G.: 2003a, *ApJ* **583**, 229.

Heitsch, F. and Zweibel, E.G.: 2003b, *ApJ* **590**, 291.
Hennebelle, P. and Pérault, M.: 2000, *A&A* **359**, 1124.
Heyer, M.H., Carpenter, J.M. and Snell, R.L.: 2001, *ApJ* **551**, 852.
Ingalls, J.G., Bania, T.M., Lane, A.P., Rumitz, M. and Stark, A.A.: 2000, *ApJ* **535**, 211.
Ingalls, J.G., Bania, T.M., Reach, W. and Carpenter, J.M., in preparation.
Jenkins, E.B. and Tripp, T.M.: 2001, *ApJS* **137**, 297.
Johnstone, D., Wilson, C.D., Moriarty-Schieven, G., et al.: 2000, *ApJ* **545**, 327.
Johnstone, D., Fich, M., Mitchell, G.F. and Moriarty-Schieven, G.: 2001, *ApJ* **559**, 307.
Joulain, K, Falgarone, E., Pineau des, Forêts G. and Flower, D.: 1998, *A&A* **340**, 241.
Lada, C.J., Lada, E.A., Clemens, D.P. and Bally, J.: 1994, *ApJ* **434**, 614.
Larson, R.B.: 1981, *MNRAS* **194**, 809.
Lazarian, A. and Pogosyan, D.: 2000, *ApJ* **537**, 720.
Lis, D.C., Pety, J., Phillips, T.G. and Falgarone, E.: 1996, *ApJ* **463**, 623.
Loren, R.B.: 1989, *ApJ* **338**, 902.
Lucas, R. and Liszt, H.S.: 2002, *A&A* **384**, 1054.
Mac Low, M.-M., Klessen, R.S.: 2004, *Review of Modern Physics* **76**, 125.
Matthews, B.C., Wilson, C.D. and Fiege, J.D.: 2001, *ApJ* **562**, 400.
Miville-Deschênes, M.-A., Joncas, G., Falgarone, E., and Boulanger, F.: 2003a, *A&A* in press.
Miville-Deschênes, M.-A., Levrier, F. and Falgarone, E.: 2003b, *ApJ* **593**, 831.
Motte, F., André, P. and Neri, R.: 1998, *A&A* **336**, 150.
Neufeld,,D.A., Kaufman, M.J., Goldsmith, P.F., Hollenbach, D.J. and Plume, R.: 2002, *ApJ* **580**, 278.
Neufeld, D.A., Bergin, E.A., Melnick, G.J. and Goldsmith, P.F.: 2003, *ApJ* **590**, 882.
Omont, A., Gilmore, G.F., Alard, C. et al.: 2003, *A&A* **403**, 975.
Ossenkopf, V. and Mac Low, M.-M.: 2002, *A&A* **390**, 307.
Ostriker, E.C., Stone, J.M., and Gammie, C.F.: 2001, *ApJ* **546**, 980.
Padoan, P., Goodman, A.A. and Juvela, M.: 2003, *ApJ* **588**, 881.
Pérault, M., Omont, A., Simon G., et al.: 1996, *A&A* **315**, L165.
Pety, J. and Falgarone, E.: 2000, *A&A* **356**, 279.
Pety, J. and Falgarone, E.: 2003, *A&A* **412**, 417.
Plume, R., Bensch, F., Howe, J.E. et al.: 2000, *ApJ* **539**, L133.
Porter, D.H., Pouquet, A. and Woodward, P.R.: 1994, *Phys. Fluids* **6**, 2133.
Price, S., Carey, S. and Egan, M.P.: 2002, *EAS* **4**, 99.
Rosolowsky, E.W., Goodman, A.A., Wilner, D.J. and Williams, J.P.: 1999, *ApJ* **524**, 887.
Sakamoto, S.: 2002, *ApJ* **565**, 1050.
Savage, B.D., Drake, J.F., Budich, W. and Bohlin, R.C.: 1977, *ApJ* **216**, 291.
Simon, R., Jackson, J.M., Clemens, D.P., Bania, T.M. and Heyer, M.H.: 2001, *ApJ* **551**, 747.
Simon, R. and Bensch, F.: 2002, *Meeting of the Center for Star Formation* Studies, Santa Cruz.
Solomon, P.M., Rivolo, A.R., Barrett, J. and Yahil, A.: 1987, *ApJ* **319**, 730.
Stanimirović, S. and Lazarian, A.: 2001, *ApJ* **551**, L53.
Stutzki, J., Bensch, F., Heithausen, A., Ossenkopf, V. and Zielinsky, M.: 1998, *A&A* **336**, 697.
Tauber, J.A., Goldsmith, P.F. and Dickman, R.L.: 1991, *ApJ* **375**, 635.
Testi, L. and Sargent, A.I.: 1999, *ApJ* **508**, 91.
Williams, J.P., Blitz, L. and McKee, C.F.: 2000, in: Mannings, V., Boss, A.P. and Russell, S.S. (eds.), *Protostars and Planets IV*, University of Arizona Press, Tucson.
Zweibel, E.G. and Josafatsson, K.: 1983, *ApJ* **270**, 511.
Zweibel, E.G. and Brandenburg, A.: 1997, *ApJ* **478**, 563.

THE TINY-SCALE ATOMIC STRUCTURE: GAS CLOUDLETS OR SCINTILLATION PHENOMENON ?

S. STANIMIROVIĆ[1], J.M. WEISBERG[2], A. HEDDEN[2], K. DEVINE[2], T. GREEN[2] and S.B. ANDERSON[3]

[1]*Radio Astronomy Lab, UC Berkeley, 601 Campbell Hall, Berkeley, CA, U.S.A.; E-mail: sstanimi@astro.berkeley.edu*

[2]*Department of Physics and Astronomy, Carleton College, Northfield, MN, U.S.A.*

[3]*Department of Astronomy, MS 18-34, California Institute of Technology, Pasadena, CA, U.S.A.*

Abstract. We present preliminary results from the recent multi-epoch HI absorption measurements toward several pulsars using the Arecibo telescope. We do not find significant variations in optical depth profiles over periods of 4 months and 9 years. The upper limits on the optical depth variations in directions of B0823+26 and B1133+16 are similar on scales of 10–20, 350 and 500 AU. The large number of non detections of the tiny scale atomic structure suggests that the AU-sized structure is not ubiquitous in the interstellar medium.

Keywords: ISM structure, pulsars

1. Introduction

For many years there have been theoretical and observational support for the existence of structure in the interstellar medium (ISM) on scales from $\sim$1 pc to 1 kpc (as summarized by Dickey and Lockman, 1990). However, it has long been believed that structure on smaller scales ($<$1 pc) would not be prominent in the ISM. Indeed, if we assume the standard thermal pressure for the cold neutral medium (CNM) $P_{\rm th} \sim 2250$ cm^{-3} K (Jenkins and Tripp, 2001), and the mean temperature for the CNM of $\sim$70 K (Heiles and Troland, 2003), then the expected volume density for the CNM clouds is about 30 cm^{-3}. With the median measured column density of 5×10^{19} cm^{-2} (Heiles and Troland, 2003), the typical expected scale length for the CNM features is $\sim$1 pc. It was therefore quite surprising to frequently find structure on AU scales in many different directions in the ISM.

The AU-sized structure is observationally probed using spatial and/or temporal variations in absorption profiles against background sources. There are several varieties of this observational technique. For the atomic medium, the following three techniques are commonly used. (1) Spatial variation of HI absorption line profiles across the extended background source (Dieter et al., 1976; Diamond et al., 1989; Davis et al., 1996; Faison and Goss, 2001) typically probes scales of a few tens to a few hundreds of AUs. (2) Comparison of optical interstellar absorption lines against binary stars and globular clusters (Meyer and Blades, 1996; Lauroesch et al., 1998) probes larger scales of 10^2–10^6 AU. (3) Time variability of HI absorption profiles against pulsars (Deshpande et al., 1992; Frail et al., 1994; Johnston et al., 2003) probes scales of tens to hundreds of AUs. In the case of the molecular medium, the fine-scale structure is usually observed through time and/or spatial variability of molecular line absorption profiles against compact extragalactic sources (Marscher et al., 1993; Moore and Marscher, 1995) and typically probes scales of tens of AUs. In this paper we concentrate on the AU-sized structure in the cold atomic medium, often referred to as the tiny-scale atomic structure (TSAS).

The direct quantities provided by observations of the TSAS are optical depth variation ($\Delta\tau$) and a particular time or spatial baseline, which can be translated into transverse size ($L_\perp$). Two additional constraints are the frequency of detection and the variability of the line shape. As the AU-sized structure was very frequently detected it was thought that it is more likely to be a general property of the ISM than the effect of some local phenomena. As noted by Heiles (1997), it is often that the line depth was found to vary rather than the line shape, suggesting that the TSAS is kinematically related to the CNM.

The conventional interpretation of the above mentioned observations is that optical depth variations are due to a blob moving in and out of the line-of-sight, whose transverse dimension is equal to $L_\perp$. Assuming a simple spherical geometry, the HI volume density of these blobs can be estimated and is typically of order of 10^4 cm^{-3}, which is very dense. The inferred thermal pressure is then of the order of 10^6 cm^{-3} K, much higher than the hydrostatic equilibrium pressure of the ISM or the standard thermal pressure of the CNM. This has been known for a long time and has caused much controversy. It is expected that such over-dense and over-pressured features should dissipate on a time scale of about 100 yrs and therefore not be common in the ISM. A variety of other questions concerning the AU-sized structure includes the following. How are AU-sized features formed and maintained in the ISM? What is the fraction of the ISM occupied by these features? Is this is a new population of interstellar clouds? How do these features relate to the continuous hierarchy of structure observed on larger spatial scales?

In order to solve this long-standing puzzle several alternative explanations were proposed. Heiles (1997) suggested that TSAS features are actually curved filaments and/or sheets that happen to be aligned along our line-of-sight. The ratio of the line-of-sight to the plane-of-the-sky length of 4–10 is required to bring the TSAS HI volume density to modest, acceptable, values. Deshpande (2000) suggested that TSAS blobs correspond to the tail of a hierarchical structure organization that exists on larger scales. They pointed out that TSAS observations were misinterpreted by associating measured $L_\perp$ with the longitudinal dimension of TSAS clouds, leading to extraordinary high volume densities. Gwinn (2001) proposed that optical depth fluctuations seen in multi-epoch pulsar observations are actually a scintillation phenomenon combined with the velocity gradient across the absorbing HI of order of 0.05–0.3 km s^{-1}AU^{-1}. Different explanations predict a different level of optical depth variations at a particular scale size. For example, Deshpande (2000) expects that optical depth variations would increase with the size of structure, while Gwinn (2001) predicts maximum variations on very small spatial scales probed by the interstellar scintillation. All suggested explanations, however, call for more observational data.

Motivated by the recent theoretical efforts in understanding the nature and origin of the TSAS we have undertaken new multi-epoch observations of HI absorption against a set of bright pulsars. We decided to observe the same sources as Frail et al. (1994) in order to enhance the number of available time baselines for comparison. This paper summarizes preliminary results from this project.

2. Observations and Data Processing

We have used the Arecibo telescope[1] to obtain new multi-epoch HI absorption measurements against six pulsars previously studied by Frail et al. (1994). For detail observing and data processing description see Stanimirovic et al. (2003). We had four observing sessions: August 2000, December 2000, September 2001 and November 2001, measuring HI absorption profiles over time intervals from less than a day to 1.25 yrs. The Caltech Baseband Recorder was used as a fast-sampling backend, with a total bandwidth of 10 MHz, recording the raw voltage data every 100 ns. The first stage of data reduction was performed at the Caltech's Center for Advance Computation and Research. The 'pulsar-on' and 'pulsar-off' spectra were accumulated by finding the pulsar pulse and extracting spectra during the pulse and between pulses, respectively. The pulsar absorption spectrum is created by generating the 'pulsar-on'–'pulsar-off' for each scan, doing frequency switching to flat the baseline, and accumulating all such spectra with a weight proportional to T_{PSR}^2, where T_{PSR} is the antenna temperature of the pulsar. Final absorption and emission spectra have velocity resolution of 0.5 km s^{-1}.

3. Results

We present here the results for three pulsars: B0823+26, B1133+16, and B2016+28, from two observing sessions in August 2000 and December 2000. In addition, we have used the third session obtained by Frail et al. (1994) to provide a long term comparison. Figure 1 shows absorption and emission spectra for PSR B2016+28 from Aug 2000 and Dec 2000, while Figure 2 compares data from December 2000 and December 1991. The top panel in each figure shows absorption spectra obtained at two different epochs, and a scaled emission spectrum to

[1]The Arecibo Observatory is part of the National Astronomy and Ionosphere Center, operated by Cornell University under a cooperative agreement with the National Science Foundation.

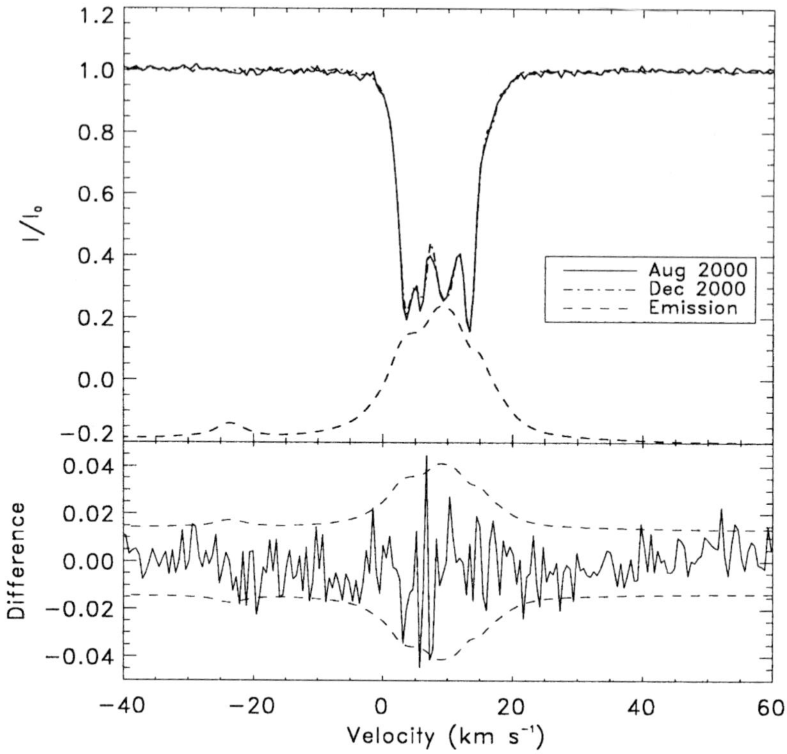

Figure 1. Top: HI absorption spectra against B2016+28 observed in August and December 2000. The HI emission spectrum, obtained in the same direction, has been multiplied by 0.0005. *Bottom:* Difference between the two absorption spectra. Contours show ± 2-σ levels where the HI brightness temperature has been taken into account.

get a feeling about how noise in the absorption spectra varies with velocity. The bottom panel in each plot shows a difference of absorption profiles at two epochs with overlaid $\pm$ 2-σ significance envelopes. To estimate the rms noise we have taken into account contribution from the sky background and, very importantly, contribution from the HI brightness temperature which increases the rms noise on line significantly (for example, by a factor of 3 in the case of B2016+28).

For all three pulsars we compared HI absorption spectra over a time span of four months (August 2000 to December 2000) and 9 yrs (December 1991 to December 2000). Except for the case of B2016+28 between August 2000 and December 2000, in all other cases we *do not* find significant change in the absorption spectra. B2016+28 shows a marginal, 2.5-σ, change in absorption spectra obtained with a four months difference. Transverse velocity of this object is 39 km s^{-1}, meaning that during the period of four months B2016+28 has traveled over a distance of 3 AU. This marginal detection could be caused by structure on scales of 3 AU.

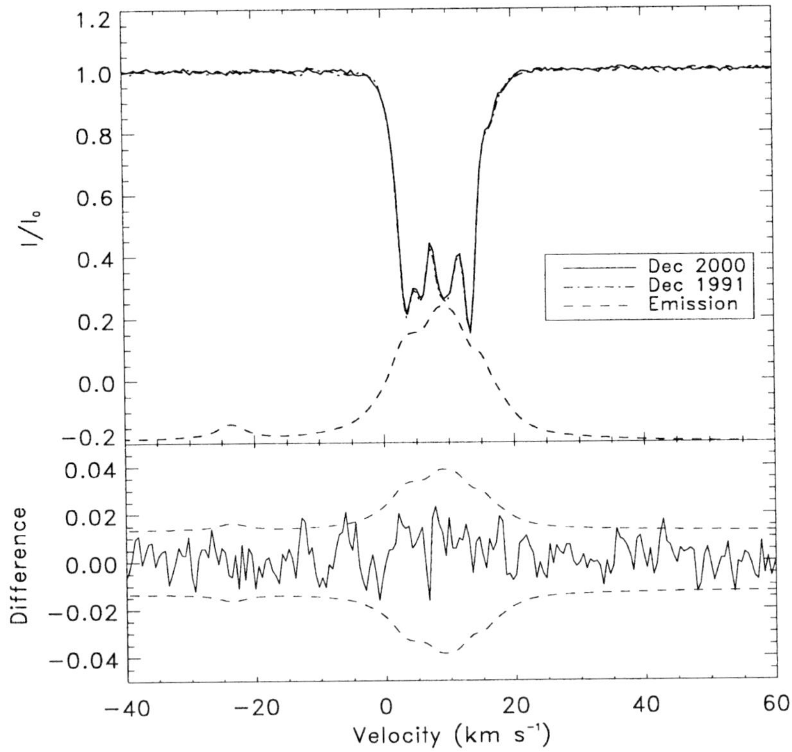

Figure 2. Comparison of HI absorption profiles against B2016+28 obtained in December 1991 and December 2000. Details are the same as in Figure 1.

Other comparisons do not show change in absorption spectra on scales of about 10–20, 70 and 350–500 AU at $\Delta\tau$ level down to about 0.02. Information on individual objects and an upper limit on $\Delta\tau$ are given in Table I (note that these are 2-σ limits on the HI line). Frail et al. (1994) found variations of 0.07 for B0823+16 and 0.04 for B1133+16 over a period of 13 months. In the case of B2016+28 they noticed even larger optical depth variations of almost one over 7 and 20 months periods. Our sensitivity is better than typical variations seen in Frail et al. (1994) and our non detections are therefore a significant result. As already pointed out by Johnston et al. (2003) there is a concern that the Frail et al.'s data have been over interpreted.

The upper limits on optical depth fluctuations set by B0823+26 and B1133+16 at scales of 10–20 and 350–500 AU are strikingly low. $\Delta\tau$ of 0.03 corresponds to 3×10^{18} cm^{-2} for the column density fluctuations of the CNM. In addition, these upper limits are almost the same for spatial scales that are more than an order of magnitude different! Deshpande (2000) predicts opacity variations as an extension of HI opacity irregularities observed on larger scales using a single power law

TABLE I

Transverse scales and maximum $\Delta\tau$ probed by the three pulsars. Note that Max. $\Delta\tau$ is a 2-σ limit estimated on the HI line

PSR	$L_\perp$ (AU)	Max. $\Delta\tau$
B0823+26	10	< 0.04
	350	< 0.04
B1133+16	20	< 0.03
	500	< 0.02
B2016+28	3	0.18
	70	< 0.2

spectrum. Using the power spectrum of opacity distribution in the direction of Cas A and extrapolating down to AU scales, they predict $\Delta\tau \sim 0.03$ at scales of about 10 AU, while at 50–100 AU $\Delta\tau$ ~0.2–0.4 is expected. This model generally expects that $\Delta\tau$ increases with spatial scales. While results for B0823+26 and B1133+16 at scales of 10–20 AU are consistent with this picture, our upper limits for $\Delta\tau$ at scales of 350 and 500 AU for the same pulsars are significantly lower from what is expected. However, these results are derived from small number statistics and we will be able to provide stronger constraints on particular theoretical models in the near future. In addition, it is important to mention that the power spectrum slope derived for the direction toward Cas A may not be applicable to directions sampled by pulsars in this study.

This large number of TSAS non detections, together with recent results by Johnston et al. (2003), is disturbing and unexpected! While previous observations were frequently detecting TSAS, in six comparisons presented here, and in many comparisons in Johnston et al. (2003), there is only one marginal detection for the case of B2016+28 and one detection for B1641-45 (Johnston et al., 2003) over a period of almost 20 yrs. This raises the question of the existence of TSAS as traced at least by multi-epoch pulsars observations. Contrary to the previous belief that the TSAS is ubiquitous in the ISM, our results indicate that TSAS may be more rare and could be related to some kind of a local phenomenon.

4. Conclusions

We have compared multi-epoch HI absorption observations toward B0823+26, B1133+16 and B2016+28 over time periods of four months and 9 yrs. Except for a marginal change in the case of B2016+28 over a 4-month period, we do not find significant changes in absorption spectra. This is very different from previous observations by Frail et al. (1994) who saw significant optical depth variations for the same pulsars over time periods of 7, 13 and 20 months. We have placed upper limits on the optical depth variations of 0.02–0.04 in directions of B0823+26 and B1133+16. In addition, the upper limit on optical depth variations appears the same for spatial scales of 10–20, 350 and 500 AU, traced by these objects. A large number of non detections of the TSAS presented here, together with recent results by Johnston et al. (2003), suggests that the TSAS is not ubiquitous in the ISM.

Acknowledgements

We are grateful to Caltech's Center for Advance Computation and Research for the use of their facilities for data storage and processing. S.S. acknowledges support by NSF grants AST-0097417 and AST-9981308.

References

Davis, R.J., Diamond, P.J. and Goss, W.M.: 1996, *MNRAS* **283**, 1105.
Deshpande, A.A.: 2000, *MNRAS* **317**, 199D.
Deshpande, A.A., McCulloch, P.M., Radhakrishnan, V. and Anantharamaiah, K.R.: 1992, *MNRAS* **258**, 19P.
Diamond, P.J., Goss, W.M., Romney, J.D., Booth, R.S., Karbela, P.N.M. and Mebold, U.: 1989, *ApJ* **347**, 302.
Dickey, J.M. and Lockman, F.J.: 1990, *A&AR* **28**, 215.
Dieter, N.H., Welch, W.J. and Romney, J.D.: 1976, *ApJ* **206**, L113.
Faison, M.D. and Goss, W.M.: 2001, *AJ* **121**, 2706.
Frail, D.A., Weisberg, J.M., Cordes, J.M. and Mathers, C.: 1994, *ApJ* **436**, 144.
Gwinn, C.R.: 2001, *ApJ* **561**, 815.
Heiles, C.: 1997, *ApJ* **481**, 193.
Heiles, C. and Troland, T.H.: 2003, *ApJ* **586**, 1067.
Heiles, C.E.: 2000, in: D.G. Finley and W.M. Goss (eds.), *Radio Interferometry: The Saga and the Science*, NRAO Workshop No.27 (Associated Universities), 7.
Jenkins, E.B. and Tripp, T.M.: 2001, *ApJS* **137**, 297.
Johnston, S., Koribalski, B., Wilson, W. and Walker, M.: 2003, *MNRAS* **341**, 941.
Lauroesch, J.T., Meyer, D.M., Watson, J.K. and Blades, J.C.: 1998, *ApJ* **507**, L89.
Marscher, A.P., Moore, E.M. and Bania, T.M.: 1993, *ApJ* **419**, L101.
Meyer, D.M. and Blades, J.C.: 1996, *ApJ* **464**, L179.
Moore, E.M. and Marscher, A.P.: 1995, *ApJ* **452**, 671.
Stanimirović, S., Weisberg, J., Dickey, J.M., de la Fuente, A., Devine, K., Hedden, A. and Anderson, S.B.: 2003, *ApJ* **592**, 953.

IS THE SUPERSONIC TURBULENCE OBSERVED IN HII REGIONS SUSTAINED BY THE PRESENCE OF MAGNETIC FIELDS?

JOHN E. BECKMAN[1,2] and MÓNICA RELAÑO[2]

[1]*Instituto de Astrofísica de Canarias, La Laguna, Tenerife, Spain; E-mail: jeb@ll.iac.es*
[2]*Consejo Superior de Investigaciones Científicas, Spain*

Abstract. We describe a set of Hα emission line profiles from populations of H II regions in nearby spiral galaxies. These are characterized by a strong Gaussian central peak, and lower intensity higher velocity wing features. From the peak we extract a non-thermal velocity component, due to the internal turbulence of the region. The plot of the widths of these non-thermal components against Hα luminosity shows a lower envelope in line width, which we assign to regions in, or close to virial equilibrium, although the region mass derived on this assumption is higher than the mass obtained by summing all known mass components. We speculate that this discrepancy, as well as the supersonicity of the line widths, can be explained by the presence of turbulent magnetic fields within the H II regions, and make a very rough estimate of the fields implied, of order a few tens of microgauss.

Keywords: H II regions, supersonic turbulence

1. The Observational Data

The observations presented here were obtained with the TAURUS Fabry–Perot line imager on the 4.2 m William Herschel Telescope (WHT) in La Palma. The data come as a "cube" in which the surface brightness of the Hα emission from a full galaxy is presented as a function of three coordinates: two in position, referenced to the galaxy centre, and one in velocity referenced to the systemic radial velocity of the galaxy's motion. From these data cubes we can extract emission line profiles from any subset of pixels, which for the present study we can select to cover any discrete H II region. A typical luminous H II region in the galaxies we have studied, subtends an angle of between 5 and 15 arcsec, and the technique allows us to obtain emission line profiles of several hundred regions per galaxy. Because flux calibration of the data cubes is subject to systematic errors, we calibrate using a continuum subtracted Hα emission map of the same object, obtained with a pair of narrow band filters using either the WHT itself, or one of the other major La Palma telescopes (3.5 m Galileo, 2.5 m NOT), which also have sub-arcsecond seeing. For the study presented here we have selected a few tens of the most isolated H II regions whose spectra contain no more than a single peak, and whose emission lines can be interpreted without ambiguity.

Astrophysics and Space Science **292:** 111–118, 2004.

2. Deriving the Non-Thermal, Turbulent, Line Widths from the Profiles

The output of the instrument can be obtained in the form of a line profile per pixel of image. These profiles can then be integrated to obtain a mean profile over any selected set of pixels, which in this case can be those of a selected H II region. A characteristic profile takes the form of a quasi–Gaussian peak, with $\sigma_{\rm obs}$ of order 20 to 30 km s^{-1} ,(large compared to the sound speed of close to 13 km s^{-1}), and with lower intensity wing features, offset by a few tens of km s^{-1} from the line centre, to the blue and to the red.

A set of spectra is shown in Figure 1, where the central peaks and wing features can be readily seen. These latter are due to expanding gas impelled by stellar winds, supernovae, and dust–coupled radiation pressure. They are of considerable interest for the dynamical development of the H II regions, and are a causal agent for the general internal turbulence of the regions, but they do not form the subject matter for the present study, and the reader is referred to Relaño et al. (2003a, b) for a description of their energetics and some simple modelling considerations. However

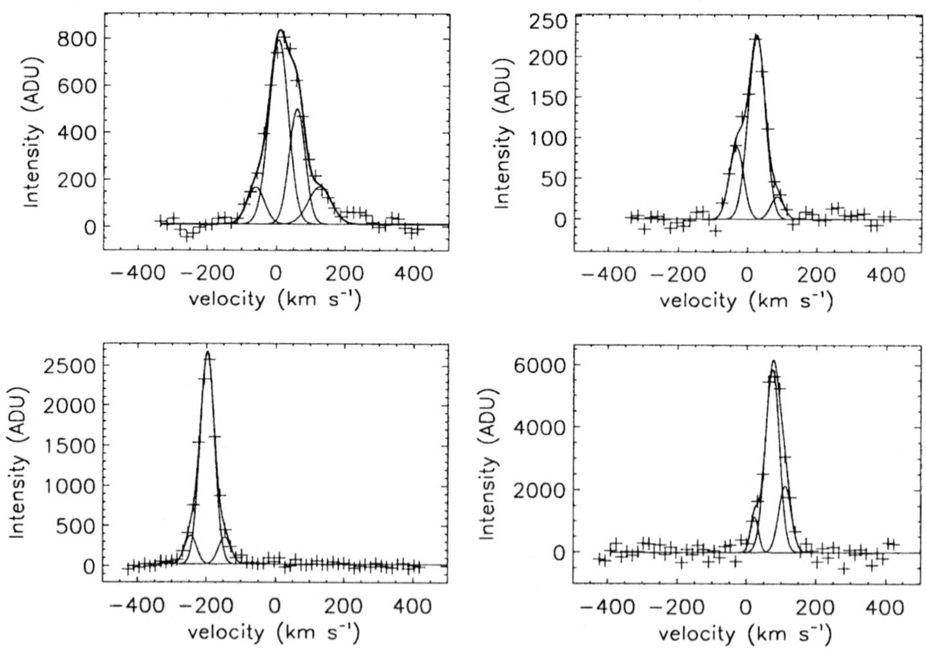

Figure 1. Integrated Hα emission line profiles for four representative H II regions in the barred spiral galaxies NGC 1530, NGC 3359 and NGC 6951, illustrating the line structure described in the text: central strong peak, ascribed to a general field of internal turbulence, plus weaker wings, ascribed to systematic expansion. The present article deals only with the peak. The upper left hand spectrum shows a subsidiary peak, well separable from the primary peak, due to a less luminous superposed H II region.

as we can see in Figure 1 most of the Hα emission as seen in the main peaks of the emission lines comes from the generalized turbulent motion within each region. We know that it is turbulence rather than bulk motion, because the line profile, and in particular the line width ($\sigma_{\rm obs}$), from a given region shows very little variation from centre to limb.

In order to obtain a turbulent $\sigma_{\rm nt}$ from an observed line profile, we have to subtract off in quadrature the following contributions to the observed $\sigma_{\rm obs}$: (a) the instrumental $\sigma_{\rm inst}$, which for the observations shown here are 17.4 km s^{-1} for NGC 1530, 16.3 km s^{-1} for NGC 3359 and 15.9 km s^{-1} for NGC 6951, (b) the thermal $\sigma_{\rm t}$, which we take as constant at 9.1 km s^{-1}, corresponding to an internal kinetic temperature of $\sim$10 000 K, and (c) the natural $\sigma_{\rm n}$, which is 3 km s^{-1}. Any excess we attribute to turbulence, so that lines with observed $\sigma_{\rm obs}$ greater than $\sim$19 km s^{-1}, the sum in quadrature of the above contributions, show significant turbulent velocities. If we go on to use an estimate for the sound speed in the gas of an H II region at the above temperature and densities characterizing these regions, (of order 1 to 10 atoms cm^{-3}, see e.g. Rozas et al. (1996)), with a value of 13 km s^{-1}, a line shows evidence for supersonic turbulence if its observed $\sigma_{\rm obs}$ is greater than $\sim$23.5 km s^{-1}, and this is the case for a major fraction of the regions observed, as we will see in detail below. One point of interest is that our $\sigma_{\rm obs}$ are systematically narrower than they would be if we had tried to fit a single gaussian peak to the observed profiles, as can easily be seen by inspecting Figure 1, and this means that the widths are systematically less than those published in some previous studies where the signal to noise ratio was not sufficient to effect a valid component separation, notably in the pioneering study of Terlevich and Melnick (1981). In spite of this, the turbulent widths are frequently supersonic.

3. Plots of Non-Thermal Line Widths Against the H II Region Luminosities

We made TAURUS observations from three nearby, nearly face–on spiral galaxies, NGC 1530, NGC 3359, and NGC 6951, from which we extracted the H II region emission line profiles and analyzed them as outlined above. A way to present the results which best displays physical relationships is to plot the non-thermal $\sigma_{\rm nt}$ against the Hα luminosity of the H II region. Plots of this kind are shown in Figures 2a and 2b for NGC 1530 and NGC 3359 respectively. They are similar in all important respects, and the figures show this; the data from third galaxy do not show any different features, so it was not necessary to include them here. The non-thermal $\sigma_{\rm nt}$ values, which we identify with the turbulent velocities, range from quite small numbers, well below 10 km s^{-1} to values in the low tens of km s^{-1}. We have not shown error bars on these graphs; however values much below 10 km s^{-1} are significantly affected by any noise in the profiles, and values below 5 km s^{-1} are not reliable. We can see several features common to the distributions in the two graphs. Firstly there is a, not surprising, tendency for $\sigma_{\rm nt}$ to increase with luminosity,

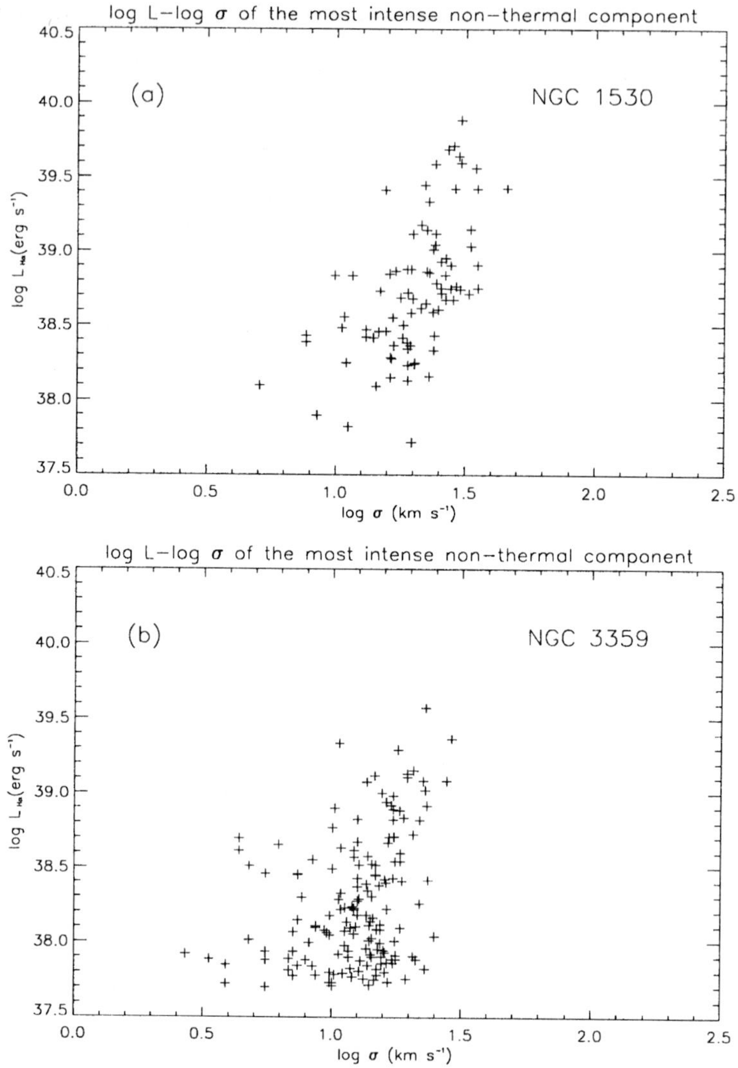

Figure 2. (**a**) Plot of the logarithmic Hα luminosity v. the logarithm of the non-thermal component of the emission line width (σ_{nt}) for the most luminous H II regions in NGC 1530. Note that many of the line widths, notably for the more luminous regions of the set, are supersonic (i.e. have values of $\sigma_{\mathrm{nt}} > 13$ km s^{-1}). (**b**) As for (a) but for a nearer barred spiral: NGC 3359.

but it is in no way monotonic. Secondly, there appears to be an upper envelope in luminosity, or a lower envelope in σ_{nt}, to the distribution. Thirdly many of the line widths appear highly supersonic.

To isolate the supersonic line widths, as well as to eliminate those line widths whose determinations are the least accurate, we have combined all the information from the three galaxies into one graph, with a lower σ_{nt} cut off at 13 km s^{-1} , shown

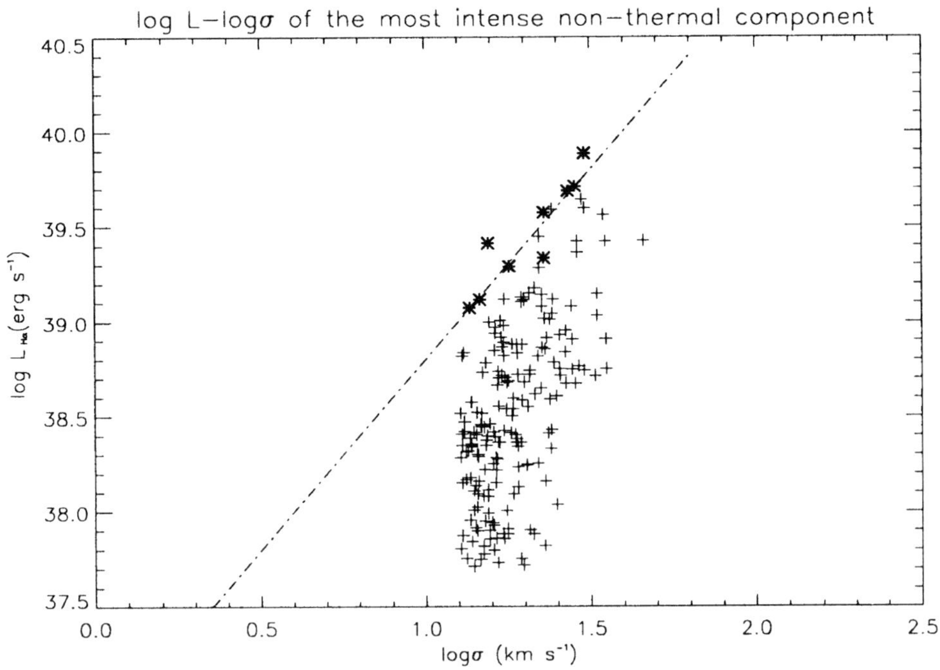

Figure 3. Luminosity–σ_{nt} diagram for the H II regions in NGC 1530, NGC 3359, and NGC 6951, with a cut–off at σ_{nt} equal to the sound speed. The values of L and σ_{nt} have been adjusted for cases with more than one identified main peak, so that the σ_{nt} refers to the major peak, and the L refers to the fraction of the total luminosity belonging to that peak, i.e. to that individual H II region. Note the lower envelope in σ_{nt}, which we hypothetically identified as representing regions in virial equilibrium, an assumption which we examine critically in the text.

in Figure 3. In this figure the luminosities of the H II regions have also been adjusted to take into account that in a minority of H II region spectra we find more than one overlapping central peak, indicating more than one region along the line of sight. In this case we have used the dominant peak as the relevant feature to measure the line width, and used a fractional value of the Hα luminosity proportional to the fractional contribution of this dominant peak to the complete spectrum. These finer details are not so important in the present context, but by including them we obtain a rather clean upper envelope to the plot, which we will comment on below, with the rest of the plot looking like a scatter diagram.

4. Are the H II Regions in Virial Equilibrium?

4.1. General Considerations

In the paper by Terlevich and Melnick (1981) referenced above, they drew the conclusion that a fit to their luminosity–σ_{nt} plot, in the log–log plane, had a slope of 4, which they claimed was evidence that the population of the H II regions as a whole is in virial equilibrium but the model they used was based on the equilibrium of stellar clusters, and their data has been superseded in quality in the CCD epoch. In an earlier piece of work in our group, Rozas et al. (1998), we showed, for the H II regions in the classical "grand design" spiral NGC 4321 (M100) that there was an upper envelope to the plot, with a slope of 2.5, which we claimed was lower than 4 because an increasing fraction of ionizing radiation escapes from luminous H II regions.

To test whether a given H II region is virialized or not, we have calculated the masses of a sample of our regions in two ways: (i) dynamically assuming virial equilibrium, using the measured $\sigma_{\rm nt}$ to calculate the internal turbulent velocities, and (ii) using the mass of the ionized gas, measured from the surface brightness in Hα of the region, and making a proportional allowance for neutral gas, via a plausible model. In the second estimate we also included the stellar mass, estimated also from the Hα luminosity via the IMF. This component was in all cases a small percentage of the total, never as much as 10%, and typically just a few percent in the most "generous" of estimates. The results can be summarized as follows: those regions with line widths away from (greater than) the "envelope value" for a given luminosity, have computed virial masses much greater than their component sum masses, typically more than an order of magnitude greater. This we interpret as indicating that they are not in virial equilibrium; their line widths are enhanced by contributions to their internal motions, from supernovae and stellar winds, which have not relaxed into equilibrium. The computed virial masses of the regions on the envelope are also greater than their component sum masses, by a factor which varies little along the envelope, and is of order 5. Although this suggests that even these regions are not in equilibrium, there is a reasonable alternative possibility which we will outline below.

4.2. Key observational evidence from NGC 604

Galactic H II regions are not good targets for optical line width studies due to strong differential dust absorption. Local group galaxies offer the best choice, and NGC 604 in M33 has been thoroughly measured by Yang et al. (1996). They made a TAURUS map in Hα of the whole region, from which we can derive an integrated non-thermal $\sigma_{\rm nt}$ of 15 km s^{-1}. However, they present a $\sigma_{\rm nt}$ analysis pixel by pixel over the whole region, showing that $\sigma_{\rm nt}$ is made up of an underlying mode $\sigma_{\rm nt}$ of 12 km s^{-1}, from the majority of pixels, averaged with much higher values emitted over limited areas (they also detect the wing features commented on above, as a separate distinct contribution, which we are not considering here). If we interpret the integrated width as containing local enhancements from non-equilibrium energy inputs, but that the mode non-thermal width is the virial width, we obtain agreement within a factor of two or three between the virial and component masses. We note that NGC 604 lies on the envelope of our luminosity–$\sigma_{\rm nt}$ diagram. The ratio between the integrated non-thermal line width and the mode value of the pixel by pixel high intensity points for NGC 604 is 0.8 and if we go on to assume that this is typical for all regions on the envelope, the discrepancy between their virial and component masses is reduced to between two and three. Although this reduction from $\sim$5 to $\sim$3 is "helpful" a discrepancy still remains. It would be of great interest to study further Local Group high luminosity H II regions such as 30 Dor in this context.

However, both the virial and the component sum estimates are subject to known possible large systematic uncertainties. For the virial estimate the effective radius of an H II region depends strongly on the radial distribution of the gas density, which is quite poorly determined. In the virial mass expression we used a uniform value of one half the luminous radius of the region for this parameter, but this needs further examination. For the component estimate, the ratio of neutral to ionized gas mass is a large and not well known parameter. We used a constant value of 10 in our estimates, based on a radiative transfer model, but this is clearly subject to major uncertainty, and a factor two here is by no means out of the question. Yang et al. (1996) used a factor, based on observations, of 6 for this ratio in the case of NGC 604.

So we conclude that our test for viriality is incomplete, but that within systematic error estimates, it is possible that the H II regions on our luminosity–σ_{nt} envelope are in virial equilibrium.

5. Supersonic Line Widths and Magnetic Fields

If in fact the turbulence in luminous H II regions, notably those on the envelope, is close to, or in equilibrium, the fact that the turbulent velocities are supersonic requires an explanation. We need to explain two points: how can supersonic velocities be sustained, and why are the values as measured? Sustaining the motions observed is not a problem; as we showed in Relaño et al. (2003b) the energy in the turbulence is quite small compared to the combined stellar wind and radiation energies from the exciting stars of the regions, without even resorting to supernova inputs. As seen in Figures 2 and 3 envelope values of the turbulent velocity range from subsonic, (which we have characterised roughly as <13 km s^{-1}) , to over 30 km s^{-1} , and these values increase systematically with region luminosity. An interpretation could be that we are observing the Alfven velocity corresponding to the coexistence of mechanical motions and turbulent magnetic fields. This explanation is quite plausible but if it is correct, we note that the condition of equipartition between mechanical and magnetic energy may not be satisfied. The observed turbulent velocities rise with luminosity, but without further careful evidence we do not know whether there is a systematic trend of increasing sound velocity in the same luminosity range, which would be necessary for equipartition to be satisfied. However speculating that the equilibrium line widths are in fact Alfvenic, (not super-Alfvenic), we can estimate the (turbulent) magnetic field strengths implied as of order a few tens of microgauss. Since we are dealing with regions containing a considerable fraction of turbulent plasma, these values which are higher that the few microgauss fields measured in diffuse clouds, are not ruled out, and could possibly account for the difference between the virial masses derived assuming purely mechanical equilibrium, and the component mass estimates. This would be the equivalent for an ionized cloud to the magnetic contribution to virial equilibrium in molecular clouds inferred by Myers et al. (1988).

Acknowledgements

The authors acknowledge grants PB97-0219 of the Spanish DGCYT and AYA2001-0435 of the Spanish Ministry of Science and Technology.

References

Myers, P. and Goodman, A.: 1988, *ApJ* **329**, 392.
Relaño, M., Beckman, J.E., Rozas, M. and Zurita, A.: 2003a, *RMAA* **15**, 205.
Relaño, M., Beckman, J.E. and Dyson, J.E.: 2003b, *RMAA* **15**, 205.
Rozas, M., Knapen, J.H. and Beckman, J.E.: 1996, *A&A* **312**, 275.
Rozas, M., Sabalisck, N., Beckman, J.E. and Knapen, J.H.: 1998, *A&A* **338**, 15.
Terlevich, R. and Melnick, J.: 1981, *MNRAS* **195**, 839.
Yang, H., Chu, Y.H., Skillman, E.D. and Terlevich, R.: 1996, *AJ* **112**, 146.

BIMA AND JCMT SPECTROPOLARIMETRIC OBSERVATIONS OF THE CO J = 2 − 1 LINE TOWARDS ORION KL/IRc2

J.M. GIRART[1], J.S. GREAVES[2], R.M. CRUTCHER[3] and S.-P. LAI[4]

[1]*Departament d'Astronomia i Meteorologia, Universitat de Barcelona, Barcelona, Catalunya, Spain; E-mail: girart@ieec.fcr.es*

[2]*Royal Observatory, Blackford Hill, Edinburgh, Scotland, U.K.*

[3]*Astronomy Department, University of Illinois, Urbana, IL U.S.A.*

[4]*Jet Propulsion Lab, California Institute of Technology, Pasadena, CA, U.S.A.*

Abstract. We present results of JCMT and BIMA CO $J = 2 - 1$ polarization observations towards the Orion KL/IRc2 high mass star forming region. The linear polarization fraction of the JCMT CO $J = 2 - 1$ spectra presents a clear decrease towards the center of the line, as expected, due to the increase of the optical depth. The position angle remains constant along the spectral line, except at the line center, where the highest optical depth and lower fractional polarization are measured. The combined BIMA and JCMT maps of the redshifted and blueshifted CO emission show a uniform polarization pattern that does not coincide with previous dust continuum observations at similar angular resolution. This suggests that the CO and dust are tracing different spatial components along the line of sight.

Keywords: polarization, magnetic fields, Orion KL/IRc2

1. Introduction

It is clear that observations of the magnetic fields in molecular clouds are very important to improve our understanding of the star formation environments and to confront theory (Crutcher et al., 1999). The properties of the magnetic fields in molecular clouds can be inferred from four different observational techniques: linear polarization of the dust emission (Greaves et al., 1999a; Lai et al., 2001; Lai et al., 2002; Matthews et al., 2002), Zeeman effect measurements (Crutcher, 1999; Levin et al., 2001), linear polarization of the molecular line emission (Greaves et al., 1999b; Girart et al., 1999), and Ion/Neutral comparison (Houde et al., 2000a; Houde et al., 2000b).

In the presence of a magnetic field, rotational levels split into magnetic sublevels. If the populations of the different magnetic sublevels are unequal net linear polarization will result. Significant linear polarization may arise in interstellar molecular spectra lines if some requirements are fulfilled (the Goldreich-Kylafis effect): an anisotropic radiation field, a similar rate of radiative and collisional transitions and an optical depth of $\tau \sim 1$ (Goldreich and Kylafis, 1981; Kylafis, 1981, 1983; Deguchi and Watson, 1984). The polarization vectors can be either parallel or perpendicular to the magnetic field direction. The first detection in molecular clouds was reported by Greaves et al. (1999a). The first mapping of the linearly polarized emission (towards NGC 1333 IRAS 4A) was obtained with the BIMA array (Girart et al., 1999). This and more recent works (Greaves et al., 2002; Lai et al., in press) show that mapping of the polarized molecular line emission complements dust polarization mapping, making it possible to extend the study of magnetic fields to regions where dust emission is too weak for polarization detection, such as outflow or extended areas of dense cores.

OMC-1 is the nearest region of OB star formation. The far-IR through mm dust polarized emission towards this molecular cloud has been studied intensively (Gonatas et al., 1990; Leach et al., 1991; Aitken et al., 1997; Schleuning et al., 1998; Coppin et al., 2000). At large scales ($\gtrsim$0.04 pc) the polarization vectors are well ordered and roughly parallel to the molecular ridge, which suggests a strong ($\sim$1 mG) and uniform magnetic field perpendicular to the ridge. The fractional polarization decreases significantly toward the Orion KL/IRc2 region. BIMA mm high angular resolution dust polarization observations show that this is due to an abrupt change in polarization direction south of the Orion hot core (Rao et al., 1998). This change in polarization direction may be due to a change in magnetic field direction. Alternatively, it could be due to the Gold effect: i.e., the grains are mechanically aligned by outflows (Lazarian, 1997).

In this paper we present the first polarization maps obtained by combining single-dish (JCMT) and interferometric (BIMA array) data. We observed the CO $J = 2 - 1$ polarized emission towards the Orion KL/IRc2 region.

2. Observations

2.1. BIMA OBSERVATIONS

The interferometric observations were carried out with the BIMA array[1] in the C configuration during December 1997. The observations were centered at the Orion IRc2 position. The digital correlator was configured to observe the CO $J = 2 - 1$ line (230.537 GHz) with a 32 channel window of 100 MHz bandwidth, which provided a spectral resolution of 4 km s^{-1}. The BIMA array polarimeter consists of a pair of quarter-wave plates placed in front of the receiver of each antenna, which enable to detect the left (L) and right (R) circularly polarized radiation. L and R are time-multiplexed on each antenna using a fast Walsh function switching pattern. This allows sampling the four Stokes parameters, I, Q, U and V on every baseline on a timescale shorter than the (u, v) cell transit time to achieve quasi-simultaneous dual polarization observations. The instrumental polarization response ("leakages") was calibrated by observing the strong polarized QSO 3C 273 and 3C 279. After leakage corrections, polarization measurements should be accurate at about 0.5% level. The BIMA polarimeter and the calibration procedures are described in more detail by Lai et al. (2001).

[1] The BIMA array is operated by the Berkeley-Illinois-Maryland Association support from the National Science Foundation.

2.2. JCMT OBSERVATIONS

The JCMT[2] observations were carried out in August 2000. The JCMT polarimeter consists of a half-wave plate in front of the heterodyne receiver RxA2. The data were taken by stepping the plate to a series of different angles, thus rotating the source plane of polarization with respect to the receiver. The instrumental polarization was subtracted from the data, with instrumental polarization residuals of about 0.3%. More information about the JCMT polarimeter can be found at Greaves et al. (2002). The CO $J = 2 - 1$ line was mapped on a 3×3 rectangular grid of points spaced $20''$, with the grid center almost coinciding with the BIMA phase center. The JCMT spectra have a spectral resolution of 0.4 km s^{-1}.

2.3. SINGLE-DISH AND INTERFEROMETER COMBINATION PROCEDURE

In order to combine the interferometer and single-dish data we use the technique of adding short and zero-spacing data from the JCMT in the region of the (u,v) plane unsampled by the BIMA observations. This was done by first, resampling the spectral coverage and resolution of the JCMT spectra to that of the BIMA data. Second, to remove the JCMT beam pattern, the JCMT map was deconvolved using the CLEAN algorithm. Then the deconvolved JCMT map was multiplied by the BIMA primary-beam response. Finally, a randomly distributed sample of visibilities was generated from the deconvolved model of the JCMT map in the (u, v) range of 0–3.5 kλ. The number of visibilities was chosen to have a JCMT visibility density similar to the BIMA data in the inner part of the (u, v) plane.

3. Results and Discussion

3.1. JCMT OBSERVATIONS

The left panel of Figure 1 shows the polarization fraction and total intensity of the JCMT CO $J = 2 - 1$ spectra towards the Orion KL/IRc2 region. Except for the central position, the polarization fraction decreases significantly toward the center of the CO line. Thus, at the line wings the polarization fraction is typically in the 2–10% range, while at the center of the line the polarization fraction decreases below the 1% value. Indeed, at some positions the polarization at the center of the line is at the level of the expected residual instrumental polarization (e.g., N, NW and SW positions). Interestingly, the polarization fraction at the central position is below 1% over the whole velocity range plotted in Figure 1. The theoretical work on molecular

[2]The James Clerk Maxwell Telescope is operated by The Joint Astronomy Centre on behalf of the Particle Physics and Astronomy Research Council of the United Kingdom, the Netherlands Organisation for Scientific Research, and the National Research Council of Canada.

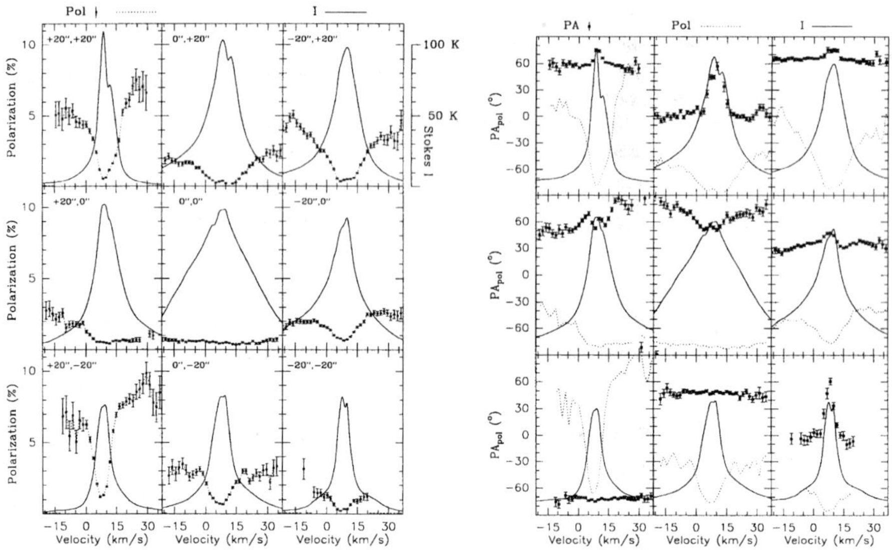

Figure 1. Left Polarization fraction (dots with error bars) and total intensity (solid line) of the CO $J = 2 - 1$ for the positions observed with the JCMT. The position of each spectrum is shown in the upper left corner of each panel. *Right* Position angle of the polarized emission (dots with error bars), polarization fraction (dotted line), and total intensity (solid line) of the CO $J = 2 - 1$ for the positions observed with the JCMT.

line polarization predicts that the polarization percentage should be maximum at an optical depth of $\sim$1, decreasing significantly for high optical depths (Kylafis, 1983). The observational results are, thus, in agreement with the theoretical predictions since the CO $J = 2 - 1$ line is clearly optically thick at the center of the line.

The right panel of Figure 1 shows the position angle of the polarization vectors for the JCMT CO $J = 2 - 1$ spectra. As a general trend, for a given spectrum the position angles of the polarization vectors remain almost constant along the CO line. This uniformity along the spectral line is only broken at some positions at the center of the line where the polarization fraction is low, $\lesssim 1\%$, and therefore where the residual instrumental polarization contribution may be significant (e.g., see top central and bottom right spectra). Taking into account that the redshifted and blueshifted CO are likely tracing different spatial regions, the uniformity of position angles along the line suggests a uniform magnetic field. This result is in agreement with the properties of the magnetic field derived from the dust polarization at large scales.

3.2. Combined JCMT and BIMA observations

Given the low polarization fraction toward the center of the CO line, we combined the BIMA and JCMT data for the velocity ranges with the highest polarization fraction and with significant polarized intensity. The JCMT and BIMA combined polarization maps presented here are centered at a velocity of $v_{\rm LSR} = -5$ (blueshifted) and 23 km s^{-1} (redshifted) with a velocity width of 16 km s^{-1}. In order to check the reliability of the combination procedure, the resulting combined maps were convolved with a Gaussian to match the angular resolution of the original JCMT data. We compared the JCMT+BIMA data with the JCMT data at the center position and at the four offset positions located 20″ from the center (the other four JCMT positions are located 28″ from the center, so slightly outside from the primary beam). At the center, east and west positions the polarization of the two data sets are in good agreement (ΔPA $\lesssim 10^\circ$) for the redshifted emission and in rough agreement (ΔPA $\simeq$ 12–20°) for the blueshifted emission. However, at the N and S positions the polarization angle differences between the two data sets are significantly higher (up to ΔPA $\simeq 37^\circ$). Incomplete sampling of the JCMT map and strong side lobes (~70% of the main beam at 18″ N and S of the center) in the BIMA dirty beam could cause this partial discrepancy. Therefore, hereafter we focuse the discussion on the polarization within a radius of $\lesssim$18″ with respect to the center, where the combined data is likely more reliable.

Figure 2 shows the total intensity map and polarization vectors of the combined BIMA and JCMT CO $J = 2 - 1$ blueshifted and redshifted emission. This figure shows that the polarization angles are well ordered. The CO polarization vectors of the blueshifted emission have an uniform distribution with all the vectors aligned in the NE-SW direction. The averaged position angle is $51^\circ \pm 8^\circ$ (see upper panel of Figure 3). The CO polarization vectors of the redshifted emission have also a uniform distribution, but there is a smooth change in the position angle direction,

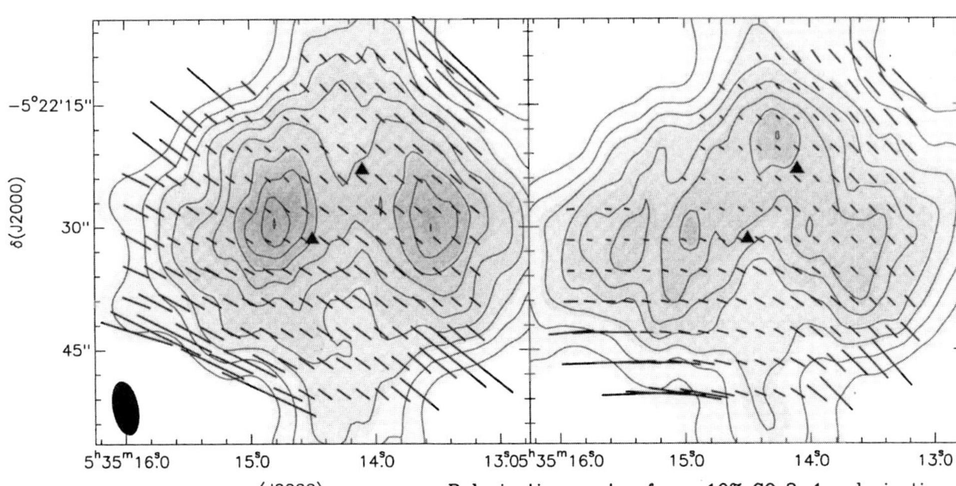

Figure 2. Polarization map of Orion KL/IRc2 from the combined BIMA and JCMT CO $J = 2 - 1$ data. The grey and contour maps show the Stokes I of the blueshifted (left panel) and the redshifted (right panel) CO $J = 2 - 1$ emission. The synthesized beam, 6.″6 × 3.″0 and PA = 10°, is shown in the bottom left corner of the left panel.

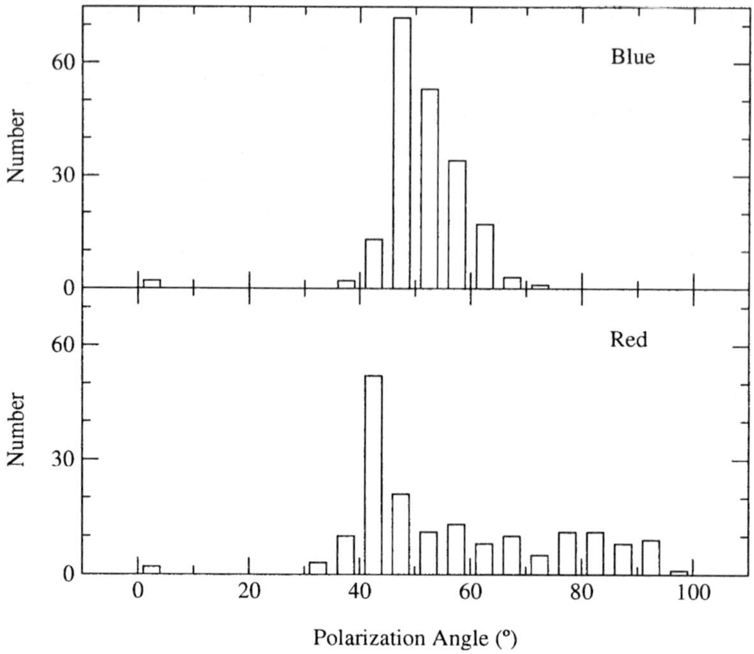

Figure 3. Polarization angle distribution of the blueshifted (upper panel) and the redshifted (bottom panel) CO $J = 2 - 1$ emission of the JCMT and BIMA combined data in Orion KL/IRc2.

from an averaged position angle of $43^\circ \pm 8^\circ$ for the western part of the map to $80^\circ \pm 10^\circ$ for the eastern part. The smoothness of the direction change can also be seen in the histogram of the polarization angle distribution (bottom panel of Figure 3).

Table I shows the polarization position angle measured at four different positions from the blueshifted and redshifted CO $J = 2-1$ emission as well from the 3.3 mm and 1.3 mm dust continuum emission. The dust measurements were obtained by Rao et al. (1998) with the BIMA array at an angular resolution similar to our data.

TABLE I

Polarization position angles (°) in Orion KL/IRc2

Offset position[a]	Dust[b]		CO $J = 2 - 1$	
	3.3 mm	1.3 mm	Blueshifted	Redshifted
(−09″, −11″)	14 ± 6	8 ± 5	49 ± 2	50 ± 3
(−01″, −03″)	94 ± 6	100 ± 5	52 ± 2	59 ± 3
(−08″, +03″)	13 ± 7	31 ± 9	48 ± 2	40 ± 3
(+08″, +19″)	52 ± 8	48 ± 8	45 ± 6	–

[a]Offset are measured in arcseconds with respect to IRc2.
[b]From Rao et al. (1998).

At the northeastern position (with respect to the IRc2) the CO and dust polarization vectors are parallel within the errors. However, for the positions closer to the center of the map (the first three rows of the table) the CO polarization vectors are neither perpendicular nor parallel to those of the dust polarization. Indeed, the difference between the CO and dust position angles is roughly 40°. This suggests that the redshifted and blueshifted CO and the dust emission trace different spatial regions along the line of sight. The redshifted and blueshifted CO maps probably trace the molecular outflow. Thus, the well ordered and uniform magnetic field inferred from the CO polarization maps imply that the molecular outflow is embedded in a region with a strong magnetic field.

Acknowledgements

This research was supported by NSF grant AST 99-81363 and AST 98-20651. JMG acknowledges partial support from MCyT grant AYA-2002-00205. RMC acknowledges support from NSF grant AST 02-05810.

References

Aitken, D.K., Smith, C.H., Moore, T.J.T., Roche, P.F., Fujiyoshi, T. and Wright, C.M.: 1997, *MNRAS* **286**, 85.
Coppin, K.E.K., Greaves, J.S., Jenness, T. and Holland, W.S.: 2000, *A&A* **356**, 1031.
Crutcher, R.M.: 1999, *ApJ* **520**, 706.
Crutcher, R.M., Troland, T.H., Lazareff, B., Paubert, G. and Kazès, I.: 1999, *ApJ* **514**, L121.
Deguchi, S. and Watson, W.D.: 1984, *ApJ* **285**, 126.
Girart, J.M., Crutcher, R.M. and Rao, R.: 1999, *ApJ* **525**, L109.
Goldreich, P. and Kylafis, N.D.: 1981, *ApJ* **243**, L75.
Gonatas, D.P., Engargiola, G.A., Hildebrand, R.H., Platt, S.R., Wu, X.D., Davidson, J.A., Novak, G., Aitken, D.K. and Smith, C.: 1990, *ApJ* **357**, 132.
Greaves, J.S., Holland, W.S., Friberg, P. and Dent, W.R.F.: 1999a, *ApJ* **512**, L139.
Greaves, J.S., Holland, W.S., Minchin, N.R., Murray, A.G. and Stevens, J.A.: 1999b, *A&A* **344**, 668.
Greaves, J.S., Holland, W.S. and Dent, W.R.F.: 2002, *ApJ* **578**, 224.
Kylafis, N.D.: 1983, *ApJ* **267**, 137.
Houde, M., Bastien, P., Peng, R., Phillips, T.G. and Yoshida, H.: 2000a, *ApJ* **536**, 857.
Houde, M., Peng, R., Phillips, T.G., Bastien, P. and Yoshida, H.: 2000b, *ApJ* **537**, 245.
Lai, S., Crutcher, R.M., Girart, J.M. and Rao, R.: 2001, *ApJ* **561**, 864.
Lai, S., Crutcher, R.M., Girart, J.M. and Rao, R.: 2002, *ApJ* **566**, 925.
Lai, S., Girart, J.M. and Crutcher, R.M.: 2003, *ApJ* **598**, 392.
Lazarian, A.: 1997, *ApJ* **483**, 296.
Leach, R.W., Clemens, D.P., Kane, B.D. and Barvainis, R.: 1991, *ApJ* **370**, 257.
Levin, S.M., Langer, W.D., Velusamy, T., Kuiper, T.B.H. and Crutcher, R.M.: 2001, *ApJ* **555**, 850.
Matthews, B.C., Fiege, J.D. and Moriarty-Schieven, G.: 2002, *ApJ* **569**, 304.
Rao, R., Crutcher, R.M., Plambeck, R.L. and Wright, M.C.H.: 1998, *ApJ* **502**, L75.
Schleuning, D.A.: 1998, *ApJ* **493**, 811.

THE MEASUREMENT OF THE ORIENTATION OF THE MAGNETIC FIELD IN MOLECULAR CLOUDS

MARTIN HOUDE[1], RUISHENG PENG[1], HIROSHIGE YOSHIDA[1], ROGER H. HILDEBRAND[2], THOMAS G. PHILLIPS[3], C. DARREN DOWELL[3,4], PIERRE BASTIEN[5], JESSIE L. DOTSON[6] and JOHN E. VAILLANCOURT[7]

[1]*Caltech Submillimeter Observatory, Hilo, HI, U.S.A.; E-mail: houde@submm.caltech.edu*
[2]*Department of Astronomy and Astrophysics and Enrico Fermi Institute and Department of Physics, University of Chicago, IL, U.S.A.*
[3]*California Institute of Technology, Pasadena, CA, U.S.A.*
[4]*Jet Propulsion Laboratory, Pasadena, CA, U.S.A.*
[5]*Département de Physique, Université de Montréal, Montréal, QC, Canada*
[6]*NASA Ames Research Center, Moffet Field, CA, U.S.A.*
[7]*Physics Department, University of Wisconsin, Madison, WI, U.S.A.*

Abstract. We discuss how the combination of polarimetry and ion-to-neutral molecular line width ratio measurements permits the determination of the magnitude and orientation of the magnetic field in the weakly ionized parts of molecular clouds. Polarimetry measurements give the field orientation in the plane of the sky and the ion-to-neutral molecular line width ratio determines the angle between the magnetic field and the line of sight. We show the first results obtained with this technique on the M17 and Orion A star-forming region using Hertz 350 μm polarimetry maps and HCO^+-to-HCN molecular line width ratios to provide the first view of the spatial orientation of the magnetic field these molecular clouds.

Keywords: ISM cloud, ISM individual (M17), ISM individual (Orion), ISM magnetic field, ISM polarization, ISM radio lines

1. Introduction

The magnetic field can be specified with three parameters: its magnitude B, the viewing angle α defining its inclination to the line of sight, and the angle β made by its projection on the plane of the sky (eastward from north). Until now, of these three quantities only the third could be measured. At submillimeter wavelengths, this can be accomplished, for example, with polarization measurements of the continuum radiation emanating from elongated dust grains that are aligned by the local magnetic field (Davis and Greenstein, 1952). The angle β is thus obtained from the angle of the polarization vector since the projection of the magnetic field vector in the plane of the sky is oriented at right angles to the polarization vector (Hildebrand, 1988). The magnitude B cannot be directly measured, only the projection of the magnetic field vector to the line of sight $B_{\rm los}$ can be obtained with Zeeman measurements (e.g., Brogan and Troland (2001)). For general cases, where the magnetic field lies out of the plane of the sky, a determination of the viewing or inclination angle α, in combination with the measurements for $B_{\rm los}$ and β, would provide a description of the magnetic field vector **B**.

As was shown by Houde et al. (2002), the determination of the viewing angle α can be accomplished through a comparison of the profile of line spectra from coexistent ion and molecular species. The ion-to-neutral line width ratio, as defined by (Houde et al., 2000a), is the fundamental parameter that, when combined with polarimetry data, holds the key to the determination of the viewing angle.

In this paper, we will summarize the technique and the first results of Houde et al. (2002) obtained for the M17 molecular cloud and present new results obtained for the Orion A region using HCO^+ and HCN spectroscopic data and extensive 350 μm continuum and polarimetry maps obtained with the Hertz polarimeter (Dowell et al., 1998) and the SHARC II camera (Dowell et al. 2003) at the Caltech Submillimeter Observatory (CSO).

2. The Ion-to-Neutral Line Width Ratio

It was shown in Houde et al. (2000a) and Houde et al. (2000b) that a comparison of the line profiles of coexistent neutral and ion species can be used to detect the presence of the magnetic field in molecular clouds. For example, if we solve the equation of motion for an isolated ion immersed in a weakly ionized plasma and undergoing elastic collisions with the numerous neutral particles which take part in flows or turbulent motions in the region under study, we find that the ion velocity is in general smaller than the average velocity of the neutrals. From this, it is possible to conclude that *in the core of molecular clouds, the width of the line profiles of molecular ions should in general be less than that of coexistent neutral molecular species*. This fact has been verified in more than twenty molecular clouds so far (see Table I of Houde et al. (2000)). Numerous examples of comparisons of neutral and ion molecular species spectra are shown in Houde et al. (2000a) and Houde et al. (2000b).

It is also possible to give upper and lower limits for the ratio of the ion-to-neutral line width ratio. If $\sigma_{l,i}$ and $\sigma_{l,n}$ are the ion and neutral line widths, respectively, as defined by the standard deviation of the line profiles we find that the ion-to-neutral line width ratio is restricted to the following range:

$$0.26 \lesssim \frac{\sigma_{l,i}}{\sigma_{l,n}} \lesssim 1, \tag{1}$$

for HCO^+.

Equation (1) can be verified by studying Figure 1 where we have plotted the ion line width against the corresponding neutral line width for every object and pair

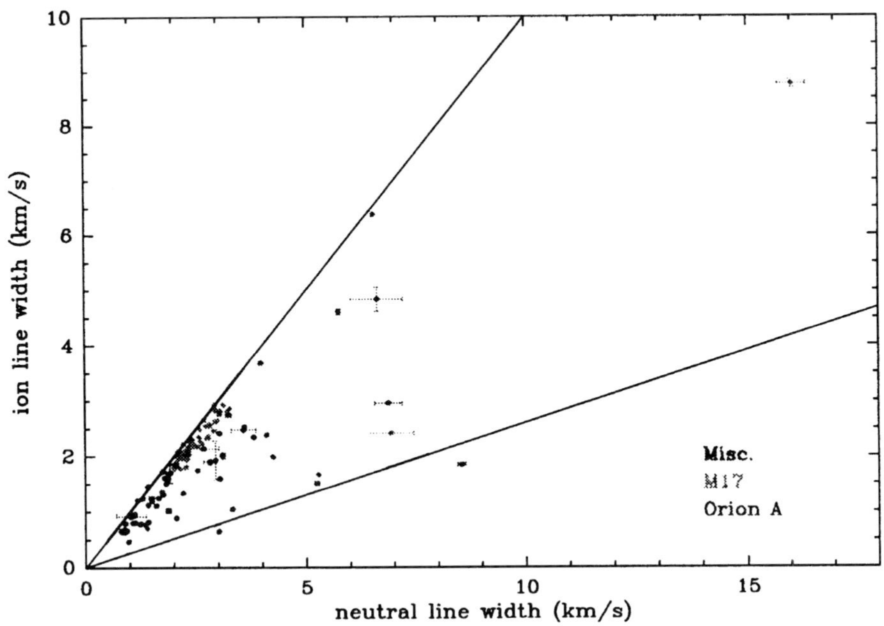

Figure 1. The ion line width vs the neutral line width from a group of miscellaneous clouds (see Table I of Houde et al. (2002)), M17 and Orion A. HCO^+ is plotted against HCN and $H^{13}CO^+$ against $H^{13}CN$.

of molecular species studied so far (HCO^+ is plotted against HCN and $H^{13}CO^+$ against $H^{13}CN$). The two straight lines correspond to the upper and lower limits discussed above. The steeper of the two (with a slope of $\simeq$1) arises when the magnetic field is oriented in a direction parallel to the line of sight while the other (with a slope of $\simeq$0.26) when the field lies in the plane of the sky. As can be readily seen from this figure, the data obtained so far is in excellent agreement with the prediction made with Eq. (1). These results have important implications for the study of the magnetic field in molecular clouds. Namely, it is possible to conclude that:

1. *In the weakly ionized regions of turbulent molecular clouds, the neutrals drive the ions.* If the opposite were true, we would expect the ion species to exhibit line profiles which would be at least as broad as those of coexistent neutral species and probably broader (Houde et al., 2001), contrary to observation.
2. The difference in the width of the line profiles of coexistent ion and neutral molecular species implies that *the coupling between ions and neutrals is poor in the core of molecular clouds*, at least at the scales probed by our observations (up to a few tenths of a parsec).
3. At the spatial resolution attained with our observations (we have a beam width of $\approx 20''$ in most cases), the diffusion between ions and neutrals can be studied

through a comparison of the width of their line profiles. It then appears from our results that *the drift speed between ions and neutrals can often be significant in the core of molecular clouds* (on the order of a km/s at the gas densities probed with the molecular species used here, i.e., $n \gtrsim 10^6 \mathrm{cm}^{-3}$).

3. The Determination of the Viewing Angle α

Since the line width ratio is maximum at approximately unity when the field is aligned with the line of sight ($\alpha = 0$) and decreases to a minimum of $\simeq 0.26$ when the field lies in the plane of the sky ($\alpha = \pi/2$), it is tempting to assume that it is a well-behaved function which decreases monotonically with increasing α. It can in fact be shown that this is true, at least when considering reasonably general models of neutral flow configurations (see Figure 2 of Houde et al. (2002) and the

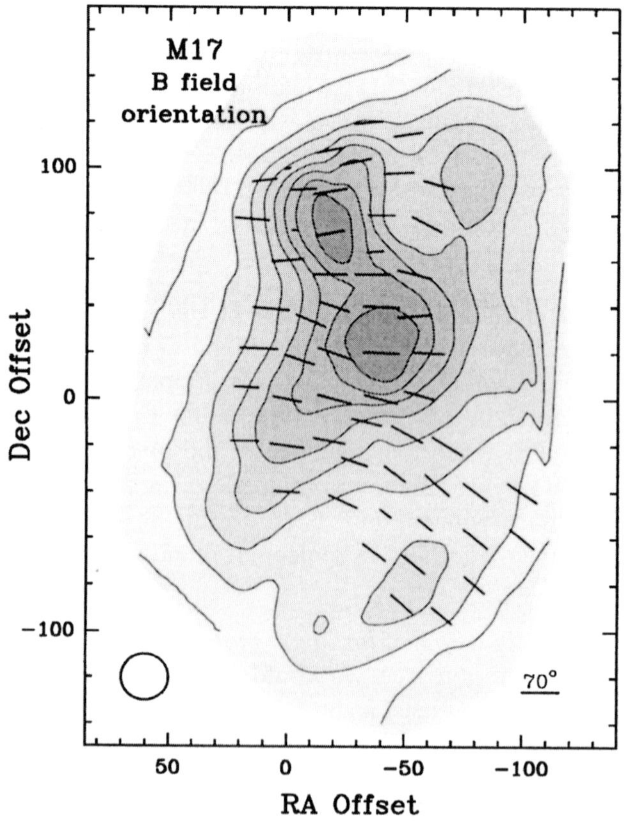

Figure 2. Orientation of the magnetic field in M17. The orientation of the projection of the magnetic field in the plane of the sky is shown by the vectors and the viewing angle is given by the length of the vectors (using the scale shown in the bottom right corner). The contours and the grey scale delineate the total continuum flux. The beam width ($\simeq 20''$) is shown in the lower left corner.

discussion that pertains to it). When studying such models one obtains a family of curves of the ion-to-neutral line width ratio as a function of the inclination angle α. The curve that best fit the physical conditions of the region under study has to somehow be identified.

To do this, we use another parameter that is a function of the orientation of the magnetic field relative to the line of sight that can be readily obtained. This parameter is the polarization level that is measured, for example, from the continuum emission from dust at submillimeter wavelengths. Since the (elongated) dust grains are presumably aligned by the magnetic field, the polarization level P detected by an observer studying a given region where the field is oriented with a viewing angle α can be expressed as

$$P = P_{\max} \sin^2(\alpha), \tag{2}$$

where $P_{\max}$ is the maximum polarization level that can be detected, i.e. when the field lies in the plane of the sky ($\alpha = \pi/2$). Equation (2) can be used in conjunction with the relevant equation for the ion-to-neutral line width ratio in order to express the latter as a function of the normalized polarization level $P/P_{\max}$. By carefully taking into account the well-known depolarization effect observed in molecular clouds toward regions of higher flux (Weintraub et al., 2000) and assigning a suitable value for $P_{\max}$, it then becomes possible to use the polarization data to constrain the neutral flow model and, in effect, calibrate the ion-to-neutral line width ratio.

4. The M17 Molecular Cloud

An extensive 350 μm polarimetry map of the M17 molecular cloud was obtained using the Hertz polarimeter (Dowell et al., 1998) (see Figure 6 of Houde et al. (2002)). We have also made maps of the same region in HCN and HCO^+ in the $J \rightarrow 4-3$ transition at the CSO, using the facility's 300–400 GHz receiver (see Figure 8 of Houde et al. (2002)). The beam size for all sets of observations was about the same at $\approx 20''$. This is a nice feature as our analysis will rest on comparisons of polarimetry and spectroscopic data across the molecular cloud.

With the polarization and spectroscopic (i.e., HCN and HCO^+) data at hand, it now becomes possible to calibrate the ion-to-neutral line width ratio. The line ratios were calculated after the ion and neutral line widths were measured at every position of the spectroscopic maps and compared with the corresponding (normalized) polarization level to achieve the calibration. Further discussion concerning this process can be found in Houde et al. (2002) and the result for M17 is illustrated in Figure 10 therein.

Following this, we can then evaluate the viewing angle α and we show in Figure 2 a map of the orientation of the magnetic field in M17 at every observed position. The angle β, made by the projection of the magnetic field in the plane of the sky, was obtained from the polarimetry data by rotating the corresponding polarization angle PA by 90° and is represented on the map by the orientation of the vectors. The viewing angle α was obtained with the now calibrated ion-to-neutral line width ratio along with Eq. (2) and can be read on the map by the length of the vectors (using the scale in the bottom right). Both angles are plotted on top of the 350 μm continuum flux obtained with Hertz. As was discussed by Houde et al. (2002), the line profiles from HCN and HCO^+ are probably saturated in some regions of the molecular cloud. This will bring some errors in the calculated line width ratios and, as the ion species has a lower critical density, the most likely effect will be an underestimation of the inclination angle. Using a pair of less-abundent molecular species would help in improving on this.

From Figure 2, we can observe some of the main features in the orientation of the magnetic field in M17. First, there is a gradual shift of some 40° in the orientation of the projection of the magnetic field in the plane of the sky from the south-west part of the map ($\beta \sim 50°$) to the north ($\beta \sim 90°$). On the other hand, the viewing angle is maximum at $\alpha \sim 65°$ in the neighborhood of the region of peak continuum emission and smoothly decreases south-westerly to a local minimum where the field is better aligned to the line of sight with $\alpha \simeq 30°$ at RA Off $\simeq -50''$, Dec Off $\simeq -50''$. The field gradually approaches the plane of the sky, once again, in the south of the map.

Most interestingly, there is an important and localized decrease in the viewing angle of roughly 30°–40° close to the position of steepest change in continuum emission where α reaches a minimum at approximately 10°. This region, located at RA Off $\simeq 0''$, Dec Off $\simeq 80''$, is also nearly coincident with the locations of H_2O and OH masers and the ultracompact HII region of the Northern Condensation, where Brogan and Troland (2001) (see their Figure 16) have obtained a value of $B_{\rm los} \simeq -300\ \mu$G using OH Zeeman measurements at 20 km/s.

It is important to realize that α does not provide us with the information concerning the direction of the magnetic field relative to the plane of the sky, this will be provided by Zeeman measurements. We know that for M17 the magnetic field is actually coming out of the plane of the sky (Brogan and Troland (2001), using their HI or OH Zeeman measurements at 20 km/s). The values of α obtained here are therefore relative to an axis directed toward the observer. Finally, taking into account that the magnetic field can be directed away from the line of sight by as much as 60° in some parts of M17, we can see that a multiplicative factor of the order of 2 has to be applied to the Zeeman measurements of Brogan and Troland (2001) in order to evaluate the magnitude of the magnetic field. They obtained a maximum value of $\approx -750\ \mu$G for $B_{\rm los}$, using HI Zeeman measurements at 20 km/s, implying that the magnitude of the field could be as high as $\approx$1.5 mG for this object.

5. Orion A

We have applied a similar procedure to data obtained in the Orion A star-forming region. The data consisting of SHARC II (Dowell et al., 2003) continuum maps

TABLE I

Mean orientation of the magnetic field in some parts of Orion A

Source	α	β
OMC-3 MMS 6	73°	46°
OMC-3 MMS 9	74°	39°
OMC-2 FIR 4	80°	31°
OMC-1	60°	115°

of the OMC-2/OMC-3 and OMC-1 regions with the polarization vectors obtained with Hertz at 350 μm are shown in Figure 3. Unfortunately, it was not possible to map this region as extensively as we did the M17 cloud with the HCN and HCO^+ lines in the $J \rightarrow 4-3$ transition. Instead we obtained spectroscopic data at 27 different positions in the OMC-3 MMS 6, OMC-3 MMS 9, OMC-2 FIR 4 and OMC-1 regions. This allows us to calculate some average of the orientation of the magnetic field at these positions as shown in Table I. From this, we can see that the orientation of the magnetic field is relatively stable in OMC-2 and OMC-3 while

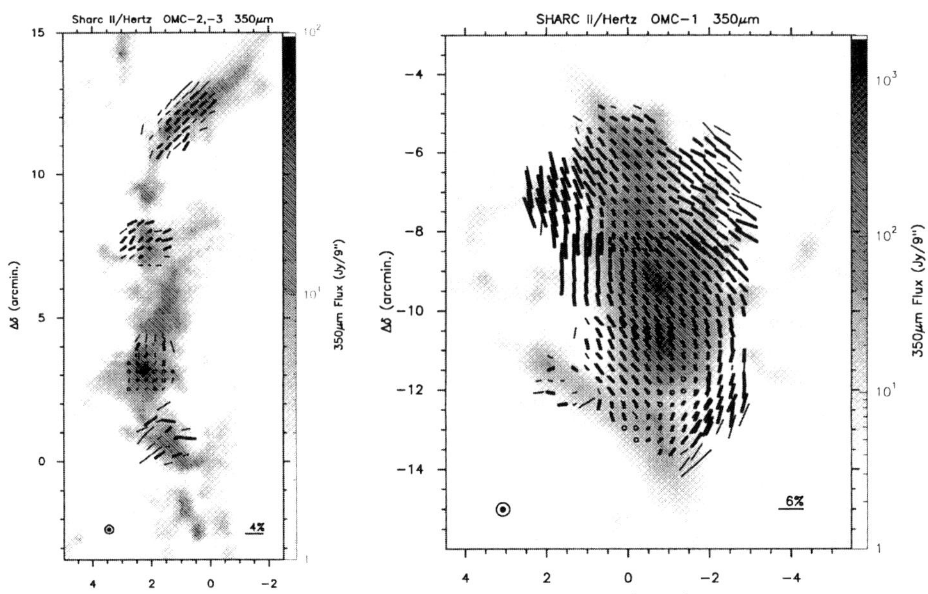

Figure 3. SHARC II continuum maps of the OMC-2/OMC-3 (left) and OMC-1 (right) regions with the polarization vectors obtained with Hertz, all at 350 μm. Thick (thin) vectors signify that $P \geq 3\sigma_P$ ($P \geq 2\sigma_P$) while circles indicate that $P + 2\sigma_P < 1\%$ and $P < 3\sigma_P$. The beam widths are shown in the lower left corner, with the solid and open circles for SHARC II at $\simeq 9''$ and Hertz at $\simeq 20''$, respectively. The reference position is at RA = 5 h 32 m 50 s, Dec = $-5°15'00''$ (B1950).

it exhibits a significant re-orientation in OMC-1. These results are discussed in more details in an upcoming paper where we infer that the magnetic field seems to be relatively unaffected by the presence of the concentrations of lower mass that characterize the OMC-2 and OMC-3 fields while it appears to be strongly perturbed in and around the massive complex of OMC-1.

Acknowledgements

The Caltech Submillimeter Observatory is funded by the NSF through contract AST 9980846 and the observations made with Hertz were supported by NSF Grants AST 9987441 and AST 0204886.

References

Brogan, C.L. and Troland, T.H.: 2001, *ApJ* **560**, 821.
Davis, L.M. and Greenstein, J.L.: 1952, *ApJ* **114**, 206.
Dowell, C.D., Hildebrand, R.H., Schleuning, D.A., Vaillancourt, J.E., Dotson, J.L., Novak, G., Renbarger, T. and Houde, M.: 1998, *ApJ* **504**, 588.
Dowell, C.D., Allen, C.A., Babu, S., Freund, M.M., Gardner, M., Groseth, J., Jhabvala, M., Kovacs, A., Lis, D.C., Moseley Jr., S.H., Phillips, T.G., Silverberg, R., Voellmer, G. and Yoshida, H.: 2003, in: T.G. Phillips and J. Zmuidzinas (eds.), *Millimeter and Submillimeter Detectors for Astronomy, Proc. SPIE*, **4855**, 73.
Hildebrand, R.H.: 1988, *QJRAS* **29**, 327.
Houde, M., Bastien, P., Peng, R., Phillips, T.G. and Yoshida, H.: 2000a, *ApJ* **536**, 857.
Houde, M., Peng, R., Phillips, T. G., Bastien, P. and Yoshida, H.: 2000b, *ApJ* **537**, 245.
Houde, M., Phillips, T. G., Bastien, P., Peng, R., and Yoshida, H.: 2001, *ApJ* **547**, 311.
Houde, M., Bastien, P., Dotson, J.L., Dowell, C.D., Hildebrand, R.H., Peng, R., Phillips, T.G., Vaillancourt, J.E. and Yoshida, H.: 2002, *ApJ* **569**, 803.
Weintraub, D.A., Goodman, A.A. and Akeson, R.L: 2000, in: V. Mannings, A.P. Boss, and S.S. Russel (eds.), *Protostars and Planets IV,* The University of Arizona Press, Tucson, p. 247.

FORMATION OF STRUCTURE AND FRAGMENTATION

FRAGMENTATION AND STRUCTURE FORMATION

ENRIQUE VÁZQUEZ-SEMADENI
Centro de Radioastronomía y Astrofísica, UNAM, Morelia, Mexico;
E-mail: e.vazquez@astroscu.unam.mx

Abstract. I present an overview of the hierarchy of structures existing in the interstellar medium (ISM) and the possible mechanisms that cause the fragmentation of one level into the next, with the formation of stars as its last step. Within this framework, I then give an overview of the contributions to this session. Numerical work addresses, at the largest scales, the shaping and formation of structures in the ISM through turbulence driven by stellar energy injection, and the resulting star formation rate as a function of mean density. At the scales of molecular clouds, results comparing observational and numerical data on the density and velocity structure of turbulence-produced cores, as well as their mass spectra, are summarized, together with existing theories of core and star formation controlled by the turbulence. Observationally, an attempt to discriminate between the standard and turbulent models of star formation is presented, finding inconclusive results, but suggesting that both turbulence and the magnetic field are dynamically important in molecular clouds and their cores. Finally, various determinations of the magnetic field strength and geometry are also presented.

Keywords: ISM, cloud formation, star formation

"Fragmentation" is a very general term used in the study of star formation (SF) and the interstellar medium (ISM) that describes the sequential breakup of a diffuse and extended mass of gas (the ISM) into ever smaller regions (clouds, clumps, cores), ultimately leading to the formation of stars. Although a single term is used to describe this iterative process, it most likely comprises a variety of physical mechanisms operating at the various levels of the sequence. The study of fragmentation thus represents an attempt to acquire a unified view of the formation of structures in the ISM, with star formation as its last step. A session on this topic therefore naturally belongs in this conference.

In order to properly discuss fragmentation, it is probably best to first make an attempt at formulating practical definitions of what we mean by this and a few other related terms. First, let us define a "structure" within the ISM as a connected region having specific values (or ranges of values) of the physical variables, with at least some of them clearly distinguishable from their values in the surroundings of the region. "Fragmentation" can then be defined as the formation of small-scale structures within a larger-scale one. This immediately implies that fragmentation is a *multi-scale* process, because it involves at least the scales of the "parent" and "daughter" structures. In particular, interstellar clouds (either HI or molecular) are essentially high-density structures embedded within larger-scale, lower-density ones. It is of course possible that, by either the same mechanism that formed them, or a different one, the small-scale structures fragment into even smaller structures, the process then being termed *hierarchical*, with many possible levels of fragmentation.

In the ISM, we can recognize a large hierarchy of structures including, from larger to smaller scales, spiral arms (of sizes $L \sim 10$ kpc), "superclouds" ($L \sim 1$ kpc, $n \sim 1$ cm^{-3}) (Elmegreen and Elmegreen, 1983), giant molecular clouds (GMCs) ($L \sim 50$ pc, $n \sim 50$ cm^{-3}), "individual" molecular clouds (MCs) ($L \sim 5$ pc, $n \sim 10^3$ cm^{-3}), clumps ($L \sim 0.5$ pc, $n \sim 10^4$ cm^{-3}) and cores ($L \lesssim 0.1$ pc, $n > 10^4$ cm^{-3}) (see, e.g., the reviews by Blitz, 1993; Elmegreen, 1993). Note, however, that, although it is only human to classify and categorize, the described "categories" really appear to merge into one another, the ISM most likely really being a continuum (see the review by Vázquez-Semadeni et al., 2003 and references therein; also, the contribution by **de Avillez** in this session[1]), with structures spanning all of the aforementioned range of scales. This contrasts with the traditional multi-phase view of the ISM (Field, et al., 1969; McKee and Ostriker, 1977).

It is interesting to compare the contents of this session with those of earlier reviews on the subject (e.g., Scalo, 1985; Elmegreen, 1993). Such a comparison shows that great progress has been made through the use of numerical simulations of the turbulent ISM, allowing a deeper understanding and an ever more quantitative determination of the role of stellar-driven turbulence in shaping the ISM (see, e.g., the review by Vázquez-Semadeni (2002), and the contributions by **de Avillez** and **Sarson et al.**), as well as the interiors of molecular clouds, and in controlling various aspects of the SF process (see the review by **Ballesteros Paredes** and the contribution by **Klessen**).

The physical mechanisms operating at the various hierarchical levels of density structure formation in the ISM include a) the gravitational (Jeans, 1902), thermal (Field, 1965; Field et al., 1969; Pikel'ner, 1968) and Parker (1966) linear instabilities, or combinations thereof (Elmegreen, 1991, 1994); b) random compressions in the globally turbulent ISM, and c) the nonlinear development of the instabilities themselves.

Specifically, the gravitational instability, parhaps aided by cooling, is thought to be responsible for the formation of the largest structures, or "superclouds" (see, e.g., the review by Elmegreen, 1993), although the Parker instability may also contribute at these scales (Franco et al., 2002). Within these, GMCs may be originated by a) gravitational instability, again aided by cooling, within their parent superclouds (e.g., Elmegreen, 1993, 1994; Wada and Norman, 1999); b) magneto-Jeans instabilities or modified swing amplification (e.g., Kim and Ostriker, 2001), and c) large-scale turbulent motions driven by stellar activity (Ballesteros Paredes et al., 1999; de Avillez, 2000; **de Avillez**; see also the review by Mac Low, 2002). Early cloudlet coagulation models (Oort, 1954; Kwan, 1979) are apparently not anymore considered as a viable GMC formation mechanism (Elmegreen, 1990), as neither is the Parker instability (**Kim, Ryu and Hong**).

[1] In this paper, I denote in **boldface** the contributions to the present session.

Regardless of their formation mechanism, GMCs, and the "individual" molecular clouds inside them, are supersonically turbulent themselves (see, e.g., the review by Blitz, 1993; and references therein), while being magnetized and roughly isothermal, with approximate equipartition between the turbulent and magnetic energies (e.g., Myers and Goodman, 1988). Thus, numerical simulations of MHD isothermal turbulent flows should constitute good models of molecular clouds. However, in this case, besides the structure-formation mechanism at work, an important parallel question is what is the driving mechanism for the turbulence at every scale, and how ubiquitous it is. This question arises because recent numerical results have shown that the turbulence, left undriven, decays rapidly, roughly in a large-scale crossing time at the large-scale velocity dispersion (Padoan and Nordlund, 1999; Mac Low et al., 1998; Stone et al., 1998).[2]

A possible driving mechanism for the turbulence in molecular clouds is dynamical ("bending mode") instabilities (Vishniac, 1994) in the compressed layers between converging flows, as in the simulations of Hunter et al. (1986), Walder and Folini (1998) and Koyama and Inutsuka (2002). It is important to note that these instabilities appear to be enhanced by the presence of cooling, so that the formation of turbulent molecular clouds at temperatures of a few tens of Kelvins out of converging streams in diffuse gas with $T > 100$ K appears particularly likely. Although this process has not been explicitely looked for in multi-scale simulations of the ISM, the adaptive-grid, high-resolution simulations presented by **de Avillez** in this session hold great promise for identifying it. In the meantime, his contribution discusses the relative values of the thermal and magnetic pressures in the various temperature and density regimes of the ISM. Other contributing driving mechanisms for the turbulence may be the variation of the background UV radiation, which induces mass flux among the warm and cool stable phases of the ISM (Kritsuk and Norman, 2002) and the "drag instability" found and discussed by **Gu, Lin and Vishniac**.

Within molecular clouds, the formation of their clumps and cores appears most likely a consequence of the supersonic turbulence within the clouds, as it appears to have a dominant role in their overall energy balance (corresponding to the standard notion that molecular clouds are in general "supported" against their self-gravity by the turbulent kinetic energy). Moreover, since stars form in dense molecular cloud cores, it is at this level of the fragmentation hierarchy that we have reached the connection to star formation (SF). However, if clumps and cores are density

[2]Cho et al. (2002) (also Lazarian, this conference) have argued that "imbalanced" turbulence, in which the turbulent excitation is not statistically uniform in space, does not decay as rapidly. This would correspond to the case where the turbulent driving occurs at localized and discrete sites in space. However, their numerical experiments were set up with rather idealized initial conditions of an excess of traveling wave packets in one direction. The more realistic simulations of Avila-Reese and Vázquez-Semadeni (2001), with the initial turbulent driving applied to models of the ISM by discrete, stellar-like sources, exhibit as rapid a decay as that of the experiments mentioned earlier. Thus, so far most numerical evidence still points toward rapid decay of undriven turbulence.

fluctuations produced by the supersonic turbulence in molecular clouds, it is natural to ask whether such a turbulent origin is compatible with the standard model of star formation (Mouschovias, 1976; Shu, 1977; Shu, et al., 1987), in which clumps and cores forming low mass stars are assumed to be in quasi-magnetostatic equilibrium, and evolving on long time scales, given by the ambipolar diffusion characteristic time. This long contraction time scale is at the basis of the low SF efficiency (SFE) in the standard model.

In recent years, a new model has been emerging, which attempts to understand such issues as the properties of molecular cloud cores, the efficiency of star formation, and even the stellar initial mass function (IMF) as a consequence of the global parameters of the turbulence in molecular clouds (Padoan, 1995; Padoan and Nordlund, 2002; Vázquez-Semadeni, et al., 2003; see also the reviews by Vázquez-Semadeni et al., 2000; Mac Low and Klessen, 2004, and **Ballesteros-Paredes**). In this model, the low SFE is a consequence of *global* (large-scale) turbulent support for the clouds, together with *local* (small-scale) collapse induced by the compressible turbulence itself, in a sort of turbulent "colander". Thus, only a small fraction of the mass in a molecular cloud can proceed to gravitational collapse, while the rest is probably dispersed in a few turbulent crossing times, due to both stellar activity and the excess turbulent energy. This latter process necessarily requires simulations of the ISM at large, in which clouds can be adequately simulated as open systems (see, e.g., the review by Vázquez-Semadeni 2002; and references therein, and the contributions by **de Avillez** and **Sarson et al.**). In particular, **Sarson et al.** show promising results towards understanding the SF rate as a function of the mean gas density (the "Schmidt law") in the ISM directly from numerical simulations.

Thus, while in the standard model the low SFE is the result of a long evolution time scale for *each* core, in the turbulent model it is a *statistical* issue, because most of the mass in a molecular cloud is supported against collapse by the turbulence, with most cores being transient, and only a small fraction of the mass being in collapsing cores. It is a task of the emerging theory to determine how this fraction is determined by the global turbulent parameters. The contribution by **Ballesteros-Paredes** reviews the accomplishments of numerical simulations of MC turbulence in producing realistic cores, and the present state of the developing theory, as well as some shortcomings. Further results of the turbulent model on the resulting mass spectra of the collapsed object in turbulent clouds are given by **Klessen**, on the basis of "smoothed particle hydrodynamics" (SPH) numerical simulations. A promising numerical code for extending SPH to the magnetic case is presented by **Price and Monaghan**.

Observationally, the contribution by **Crutcher** makes an attempt at discerning between the two models, concluding that *both* turbulence and magnetic fields are important in MCs and SF, but finding that the presently available evidence is still insufficient to discern between the competing models. The contributions by **Lai, Velusamy and Langer** and by **Wolf, Launhardt and Henning** present magnetic field determinations using interesting novel techniques. As mentioned by **Crutcher**, a large database of this kind of measurements is needed in order to provide solid observational constraints as to which model better describes the star formation process as a part of the hierarchical fragmentation sequence from the largest scales in the ISM.

All in all, the contents of this session reflect the exciting advances of recent years in both observations and theoretical understanding of the hierarchical structure in the ISM and the SF process, and point towards the necessary steps for the future.

Acknowledgements

Useful comments from J. Ballesteros-Paredes are gratefully acknowledged. This work has received financial support from CONACYT Grant 36571-E.

References

Avila-Reese, V.A. and Vázquez-Semadeni, E.: 2001, *ApJ* **553**, 645.
de Avillez, M.A.: 2000, *MNRAS* **315**, 479.
Ballesteros-Paredes, J., Vázquez-Semadeni, E. and Scalo, J.: 1999, *ApJ* **515**, 286.
Blitz, L.: 1993, in: E. Levy and J.I. Lunine (eds.), *Protostars and Planets III*, University of Arizona, Tucson, 125.
Cho, J., Lazarian, A. and Vishniac, E.T.: 2002, *ApJ* **564**, 291.
Elmegreen, B.G. and Elmegreen, D.M.: 1983, *MNRAS* **203**, 31.
Elmegreen, B.G.: 1990, *ApJ* **357**, 125.
Elmegreen, B.G.: 1991, *ApJ* **378**, 139.
Elmegreen, B.G.: 1993, in: E. Levy and J.I. Lunine (eds.), *Protostars and Planets III*, University of Arizona, Tucson, p. 97.
Elmegreeen, B.G.: 1994, *ApJ* **433**, 39.
Field, G.B.: 1965, *ApJ* **142**, 531.
Field, G.B., Goldsmith, D.W. and Habing, H.J.: 1969, *ApJ* **155**, L149.
Franco, J., Kim, J., Alfaro, E.J. and Hong, S.S.: 2002, *ApJ* **570**, 647.
Hunter, J., Sandford, M.T., Whitacker, R.W. and Klein, R.I.: 1986, *ApJ* **305**, 309.
Jeans, J.H.: 1902, *Phil. Trans. A* **199**, 1.
Kim, W.-T. and Ostriker, E.C.: 2001, *ApJ* **559**, 70.
Koyama, H. and Inutsuka, S.-I.: 2002, *ApJL* **564**, L97.
Kritsuk, A. and Norman, M.L.: 2002, *ApJL* **569**, L127.
Kwan, J.: 1979, *ApJ* **229**, 567.
Mac Low, M.-M., Klessen, R.S., Burkert, A. and Smith, M.D.: 1998, *Phys. Rev. Lett.* **80**, 2754.
Mac Low, M.-M.: 2002, in: *From Observations to Self-Consistent Modeling of the Interstellar Medium* (astro-ph/0211616).
Mac Low, M.-M. and Klessen, R.S.: 2004, *Rev. Mod. Phys.* **76**, 125.
McKee, C.F. and Ostriker, J.P.: 1977, *ApJ* **218**, 148.
Mouschovias, T.C.: 1976, *ApJ* **207**, 141.
Myers, P.C. and Goodman, A.A.: 1988, *ApJL* **326**, L27.
Oort, J.H.: 1954, *Bull. Astron. Inst.* Netherlands **12**, 177.
Padoan, P.: 1995, *MNRAS* **277**, 377.
Padoan, P. and Nordlund, Å: 1999, *ApJ* **526**, 279.
Padoan, P. and Nordlund, Å: 2002, *ApJ* **576**, 870.

Parker, E.N.: 1966, *ApJ* **145**, 811.
Pikel'ner, S.B.: 1968, *Sov. Astron* **11**, 737.
Scalo, J.M.: 1985, in: D.C. Black and M.S. Matthews (eds.), *Protostars & Planets II*, University of Arizona, Tucson, p. 201.
Shu, F.H.: 1977, *ApJ* **214**, 488.
Shu, F.H, Adams, F.C. and Lizano, S.: 1987, *ARAA* **25**, 23.
Stone, J.M., Ostriker, E.C. and Gammie, C.F.: 1998, *ApJL* **508**, L99.
Vázquez-Semadeni, E., Ostriker, E.C., Passot, T., Gammie, C.F. and Stone, J.M.: 2000, in: V. Mannings, A. Boss and S. Russell (eds.), *Protostars and Planets IV*, University. of Arizona Press, Tucson p. 3.
Vázquez-Semadeni, E.: 2002, in: R. Taylor, T. Landecker, and A. Willis (eds.), *Seeing Through the Dust: The Detection of HI and the Exploration of the ISM in Galaxies*, San Francisco, ASP, p. 155.
Vázquez-Semadeni, E., Gazol, A., Passot, T. and Sánchez-Salcedo, F.J.: 2003, in: E. Falgarone and T. Passot (eds.), *Turbulence and Magnetic Fields in Astrophysics* Springer, Lecture Notes in Physics, p. 213.
Vázquez-Semadeni, E., Ballesteros-Paredes, J. and Klessen, R.S.: 2003, *ApJL* **585**, L134.
Vishniac, E.T.: 1994, *ApJ* **428**, 186.
Wada, K. and Norman, C.A.: 1999, *ApJL* **516**, L13.
Walder, R. and Folini, D.: 1998, *A&A* **330**, 21.

TURBULENT FRAGMENTATION AND STAR FORMATION

JAVIER BALLESTEROS-PAREDES

Instituto de Astronomía, UNAM, Morelia, Mexico; E-mail: j.ballesteros@astrosmo.unam.mx

Abstract. We review the main results from recent numerical simulations of turbulent fragmentation and star formation. Specifically, we discuss the observed scaling relationships, the "quiescent" (subsonic) nature of many star-forming cores, their energy balance, their synthesized polarized dust emission, the ages of stars associated with the molecular gas from which they have formed, the mass spectra of clumps, and the density and column density probability distribution function of the gas. We then give a critical discussion on recent attempts to explain and/or predict the star formation efficiency and the stellar initial mass function from the statistical nature of turbulent fields. Finally, it appears that turbulent fragmentation alone cannot account for the final stages of fragmentation: although the turbulent velocity field is able to produce filaments, the spatial distribution of cores in such filaments is better explained in terms of gravitational fragmentation.

Keywords: ISM: fragmentation, clouds, turbulence, kinematics and dynamics, stars: formation

1. Introduction

It is well known that stars form in dense cores within molecular clouds, but the physical processes that control the formation of low-mass stars within these cores are not well understood yet. The "standard" scenario assumes that low-mass protostellar cores are in quasi-static equilibrium supported against gravitational collapse by a combination of magnetic and thermal pressures (see, e.g., Shu et al., 1987). In this scenario, a core forms stars once magnetic support is lost through the process known as "ambipolar diffusion": neutral gas particles slowly drift through the ions, which are held up by the magnetic field, allowing the core to eventually attain the "critical" mass-to-flux ratio, beyond which the gravitational energy exceeds the magnetic energy and collapse sets in from the inside-out.

The more recent turbulent scenario, on the other hand, suggests that cores are formed by compressible motions in the turbulent velocity field of the cloud (Sasao, 1973; Elmegreen, 1993; Padoan, 1995; Ballesteros-Paredes et al., 1999a; Klessen et al., 2000). Those cores with an excess of gravitational energy collapse rapidly to form stars, while the others with sufficiently large internal or kinetic energies re-expand once the turbulent compression subsides (Taylor et al., 1996; Vázquez-Semadeni et al. 2002b).

Although the standard scenario has explained many observational features of protostellar cores, it does not address the problem of how the quasi-static cores can form and persist within a turbulent environment. On the other hand, the turbulent scenario naturally incorporates the process of core formation, and is increasingly

Astrophysics and Space Science **292:** 193–205, 2004.

succesful in reproducing the observed core properties. In this paper, we will review the main results of the turbulent scenario, the observational facts that numerical works have successfully reproduced, the simulations' predictions concerning the star formation efficiency and the attempts to explain the initial mass function from statistical considerations of a turbulent field. Finally, we discuss the final stages of fragmentation, which appear to be gravitational rather than turbulent.

2. Turbulent Fragmentation

The process through which a chaotic velocity field produces a clumpy density structure in the gas within in a few dynamical timescales is referred to as "turbulent fragmentation".

It is generally accepted that small-scale turbulence can provide support against global gravitational collapse (e.g., Chandrasekhar and Fermi 1953). In fact, if only small-scale modes are considered, and if they are more or less isotropic, they provide extra pressure. In this case, an "effective" Jeans length can be defined in terms of the "effective" sound speed, $c_{\rm eff}^2 = c_{\rm s}^2 + \frac{1}{3}u_{\rm rms}^3$, where $u_{\rm rms}$ is the turbulent velocity dispersion, which in principle depends on the scale (Bonazzola et al., 1987). Thus, if the scale of a system is smaller than the modified Jeans length, it is said that turbulence inhibits collapse. This is valid only if turbulence is isotropic, and operates on small scales. In reality, interstellar turbulence is a multiscale phenomenon, and its largest scales are in fact comparable to the size of the system (Ballesteros-Paredes et al., 1999a; Brunt, 2003), and contain most of the kinetic energy. If this occurs, the large- and intermediate-scale modes of turbulence can produce fragmentation, while the small-scale modes can provide additional pressure. The structures produced by the fragmentation (cores) can either proceed to collapse, or redisperse and merge back into the ambient medium, depending on the balance between the energies within them (Sasao, 1973; Elmegreen, 1993; Ballesteros-Paredes et al., 1999a). Thus, supersonic turbulence plays a dual role, providing support for large structures while simultaneously inducing their fragmentation into smaller-scale subunits. In what follows, we review several predictions and comparisons with observations from numerical simulations of turbulent fragmentation.

3. Results

Turbulence decays fast. Molecular clouds (MC) are known to be turbulent, with motions that are supersonic at scales $\geq$0.1 pc (Larson, 1981). Early concerns that turbulence should be dissipated rapidly in shocks (Goldreich and Kwan, 1974) were shortly afterwards argued to be inapplicable if turbulence consisted of MHD waves (Arons and Max, 1975). However, an important result found in numerical simulations is that the original concerns were well found, and that strong MHD turbulence decays rapidly even in the presence of strong magnetic fields (Padoan and Nordlund, 1999; Mac Low et al., 1998; Stone et al., 1998; Mac Low, 1999; Avila-Reese and Vázquez-Semadeni, 2001; see, however, Cho et al., 2002 for an alternative viewpoint). Thus, it appears that MC turbulence must be continually driven during the entire lifetime of the clouds (which, on the other hand, probably does not exceed several Myr– Ballesteros-Paredes et al., 1999b). The driving mechanism, however, is probably not restricted to the stellar activity internal to the clouds. Turbulence in clouds may be generated by the very compressive process that forms the cloud, through bending-mode instabilities aided by cooling (Folini and Walder, 1998; Klein and Woods, 1998; Koyama and Inutsuka, 2002), as proposed by Ballesteros-Paredes et al., (1999a) and Vázquez-Semadeni et al. (2003).

Energy balance. An evident result from numerical simulations is that the structures formed in those models evolve dynamically because they are part of a globally supersonic turbulent flow. Then, turbulence continuously reshapes their structure, producing changes of mass, momentum and energy in them (Ballesteros-Paredes et al., 1999a). Clouds in simulations are found to be roughly in energy equipartition, but not in virial *equilibrium*, since the time derivatives of the moment of inertia and the surface terms in their virial balance are at least as important as the volume energy terms[1] (Ballesteros-Paredes and Vázquez-Semadeni, 1995; Ballesteros-Paredes and Vázquez-Semadeni, 1997; see Ballesteros-Paredes, 2003 for a review). The fact that the surface terms are large implies the existence of strong fluxes of all physical quantities (mass, energy and momentum) across cloud "boundaries" which, on the other hand, are rather arbitrary.

Larson's relations. Numerical simulations including turbulence have been helpful to understand the origin of the observed scaling relationships, $\Delta v \propto R^{1/2}$ and $\langle \rho \rangle \propto R^{-1}$ (Larson, 1981). It is more or less accepted that the first relationship is a direct consequence of the turbulent power spectrum of a field dominated by shocks. However, it has been suggested that the second relationship may be an artifact of the observations (Kegel, 1989; Scalo, 1990). In fact, it has been shown from numerical simulations that the relationship does not exist in real space (Vázquez-Semadeni et al., 1997; Ballesteros-Paredes and Mac Low, 2002), but it appears in the observational (position–position–velocity) space as a consequence of the thresholding done in the process of defining clouds or clumps (Ballesteros-Paredes and Mac Low, 2002).

Fast cloud and star formation. If, as suggested by simulations, clouds are far from virial equilibrium and supersonic turbulence decays fast, then probably cloud formation can be faster than earlier estimates, which were ~20–30 Myr (Blitz

[1]This does not mean that "Virial Mass" estimates are wrong. In fact, those estimations are based on equipartition between gravitational and kinetic energies which, as mentioned, are roughly in equipartition. Instead, "Virial Mass" estimations should be termed "Equipartition Mass" estimations.

and Shu, 1980). This agrees with the suggestions by, among others, Sasao (1973), Hunter et al. (1986), Elmegreen (1993) and Ballesteros-Paredes et al. (1999a), that clouds are turbulent density fluctuations within the global ISM flow.

Ballesteros-Paredes et al. (1999b) have shown that the global turbulence driven by star formation events can produce locally the gravitationally unstable clouds with characteristics similar to those found in Taurus. They argue that rapid assembly of molecular clouds by large-scale streams is a very plausible mechanism to form clouds, in which, furthermore, stars older than $\sim$5 Myr are not found.[2] Indeed, Hartmann et al. (2001) tabulate the ages of stars in 13 nearby star-forming regions. For those regions with stars older than $\sim$5 Myr, there is no molecular gas associated, suggesting that the time scales for both cloud and star formation are short.

Isolated and clustered modes of star formation. In numerical simulations, the structure of a turbulent molecular cloud depends on the largest scale of the turbulent motions (Klessen et al., 2000; Heitsch et al., 2001a) . In particular, for turbulence driven at large scales, "clustered" collapse occurs predominantly in filaments in a more or less efficient way, while for small-scale driven turbulence, a more "isolated" mode of star formation occurs (see Mac Low and Klessen, 2003, for a review). However, a potentionally confusing point is worth noting here. For the star formation community, a clustered region is one in which many stars are formed in very compact regions. For instance, while the Orion nebula has a star density of $\sim 10^4$ to 10^5 stars per cubic pc, the Taurus molecular cloud contains only about 200 T Tauri stars. Thus, Orion is conventionally considered to belong to the clustered mode, and Taurus to the isolated one. However, from the point of view that Taurus is a coherent filamentary region, apparently due to an external compression (Ballesteros-Paredes et al., 1999b) that has formed stars almost simultaneously, it would belong to the "clustered" mode as defined by Klessen et al. (2000).

Existence of quiescent cores. Observational evidence suggests that low-mass stars formed from molecular cloud cores of which the column density profiles may resemble Bonnor–Ebert (Ebert, 1955; Bonnor, 1956) equilibrium spheres, and velocity dispersions are transonic, or even subsonic (Myers, 1983; Jijina et al., 1999;

[2]Note that Palla and Stahler (2000, 2000) argue that nearby MC have been forming stars in the last 10 Myr or more, with a recent burst of star formation. However, Hartmann (2003) argues that their conclusions are skewed by a statistically small sample of stars with masses larger than $1 M_{\odot}$, and by biases in their birthline age corrections. He notes that the picture by Palla and Stahler, 2002 (a) requires the last 1–2 Myr to be a special epoch for most MCs; (b) implies that most MCs are forming stars at extremely low rates, if any; and that (c) the apparently oldest stars are systematically higher in mass, implying that for most of a typical MC's star-forming history, the Initial Mass Function was strongly skewed. Item (a) seems implausible, and items (b) and (c) are contradicted by observations (see Hartmann, 2003 and references therein). A more plausible explanation of the observations, he argues, is that the "tail" of older stars is really the result of including older foreground stars, as well as problems with the isochrone calibration in the higher mass stars.

Caselli et al., 2002) . For this reason such cores are usually termed "quiescent". In the standard scenario of star formation (Shu et al., 1987) quiescent structures are naturally explained as consequences of the quasi-static contraction process. In the turbulent scenario, however, protostellar cores are transient features generated by the dynamical flow in the cloud. In a recent paper, Ballesteros-Paredes et al. (2003) demonstrated that indeed, transient, dynamic cores have an angle-averaged column density structure that often resembles hydrostatic Bonnor–Ebert profiles, such as observed cores (Alves et al., 2001; Evans et al., 2001; Shirley et al., 2000 and Johnstone et al., 2000). They argue that this kind of analysis in observational work is not an unambiguous test of hydrostatic equilibrium.

Furthermore, Klessen et al. (2003) investigate the velocity structure of these cores. They exhibit sub- and transonic velocity dispersions and, in self-gravitating cases are in approximate equipartition between the kinetic and gravitational energies. They thus closely resemble the observed population of so-called quiescent low-mass cores. The fact that dynamically evolving cores in highly supersonic turbulent flows can have subsonic velocity dispersions is a direct consequence of a turbulent energy spectrum that decreases with decreasing length scale. Note that this also implies that the quiescent nature of the cores is not necessarily indicative that they constitute the dissipative scale of turbulence, as is sometimes suggested (e.g., Goodman et al., 1998). The observed properties of quiescent low-mass protostellar cores in molecular clouds thus do neither directly imply the existence of strong magnetic fields for support against collapse nor quasi-static evolution through ambipolar diffusion.

Polarized thermal dust emission. Other studies have focused on the polarized emission of dust from numerical simulations. The main results are, first, that cores in the magnetized simulations exhibit degrees of polarization between 1 and 10%, regardless of whether the turbulence is sub- or super-Alfvénic (Padoan et al., 2001a); second, that submillimeter polarization maps of quiescent cores do not map the magnetic fields inside the cores for visual extinctions larger than $A_V \sim 3$ (Padoan et al., 2001a); third, that the Chandrasekhar–Fermi method of estimating the mean magnetic field in a turbulent medium underestimates very weak magnetic fields by factors up to 150 (Heitsch et al., 2001b); and fourth, that limited telescope resolution leads to a systematic overestimation of the field strengths, as well as to an impression of more regular, structureless magnetic fields (Heitsch et al., 2001b).

Density and column density PDF of turbulent gas. Several numerical studies have investigated the statistical distribution of the density field. The volume- (mass-) weighted probability distribution function (PDF) of the density $f(\rho)d\rho$ is defined as the fractional volume (mass) occupied by gas in the density range $(\rho, \rho + d\rho)$.

It has been shown that, for a polytropic gas (i.e., for a gas where $P \propto \rho^{\gamma}$) the volume-weighted density PDF depends on the effective polytropic exponent γ of the fluid. In particular, for an isothermal flow, ($\gamma = 1$), the PDF is lognormal

(Vázquez-Semadeni, 1994; Padona et al., 1997; Passot and Vázquez-Semadeni, 1998; Scalo et al., 1998; Ostriker et al., 2001 and Li et al., 2003), because the density fluctuations are the result of a stochastic multiplicative process, and so the distribution of its logarithm is Gaussian according to the Central Limit Theorem (Passot and Vázquez-Semadeni 1998). In the more general polytropic case ($\gamma \neq 1$), Passot and Vázquez-Semadeni (1998) extended this result and showed that the PDF should approach a power law at high (respectively low) densities for $\gamma < 1$ (respectively $\gamma > 1$). Subsequently, other workers seem to have found somewhat different numerical results. Wada and Norman (2001) claim that a lognormal density PDF also appears in their non-isothermal simulations of galactic disks, while Li et al. (2003) report that the shape of the PDF may depend on the scale in which the turbulence is driven (see also Klessen, 2000). However, the discrepancy may be apparent. The lognormal PDF of Wada and Norman (2001) appears only at the highest densities in their simulations, where they are nearly isothermal, while the result of Li et al. (2003) may be due to a different choice of the turbulent driving, as already noted by Passot and Vázquez-Semadeni (1998) and by Klessen (2000).

Finally, Passot and Vázquez-Semadeni (2003) have investigated the density PDF in the magnetic isothermal case, together with the role of the magnetic pressure, finding that the latter is only poorly correlated with the density, because of the different scalings of the magnetic field strength with the density fluctuations for the different types of nonlinear MHD waves. This implies that, rather than providing a restoring force, the magnetic pressure gradient acts more as a random forcing, and therefore in general has little effect on the density PDF, which in most cases retains its lognormal shape in the isothermal case, as also found by Ostriker et al. (2001).

Since direct measurements of the density field in the ISM are not available, it is important to understand the PDF of the column density field. For that purpose, Ostriker et al. (2001) measured the PDF of the column density for numerical simulations, finding that this distribution has a similar shape as that of the underlying density field, but with smaller mean and width. However, Vázquez-Semadeni and García (2001) argued that if the events are separated by distances larger than a correlation length, then the distribution of the density field should approach a Gaussian shape, although the convergence is very slow in the high-density tail, so that the column density PDF is not expected to exhibit a unique functional shape, but to transit instead from a lognormal to a Gaussian form, as the ratio of the column length to the decorrelation length increases. This result, then, explains the nearly lognormal distribution found by Ostriker et al. (2001).

Blitz and Williams (1997) performed a comparison between column density PDFs from models and observations by analyzing the PDF of the antenna temperature T_A of ^{13}CO for Taurus, assuming that T_A is proportional to the column density of the cloud. In their Figure 1, it is clear that the central region behaves closely to a lognormal distribution. However, at larger T_A the distribution found for Taurus has an extended tail. This result may be due to the fact that in Taurus, self-gravity has produced high-density collapsing cores. This is in fact apparent in the evolution of the volumetric density PDF shown by Klessen (2000) (see also Figure 9 in Li et al., 2003), it is clear that self-gravity allows for the production of such tails in the volumetric density PDF. Thus, such tails should appear also in the column density PDF.

Mass spectra. The mass distribution of clouds and clumps in turbulent simulations has been studied by several workers (Vázquez-Semadeni, et al., 1997; Klessen, 2001; Ballesteros-Paredes and Mac Low, 2002 and Gammie et al., 2003). Their results generally give a mass spectrum of the form $dN/dM \propto M^{\alpha}$, with $-2.5 \lesssim \alpha \lesssim -1.5$. Since some of the results are discussed in the chapter by Klessen (this volume), here we just comment that several studies have recently tried to relate the mass spectrum of compact cores with the slope of the mass distribution of the newborn stars (IMF). However, this appears questionable because (a) the mass distribution of cores changes with time as clumps merge together (Klessen, 2001); (b) cores generally produce a cluster of stars, not a single one, and so the relation between the masses of cores and of individual stars is unclear, and (c) many other processes are at play in determining a star's final mass, such as competitive accretion (see, e.g., Bate, this volume) and stellar feedback through winds and outflows. In other words, a core with mass M will not necessarily form, in general, a star of the same mass, or even proportional to it.

The star formation efficiency. Studies by Padoan (1995) and Klessen et al. (2000) have proposed that turbulent fragmentation has an important role in determining the star formation efficiency (SFE). Moreover, the latter authors found that the efficiency in numerical simulations (measured as the fraction of mass in collapsed objects) increases as the energy injection scale is increased.

As remarked by Padoan (1995), the compressible cascade regime must end at a sonic scale, so that the turbulence within structures smaller than this size is subsonic and thus cannot produce any subfragmentation. Thus, in this scenario, subsonic cores are the natural end of the compressible part of the cascade in the ISM.[3] Moreover, when turbulence becomes subsonic, it ceases to be a stronger source of global support than thermal pressure, and the Jeans criterion can be applied to determine whether a core is gravitationally unstable and should proceed to collapse (Podoan, 1995), or else rebound and merge back into its environment (Elmegreen, 1993; Taylor et al., 1996; Ballesteros-Paredes et al., 1999a and Vázquez-Semadeni, et al., 2002b). This effectively sets an upper limit to the SFE because if the cloud is globally supported by its internal turbulence, only massive

[3]This does not necessarily imply that the dissipative (or "inner") scale of the turbulence has been reached, but only that at this point, the cascade can turn into an incompressible one. Similarly, it does not imply that there is a unique sonic scale. A large scatter in the velocity dispersion of structures with a given size is observed in simulations (e.g., Vázquez-Semadeni et al., 1997; Vázquez-Semadeni et al., 2003).

enough clumps among those with sizes λ smaller than the typical scale where the turbulence becomes subsonic, λ_s are susceptible of local collapse (Padoan, 1995). In this sense, star-forming cores constitute the intersection of the set of subsonic regions with the high-mass, super-Jeans tail of the mass distribution (Padoan, 1995; Ballesteros-Paredes and Vázquez-Semadeni, 1997; Padoan and Nordlund, 2002; Vázquez-Semadeni, et al., 2003). In particular, Padoan (1995) gave a semiphenomenological theory which predicted the SF efficiency and mass distribution of protostellar cores in turbulent MCs from the turbulent flow parameters, using the Larson scaling relations as empirical input. In that work, however, he used a questionable transformation between density and Jeans mass (see Scalo et al., 1998 for a discussion).

In this scenario, since progressively smaller density peaks contain progressively smaller fractions of the mass, one can expect the SFE to decrease with decreasing sonic scale, the latter being a function of the turbulent flow parameters (Vázquez-Semadeni et al., 2003). Indeed, the latter authors found that the star formation efficiency in numerical simulations scales with λ_s as $e^{-\lambda_0/\lambda_s}$, where λ_0 is a reference scale, which for the set of simulations fitted, has a value of $\sim$0.02 pc.

The initial mass function from turbulent fragmentation. Interesting attempts to explain the Initial Mass Function (IMF) from turbulent models have been performed by Padoan et al. (1997) and Padoan and Nordlund (2002). In particular, in the latter work, the authors propose a model in which the power spectrum of the turbulence is a power law, the cores are formed by shocks, and the typical size of a dense core scales as the thickness of the postshock gas; then the number of gravitationally unstable cores is given by

$$N(m)d\log m \propto m^{-3/(4-\beta)}\left(\int_0^m p(m_j)dm_j\right) d\log m, \tag{1}$$

where β is the exponent of the energy power spectrum of the turbulence, $E(k) \propto k^{-\beta}$, and the integral of $p(m_j)$ is the fraction of cores of mass m with gravitational energy larger than the thermal energy. In particular, (Padoan and Nordlund, 2002) argue that the value of $\beta = 1.8$, which is close to the value found in their numerical simulations, and is close to the average value inferred from observations of the velocity dispersion–size relationship, gives a mass dependence in Eq. (1) consistent with the Salpeter IMF.

Caveats in the derivation of the SFE and the IMF. As interesting and promising as it is, the model by Padoan and Nordlund (2002) is based on strong assumptions which we briefly discuss here in order to asses their feasibility. First, the number of cores as a function of scale, $N(L)$ (their Eq. (17)) is derived from scaling considerations as in Elmegreen (1997). Thus, it is independent of estimates of core properties (Eqs. (1)–(3), (11) in Padoan and Nordlund, 2002), while ideally, they should be self-consistent. Second, the derivation of estimate of $N(L)$ is based on assuming that a large-scale (L_2) simulation can be thought of as made of $(L_2/L_1)^3$ space-filling simulations of scale L_1, thus not taking into account intermittency of the density and velocity gradient fields. Instead, it assumes that the small scales are uniformly distributed in space without belonging to localized larger-scale ones. Thus, it will be probably neccessary to do a correction analogous to Kolmogorov (1962) or β-model intermittency corrections to original Kolmogorov (1941) model. This might alter the scaling. Third, their Eq. (24), which gives the number of collapsing cores, assumes that all collapsing cores are subsonic and indivisible by turbulence. Specifically, if a core with mass $m_1 < m_2$ is contained *within* a core of mass m_2, their Eq. (24) will predict two stars of masses $\propto m_1$ and $\propto m_2$, implying that the same mass used to form the first star, is also used to form the second one. Moreover, recent numerical simulations of the collapse of molecular cloud cores by (Goodwin et al., 2003) have found that the number and properties of objects formed in an individual core depends in a chaotic way on the details of the initial velocity field (assumed subsonic, with Mach ~0.3), showing that even subsonic cores fragment into a random number of stars. Fourth, in their treatment, competitive accretion and stellar feedback is not considered as agents that might change in a substantial way the IMF. However, the interaction of protostellar cores and their competition for mass growth from their surroundings are important processes shaping the distribution of final star properties (see the reviews by Mac Low and Klessen (2003), and Larson (2003)).

Finally, it is worth mentioning that in their treatment, both Padoan and Nordlund (2002) and Vázquez-Semadeni, et al. (2003) consider that the collapsing regions are those with subsonic turbulence and masses larger than the Jeans mass. However, there are examples in our Galaxy of compact ($R \leq 0.1$ pc), massive ($M \sim 10^4 M_\odot$) collapsing cores with velocity dispersions of the order of $\delta v \sim$ 4–12 km s^{-1} (see, e.g., Kurtz et al., 2000; Molinari et al., 2002 and Faundez et al., 2003). Although it is likely that they might suffer from internal turbulent fragmentation, it is not clear whether there will be one or several (and how many) sites of collapse in them. In fact, ~20% of the objects observed by Faundez et al., (2003) (e.g., IRAS 16272-4837) seem to be good candidates for isolated, massive star-formation regions. Thus, these still need to be accommodated by the emerging theory of turbulent star formation.

Gravitational fragmentation on a filament produced by large-scale streams. One important question is whether there is again a transition from turbulent fragmentation to gravitational fragmentation at some small scale, and if so, what is that scale.

In the particular case of Taurus, once turbulence has induced the formation of the filamentary structure of the gas (Ballesteros-Paredes et al., 1999b), gravitational fragmentation in the filaments should be considered as a mechanism for forming cores and stars. In fact, following Larson (1985), Hartmann (2002) has shown that the Jeans length and the free-fall timescale for a filament, are given by

$$\lambda_J = 1.5\, T_{10}\, A_V^{-1}\ \mathrm{pc}, \tag{2.1}$$

$$\tau \sim 3.7\, T_{10}^{1/2}\, A_V^{-1}\ \mathrm{Myr}, \tag{2.2}$$

where T_{10} is the temperature in units of 10 K, and A_V is the visual extinction through the center of the filament. By using a mean visual extinction of starless cores $A_V \sim 5$ (Onishi et al., 1998), Eq. (2.1) gives a characteristic Jeans length of $\lambda_J \sim 0.3$ pc, and collapse should occur in about 0.74 Myr. Hartmann (2002) argues that, since the number of young stellar objects per length scale in Taurus is about $3 - 4\ \mathrm{pc}^{-1}$, then self-gravity is working and producing the cores and stars.

In order to test this idea in the simulations, we analyze an SPH calculation performed by Klessen (see his contribution in this volume for more details). This simulation has been performed without gravity until a particular, well-defined elongated structure is formed. At this time, self-gravity is turned on. The results are in good agreement with Hartmann (2002). In Figure 1, we show eight column density frames of the simulation. The first frame shows the structure just before self-gravity is turned-on. We see that the filament forms cores in a fraction of a Myr. The timestep between frames is 0.1 Myr. The mean surface density for the filament is 0.033 g cm^{-2}, corresponding to a visual extinction of $\sim$7.5. Using Eqs. (2.1) and (2.2), this value gives a Jeans length of $\lambda_J \sim 0.2$ pc, and a collapsing timescale of $\tau \sim 0.5$ Myr. Note from Figure 1 that the first cores appear roughly at $\tau \sim 0.3$ Myr, although the final structure of collapsed objects is clearly defined at $t = 0.5$ Myr. Moreover, the typical separation between the cores corresponds roughly to the Jeans length, which is indicated by a small bar in the figures.

We then conclude that even though the filament is not in hydrostatic equilibrium, gravitational fragmentation may produce the structures seen either in the

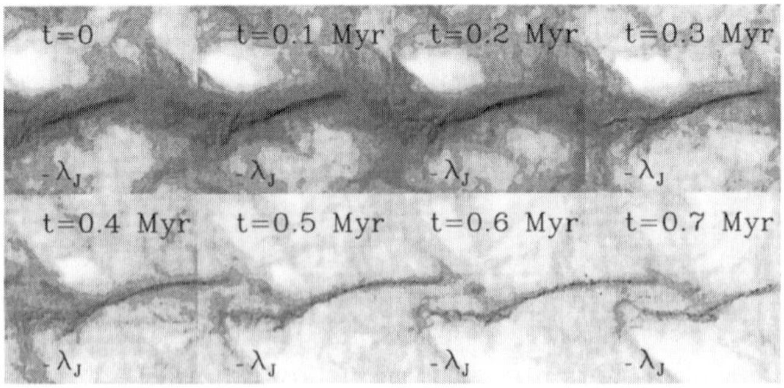

Figure 1. Evolution of the column density of an SPH simulation (courtesy of R. Klessen). The filament in the first frame (before self-gravity is turned-on) shows that turbulence is responsible for forming this kind of structures. The small bar in the bottom-left of each frame denotes the Jeans length (Eq. 2.1) at this time. At later times, self-gravity is turned on and the filament suffers gravitational fragmentation on a free-fall timescale (Eq. 2.2). The size of each box is 6 pc.

simulations, and in actual clouds. This means that turbulence causes first the production of the filament, and then self-gravity takes over to form the collapsed objects. However, since (a) the filament is not a perfect cylinder, (b) the collapsed objects are not perfectly equally spaced, and (c) there is a range in timesteps in which the cores are formed (between $t \sim 0.3$ and 0.6 Myr), it becomes clear that the final properties of the star-forming regions still depend on the initial conditions induced by the global turbulence. In a similar way, the final mass of the stars may still depend on the accretion history from the surrounding cloud material (see chapter by Klessen, this volume).

4. Conclusions

This review shows that numerical simulations of turbulent fragmentation have been quite successful in reproducing a number of observational properties of molecular cloud cores, while simultaneously making a number of predictions about the behavior of turbulence in molecular clouds and their cores, as well as providing an explanation for another set of observed features. Although much work remains to be done, and a full theory is still elusive, the progress made is highly encouraging, and makes the dynamical picture of star formation a plausible alternative to the ambipolar-diffusion-mediated magnetostatic-equilibrium standard model.

Acknowledgements

I want to thank Enrique Vázquez-Semadeni and Ralf Klessen for fruitful discussions and careful reading of this manuscript; the Scientific Organizing Committee for their invitation; Paolo Padoan for interesting (and still ongoing discussions) on the derivation of the IMF from the mass distribution of clumps formed by the turbulent field; and support from Conacyt's Grant I39318-E.

References

Arons, J. and Max, C.E.: 1975, *ApJL* **196**, L77.
Alves, J., Lada, C.J., and Lada, E.A.: 2001, *Nature* **409**, 159.
Avila-Reese, V. and Vázquez-Semadeni, E.: 2001, *ApJ* **553**, 645.
Ballesteros-Paredes, J.: in: M. Avillez and D. Breitschwerdt eds. *From Observations to Self-Consistent Modeling of the Interstellar Medium*, Astrophysics and Space Science, in press.
Ballesteros-Paredes, J., Hartmann, L., and Vázquez-Semadeni, E.: 1999b, *ApJ* **527**, 285.
Ballesteros-Paredes, J., Klessen, R.S., and Vázquez-Semadeni, E.: 2003. *ApJ* in press.
Ballesteros-Paredes, J. and Mac Low, M.: 2002, *ApJ* **570**, 734.
Ballesteros-Paredes, J. and Vázquez-Semadeni, E.: 1995, *Revista Mexicana de Astronomia y Astrofisica Conference Series* **3**, 105.
Ballesteros-Paredes, J. and Vázquez-Semadeni, E.: 1997, *Star Formation Near and Far*, 81.
Ballesteros-Paredes, J., Vázquez-Semadeni, E., and Scalo, J.: 1999a, *ApJ* **515**, 286.
Blitz, L. and Shu, F.H.: 1980, *ApJ* **238**, 148.

Blitz, L. and Williams, J.P.: 1997, *ApJL* **488**, L145.
Bonazzola, S., Heyvaerts, J., Falgarone, E., Perault, M., and Puget, J.L.: 1987, *AA* **172**, 293.
Bonnor, W.B.: 1956, *MNRAS* **116**, 351.
Brunt, C.: 2003. *ApJ* **583**, 280.
Caselli, P., Benson, P.J., Myers, P.C. and Tafalla, M.: 2002, *ApJ* **572**, 238.
Chandrasekhar, S. and Fermi, E.: 1953, *ApJ* **118**, 116.
Cho, J., Lazarian, A. and Vishniac, E.: 2002, astro-ph/0205286.
Faundez, S., Bronfman, L., Garay, G., Chini, R., May, J., and Nyman, L.: 2003, AA submitted.
Walder, R. and Folini, D.: 1998, *AA* **330**, L21.
Ebert, R.: 1955, *Zeitschrift für Astrophysik* **36**, 222.
Elmegreen, B.G.: 1993, in: *Protostars and Planets III*, p. 97.
Elmegreen, B.G.: 1997, *ApJ* **486**, 944.
Evans, N.J., Rawlings, J.M.C., Shirley, Y.L. and Mundy, L.G.: 2001, *ApJ* **557**, 193.
Gammie, C.F., Y.-T. Lin, Y.-T., Stone, J.M., and Ostriker, E.C.: 2003. *ApJ* in press.
Goldreich, P. and Kwan, J.: 1974, *ApJ* **189**, 441.
Goodman, A.A., Barranco, J.A., Wilner, D.J. and Heyer, M.H.: 1998, *ApJ* **504**, 223.
Goodwin, S.P., Whitworth, A.P. and Ward-Thompson, D.: 2004, *A&A* **414**, 633.
Hartmann, L.: 2002, *ApJ* **578**, 914.
Hartmann, L.: 2003, *ApJ* **585**, 398.
Hartmann, L., Ballesteros-Paredes, J. and Bergin, E.A.: 2001, *ApJ* **562**, 852.
Hunter, J.H., Sandford, M.T., Whitaker, R.W. and Klein, R.I.: 1986, *ApJ* **305**, 309.
Heitsch, F., Mac Low, M., and Klessen, R.S.: 2001a, *ApJ* **547**, 280.
Heitsch, F., Zweibel, E.G., Mac Low, M., Li, P. and Norman, M.L.: 2001b, *ApJ* **561**, 800.
Jijina, J., Myers, P.C. and Adams, F.C.: 1999, *ApJS* **125**, 161.
Johnstone, D., Wilson, C.D., Moriarty-Schieven, G., Joncas, G., Smith, G., Gregersen, E. and Fich, M.: 2000, *ApJ* **545**, 327.
Kegel, W.H.: 1989, *AA* **225**, 517.
Klein, R.I. and Woods, D.T.: 1998, *ApJ* **497**, 777.
Klessen, R.S.: 2000, *ApJ* **535**, 869.
Klessen, R.S.: 2001, *ApJ* **556**, 837.
Klessen, R.S., Ballesteros-Paredes, J., Vázquez-Semadeni, E. and Durán-Rojas, C.: 2003. *ApJ*, submitted.
Klessen, R.S., Heitsch, F., and Mac Low, M.: 2000, *ApJ* **535**, 887.
Kolmogorov, A.N.: 1941, *Dokl. Akad. Nauk SSSR* **30**, 301 (reprinted in *Proc. R. Soc. London A* **434**, 9 [1991]).
Kolmogorov, A.N.: 1962, *J. Fluid Mech.* **13**, 82.
Koyama, H. and Inutsuka, S.: 2002, *ApJL* **564**, L97.
Kurtz, S., Cesaroni, R., Churchwell, E., Hofner, P. and Walmsley, C.M.: 2000, in: *Protostars and Planets IV*, p. 299.
Larson, R.B.: 1981, *MNRAS* **194**, 809.
Larson, R.B.: 1985, *MNRAS* **214**, 379.
Larson, R.B.: 2003, *Rep. Prog. Phys.*, submitted.
Li,Y., Klessen, R.S. and Mac Low, M.-M.: 2003. *ApJ* **592**, 975.
Mac Low, M.: 1999, *ApJ* **524**, 169.
Mac Low, M.-M. and Klessen, R.S.: 2003, *Rev. Mod. Phys.* (astro-ph/0301093).
Mac Low, M., Klessen, R.S., Burkert, A. and Smith, M.D.: 1998, *Physical Review Letters* **80**, 2754.
Molinari, S., Testi, L., Rodríguez, L.F. and Zhang, Q.: 2002, *ApJ* **570**, 758.
Myers, P.C.: 1983, *ApJ* **270**, 105.
Onishi, T., Mizuno, A., Kawamura, A., Ogawa, H. and Fukui, Y.: 1998, *ApJ* **502**, 296.

Ostriker, E.C., Stone, J.M. and Gammie, C.F.: 2001, *ApJ* **546**, 980.
Padoan, P.: 1995, *MNRAS* **277**, 377.
Padoan, P., Goodman, A., Draine, B.T., Juvela, M., Nordlund, Å. and Rögnvaldsson, Ö. E.: 2001, *ApJ* **559**, 1005.
Padoan, P. and Nordlund, Å.: 1999, *ApJ* **526**, 279.
Padoan, P. and Nordlund, Åke 2002, *ApJ* **576**, 870.
Padoan, P., Nordlund, A. and Jones, B.J.T.: 1997, *MNRAS* **288**, 145.
Palla, F. and Stahler, S.W.: 2000, *ApJ* **540**, 255.
Palla, F. and Stahler, S.W.: 2002, *ApJ* **581**, 1194.
Passot, T. and Vázquez-Semadeni, E.: 1998, *Phys. Rev. E* **58**, 4501.
Passot, T. and Vázquez-Semadeni, E.: 2003, *AA* **398**, 845.
Sasao, T.: 1973, *PASJ* **25**, 1.
Scalo, J.: 1990, ASSL Vol. 162: Physical Processes in Fragmentation and Star Formation, 151.
Scalo, J., Vázquez-Semadeni, E., Chappell, D., and Passot, T.: 1998, *ApJ* **504**, 835.
Shirley, Y.L., Evans, N.J., Rawlings, J.M.C., and Gregersen, E.M.: 2000, *ApJS* **131**, 249.
Shu, F.H., Adams, F.C., and Lizano, S.: 1987, *ARAA* **25**, 23.
Stone, J.M., Ostriker, E.C., and Gammie, C.F.: 1998, *ApJL* **508**, L99.
Taylor, S.D., Morata, O., and Williams, D.A.: 1996, *AA* **313**, 269.
Vázquez-Semadeni, E., Ballesteros-Paredes, J. and Rodriguez, L.F.: 1997, *ApJ* **474**, 292.
Vázquez-Semadeni, E.: 1994, *ApJ* **423**, 681.
Vázquez-Semadeni, E. and García, N.: 2001, *ApJ* **557**, 727.
Vázquez-Semadeni, E., Ballesteros-Paredes, J. and Klessen, R.S.: 2003, *ApJL* **585**, L131.
Vázquez-Semadeni, E., Shadmehri, M. and Ballesteros-Paredes, J.: 2002. *ApJ* submitted.
Wada, K. and Norman, C.A.: 2001, *ApJ* **547**, 172.

MHD SIMULATIONS OF THE ISM: THE IMPORTANCE OF THE GALACTIC MAGNETIC FIELD ON THE ISM "PHASES"

MIGUEL DE AVILLEZ[1] and DIETER BREITSCHWERDT[2]

[1]*Department of Mathematics, University of Évora, R. Romão Ramalho 59, Evora, Portugal; E-mail: mavillez@galaxy.lca.uevora.pt*

[2] *Max-Planck-Institut für Extraterrestrische Physik, Giessenbachstraße, Postfach 1312, Garching, Germany*

Abstract. We have carried out 1.25 pc resolution MHD simulations of the ISM, on a Cartesian grid of $0 \leq (x, y) \leq 1$ kpc size in the galactic plane and $-10 \leq z \leq 10$ kpc into the halo, thus being able to fully trace the time-dependent evolution of the galactic fountain. The simulations show that large scale gas streams emerge, driven by SN explosions, which are responsible for the formation and destruction of shocked compressed layers. The shocked gas can have densities as high as 800 cm^{-3} and lifetimes up to 15 Myr. The cold gas is distributed into filaments which tend to show a preferred orientation due to the anisotropy of the flow induced by the galactic magnetic field. Ram pressure dominates the flow in the unstable branch $10^2 < T \leq 10^{3.9}$ K, whereas for $T \leq 100$ K (stable branch) magnetic pressure takes over. Near supernovae thermal and ram pressures determine the dynamics of the flow. Up to 80% of the mass in the disk is concentrated in the thermally unstable regime $10^2 < T \leq 10^{3.9}$ K with ~30% of the disk mass enclosed in the $T \leq 10^3$ K gas. The hot gas in contrast is controlled by the thermal pressure, since magnetic field lines are swept towards the dense compressed walls.

Keywords: Magneto-hydrodynamics, Galaxy: disk, ISM: general, ISM: kinematics and dynamics, ISM: structure

1. Introduction

Modelling the ISM with high spatial resolution allows us to tackle a set of problems simultaneously, encompassing both large and small scales, provided that the appropriate grid size, resolution and numerical tools (e.g. adaptive mesh refinement) are used. To name just the most important ones, global modelling yields information on the formation and lifetimes of molecular clouds, how star-forming regions are influenced by large-scale flows in the ISM, and which dynamical rôle SNe and superbubbles play in triggering local and global star formation. In this paper we investigate the effects of the magnetic field on the dynamics and evolution of the cold gas in the disk, the relative importance of the field for the ISM "phases," and its influence on the support of shocked compressed layers.

2. Model and Simulations

In the current work we report on kpc-scale simulations of the ISM on a cartesian grid of $0 \leq (x, y) \leq 1$ kpc size in the Galactic plane and $-10 \leq z \leq 10$ kpc into the

Astrophysics and Space Science **292:** 207–214, 2004.

halo, using a modified version of the 3D SN driven ISM model of Avillez (2000) coupled to a three-dimensional MHD code that uses adaptive mesh refinement (AMR) in a block-based structure in combination with Message Passing Interface (MPI) suitable for massive parallel computations.

The modified model includes a fixed gravitational field provided by the stars in the disk, radiative cooling assuming optically thin gas in collisional ionization equilibrium, and uniform heating due to starlight. The radiative cooling function is a tabulated version of that shown in Figure 2 of Dalgarno and McCray (1972) with an ionization fraction of 0.1 at temperatures below 10^4 K and a temperature cutoff at 10 K. Background heating due to starlight varies with z as described in Wolfire et al. (1995); in the midplane, at $z = 0$, it is chosen to initially balance radiative cooling at 9000 K. With the inclusion of background heating the gas becomes thermally stable at $T \leq 100$ K and $10^{3.9} < T \leq 10^{4.2}$ K.

The prime sources of mass, momentum and energy are SN whose setup is similar to that described in Avillez (2000) with some modifications. In the present work SNe type Ia with a scale height of 325 pc are also included. In regions with a density and temperature thresholds of 10 cm^{-3} and 100 K, respectively, an initial mass function is applied to determine the number of OB stars and their masses, forming an OB association. In agreement with observations 60% of these stars explode within the association, while the remaining lowest mass stars occur in the field. The location of the field stars is determined kinematically by attributing to each star a random direction and a velocity. The associations are allowed to form in a layer with a scale height of 46 pc, while the field stars occur in a layer with a scale height of 90 pc. The time interval between the explosions of all OB stars is determined by their main sequence life time.

The interstellar gas is initially setup with a density stratification distribution that includes the cold, cool, warm, ionized and hot gas "phases" in the Galaxy as described by Ferrière (1998). In these simulations the Galactic SN rate has been used and the canonical energy of explosion is 10^{51} erg for all types of SNe. At the beginning of the simulations the uniform field components along the three axes are given by $\vec{B}_u = (B_{u,0}(n(z)/n_0)^{1/2}, 0, 0)$, where $B_{u,0} = 3$ μG is the field strength and $n(z)$ is the number density of the gas as a function of distance from the Galactic midplane and $n_0 = 1$ cm^{-3} is the average midplane density. The random field component is set to zero in the beginning of the simulations. This component is built up during the first millions of years of evolution as a result of turbulent motions, mainly induced by SN explosions. Self-gravity is not included in these calculations.

The computational grid has a resolution of 10 pc, except in the layer between -500 and 500 pc, where three levels of AMR are used, yielding a finest level resolution of 1.25 pc. Periodic boundary conditions are applied along the four vertical boundary faces, while outflow boundary conditions are imposed at the top ($z = 10$ kpc) and bottom ($z = -10$ kpc) boundaries. All the simulations were evolved for 400 Myr corresponding to several sound crossing times along the fountain as required for the establishment of the duty disk-halo-disk cycle and the global dynamical equilibrium.

3. Results

3.1. Global Evolution

The initial evolution of the magnetized disk is similar to that seen in the HD runs (Avillez, 2000; Avillez and Breitschwerdt, 2004a), that is, the initial stratified distribution does not hold for long as a result of the lack of equilibrium between gravity and (thermal, kinetic and turbulent) pressure during the "switch-on phase" of SN activity. As a consequence the gas in the upper and bottom parts of the grid collapses into the midplane, leaving low density material in its place. However, in the MHD run it takes a longer time for the collapse to be completed as a result of the magnetic pressure and tension forces. As soon as the system has collapsed and enough SNe have gone off in the disk building up the required pressure support, transport into the halo is not prevented, although the escape of the gas takes a few tens of Myr to occur. The crucial point is that a huge thermal overpressure due to combined supernova explosions can sweep the magnetic field into dense filaments and punch holes into the extended warm and ionized HI layers. Once such pressure release valves have been set up, there is no way from keeping the hot overpressured plasma to follow the pressure gradient into the halo. As a consequence the duty disk-halo-disk cycle of the hot gas is fully established, which combined to the input of energy into the ISM by SNe, diffuse heating, energy lost by radiative cooling and magnetic pressure leads the system to evolve into a dynamical equilibrium state within some 200 Myr (Kahn, 1981).

Figure 1 shows slices of the 3D data cube of the density and magnetic field distributions in the Galactic midplane. The highest density ($n \geq 100$ cm^{-3}) gas tends to be confined into shocked compressed layers that form in regions where several large scale streams of convergent flow (associated with laminar flows driven by SNe) occur. The compressed regions are filamentary in structure, tend to be aligned with the local field and are associated with the highest field strengths[1]. The formation time of these high density structures depends on how much mass is carried by the convergent flows, how strong compression is and on the rate of cooling of the compressed region. The compressed layers have on average lifetimes of 10–15 Myr. The streams that promote the formation of the shocked dense layers are also responsible for their destruction. The filamentary structures are thicker in these simulations than in similar HD simulations, because the compression of a shock wave at given Mach number must go into increase of thermal and magnetic pressure in the MHD case and only into thermal in HD.

[1] A similar result is seen in similar HD simulations. However, there is no preferable orientation for the filaments.

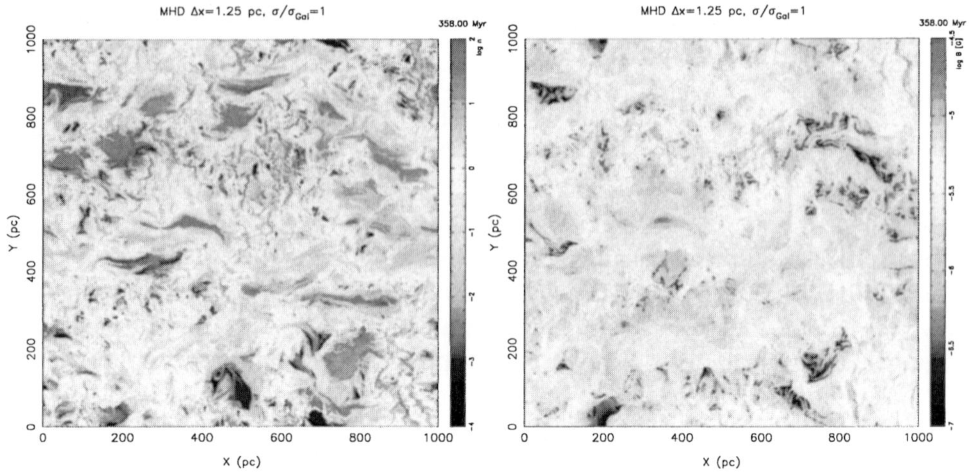

Figure 1. Density and magnetic field distribution in the Galactic midplane after 374 Myr of disk evolution. The resolution of the finest AMR level is 1.25 pc.

3.2. Field Dependence with Density

The simulations started with a fixed value for the magnetic field varying with z and having a midplane value of 3×10^{-6} G. During the evolution of the system thermal and dynamical processes broaden the distribution of the field strength in such a way that after the global dynamical equilibrium has been set up the field strength in the disk spans two orders of magnitude from 10^{-7} to 10^{-5} G as can be seen in Figure 2. The figure shows a scatter plot of the field strength as function of density in the simulated disk after 400 Myr of evolution.

Both panels in the figure show a large scatter in the field for the same density. This lack of correlation suggests that the field may not follow the "classical" relation of $B \sim \rho^{1/2}$. This essentially means that the equation of state is considerably softer than in the pure magnetic case and hence the effective adiabatic index γ must be well below unity. This may be attributed to the variety of processes that affect the energy balance of the magnetized plasma, such as turbulence, heating and cooling. Specifically the equation of state becomes stiffer, if the dependence of the heating rate on density is steeper than that of the cooling rate.

The large scatter in the field seen in Figure 2 also suggests that the field is being driven by the inertial motions, rather than it being the agent determining the motions. In the latter case the field would not be strongly distorted, and it would direct the motions predominantly along the field lines. The high field variability is also seen in the right panel of Figure 1, which shows a highly turbulent field, that seems to be uncorrelated with the density.

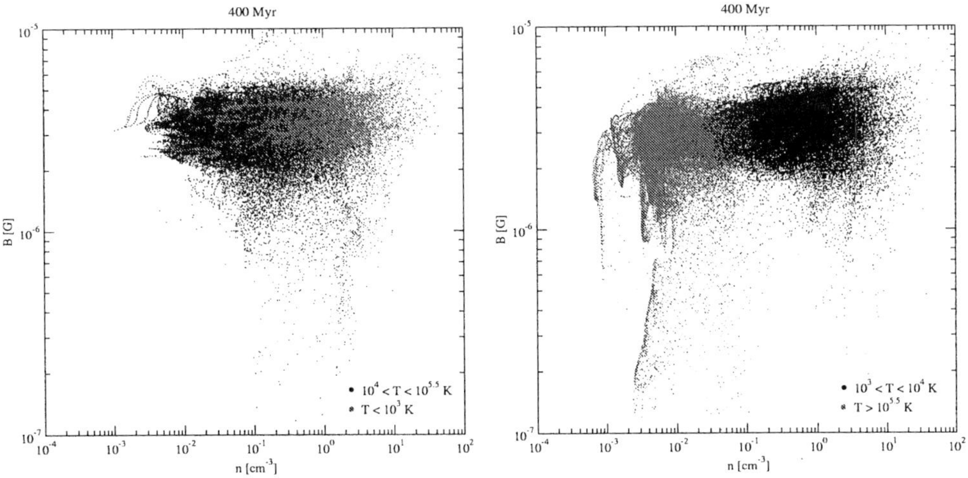

Figure 2. Scatter plot of the magnetic field strength as a function of density for the $T \leq 10^3$ (red) and $10^4 < T \leq 10^{5.5}$ K (black) regimes (left panel) and $10^3 < T \leq 10^4$ (black) and $T > 10^{5.5}$ K (red) regimes (right panel) at 400 Myr of evolution.

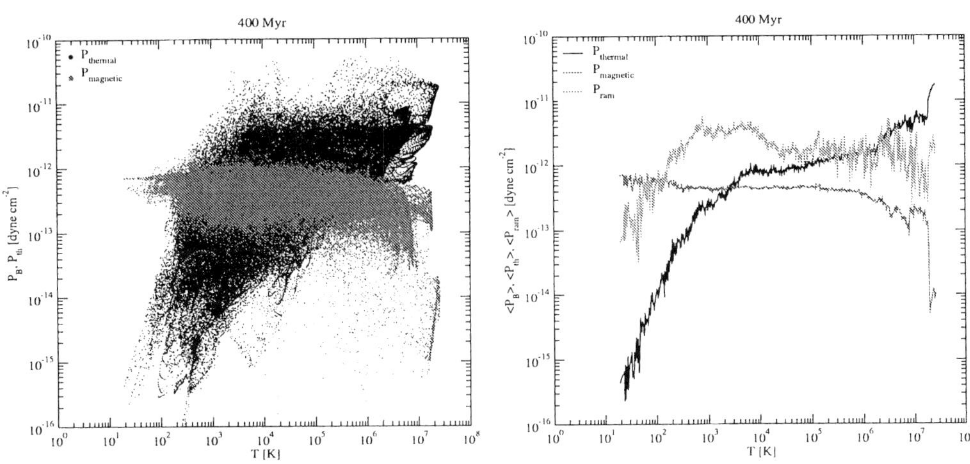

Figure 3. Left panel shows a scatter plot of the magnetic (red) and thermal (black) pressures as functions of temperature at 400 Myr. Right panel shows the average magnetic (red), thermal (black) and ram (green) pressures as functions of temperature at 400 Myr. This plot is obtained by calculating the average pressure in a certain temperature bin. The red and black curves represent the average values of the pressures seen in the left panel.

3.3. Pressure Variation with Temperature

Left panel of Figure 3 shows the scatter points of the thermal (black) and magnetic (red) pressures as a function of temperature, while the right panel shows the average magnetic, thermal and ram pressures as function of temperature. From the figure it is seen that all the $T \leq 10^2$ K gas has $P_B > P_{\rm ram} \gg P_{\rm th}$, demonstrating that magnetically dominated regions do exist, while the hot gas has the highest thermal pressure and the lowest magnetic pressure. For $T < 10^2$ this relation does not hold, as α varies between -0.06 and 0.085 (Avillez and Breitschwerdt, 2004) indicating that the magnetic and thermal pressures are independent. A result that is consistent with the almost zero variation of magnetic pressure with temperature seen in the right panel of Figure 3.

The figure also shows that for $10^2 < T < 10^6$ K ram pressure dominates over magnetic and thermal pressures, that is, ram pressure determines the dynamics of the flow, and therefore, the magnetic pressure does not act as a restoring force as it was already suggested by the lack of correlation between the field strength and the density (for a detailed discussion see Passot and Vázquez-Semadeni, 2003). The alternating dominance between magnetic and ram pressures seen for $10^6 < T < 3 \times 10^6$ K results from the fact that about half of the explosion energy in SNe becomes kinetic, and therefore, ram pressure becomes also important near the energy sources.

The dynamical picture that emerges from these simulations is that thermal pressure gradients dominate mostly in the neighbourhood of SNe, which drive motions whose ram pressures are dominant over the mean thermal pressure (away from the energy sources) and the magnetic pressure. The magnetic field is dynamically important at low temperatures. The field does not act only by its pressure gradient, but it acts *dynamically* also by magnetic tension forces, $1/4\pi(\vec{B}\nabla)\vec{B}$. Whereas magnetic pressure is isotropic, magnetic tension is always *along* the lines of force. Equilibrium only exists in force-free configurations for which $\nabla \times \vec{B} = 0$. However, if there is a thermal pressure then it is practically impossible to balance the magnetic stresses and most of the configurations will be unstable. In other words, in a dynamical system like the ISM, which includes systematic gas motions, $\vec{j} \times \vec{B}$-forces are inevitable, and hence the magnetic field can be dominant. Thus, regions with a high field strength such as the cold clouds should be magnetically controlled.

3.4. Distribution of the ISM Mass

Up to 80% of the ISM mass in the simulated disk is concentrated in the unstable regime at $10^2 < T \leq 10^{3.9}$ K (orange curves in Figure 4) where $\sim$30% of the ISM mass is concentrated in $10^2 < T < 10^3$ K gas (dashed orange curve), while the $10^3 < T \leq 10^{3.9}$ K gas (solid orange line) encloses some 50% of the mass. The stable regimes $T \leq 10^2$ K and $10^{3.9} < T \leq 10^{4.2}$ K enclose $\sim$1% and $\sim$10% of the disk mass, respectively. Note that most of the cold gas is concentrated in the unstable branch at temperatures $T \leq 10^3$ K and has densities up to 100 cm^{-3} while gas with highest density has $T > 10^3$ K. The hot gas amounts to less than 1%, while the unstable regime with $10^{4.2} < T \leq 10^{5.5}$ encloses $\sim$10% of the total mass.

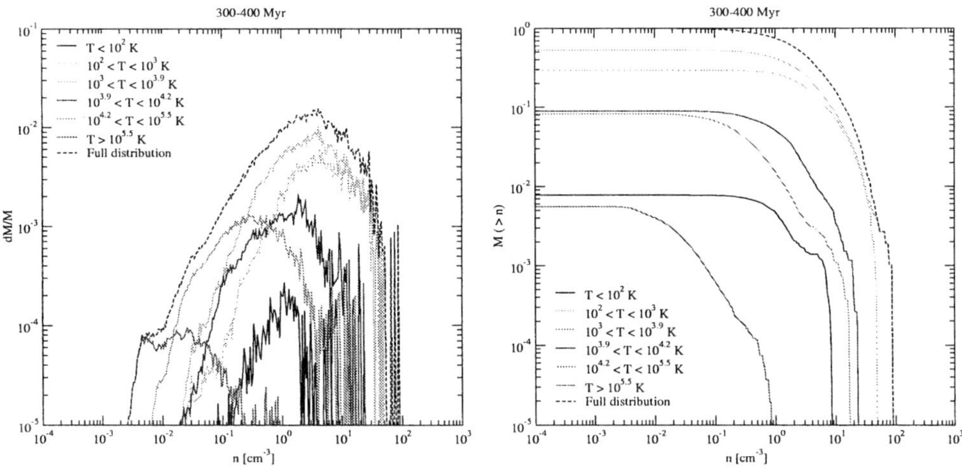

Figure 4. Averaged mass weighted (left panel) and cumulative (right panel) histograms of the density in the Galactic disk calculated between 300 and 400 Myr using 101 snapshots with 1 Myr of interval. The black and blue solid lines represent gas in the thermally stable regimes at $T \leq 10^2$ K and $10^{3.9} < T \leq 10^{4.2}$ K, the orange solid and dashed lines represent the thermally unstable gas with $10^2 < T \leq 10^3$ K and $10^3 < T \leq 10^{3.9}$ K, respectively. The gas with $10^{4.2} < T \leq 10^{5.5}$ K is represented by the green line, while the hot ($T > 10^{5.5}$ K) gas is represented by the red line.

4. Summary

The highest resolution MHD simulation of the ISM carried out to date, discussed in this paper, shows that large scale streams driven by SNe are responsible for the formation and destruction of high density clouds in shocked compressed layers. The clouds formed within the shocked gas can have densities as high as 800 cm^{-3} and lifetimes up to 15 Myr. Ram pressure dominates most of the coolest flows (in the unstable branch $10^2 < T \leq 10^{3.9}$ K) except at temperatures below 100 K (in the stable branch), where magnetic pressure takes over.

Most of the mass in the disk is concentrated in the thermally unstable regime for $T \leq 10^{3.9}$ K with some 30% of the mass enclosed in gas with $T \leq 10^3$ K. This is consistent with observations and numerical simulations discussed by Heiles (2001), Gazol et al. (2001) and Kritsuk and Norman (2002). However, these results are certainly in disagreement with classical ISM theories (McKee and Ostriker, 1977), as from our simulations up to 90% of the ISM mass is found to be in the *thermally unstable* regimes with $10^2 < T \leq 10^{3.9}$ K (~80%) and $10^{4.2} < T \leq 10^{5.5}$ K (~10%), where it should not exist, if it were not for the importance of dynamical processes. This calls into question the standard paradigm of an ISM distributed over three phases and being in pressure equilibrium.

Acknowledgements

We acknowledge fruitful discussions with the referee, E. Vázquez-Semadeni, and M.-M. Mac Low. MAA thanks A.I. Gómez de Castro for the financial support.

References

Avillez, M.A.: 2000, *MNRAS* **315**, 479.
Avillez, M.A. and Breitschwerdt, D.: 2004a, *Ap&SS* **289**, 479.
Avillez, M.A. and Breitschwerdt, D.: 2004b, *A&A* (submitted).
Dalgarno, A. and McCray, R.A.: 1972, *ARA&A* **10**, 375.
Ferrière, K.M.: 1998, *ApJ* **503**, 700.
Gazol, A., Vázquez-Semadeni, E., Sanchez-Salcedo, J. and Scalo, J.: 2001, *ApJ* **557**, 121.
Heiles, C.: 2001, *ApJ* **551**, L105.
Kahn, F.D.: 1981, in: F.D. Kahn (ed.), *Investigating the Universe*, Reidel, Dordrecht, p. 1.
Kritsuk, A. and Norman, M.L.: 2002, *ApJ* **569**, L127.
McKee, C.F. and Ostriker, J.P.: 1977, *ApJ* **218**, 148.
Passot, T. and Vázquez-Semadeni, E.: 2003, *A&A* **398**, 845.
Wolfire, M.G., McKee, C.F., Hollenbach, D., Tielens, A.G.G.M. and Bakes, E.L.O.: 1995, *ApJ* **443**, 152.

COMMENTS ON GRAVOTURBULENT STAR FORMATION

RALF S. KLESSEN
Astrophysikalisches Institut Potsdam, An der Sternwarte 16, 14482 Potsdam, Germany;
E-mail: rklessen@aip.de

Abstract. Understanding the star formation process is central to much of modern astrophysics. Stellar birth is intimately linked to the dynamical behavior of the parental gas cloud. Gravoturbulent fragmentation determines where and when protostellar cores form, and how they contract and grow in mass via accretion from the surrounding cloud material to build up stars. Supersonic turbulence can provide support against gravitational collapse on global scales, whereas at the same time it produces localized density enhancements that allow for collapse on small scales. The efficiency and timescale of stellar birth in Galactic molecular clouds strongly depend on the properties of the interstellar turbulent velocity field, with slow, inefficient, isolated star formation being a hallmark of turbulent support, and fast, efficient, clustered star formation occurring in its absence.

Keywords: star formation, turbulence

1. Introduction

Stars are born in interstellar clouds of molecular hydrogen and form by gravitational collapse of shock-compressed density fluctuations generated from the supersonic turbulence ubiquitously observed in molecular clouds (e.g. Hunter and Fleck, 1982; Elmegreen, 1993; Padoan, 1995; Ballesteros-Paredes et al., 1999a,b, 2003; Klessen et al., 2000; Padoan and Nordlund, 2002). Once gas clumps become gravitationally unstable, collapse sets in and the central density increases until a protostellar object forms which continues to grow in mass via accretion from the infalling envelope (for the first consistent protostellar collapse calculation from cloud to stellar densities see Wuchterl and Klessen, 2001; Wuchterl and Tscharnuter, 2003). As stars typically form in small aggregates or larger clusters (e.g. Adams and Myers, 2001) the interaction of protostellar cores and their competition for mass growth from their surroundings are important processes shaping the final star properties (see the reviews by Larson, 2003; and Mac Low and Klessen, 2004).

Altogether star formation can be seen as a two-phase process: First, *turbulent fragmentation* leads to transient clumpy molecular cloud structure, with some of the density fluctuation exceeding the critical mass and density for *gravitational* contraction. This sets the stage for the second phase of star formation, the *collapse of individual protostellar clumps* building up central protostars. To fully understand the star formation process and all its phases in detail is one of the fundamental challenges in astronomy. Describing the first phase of star formation requires the three-dimensional dynamical modeling of molecular cloud formation,

Astrophysics and Space Science **292:** 215–223, 2004.

evolution and fragmentation. In this contribution, we report results from smoothed-particle hydrodynamics calculations. The numerical method and its application to (driven or decaying) interstellar turbulence is introduced in Section 2 and Section 3. We demonstrate how supersonic turbulence may influence the properties of forming (proto)stellar aggregates and clusters, and discuss the spatial distribution and timescale of star formation in Section 4 and the expected mass spectra for gas clumps and protostars in Section 5.

2. SPH—A Particle-Based Flexible Method to Hydrodynamics

An important prerequisite for adequately describing the density and velocity structure of cores in numerical models of turbulent molecular cloud evolution is the ability to resolve high density contrasts at arbitrary locations. Smoothed particle hydrodynamics (SPH) is probably the best method currently available for this purpose. It is a Lagrangian method, where the fluid is represented by an ensemble of particles and thermodynamical observables are obtained by averaging over an appropriate subset of the SPH particles (Benz, 1990; Monaghan, 1992). High density contrasts are resolved by simply increasing the particle concentration where needed. The simulations presented here concentrate on subregions within a much larger cloud, therefore periodic boundary conditions are adopted (Klessen, 1997). For a detailed discussion of the performance and convergence properties of the method and tests against analytic models and other numerical schemes in the context of turbulent supersonic astrophysical flows the reader is referred to Bate and Burkert (1997), Mac Low et al. (1998), Lombardi et al. (1999), Klessen and Burkert (2000, 2001) and Klessen et al. (2000).

3. Models of Driven Supersonic Turbulence

The large observed linewidths in molecular clouds imply the presence of supersonic velocity fields that carry enough energy to counterbalance gravity on global scales (see, e.g., the review by Williams et al., 2000). However, it is known that turbulent energy dissipates rapidly, roughly on the free-fall timescale (Mac Low et al., 1998; Stone et al., 1998; Padoan and Nordlund, 1999). Unlike previously thought, this is independent of the presence of magnetic fields. The fields are therefore taken as being dynamically unimportant in the current models, and are not included. They may, however, play a role in retarding the late stages of protostellar collapse (see Crutcher, this volume) and removing angular momentum by magnetic breaking (e.g. Mouschovias and Paleologou, 1979, 1980). These effects should be considered in more consistent models in the future.

To prevent or considerably postpone global collapse, turbulence must be continuously replenished. This is achieved by applying a Gaussian driving scheme, that inserts kinetic energy in a specified range of wavenumbers k. We choose $1 \leq k \leq 2$, $3 \leq k \leq 4$ and $8 \leq k \leq 8$, corresponding to sources that act on large, intermediate and small scales, respectively. The energy input at each timestep is adjusted to reach a constant level of the total kinetic energy sufficient to stabilize the cloud as a whole (Mac Low, 1999). Furthermore one molecular cloud model is considered where turbulence is assumed to have decayed, leaving behind Gaussian density fluctuations that begin to contract on all scales. For details see Klessen (2001a,b).

The models presented here are computed in normalized units. If scaled to mean densities of $n(H_2) = 10^4\,cm^{-3}$, a value typical for star-forming molecular cloud regions (e.g. ρ-Ophiuchus: see Motte et al., 1998) and a temperature of 11.4 K (a sound speed $c_s = 0.2\,km\,s^{-1}$), then the total mass contained in the computed volume is $413\,M_\odot$, corresponding to 120 thermal Jeans masses within a cube of 0.89 pc side length.

4. Spatial Distribution and Timescale of Star Formation

Stars form from gravoturbulent fragmentation of molecular cloud material. Supersonic turbulence that is strong enough to counterbalance gravity on global scales will usually provoke *local* collapse (e.g. Sasao, 1973 for an early analysis; or Vázquez-Semadeni et al., 2000; Mac Low and Klessen, 2003; Larson, 2003 for recent reviews on ISM turbulence and star formation). Turbulence establishes a complex network of interacting shocks, where regions of high-density build up at the stagnation points of convergent flows. These gas clumps can be dense and massive enough to become gravitationally unstable and collapse when the local Jeans length becomes smaller than the size of the fluctuation. However, the fluctuations in turbulent velocity fields are highly transient. If they are not strongly self-gravitating they disperse naturally once the converging flow fades away (Vazquez-Semadeni et al., 2002). But also clumps that are dominated by gravity may get disrupted by the passage of a new shock front if it is strong enough (Klein et al., 1994; Mac Low et al., 1994).

For local collapse to result in the formation of stars, Jeans unstable, shock-generated, density fluctuations therefore must collapse to sufficiently high densities on time scales shorter than the typical time interval between two successive shock passages. Only then do they "decouple" from the ambient flow pattern and survive subsequent shock interactions. The shorter the time between shock passages, the less likely these fluctuations are to survive. The overall efficiency of star formation depends strongly on the wavelength and strength of the driving source (Klessen et al., 2000; Heitsch et al., 2001) which both regulate the amount of gas available for collapse on the sonic scale where turbulence turns from supersonic to subsonic (Vázquez-Semadeni et al., 2003).

The velocity field of long-wavelength turbulence is dominated by large-scale shocks that are very efficient in sweeping up molecular cloud material, thus creating massive coherent structures. These exceed the critical mass for gravitational collapse by far, because the velocity dispersion within the shock compressed region is much smaller than in the ambient turbulent flow. The situation is similar to localized turbulent decay, and quickly a cluster of protostellar cores builds up. Both decaying and large-scale turbulence lead to a *clustered* mode of star formation. Prominent examples are the Trapezium Cluster in Orion with a few thousand young stars, but also the Taurus star forming region which is historically considered as a case of isolated stellar birth. Its stars, however, have formed almost simultaneously within several coherent filaments which apparently are created by external compression (see Ballesteros-Paredes et al., 1999b). This renders it a clustered star forming region in the sense of the above definition.

The efficiency of gravoturbulent fragmentation is reduced if the sonic scale decreases, either by varying $M_{\rm rms}$ or the driving scale. In the latter case the network of interacting shocks is very tightly knit, hence, protostellar cores form independently of each other at random locations and at random times. There are no coherent structures with multiple Jeans masses. Individual shock generated clumps are of low mass and the time interval between two shock passages through the same point in space is small. Hence, collapsing cores are easily destroyed again. Altogether star formation is inefficient. This scenario then corresponds to an *isolated* mode of star formation. Stars that truly form in isolation are, however, very rarely observed—most young stars are observed in clusters or at most loose aggregates. From a theoretical point of view, there is no fundamental dichotomy between these two modes of star formation, they rather define the extreme ends in the continuous spectrum of the properties of turbulent molecular cloud fragmentation.

This is visualized in Figures 1 and 2. It compares the density evolution in models of decayed, large-wavelength ($k = 1 \ldots 2$), intermediate ($k = 3 \ldots 4$), and small-scale turbulence ($k = 7 \ldots 8$). The upper panes shows the systems at $t = 0$ corresponding to the phase of fully developed turbulence just before gravity is "switched on". For the Gaussian case it shows the initial random fluctuation field. The lower panel describes the four models some time after the first protostars have formed.

Figure 3 shows the mass accretion history of individual protostellar cores for the three turbulent models, together with the distribution of core formation times. Star formation efficiency is high in the long-wavelength model, all cores form within two free-fall times, whereas in the short-wavelength model the efficiency is low and core formation continues for over 15 free-fall timescales.

5. Mass Spectra of Clumps and Protostellar Cores

The dominant parameter determining stellar evolution is the mass. It is therefore important to investigate the relation between the masses of molecular clumps, protostellar cores and the resulting stars. Figure 4 plots, for all four models, the mass distribution of all gas clumps, of the subset of Jeans critical clumps, and of collapsed

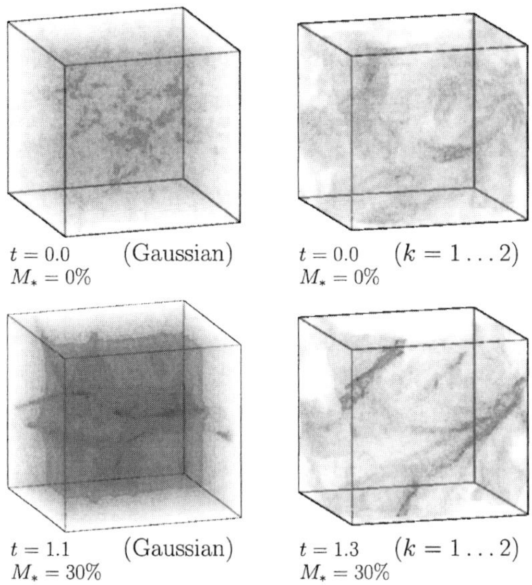

Figure 1. Comparison of the gas distribution in models of decayed turbulence leading to a Gaussian field (left column) and large-scale turbulence (second column). The upper panel depicts the initial stage when gravity is "turned on", the lower panel shows system after the first collapsed cores (protostars) have formed and accumulated roughly 30% of the total mass. The location of the newly formed protostars is indicated by dark dots. Note the different time intervals needed to reach similar evolutionary phases. The time unit is normalized to the global free-fall time of the system. Compare with Figure 2.

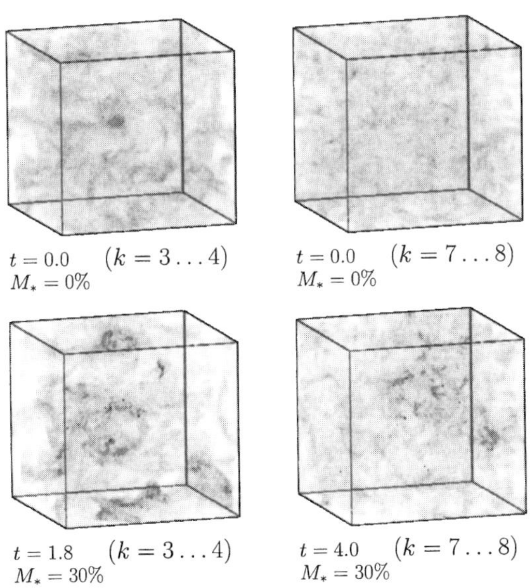

Figure 2. Same as Figure 1, but now comparing the gas distribution in intermediate-wavelength turbulence (third column) and small-wavelength turbulence (right column).

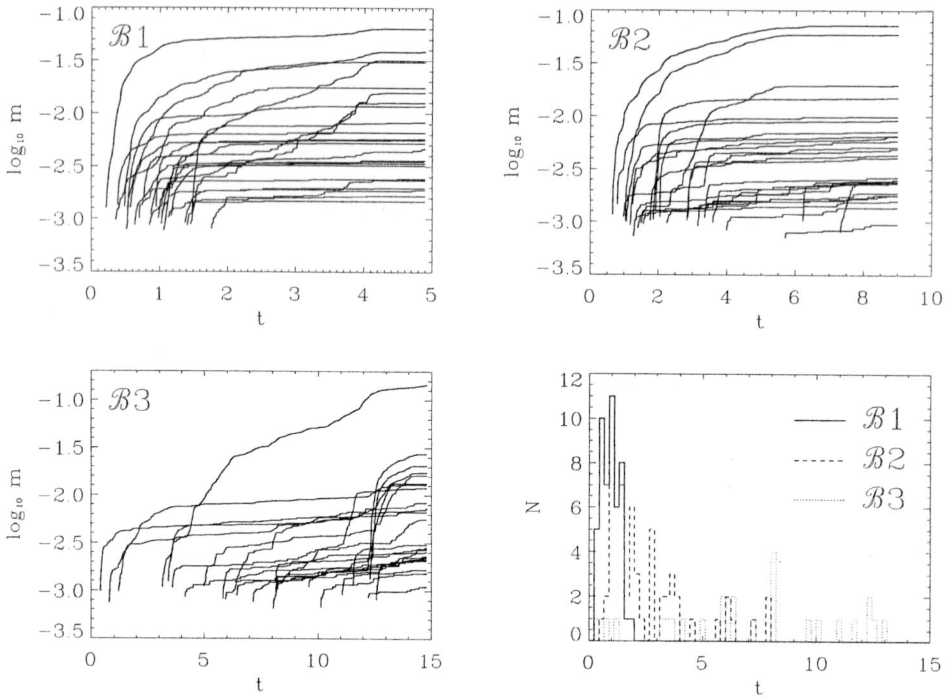

Figure 3. Mass growth history of protostellar cores (only every second core is depicted): long-wavelength ($\mathcal{B}1h$), intermediate- ($\mathcal{B}2h$) and small-scale turbulence ($\mathcal{B}3h$). The lower right plot shows the distribution of the formation times of the cores. Time is given in units of the free-fall time and masses are scaled to the total mass in the system.

cores (which we can identify as protostars in fully resolved calculations). Four different evolutionary phases are shown, initially just when gravity is "switched on", and after turbulent fragmentation has lead to the accumulation of $M_* \approx 5\%$, $M_* \approx 30\%$ and $M_* \approx 60\%$ of the total mass in protostars.

In the initial, completely pre-stellar phase gas clumps are hardly more massive than a few $\langle M_J \rangle$. The mass spectrum (thin line) is very steep (about Salpeter slope or less), but has a break and gets shallower below $M \approx 0.4\,\langle M_J \rangle$ with slope -1.5. Beyond the SPH resolution limit, the spectrum strongly declines, as smaller-mass objects cannot be described adequately by the method.

Gravity strongly modifies the distribution of clump masses during the later evolution. As clumps merge and grow bigger, the mass spectrum becomes flatter and extends toward larger masses. Consequently the number of objects exceeding the Jeans limit increases. This is most evident in the Gaussian model of decayed turbulence, where the velocity field is entirely determined by gravitational contraction on all scales. The clump mass spectrum in intermediate phases of the evolution (i.e. when the first protostars form but the overall gravitational potential is still dominated by non-accreted gas) exhibits a slope -1.5 similar to the observed one.

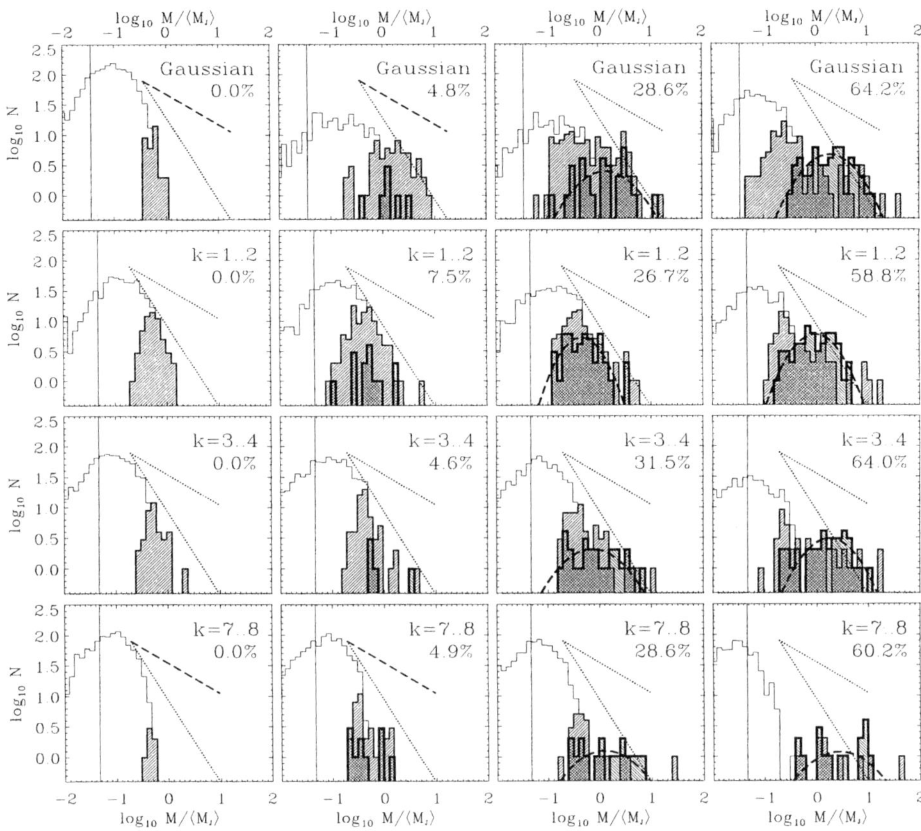

Figure 4. Mass spectra of protostars (hatched thick-lined histograms), of gas clumps (thin lines), and of the subset of Jeans unstable clumps (thin lines, hatched distribution). Masses are binned logarithmically and normalized to the average Jeans mass $\langle M_J \rangle$. The left column gives the initial state of the system, the second column shows the mass spectra when roughly $M_* \approx 5\%$ of the mass is in dense cores (i.e. protostars), the third column when this fraction is about 30%, and the right column when $M_* \approx 60\%$. For comparison with power–law spectra ($dN/dM \propto M^{\nu}$), the typical slope $\nu = -1.5$ of the observed clump mass distribution, and the Salpeter slope $\nu = -2.33$ for the high-mass end of the IMF, are indicated by the long dashed and by the dotted lines in each plot. The best log-normal fit to the protostellar mass distribution in the models is indicated by thick dashed lines in the last two columns. The vertical line shows the SPH resolution limit.

The slope also depends on the Mach number of the flow. If the kinetic energy strongly exceeds the gravitational one, the situation is similar to purely hydrodynamic turbulence and the clump mass spectrum remains steep as seen in the left panel of Figure 4.

The mass spectrum also depends on the wavelength of the dominant velocity modes. Small-scale turbulence does not allow for massive, coherent and strongly self-gravitating structures. Together with the short interval between shock passages, this prohibits efficient merging and the build up of a large number of massive clumps.

Only few clumps become Jeans unstable and collapse to build up protostars. These form independent of each other at random locations and times, and typically do not interact. The clump mass spectrum remains steep similar to the case without gravity. Increasing the driving wavelength leads to more coherent and rapid core formation and subsequently to a larger number of protostars.

Long-wavelength turbulence or turbulent decay leads to a core mass spectrum that is well approximated by a *log-normal*. It roughly peaks at the *average thermal Jeans mass* $\langle M_{\rm J}\rangle$ of the system (see Klessen and Burkert, 2000, 2001) and is comparable in width with the observed IMF (Kroupa, 2002). It is also in agreement with mass spectra of dense prestellar cores typical several star forming regions (ρ-Ophiuchus: Motte et al., 1998; Johnstone et al., 2000; NGC 2068/2071: Motte et al., 2001; Orion: Johnstone et al., 2001). The log-normal shape of the mass distribution may be explained by invoking the central limit theorem (e.g. Zinnecker, 1984), as protostellar cores form and evolve through a sequence of highly stochastic events.

To find the mass peak at $\langle M_{\rm J}\rangle$ may be somewhat surprising given the fact that the local Jeans mass strongly varies between different clumps. In a statistical sense the system retains knowledge of its mean properties. The total width of the core distribution is about two orders of magnitude in mass and is approximately the same for all four models.

However, the spectrum for intermediate and short-wavelength turbulence, i.e. for isolated star formation, is too flat (or equivalently too wide) to be comparable to the observed IMF. This appears to point toward the importance of mutual protostellar interactions during their formation and main mass-growth phase. In essence, collisional dynamics contributes additional stochasticity to the underlying randomness of pure turbulent flows. For a further discussion of the theory of IMF from gravoturbulent fragmentation see Ballesteros-Paredes (2004) and references therein.

Acknowledgements

I thank Javier Ballesteros-Paredes, Peter Bodenheimer, Fabian Heitsch, Mordecai-Mark Mac Low, and Enrique Vázquez-Semadeni for many stimulating discussions and fruitful collaboration. I acknowledge support by the Emmy Noether Program of the Deutsche Forschungsgemeinschaft (Grant no. KL1358/1).

References

Adams, F.C. and Myers, P.C.: 2001, *ApJ* **553**, 744.
Bate, M.R. and Burkert, A.: 1997, *MNRAS* **288**, 1060.
Ballesteros-Paredes, J., Hartmann, L. and Vázquez-Semadeni, E.: 1999b, *ApJ* **527**, 285.
Ballesteros-Paredes, J., Klessen, R.S. and Vázquez-Semadeni, E.: 2003, *ApJ* **592**, 188.
Ballesteros-Paredes, J., Vázquez-Semadeni, E. and Scalo, J.: 1999a, *ApJ* **515**, 286.

Benz, W.: 1990, in: J.R. Buchler (ed.), *The Numerical Modeling of Nonlinear Stellar Pulsations*, Kluwer Dordrecht p. 269.
Elmegreen, B.G.: 1993, *ApJ* **419**, L29.
Heitsch, F. Mac Low, M.-M. and Klessen, R.S.: 2001, *ApJ* **547**, 280.
Hunter, J.H. and Fleck, R.C.: 1982, *ApJ* **256**, 505.
Johnstone, D., Fich, M., Mitchell, G.F. and Moriarty-Schieven, G.: 2001, *ApJ* **559**, 307.
Johnstone, D., Wilson, C.D., Moriarty-Schieven, G., Joncas, G., Smith, G., Gregersen, E. and Fich, M.: 2000, *ApJ* **545**, 327.
Klein, R.I., McKee, C.F. and Colella, P.: 1994, *ApJ* **420**, 213.
Klessen, R.S.: 1997, *MNRAS* **292**, 11.
Klessen, R.S.: 2001a, *ApJ* **550**, L77.
Klessen, R.S.: 2001b, *ApJ* **556**, 837.
Klessen, R.S. and Burkert, A.: 2000, *ApJS* **128**, 287.
Klessen, R.S. and Burkert, A.: 2001, *ApJ* **549**, 386.
Klessen, R.S., Heitsch, F. and Mac Low, M.-M.: 2000, *ApJ* **535**, 887.
Kroupa, P.: 2002, *Science* **295**, 82.
Larson, R.B.: 2003, *Rep. Prog. Phys.* submitted (astro-ph/0306595).
Lombardi, J.C., Sills, A., Rasio, F.A. and Shapiro, S.L.: 1999, *JCP* **152**, 687.
Mac Low, M.-M.: 1999, *ApJ* **524**, 169.
Mac Low, M.-M. and Klessen, R.S.: 2004, *Rev. Mod. Phys.* **76**, 125.
Mac Low, M.-M., Klessen, R.S., Burkert, A. and Smith, M.D.: 1998, *Phys. Rev. Lett.* **80**, 2754.
Mac Low, M.-M., McKee, C.F., Klein, R.I., Stone, J.M. and Norman, M.L.: 1994, *ApJ* **433**, 757.
Monaghan, J.J.: 1992, *ARAA* **30**, 543.
Mouschovias, T.C. and Paleologou, E.V.: 1979, *ApJ* **230**, 204.
Mouschovias, T.C. and Paleologou, E.V.: 1980, *ApJ* **237**, 877.
Motte, F., André, P. and Neri, R.: 1998, *A&A* **336**, 150.
Motte, F., André, P., Ward-Thompson, D. and Bontemps, S.: 2001, *A&A* **372**, L41.
Padoan, P.: 1995, *MNRAS* **277**, 337.
Padoan, P. and Nordlund, Å.: 1999, *ApJ* **526**, 279.
Padoan, P. and Nordlund, Å.: 2002, *ApJ* **576**, 870.
Stone, J.M., Ostriker, E.C. and Gammie, C.F.: 1998, *ApJ* **508**, L99.
Vázquez-Semadeni, E., Ballesteros-Paredes, J. and Klessen, R.S.: 2003, *ApJ* **585**, L131.
Vázquez-Semadeni, E., Shadmehri, M. and Ballesteros-Paredes, J.: 2002, *ApJ* submitted (astro-ph/0208245).
Vázquez-Semadeni, E., Ostriker, E.C., Passot, T., Gammie, C.F. and Stone, J.M.: 2000, in: V. Mannings, A. Boss, and S. Russell (eds.), *Protostars and Planets IV*, University of Arizona Press. Tucson, p. 3.
Williams, J.P., Blitz, L. and McKee, C.F.: 2000, in: V. Mannings, A. Boss and S. Russell (eds.), *Protostars and Planets IV*, Univ. of Arizona Press, Tucson, p. 97.
Wuchterl, G. and Tscharnuter, W.M.: 2003, *A&A* **398**, 1081.
Wuchterl, G. and Klessen, R.S.: 2001, *ApJ* **560**, L185.
Zinnecker, H.: 1984, *MNRAS* **210**, 43.

WHAT DRIVES STAR FORMATION?

RICHARD M. CRUTCHER
Department of Astronomy, University of Illinois, 1002 W. Green Street, Urbana, IL, U.S.A.; E-mail: crutcher@uiuc.edu

Abstract. Current theoretical models for what drives star formation (especially low-mass star formation) are: (1) magnetic support of self-gravitating clouds with ambipolar diffusion removing support in cores and triggering collapse and (2) compressible turbulence forming self-gravitating clumps that collapse as soon as the turbulent cascade produces insufficient turbulent support. Observations of magnetic fields can distinguish between these two models because of different predictions in three areas: (1) magnetic field morphology, (2) the scaling of field strength with density and non-thermal velocities, and (3) the mass to magnetic flux ratio, M/Φ. We first discuss the techniques and limitations of methods for observing magnetic fields in star formation regions, then describe results for the L1544 prestellar core as an exemplar of the observational results. Application of the three tests leads to the following conclusions. The observational data show that *both* magnetic fields and turbulence are important in molecular cloud physics. Field lines are generally regular rather than chaotic, implying strong field strengths. But fields are not aligned with the minor axes of oblate spheroidal clouds, suggesting that turbulence is important. Field strengths appear to scale with non-thermal velocity widths, suggesting a significant turbulent support of clouds. Giant Molecular Clouds (GMCs) require mass accumulation over sufficiently large volumes that they would likely have an approximately critical M/Φ. Yet H I clouds are observed to be highly subcritical. If self-gravitating (molecular) clouds form with the subcritical M/Φ of H I clouds, the molecular clouds will be subcritical. However, the observations of molecular cloud cores suggest that they are approximately critical, with no direct evidence for subcritical molecular clouds or cloud envelopes. Hence, the observations remain inconclusive in deciding between the two extreme-case models of what drives star formation. What is needed to further advance our understanding of the role of magnetic fields in the star formation process are additional high sensitivity surveys of magnetic field strengths and other cloud properties in order to further refine the assessment of the importance of magnetic fields in molecular cores and envelopes.

Keywords: ISM: molecular clouds, magnetic fields, polarization, stars: formation

1. Introduction

Understanding star formation is one of the outstanding challenges of modern astrophysics. However, in spite of significant progress in recent years, there remain unanswered fundamental questions about the basic physics of star formation. In particular, what drives the star formation process? The prevailing view for the past 20 years has been that self-gravitating dense clouds are supported against collapse by magnetic fields. However, only ions are frozen into the magnetic field, while the neutral material (by far the majority of the mass) can contract gravitationally unaffected directly by the magnetic field. Because neutrals will collide with ions in this process, there will be support against gravity for the neutrals as well as the

Astrophysics and Space Science **292:** 225–237, 2004.

ions. But there will be a drift of neutrals into the core without a significant increase in the magnetic flux; this is ambipolar diffusion. Eventually the core mass will become sufficiently large that the magnetic field can no longer support the core, and dynamical collapse and star formation can proceed. The other extreme from the magnetically dominated star formation scenario is that molecular clouds are intermittent phenomena in an interstellar medium dominated by turbulence, and the problem of cloud support for long-time periods is irrelevant. Clouds form and disperse by the operation of compressible turbulence, with clumps sometimes achieving sufficient mass to become self-gravitating. Even if the turbulent cascade has resulted in turbulence support, turbulence then dissipates rapidly, and the cores collapse to form stars. Hence, there are two competing models for driving the star formation process. The issue of what drives star formation is far from settled, on either observational or theoretical grounds.

In this paper, we discuss how observations of magnetic fields in molecular clouds can distinguish between these models. In Section 2 we describe predictions of the two models that may be tested via observations of magnetic fields. A brief review of techniques for observing magnetic fields in molecular clouds is given in Section 3. Section 4 presents an exemplar of the observational results, the L1544 prestellar core. In Section 5, we apply the tests and discuss the results. Finally, in Section 6 we summarize the conclusions and suggest what new observations are necessary to answer more definitively the question—what drives star formation?

2. Model Predictions and Observational Tests

The turbulence and ambipolar diffusion models for driving star formation have different predictions for magnetic field morphology and magnetic field strength, which form the basis for tests of the two models using observations of magnetic fields. Of course, it is clear that there are both magnetic fields and turbulence in real clouds. To sharpen the distinctions between the two models, here we will consider only turbulence models in which magnetic fields are negligibly weak and magnetic support/ambipolar diffusion models without turbulence. There are three tests of the two extreme-case models based on observations of magnetic fields that we will discuss here.

2.1. Test 1: Magnetic Field Morphology

Ambipolar diffusion: Because the magnetic field is relatively strong, magnetic field lines should be smooth, without being dominated by irregular structure. Clouds will be thin disks or oblate spheroids, because thermal pressure provides the only support along field lines. Therefore, the field lines should be parallel to the minor axes of clouds. Finally, an original morphology with parallel magnetic field lines will be transformed into an hourglass morphology because it is the tension of the bent field lines that provides support. This hourglass morphology will become still more dominant as ambipolar diffusion allows contraction of the disk perpendicular to the magnetic field, drawing in field lines.

Turbulence: Because the magnetic energy is much less than the turbulent energy, field lines will not be smooth but chaotic, with random small-scale structure. Simulations (Balsara et al., 2001; Padoan et al., 2001), also show that matter may flow onto cores along filaments with the field being stretched along the filaments, which would predict that the field will tend to be along the major axis of elongated low-density structures connected with higher-density cores. The fields inside cores may be more regular, especially as turbulence decays, but with no preferred orientation with respect to density structures.

2.2. Test 2: B versus ρ and σ

The relationship between magnetic field strength and the volume density of gas is usually parameterized as $B \propto \rho^{\kappa}$. However, a dependence on the velocity dispersion σ may also be significant if turbulence plays a role in cloud support.

Ambipolar diffusion: Ambipolar diffusion driven star formation has been widely investigated, e.g., Mouschovias (1976), Nakano (1984), Shu et al. (1987), Mouschovias and Ciolek (1999), Ciolek and Basu (2000). Clouds start subcritical, with little change in B as ambipolar diffusion gradually increases the mass to magnetic flux ratio in the core. Therefore κ for the core is small, ~0.1–0.2, in the early stages of evolution. Once the core becomes supercritical and rapid collapse ensues, the magnetic field is dragged inward with the collapsing gas, and κ gradually approaches 0.5; a typical value during the core collapse phase is $\kappa \approx 0.4$. Little change in B takes place in the envelope during the entire process, however, there is essentially no predicted correlation of B with ρ in the envelope.

Turbulence: The correlation of field strength with other parameters has been discussed by several authors, e.g., Padoan and Nordlund (1999), Ostriker et al. (2001), Ballesteros-Paredes and MacLow (2002), Passot and Vazquez-Semadeni (2003). In the low-density case before gravity becomes important no relation is expected between the magnetic field strength and other parameters such as density, because clouds may be continually changing shape and exchanging mass and energy with their surroundings. Once gravity becomes important, a rough equipartition of energies is achieved. Although clouds will generally not be in equilibrium, we can extract the form of the correlations if we assume virial equilibrium between gravity and turbulence ($3GM^2/5R = 3M\sigma^2/2$). Then $\rho R^2 \propto \sigma^2$. Flux freezing ($M \propto \Phi$) gives $\rho R \propto B$, so $B \propto \sigma\rho^{1/2}$ is predicted. Hence, although both models predict $\kappa \approx 0.5$, if turbulence is significant in cores an additional dependence of B on σ is predicted.

2.3. Test 3: mass-to-flux ratio

The ratio of the mass in a magnetic flux tube to the magnitude of the magnetic flux is a crucial parameter for the magnetic support/ambipolar diffusion model. The critical value for the mass that can be supported by magnetic flux Φ is $M_{B\text{crit}} = \Phi/2\pi\sqrt{G}$ (Nakano and Nakamura, 1978); the precise value of the numerical coefficient is slightly model dependant, e.g., Mouschovias and Spitzer (1976). It is convenient to state M/Φ in units of the critical value, and to define $\lambda \equiv (M/\Phi)_{\text{actual}}/(M/\Phi)_{\text{crit}}$. Inferring λ from observations is possible if the column density N and the magnetic field strength B are measured:

$$\lambda = \frac{(M/\Phi)_{\text{observed}}}{(M/\Phi)_{\text{crit}}} = \frac{mNA/BA}{1/2\pi\sqrt{G}} = 7.6 \times 10^{-21}\frac{N(H_2)}{B} \tag{1}$$

where $m = 2.8\,m_H$ allowing for He, $N(H_2)$ is in cm^{-2} and B is in μG.

Ambipolar diffusion: Clouds are initially subcritical, $\lambda < 1$. Ambipolar diffusion is fastest in shielded, high-density cores, so cores become supercritical, and rapid collapse ensues. The envelope continues to be supported by the magnetic field. Hence, the prediction is that λ must be <1 in cloud envelopes (models typically have $\lambda \sim$0.2–0.5), whereas in collapsing cores λ becomes slightly >1. Hence, this model tightly constrains λ.

Turbulence: The turbulent model imposes no direct constraints on λ, although strong magnetic fields would resist the formation of gravitationally bound clouds by compressible turbulence. Also, if magnetic support is to be insufficient to prevent collapse of self-gravitating clumps that are formed by compressible turbulence, the field must be supercritical, $\lambda > 1$. λ may take any value >1, although of course for the turbulence model with very weak magnetic fields that we are considering, clouds will be highly supercritical.

3. Observational Techniques

3.1. Zeeman Splitting

The Zeeman effect is the standard technique for directly measuring magnetic field strengths in molecular clouds. Stokes V spectra reveal the direction and magnitude of the line-of-sight component B_{los}; with current sensitivities, the plane-of-sky component of **B** cannot be measured, e.g. Crutcher et al. (1993).

It is possible to correct statistically for the fact that only one component of **B** is measured, i.e., $B_{\text{los}} = |\mathbf{B}|\cos\theta$, where θ is the angle between the line-of-sight and the magnetic field. For a large number of clouds whose magnetic fields are randomly oriented with respect to the observed line-of-sight,

$$\overline{B}_{\text{los}} = \int_0^{\pi/2} |\mathbf{B}|\cos\theta\sin\theta d\theta = \frac{1}{2}|\mathbf{B}|. \tag{2}$$

If **B** is strong, clouds will have a disk morphology with **B** along the minor axis. To properly measure λ, one needs B and N along a flux tube, i.e., parallel to the minor axis. As noted by Crutcher (1999), the path length through a disk will be too long by $1/\cos\theta$ and N will be overestimated, whereas $|\mathbf{B}|$ will be underestimated by $\cos\theta$. Statistically,

$$\overline{M/\Phi} = \int_0^{\pi/2} (M/\Phi)_{\rm obs} \cos^2\theta \sin\theta d\theta = \frac{1}{3}(M/\Phi)_{\rm obs}. \tag{3}$$

The 1/3 correction factor is only appropriate for the strongly magnetized case, where clouds have a disk-like morphology. We will discuss λ_C obtained by applying this correction factor of 1/3 to the directly observed λ, because a disk morphology is the prediction of the ambipolar diffusion model; the 1/3 correction must be applied to test this model.

A bias can occur if the angular resolution is poor and $B_{\rm los}$ is not mapped over the cloud. For example, if rotational twisting of a poloidal field takes place, the field may retain a significant toroidal component even after magnetic braking has reduced the rotation of a cloud to very small values. Since a toroidal field will have $B_{\rm los}$ with opposite signs on the two sides of the cloud, partial cancellation of the Zeeman effect will take place unless the field is spatially resolved. An example of reversal of $B_{\rm los}$ over a molecular cloud is W3 (Roberts et al., 1993). Also, if there are multiple self-gravitating clumps within a single flux tube, due to fragmentation within a cloud, each clump may have a subcritical λ even though λ may appear to be supercritical if the masses of the clumps are added together by the observations. Both effects bias the observations toward supercritical λs.

3.2. Dust Polarization

Linear polarization at optical, infrared, or millimeter wavelengths arises from elongated grains with their short axes aligned, generally with **B** (Lazarian, 2003). When emission from dust grains is observed, the position angle of maximum emission will be perpendicular to the magnetic field direction projected onto the sky, $B_{\rm pos}$; however, the degree of polarization is not sensitive to the strength of the magnetic field.

In the early days of interstellar polarization studies Chandrasekhar and Fermi (1953) suggested that analysis of the small-scale randomness of magnetic field lines could yield estimates of the field strengths. The method depends on the fact that turbulent motions will lead to irregular magnetic fields (because under interstellar conditions fields will be frozen into the matter). There will therefore be a perturbed or MHD-wave component to the field that should show up as an irregular scatter in polarization positions angles relative to those that would be produced by a regular magnetic field. The stronger the regular field, the more it resists being irregularized by turbulence. They showed that the magnitude of the irregularity of field lines could yield the regular field strength in the plane of the sky:

$$B_{\rm pos} = Q\sqrt{4\pi\rho}\,\frac{\delta V}{\delta\phi} \approx 9.3\sqrt{n(H_2)}\,\frac{\Delta V}{\delta\phi}\ \mu G, \tag{4}$$

where $\rho = mn(H_2)$ is the gas density, δV is the velocity dispersion, $\delta\phi$ is the dispersion in polarization position angles in degrees, Q is a factor of order unity, $n(H_2)$ is the molecular hydrogen density in molecules cm^{-3}, and $\Delta V = \sqrt{8 ln 2}\, \delta V$ is the FWHM line width in $km\, s^{-1}$. Here we have used $Q = 0.5$, a calibration based on study of simulations of interstellar clouds by Ostriker et al. (2001), but see also Heitsch et al. (2001) and Padoan et al. (2001). These simulations found that this method could yield reliable results in molecular clouds if $\delta\phi < 25^\circ$; Heitsch et al. (2001) discussed an alternative form of the method that appears to allow the method to be applied even when $\delta\phi$ is large. One should note that while fluctuations in the field along the line of sight will be smoothed out by the polarization measurements, the calibration by the simulations referred to above include this in the Q factor.

The Chandrasekhar–Fermi method of estimating B is a statistical one that may be in error by $\sim$2 for an individual cloud. Also, similarly to Zeeman observations, in general N (and M) will be overestimated by $1/\cos\theta$ if the sight line has an angle θ to the minor axis, whereas $|\mathbf{B}|$ (and Φ) is underestimated by $\sin\theta$ (rather than by $\cos\theta$ in Zeeman observations). Again statistically,

$$\overline{M/\Phi} = \int_0^{\pi/2} \frac{M\cos\theta}{\Phi/\sin\theta} \sin\theta d\theta = \frac{1}{3}(M/\Phi)_{\text{obs}}. \tag{5}$$

3.3. Spectral-line Linear Polarization

Linear polarization may also arise in radio-frequency spectral lines formed in the interstellar medium, even when Zeeman splitting is negligible. This Goldreich–Kylafis effect (e.g., Kylafis, 1983) may be used to probe magnetic field morphologies in molecular clouds. The direction of the polarization can be either parallel or perpendicular to the magnetic field, depending on the relationship between the line-of-sight, the direction of the magnetic field, and the direction of a velocity gradient that produces the anisotropic line optical depth that is required to produce linear polarization. Although the theory makes specific predictions for whether the field is parallel or perpendicular to the line polarization, in general the observations do not provide all of the necessary information. This ambiguity is unfortunate, but if structure in a cloud causes a flip by 90° in the polarization direction, it would easily be recognized and not confused with random magnetic fields. It, therefore, is a valuable tool in the measurement of magnetic field direction and in the degree of randomness of the field. As for dust polarization, the Chandrasekhar–Fermi method may be applied to maps of spectral-line linear polarization to estimate field strengths.

4. L1544—An Exemplar of Observational Results

L1544 is a dark cloud in Taurus with a prestellar core that has been studied by single-dish (Tafalla et al., 1998) and interferometer spectroscopy (Williams et al., 1999).

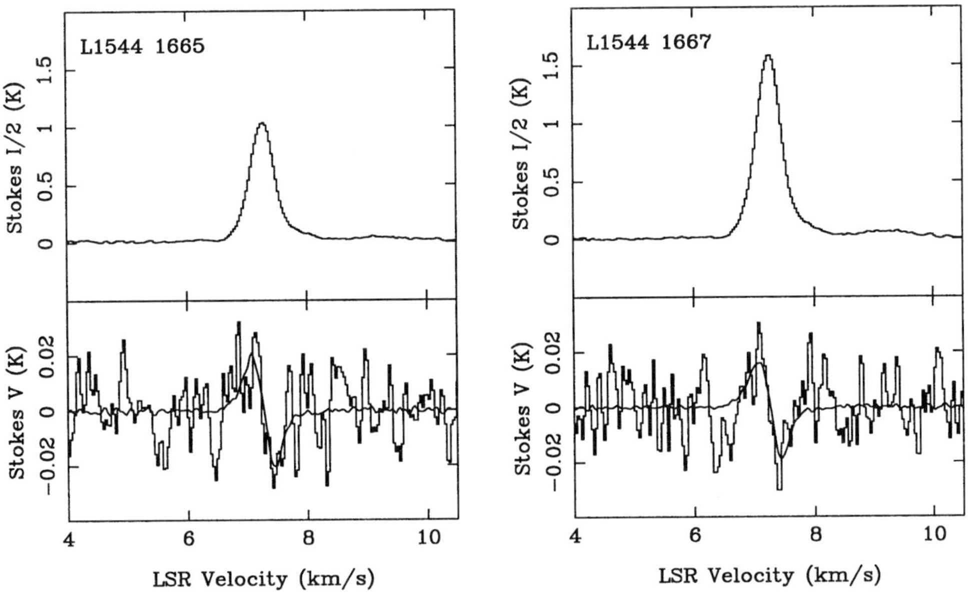

Figure 1. L1544 1665 and 1667 MHz Arecibo OH spectra. Observed data are histogram plots. The heavy line superposed on Stokes V is the fit of dI/dv to the observed V spectra. Note that the Zeeman splitting factor of the 1665 MHz line is 5/3 that of the 1667 MHz line. Obtaining the same B_{los} from the two lines lends confidence to the validity of the result. The result is $B_{los} = +10.8 \pm 1.7\ \mu$G.

These studies have suggested that L1544 is contracting. Information on the magnetic field includes OH Zeeman observations with the Arecibo telescope (Crutcher and Troland, 2000) and dust polarization mapping with the JCMT SCUBA polarimeter (Ward-Thompson et al., 2000; Crutcher et al., 2004). This cloud is therefore a good exemplar of observational results.

Figure 1 shows the Arecibo OH Zeeman spectra, which imply $B_{los} = 10.8 \pm 1.7\ \mu G$, $\Delta V_{NT} \approx 0.5$ km s^{-1}, and $N(OH) \approx 3.9 \times 10^{14}$ cm^{-2}. Since OH/H $\approx 4 \times 10^{-8}$ in dark clouds, $N(H_2) \approx 4.8 \times 10^{21}$ cm^{-2}. With a mean radius of the OH emitting region $\bar{r} \approx 0.08$ pc, $n(H_2) \approx 1 \times 10^4$ cm^{-3}. The directly observed $\lambda \approx 3.4$; corrected for geometry bias, $\lambda_C \approx 1.1$, or about critical.

Figure 2 shows the SCUBA dust intensity and polarization map. The field is fairly uniform in direction, with a position angle dispersion $\delta\phi \approx 13^\circ$, but the direction of the field in the plane of the sky is at an angle of $29^\circ \pm 7^\circ$ to the minor axis. This difference between the magnetic field direction and the minor axis of the core is in conflict with symmetric models, such as that of Ciolek and Basu (2000), that rely only on thermal and static magnetic pressure to balance gravity, because the minor axis projected onto the sky should lie along B_{pos}. However, projection effects can produce the observed position angle difference if the core has a more complicated shape, such as triaxial (Basu, 2000). The initial conditions of cloud formation and turbulence may produce the more complicated

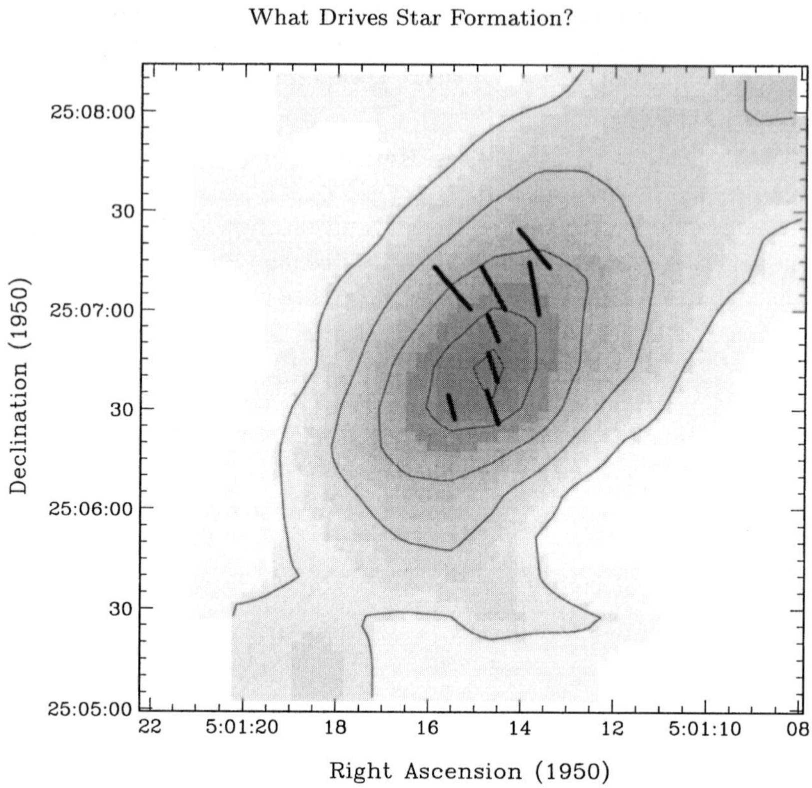

Figure 2. Dust continuum emission at λ850 μm from the L1544 prestellar core. The Stokes I map is shown as a gray scale with contours overlaid, with contours at 50, 100, 150, and 200 mJy beam^{-1}. The direction of **B** in the plane of the sky is shown by lines at positions where a measurement of the polarized flux above the $2 - \sigma$ level was achieved. The length of the lines is proportional to the percentage polarization.

shapes (Gammie et al., 2003). Crutcher et al. (2003) inferred physical parameters of the core from the SCUBA data: $\overline{r} \approx 0.021$ pc, $N(H_2) \approx 4.2 \times 10^{22}$ cm^{-2}, and $n(H_2) \approx 4.9 \times 10^5$ cm^{-3}. Using the N_2H^+ line data of Caselli et al. (2002) that gave $\Delta V_{NT} \approx 0.28$ km s^{-1}, Crutcher et al. (2004) applied the Chandrasekhar–Fermi technique and found $B_{\rm pos} \approx 140$ μG. Then $\lambda \approx 2.3$; corrected for geometry bias, $\lambda_C \approx 0.8$, or about critical.

The λs found from the OH Zeeman and the dust polarization techniques are essentially in agreement, but it is clear that very different regions are sampled by the two methods. The region sampled by OH has 4 times the radius, 0.1 the column density, and 0.02 the volume density of the region sampled by the dust emission. Therefore, the data probe separately the envelope and the core regions of the cloud. Taking the numbers at face value, the mass-to-flux ratio is approximately critical and roughly unchanged from envelope to core, and the field strength has scaled as $\rho^{0.66}$ or $\rho^{0.51}\sigma$. However, the Zeeman effect measures $B_{\rm los}$ and dust emission measures B_{pos}, and we do not know the inclination of **B** to the line of sight. The statistically corrected results may be incorrect for a single cloud. A large sample must be analyzed in order for the statistical corrections to yield trustworthy results.

5. Test Results

Rather than attempting an exhaustive citation and description of the entire database of observational results on magnetic fields in molecular clouds, we will cite specific examples that sharpen the focus on the tests of the two models.

5.1. Test 1—Field Morphology

Maps of dust and spectral-line linear polarization and of the Zeeman effect generally show a quite smooth field morphology; examples include the L183 prestellar core (Crutcher et al., 2003), NGC1333 IRAS4 (Girart et al., 1999), OMC3 (Matthews and Wilson, 2000), and OMC1 (Schleuning, 1998). An hourglass morphology is sometimes seen (Schleuning, 1998; Girart et al., 1999). A regular field dominating a random field and an hourglass morphology toward cores are predictions of the magnetic support model. Even though fairly small, the dispersion in polarization position angles is often greater than observational errors (Crutcher et al., 2003), implying that turbulence is producing an irregular component to **B**.

5.2. Test 2—B versus ρ and σ

At low densities $n \sim 0.1$–$100\ \mathrm{cm}^{-3}$, it has been clear for some time that there is no correlation of B with ρ (Troland and Heiles, 1986). At higher densities the available data cover $\rho > 300\ \mathrm{cm}^{-3}$. Unfortunately, for the OH absorption-line results of Bourke et al. (2001) and for the unfinished Arecibo dark-cloud OH emission results of Troland and Crutcher (2004), values of the volume density ρ are not yet available. The available data set is therefore that of Crutcher (1999). He showed that $B \propto \rho^{\kappa}$, $\kappa = 0.47 \pm 0.08$. Basu (2000) showed that a better correlation was obtained by fitting B vs. $\sigma\sqrt{\rho}$; the correlation coefficient went from 0.84 to 0.95.

Both the turbulence and magnetic support models predict little or no correlation between B and ρ at low densities, as observed. At higher densities, $\kappa \approx 0.5$ is the signature of flux freezing together with gravitational contraction, and does not discriminate between the two models. However, $B \propto \sigma$ is a prediction of a significant turbulent component to the support of clouds.

5.3. Test 3—M/Φ

Because molecular clouds must form from more diffuse gas, study of λ in diffuse gas is relevant to star formation. Whether clouds formed from the diffuse interstellar medium will be subcritical or supercritical depends on the initial volume being considered. Hartmann et al. (2001) pointed out that for the typical diffuse ISM values $n(H) \approx 1\,\text{cm}^{-3}$ and $B \approx 5\,\mu\text{G}$, if the formation of molecular clouds involves mass accumulation along field lines over distances greater than ~430 pc, Eq. (1) implies that the resulting molecular cloud will be supercritical. Accumulation over a flux tube with $R \approx 100$ pc and $L \approx 430$ pc would result in $\lambda = 1$ and $M \approx 3 \times 10^5$ $M_\odot$—a typical GMC mass.

An intermediate step would be the formation of H I clouds, with $n(H\ I) \sim 50$ cm^{-3}. Heiles (2004) reported recent results from H I absorption-line Zeeman observations toward background extragalactic continuum sources. His plot of B_{los} vs. $N(H\ I)$ showed that for all of the detections, the field was highly subcritical. Moreover, almost all of the non-detections were also consistent with $\lambda < 1$.

In the ambipolar diffusion model the envelopes of dark clouds are the regions where λ remains essentially unchanged while ambipolar diffusion drives M/Φ supercritical in the core. Hence, dark clouds envelopes provide a crucial test of magnetic support models—M/Φ *must* be subcritical in these regions. Dark-cloud envelopes were observed by Crutcher et al. (1993); the result was $\bar{\lambda} \approx 1 - 2$. If correct, this result would be fatal for magnetic support models. However, the observations had poor angular resolution ($18'$) with no mapping, and the caveats discussed above (Section 3.1) may have biased these results to larger λ.

Figure 3 shows all of the λ_C available from Zeeman detections and sensitive limits for molecular clouds (Crutcher, 1999; Bourke et al., 2001; Troland and Crutcher, 2004), plus the Chandrasekhar–Fermi results for three prestellar cores (Crutcher et al., 2003). Although a statistical correction of 1/3 for geometrical bias (Sections 3.1, 3.2) has been applied to each point, so that statistically this plot should be valid, for any individual point the true λ could be higher or lower than the plotted λ_C. A large part of the scatter in Figure 3 may therefore be due to geometrical projection effects. The observations are distributed roughly equally above and below the $\lambda_C = 1$ line that divides subcritical and supercritical M/Φ ratios for disk geometries. Therefore, the data suggest that $\bar{\lambda} \approx 1$; that is, the typical mass to magnetic flux ratio is approximately critical. There is a slight indication that for large column densities, $\bar{\lambda}$ may be supercritical, and for small column densities, subcritical. The H I data of Heiles (2004) (with subcritical λs) lie outside and to the left of this figure, at smaller column densities.

On the other hand, if clouds are not disks then the geometrical correction factor is only 1/2. This correction would place the dividing line between subcritical and supercritical clouds at $\log \lambda_C = -0.18$ in Figure 3, and the majority of clouds would be supercritical.

5.4. Test Results Summary

Test 1. Magnetic field morphologies do not appear to be highly random, consistent with strong magnetic field strengths. Morphologies are sometimes consistent with hourglass shapes predicted by strong magnetic fields and gravitational contraction.

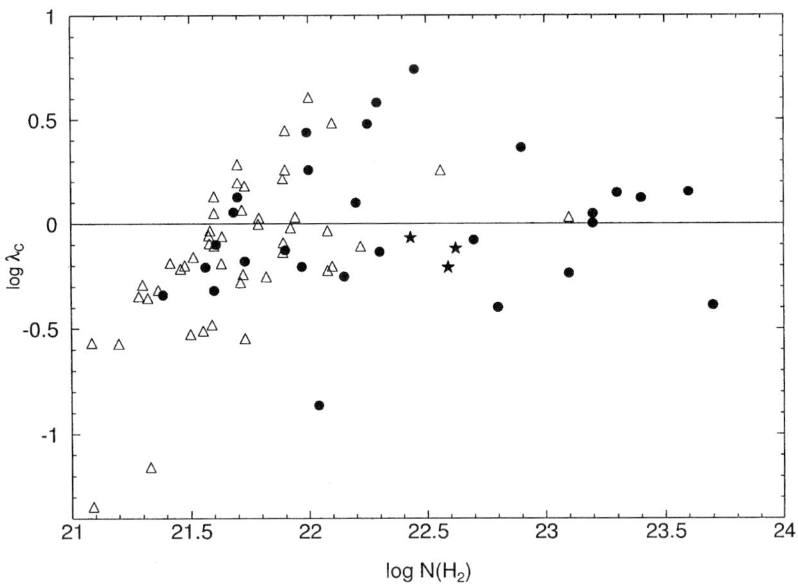

Figure 3. Mass to magnetic flux for molecular cloud data. λ_C are the observed mass to magnetic flux ratios, divided by 3 to correct for projection bias, in units of the critical values. Dots are for Zeeman data with $B_{\rm los} > 3\sigma(B_{\rm los})$, stars are for Chandrasekhar–Fermi estimates of $B_{\rm pos}$, and triangles are lower limits plotted at $B_{\rm los} = 3\sigma(B_{\rm los})$. $\lambda_C > 1$ is supercritical, $\lambda_C < 1$ is subcritical.

However, fields projected onto the sky do not align with the projected minor axes of clouds, suggesting a triaxial cloud morphology that could be produced by cloud formation via compressible turbulence.

Test 2. The observations are that $B \propto \rho^{\kappa}\sigma$, with κ close to zero at small densities and approaching 1/2 at larger densities. The scaling of B with ρ is predicted by contraction and flux freezing, and is not diagnostic for the two models. However, B scaling linearly with the non-thermal velocities is a prediction of turbulent support being significant.

Test 3. Magnetic support requires that the mass to magnetic flux ratio M/Φ must be subcritical in the diffuse gas from which molecular clouds form and in the envelopes of clouds; turbulence models require that M/Φ be supercritical, at least in self-gravitating cores, but make no further detailed prediction. Accumulation of sufficient mass to form GMCs must take place over distances such that $\lambda \approx 1$ in GMCs. However, the data show that M/Φ is subcritical in H I clouds. In molecular clouds $\lambda \approx 1$, with slight evidence for an increase in λ with density, in agreement with ambipolar diffusion.

6. Conclusions

One conclusion is that the data remain inconclusive in deciding between the two extreme models of star formation. However, a strong conclusion does come from the observations: *both* turbulence and strong magnetic fields are important in the physics of molecular clouds. There does not seem to be a single driver of star formation.

To form giant molecular clouds, mass must be accumulated over large volumes of the diffuse interstellar medium. As a result, it appears that GMCs will end up being approximately magnetically critical. However, H I clouds are observed to be subcritical. If these H I clouds accumulate to form molecular clouds (and flux freezing holds), molecular clouds will be formed with subcritical mass-to-flux ratios. Yet the observations of molecular cloud cores are consistent with $\lambda \approx 1$. Although this is predicted to occur in cores as a result of ambipolar diffusion when the original mass-to-flux ratio was subcritical, the observations have not found direct evidence of subcritical clouds and especially of subcritical envelopes. Whether λ is slightly greater than or slightly less than 1 is crucial for whether magnetic fields provide support that is removed by the action of ambipolar diffusion. The present data are not sufficiently accurate to discriminate between these cases because λ in molecular clouds is observed to be (frustratingly) close to 1.

Tests of the role of magnetic fields are of necessity statistical in nature and must involve data covering a wide range of physical parameters. What is needed are additional high sensitivity surveys of magnetic field strengths and other cloud properties in order to assess the relative importance of magnetic fields in molecular cores and envelopes. These surveys should include extended polarization maps that allow a determination of how the field in the cores connects with that in its surroundings. But the single most crucial observation that is needed is the measurement of magnetic field strengths in a statistically significant sample of molecular cloud envelopes. High resolution mapping of envelopes of cores to resolve possible toroidal fields should remove the bias that may exist if a toroidal component of **B** is significant. If envelopes are subcritical, with a large sample at least some of the fields must be nearly along the line-of-sight, and Zeeman observations will detect unambiguous subcritical mass-to-flux ratios. Clear evidence for an increase in the mass-to-flux ratio from envelope to core would then verify the ambipolar diffusion model. Alternatively, if future observations show that this is not found, turbulence-driven star formation (although with dynamically important magnetic fields) would be favored. In any case, present data make it clear that both strong magnetic fields and turbulence play significant roles in star formation.

Acknowledgements

This work was partially supported by NSF grant AST 02-05810.

References

Ballesteros-Paredes, J. and MacLow, M.-M.: 2002, *ApJ* **570**, 734.
Balsara, D., Ward-Thompson, D. and Crutcher, R.M.: 2001, *MNRAS* **327**, 715.
Basu, S.: 2000, *ApJ* **540**, L103.

Bourke, T.L., Myers, P.C., Robinson, G. and Hyland, A.R.: 2001, *ApJ* **554**, 916.
Caselli, P., Benson, P.J., Myers, P.C. and Tafalla, M.: 2002, *ApJ* **572**, 238.
Chandrasekhar, S. and Fermi, E.: 1953, *ApJ* **118**, 113.
Ciolek, G.E. and Basu, S.: 2000, *ApJ* **529**, 925.
Crutcher, R.M., Troland, T.H., Goodman, A.A., Heiles, C., Kazès, I. and Myers, P.C.: 1993, *ApJ* **407**, 175.
Crutcher, R.M.: 1999, *ApJ* **520**, 706.
Crutcher, R.M. and Troland, T.H.: 2000, *ApJ* **537**, L139.
Crutcher, R.M., Nutter, D., Ward-Thompson, D. and Kirk, J.M.: 2004, *ApJ* **600**, 279.
Gammie, C.F., Lin, Y.-T., Stone, J.M. and Ostriker, E.C.: 2003, *ApJ* **592**, 203.
Girart, J.M., Crutcher, R.M. and Rao, R.: 1999, *ApJ* **525**, L169.
Hartmann, L., Ballesteros-Paredes, J. and Bergin, E.A.: 2001, *ApJ* **562**, 852.
Heiles, C.: 2004, *Ap&SS* **292**, 77.
Heitsch, F., Zweibel, E.G., MacLow, M.-M., Li, P.S. and Norman, M.L.: 2001, *ApJ* **561**, 800.
Kylafis, N.D.: 1983, *ApJ* **275**, 135.
Lazarian, A.: 2003, *JQSRT* **79**, 881.
Matthews, B.C. and Wilson, C.D.: 2002, *ApJ* **531**, 868.
Mouschovias, T.Ch.: 1976, *ApJ* **207**, 141.
Mouschovias, T.Ch. and Spitzer, L.: 1976, *ApJ* **210**, 326.
Mouschovias, T.Ch. and Ciolek, G.E.: 1999, in: C.J. Lada and N.D. Kylafis (eds.), *The Origin of Stars and Planetary Systems* Kluwer Academic Publishers, Dordrecht, p. 305.
Nakano, T. and Nakamura, T.: 1978, *PASJ* **30**, 681.
Nakano, T.: 1984, *Fund. Cosmic Phys.* **9**, 139.
Ostriker, E.C., Stone, J.M. and Gammie, C.F.: 2001, *ApJ* **546**, 980.
Padoan, P. and Nordlund, Å.: 1999, *ApJ* **526**, 279.
Padoan, P., Goodman, A., Draine, B.T., Juvela, M., Nordlund, Å. and Rögnvaldsson, Ö.E.: 2001, *ApJ* **559**, 1005.
Passot, T. and Vázquez-Semadeni, E.: 2003, *A&A* **398**, 845.
Roberts, D., Crutcher, R.M., Troland, T.H. and Goss, W.M.: 1993, *ApJ* **412**, 675.
Schleuning, D.A.: 1998, *ApJ* **493**, 811.
Shu, F.H., Adams, F.C. and Lizano, S.: 1987, *ARA&A* **25**, 23.
Tafalla, M., Mardones, D., Myers, P.C., Caselli, P., Bachiller, R. and Benson, P.J.: 1998, *ApJ* **504**, 900.
Troland, T.H. and Heiles, C.: 1986, *ApJ* **301**, 339.
Troland, T.H. and Crutcher, R.M.: 2004, submitted.
Ward-Thompson, D., Kirk, J.M., Crutcher, R.M., Greaves, J.S., Holland, W.S. and André, P.: 2000, *ApJ* **537**, L135.
Williams, J.P., Myers, P.C., Wilner, D.J. and Di Francesco, J.: 1999, *ApJ* **513**, L61.

EVOLUTION OF MAGNETIC FIELDS IN BOK GLOBULES?

SEBASTIAN WOLF[1,2], RALF LAUNHARDT[2] and THOMAS HENNING[2]
[1]*California Institute of Technology, 1201 East California Blvd, Pasadena, CA, USA;*
E-mail: swolf@astro.caltech.edu
[2]*Max-Planck-Institut für Astronomie, Königstuhl 17, Heidelberg, Germany*

Abstract. We study the influence and structure of the magnetic field in the early phases of low-mass star formation using polarization maps of Bok globules at a wavelength of 850 μm, obtained with the Submillimeter Common-User Bolometer Array (SCUBA) at the James Clerk Maxwell Telescope (JCMT). We discuss observations of the following sources: CB 26—a small globule with a nearly dispersed dense core and a young and large circumstellar disk, CB 54—a large globule with a massive dense core and a deeply embedded young stellar cluster, and B 335, CB 230, and CB 244—three nearby, relatively isolated small globules with low-mass protostellar cores.

We find strongly aligned polarization vectors in the case of CB 26, B 335, and CB 230, whereas the vector orientations in the case of CB 54 and CB 244 are more or less randomly distributed. The degree of polarization, amounting to several percent, was found to decrease toward the center in each source. Assuming dichroic emission by aligned non-spherical grains as the polarization mechanism, where the magnetic field plays a role in the alignment process, we derive magnetic field strengths and structures from the observed polarization patterns.

We compare the magnetic field topology with the morphology and outflow directions of the globules. In the Class 0 sources B 335, CB 230, and CB 244, the magnetic field is oriented almost perpendicular to the ouflow direction. In contrast, the inclination between outflow axis and magnetic field direction is much more moderate (36°) in the more evolved Class I source CB26.

Keywords: magnetic fields, polarization, individual objects: CB 26, CB 54, CB 230, CB 244, B 335, magnetic fields, submillimeter

1. Introduction

Bok globules are small, relatively isolated, simply-structured molccular clouds and therefore excellent objects to study the earliest processes of star formation, in particular the interplay between protostellar collapse, fragmentation, and magnetic fields. Signs of star formation in many Bok globules include bipolar molecular outflows (e.g., Yun and Clemens, 1994a), infrared colors, and submillimeter properties that are consistent with Class 0 protostars or embedded Class I sources (Yun and Clemens, 1994b; Launhardt and Henning, 1997).

To study one of the key parameters of the star formation process—namely the magnetic field—submillimeter polarization measurements represent a powerful technique. Assuming emission by aligned nonspherical grains as the dominating polarization mechanism, where the magnetic field plays a role in the alignment

Astrophysics and Space Science **292:** 239–246, 2004.

process, magnetic field strengths and structures can be derived from the submillimeter polarization pattern (e.g. Wolf et al., 2003; Henning et al., 2001).

2. Observations and Data Reduction

We observed the Bok globules CB 26, CB 54, B 335, CB230, and CB 244 for which brief descriptions are given in the following: CB 26 is a small, slightly cometary-shaped Bok globule at $D \sim 140$ pc. The globule contains a nearly dispersed dense core and a young and large, perhaps early protoplanetary, disk (Launhardt and Sargent, 2001). CB 54 is a large Bok globule associated with the molecular cloud BBW 4 at $D \sim 1.1$ kpc (Brand and Blitz, 1993) and the reflection nebula LBN 1042. The globule contains a massive dense core of $M_{\mathrm{H}} \sim 100\,\mathrm{M}_{\mathrm{sun}}$, which is associated with a bipolar molecular outflow (Launhardt and Henning, 1997; Yun and Clemens, 1994a). At near-infrared wavelengths, a small, deeply embedded young stellar cluster becomes visible. B 335 is an isolated, nearly spherical Bok globule at a distance of ~250 pc (Tomita et al., 1979; Frerking et al., 1987). The deeply embedded Class 0 protostar drives a collimated bipolar outflow with a dynamical age of $\sim 3 \times 10^4$ yr (e.g. Chandler et al., 1990; Chandler and Sargent, 1993). CB230 is a small, bright-rimmed Bok globule at a distance of 400 ± 100 pc associated with the Cepheus Flare molecular cloud complex. The globule contains a protostellar double core with $10''$ separation (east–west). The western source with the more massive accretion disk and envelope drives a large-scale collimated molecular outflow of dynamical age $\sim 2 \times 10^4$ yr (Yun and Clemens, 1994a). The eastern source drives a weaker outflow that is probably not aligned with the large-scale outflow (Launhardt et al., in prep.). CB 244 is an isolated Bok globule located at a distance of 180 pc and probably associated with the Lindblad ring (Kun, 1998). It contains two dense cores separated by $\sim 90''$. The more prominent south-eastern core is associated with a bipolar molecular outflow with a dynamical age of $\sim 10^4$ yr (Yun and Clemens, 1994a; Launhardt et al., 1997).

The observations were performed at the 15-m James Clerk Maxwell Telescope (JCMT) between 1 and 6 March, 2000 and between 10 and 14 September, 2001. The effective beam size (HPBW) is $\sim 14.7''$ at 850 μm. Polarimetry was conducted with the Submillimeter Common-User Bolometer Array (SCUBA; Holland et al., 1999) and its polarimeter, SCU-POL, using the 350–850 μm achromatic half-waveplate. Flat-fielding, extinction correction, sky-noise and instrumental polarization removal have been performed with the data reduction package SURF (Jenness and Lightfoot, 1998). The polarization was computed using the POLPACK data reduction package (Berry and Gledhill, 1999).

We restrict the polarization analysis to the regions in which the total flux density per beam is above five times the rms in the maps (measured outside the central sources). We do not use polarization vectors derived at positions where the scatter

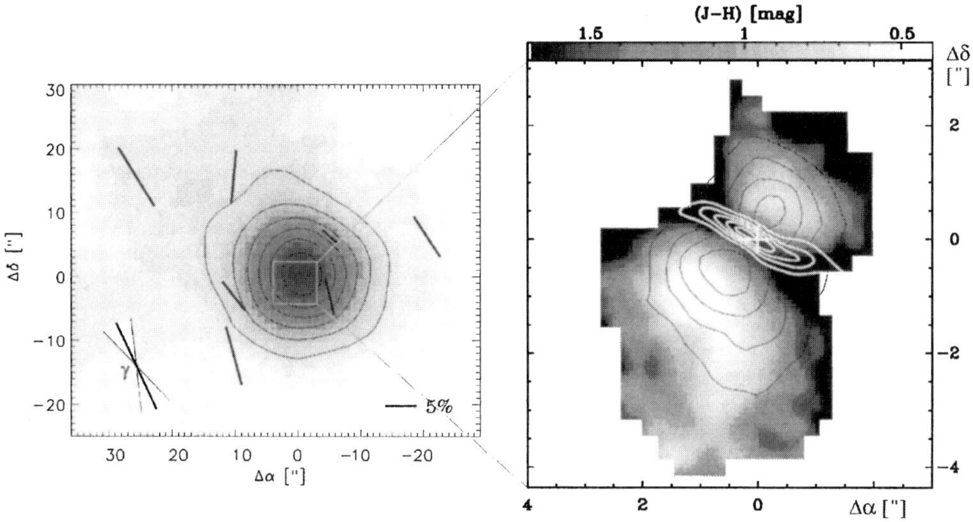

Figure 1. Left: SCUBA intensity map (850 μm) of the Bok globule CB 26 with overlayed iso-intensity contour lines (contour levels: 0.15%, 0.30%, ..., 0.90% of the maximum intensity) and polarization pattern. Furthermore, the mean direction of the polarization $\bar{\gamma}(\pm\sigma_{\bar{\gamma}})$ is shown. Right: Central part of the Bok globule CB 26 (from Launhardt and Sargent, 2001). J–H color map of the bipolar near-infrared reflection nebula (black contour lines: K band emission). The white contour lines show the proto-planetary disk discovered by Launhardt and Sargent (2001; contour levels: 4, 11,..., 32 mJy arcsec^{-2}; obtained with OVRO at 1.3 mm - beam width: $0.58'' \times 0.39''$). (*From* Wolf et al., 2003).

of the total flux density obtained through independent measurements was larger than 20% of the average total flux density at that point and exclude polarization vectors with $P_l/\sigma(P_l) < 3$, where $\sigma(P_l)$ is the standard deviation of the polarization degree. The resulting polarization maps are shown in Figures 1 and 2.

3. Polarization Versus Intensity

As found in previous polarization measurements toward other star-forming cores (see, e.g., Matthews and Wilson, 2002; Houde et al., 2002; Glenn et al., 1999), the degree of polarization decreases toward regions of increasing intensity (see Figure 3). Possible explanations for this behavior are: (a) increased densities in the cores, which cause an increase of the collisional disalignment rate of the grains toward the centers of the cores, (b) insufficient resolution of the field structure associated with the core collapse, and (c) spherical grain growth in the denser regions which would result in unpolarized re-emission by the dust (Weintraub et al., 2000). Because of the large density contrast between the unresolved central condensations and the envelopes, the increasingly poor grain alignment as density increases is expected to be the main effect.

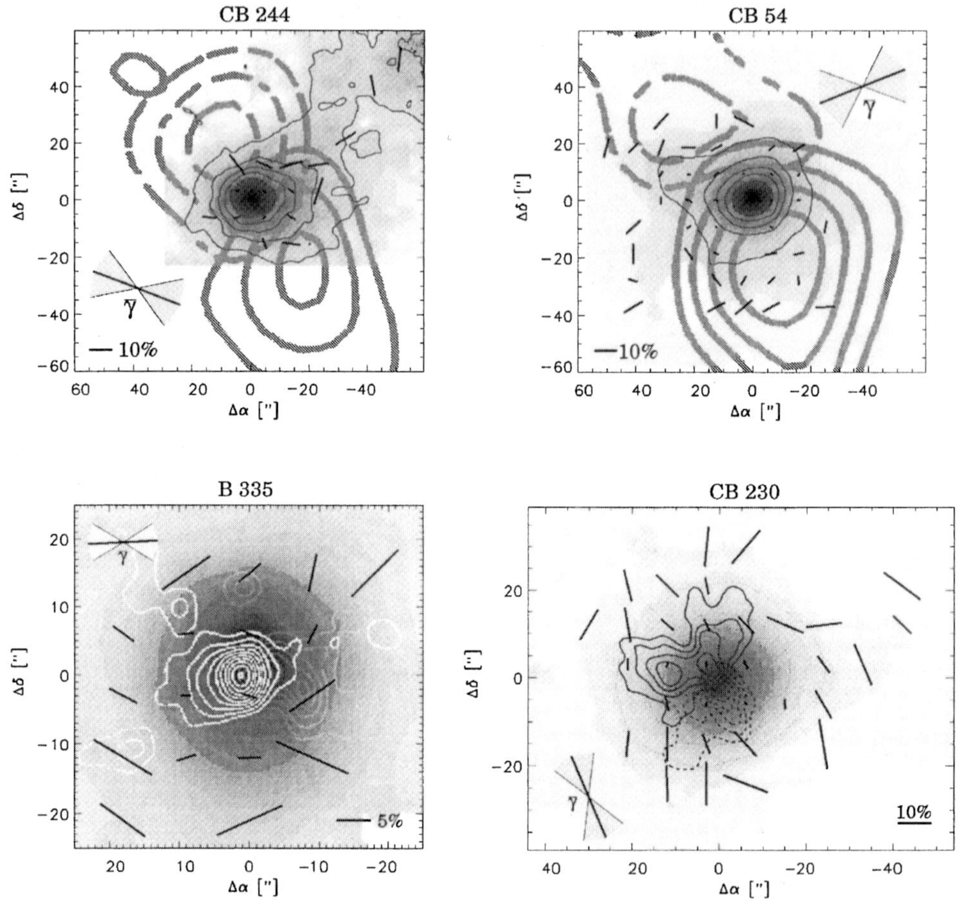

Figure 2. SCUBA intensity maps (850 μm) with overlayed iso-intensity contour lines (contour levels: 0.15%, 0.30%, ..., 0.90% of the maximum intensity) and polarization patterns of the Bok globules CB 244, CB 54, B 335, and CB 230. Furthermore, the solid/dashed contour lines represent the spatially well-separated blue-shifted/red-shifted outflow lobe seen in either the ^{12}CO(1-0) or ^{13}CO(1-0) spectral channel maps of CB 244 (blue/red-shifted outflow contour lines begin with 0.5 K km s^{-1}/0.6 K km s^{-1} and are stepped by 0.15 K km s^{-1}/0.2 K km s^{-1}; from Yun and Clemens, 1994a), CB 54 (blue/red-shifted outflow contour lines begin with 1.7 K km s^{-1}/0.45 K km s^{-1} and are stepped by 0.3 K km s^{-1}/0.15 K km s^{-1}; from Yun and Clemens, 1994a), B 335 (contours are spaced at 2σ intervals of 200 mJy beam^{-1}; from Chandler and Sargent, 1993), and CB 230 (blue lobe: $v_{\rm LSR}$ = 1.6...2.8 km s^{-1}—solid contours, red lobe: $v_{\rm LSR}$ = 3.0, ...,4.2 km s^{-1}—dashed contours. The step width amounts to 0.3 km s^{-1} for both lobes; from Launhardt, 2001) are overlayed. (*From* Wolf et al., 2003).

Furthermore, the observations of B 335 and CB 230 not only show a decrease of the polarization degree, but also that the orientation of the polarization vectors does not change remarkably toward the dense cores. This supports the hypothesis that despite the decrease of the grain alignment rate, the magnetic field structure in the cores of these objects is not seriously disturbed, and thus still represents

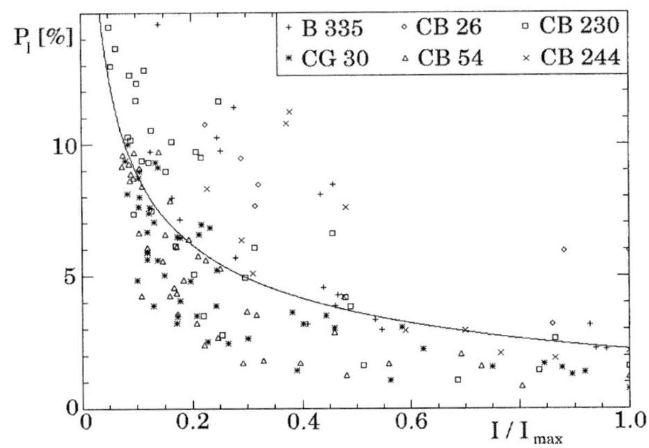

Figure 3. Scatter diagram showing the distribution of the linear polarization degree P_l vs. intensity I across the Bok globules. Additionally, data obtained for the Bok globule CG 30 have been taken into account. The solid line represents the best fit of the function of the form $P_l = a_0 + a_1(\frac{I}{\max(I)})^{a_2}$, where P_l is the degree of linear polarization, I is the measured intensity, and a_0, a_1, and a_2 are constant quantities.

the primordial field. In the case of CB 244 and CB 54 on the other hand, the orientation of the polarization vectors and therefore the structure of the magnetic field in the cores is irregular and we assume that it can no longer be accounted for being representative for the primordial field (see Henning et al., 2001 and Wolf et al., 2003 for histograms of the position angles derived from the polarization maps shown in Figures 1 and 2).

4. Magnetic Field Strength and Structure

Polarized thermal emission by aligned non-spherical grains is the main source of polarized submillimeter radiation in Bok globules (e.g., Weintraub et al., 2000; Greaves et al., 1999). Irrespective of the alignment mechanism, charged rotating interstellar grains would have a substantial magnetic moment, leading to a rapid precession of the grain angular momentum around the magnetic field direction (Draine and Weingartner, 1997). This implies a net alignment of the grains with the magnetic field. Based on the work by Chandrasekhar and Fermi (1953), the dispersion of polarization position angles is thought to be inversely proportional to the magnetic field strength

$$B = \sqrt{\frac{4\,\pi}{3}\,\rho_{\rm Gas}} \cdot \frac{\upsilon_{\rm turb}}{\sigma_{\bar{\gamma}}} \,. \tag{1}$$

Here, B is the magnetic field strength in units of G, $\rho_{\rm Gas}$ is the gas density (in units of g cm^{-3}), $\upsilon_{\rm turb}$ the rms turbulence velocity (in units of cm s^{-1}), and $\sigma_{\bar{\gamma}}$ the standard

TABLE I
Masses, gas densities, polarization, and magnetic field strengths of the envelopes.

Object	ρ_{Gas} (g cm^{-3})	v_{turb} (km s^{-1})	$\sigma_{\bar{\gamma}}$ (°)	B (μG)
B 335 (CB 199)	8.6E-18	0.14 ± 0.02[a]	$35.8^{+14.6}_{-9.1}$	134^{+46}_{-39}
CB 230 (L 1177)	3.6E-18	0.29 ± 0.04[b]	$29.8^{+8.8}_{-6.1}$	218^{+56}_{-50}
CB 244 (L 1262, SE core)	8.0E-18	≈0.29[c]	$33.1^{+19.2}_{-10.4}$	257^{+111}_{-91}
CB 26 (L 1439)	8.6E-19	0.25[d]	$18.9^{+16.7}_{-7.3}$	144^{+91}_{-68}
CB 54	3.4E-19	0.65[c]	$42.7^{+11.1}_{-8.0}$	104^{+24}_{-21}

(a) Frerking et al., 1987. (b) Wang et al., 1995; Launhardt et al., 1996. (c) Wang et al., 1995. (d) RMS turbulence velocity of a large sample of nearby star-forming Bok globules derived from $C^{18}O$ (J = 2-1)

deviation to the mean orientation angle $\bar{\gamma}$ of the polarization vectors (in units of radians). The resulting magnetic field strengths are in the range of ≈100–260 μG (see Table I).

The comparison between the magnetic field structure and morphologic features of the globules, such as their elongation and outflow direction, reveals following correlations: First, the dense cores in B 335 and CB 230 are slightly elongated whereby the major axes are oriented almost perpendicular to the direction of the bipolar molecular outflows. Second, in the Class 0 source harbouring Bok globules B 335, CB 230, and CB 244, the outflows are oriented almost perpendicular to the mean direction of the magnetic field as seen in projection on the plane of the sky. In contrast, the directions of the magnetic field and ouflow are aligned more parallel in the massive globule core CB 54 and in CB 26 with an embedded more evolved Class I source. We want to remark that the statistical significance of the latter is low because of the large scattering of polarization directions measured in CB 54 and the small number of polarization vectors and low-spatial resolution in the case of in CB 26.

5. Conclusions

We presented 850 μm polarimetric observations of dense envelopes around the very high-density protostellar condensations in five selected Bok globules. In each object we found polarization degrees of several percent that decrease toward the cores of the globules. Because the functional dependence of this behavior between the intensity and polarization is very similar for each globule, the optical properties of the grains do not seem to play a key role for the observed polarization decrease, but merely the coupling of the magnetic field to the grains.

Using the formalism by Chandrasekhar and Fermi (1953) we derive an estimate of the mean magnetic field strengths in these globules on the order of several hundred μG. These magnetic field strengths are well above those of the interstellar medium (see, e.g., Myers et al., 1995) but comparable to typical magnetic field strengths found in molecular clouds, pre-protostellar cores, and other star-forming regions.

We find the direction of the polarization vectors/magnetic field, the elongation of the Bok globules, and the orientation of the outflows (in the case of CB 26: potential outflow direction along the lobes) to be related to one another. In particular, we find the mean magnetic field direction of the Bok globules with embedded Class 0 sources—B 335, CB 230, and CB 244—to be oriented almost perpendicular to the outflow axis. In contrast to this, the magnetic field is nearly aligned with the potential outflow direction of CB 26, harboring a more evolved Class I source.

Our findings suggests that the mean magnetic field direction in a protostellar envelope changes from initially perpendicular to the outflow axis to a more moderate inclination at the end of the main accretion/ouflow phase. However, a larger sample of protostellar sources in different evolutionary stages has to be investigated to confirm this preliminary conclusion.

Acknowledgements

SW acknowledges support through the HST Grant GO 9160, through the NASA Grant NAG5-11645, and through the Emy Noether Grant WO 857/2-1 of the German Research Foundation. JCMT is operated by the Joint Astronomical Centre on behalf of the UK Particle Physics and Astronomy Research Council.

References

Berry, D.S. and Gledhill, T.M.: 1999, Starlink User Note 223.
Brand, J. and Blitz, L.: 1993, *A&A* **275**, 67.
Chandler, C.J., Gear, W.K., Sandell, G., Hayashi, S., Duncan, W.D., Griffin, M.J. and Hazella, S.: 1990, *MNRAS* **243**, 330.
Chandler, C.J. and Sargent, A.I.: 1993, *ApJ* **414**, L29.
Chandrasekhar, S. and Fermi, E.: 1953, *ApJ* **118**, 113.
Draine, B.T. and Weingartner, J.C.: 1997, *ApJ* **480**, 633.
Fiege, J.D. and Pudritz, R.E.: 2000a, *ApJ* **534**, 291.
Fiege, J.D. and Pudritz, R.E.: 2000b, *MNRAS* **311**, 105.
Frerking, M.A., Langer, W.D. and Wilson, R.W.: 1987, *ApJ* **313**, 320.
Glenn, J., Walker, C.K. and Young, E.T.: 1999, *ApJ* **511**, 812.
Greaves, J.S., Holland, W.S., Minchin, N.R., Murray, A.G. and Stevens, J.A.: 1999, *A&A* **334**, 668.
Henning, Th., Wolf, S., Launhardt, R. and Waters, R.: 2001, *ApJ* **561**, 871.
Hirano, N., Kameya, O., Nakayama, M. and Takakubo, K.: 1988, *ApJl* **327**, L69.
Holland, W.S., Robson, E.I., Gear, W.K., Cunningham, C.R., Lightfoot, J.F., Jennes, T., Ivison, R.J., Stevens, J.A., Ade, P.A.R., Griffin, M.J., Duncan, W.D., Murphy, J.A. and Naylor, D.A.: 1999, *MNRAS* **303**, 659.

Houde, M., Bastien, P., Dotson, J.L., Dowell, C.D., Hildebrand, R.H., Peng, R., Phillips, T.G., Vaillancourt, J.E. and Yoshida, H.: 2002, *ApJ* **569**, 803.

Jenness, T. and Lightfoot, J.F.: 1998, in: R. Albrecht, R.N. Hook and H.A. Bushouse (eds.), *ASP Conference Series 145, Astronomical Data Analysis Software and Systems VII*, ASP, San Francisco, p. 216.

Keene, J., Davidson, J.A., Harper, D.A., Hildebrand, R.H., Jaffe, D.T., Loewenstein, R.F., Low, F.J. and Pernic, R.: 1983, *ApJl* **274**, L43.

Kun, M.: 1998, *ApJs* **115**, 59.

Launhardt, R.: 1996, Ph.D. Thesis, University of Jena.

Launhardt, R.: 2001, in: Reipurth, B. and Zinnecker, H. (eds.), *Birth and Evolution of Binary Stars, Proceedings of IAU Symposium No. 200*, p. 117.

Launhardt, R., Evans, N.J., Wang, Y., Clemens, D.P., Henning, Th. and Yun, J.L.: 1998, *ApJSS* **119**, 59.

Launhardt, R. and Henning, Th.: 1997, *A&A* **326**, 329.

Launhardt, R., Mezger, P.G., Haslam, C.G.T., Kreysa, E., Lemke, R., Sievers, A. and Zylka, R.: 1996, *A&A* **312**, 569.

Launhardt, R. and Sargent, A.: 2001, *ApJ* **562**, L173.

Launhardt, R., Ward-Thompson, D. and Henning, Th.: 1997, *MNRAS* **288**, L45.

Matthews, B.C. and Wilson, C.D.: 2002, *ApJ* **574**, 822.

Myers, P.C., Fuller, G.A., Goodman, A.A. and Benson, P.J.: 1991, *ApJ* **376**, 561.

Myers, P.C., Goodman, A.A., Gusten, R. and Heiles, C.: 1995, *ApJ* **442**, 177.

Ryden, B.S.: 1996, *ApJ* **471**, 822.

Tomita, Y., Saito, T. and Ohtani, H.: 1979, *PASJ* **31**, 407.

Wang, Y., Evans, N.J.II, Zhou, S. and Clemens, D.P.: 1995, *ApJ* **454**, 217.

Weintraub, D.A., Goodman, A.A. and Akeson, R.L.: 2000, in: V. Mannings, A.P. Boss and S.S. Russell (eds.), *Protostars and Planets IV*, Univ. Arizona Press, Tucson, p. 247.

Wolf, S., Launhardt, R. and Henning, Th.: 2003, *ApJ* **592**, 233.

Yun, J.L. and Clemens, D.P.: 1994a, *ApJS* **92**, 145.

Yun, J.L. and Clemens, D.P.: 1994b, *AJ* **108**, 612.

HIGH ANGULAR RESOLUTION STUDY OF THE ION AND NEUTRAL SPECTRA AS A PROBE OF THE MAGNETIC FIELD STRUCTURE

SHIH-PING LAI, T. VELUSAMY and W.D. LANGER

Jet Propulsion Laboratory, California Institute of Technology, Pasadena, CA 91109, U.S.A.; E-mail: velusamy@jpl.nasa.gov

Abstract. We present the observations and analysis of the spectral linewidths of neutral and ion species across the DR21(OH) core to investigate the cyclotron interaction as a probe of magnetic fields. It has been suggested that under average interstellar field strengths the cyclotron interaction between ions and magnetic fields is strong enough to narrow the linewidth and suppress the line wings in the ion spectra. We have obtained high spatial resolution spectral line images of $H^{13}CN$ and $H^{13}CO^+$ at 3 mm in DR21(OH) with the Millimeter Array of the Owens Valley Radio Observatory. Our results show that in the eastern parts of the DR21(OH) core the spectral linewidth of the ions is narrower than that of the neutrals. We use our results, along with the existing Zeeman and dust/CO polarization data on small scales, to derive the 3D magnetic field structure. We obtain a field strength of 0.44 ± 0.12 mG with inclination of 36° to the line of sight, directed toward the observer, and a position angle of −75° in the plane of the sky.

Keywords: star formation, magnetic fields

1. Introduction

Measurements of magnetic fields in the environment of star forming regions (see the review article by Crutcher in this journal) are mostly carried out with observations of Zeeman splitting (Crutcher, 1999), the polarimetry of dust continuum (Dotson et al., 2000; Lai et al., 2001, 2002), and the polarimetry of CO emission (Girart et al., 1999; Lai et al., 2003a). The Zeeman observations only give the field *strength* along the line of sight, and the polarimetry of dust continuum and CO emission yield the field *direction* in the plane of sky. As a result, these observations do not determine the full strength and orientation of the magnetic fields in 3D. To characterize the magnetic field environment, we need to measure directly the magnetic field strength in the plane of sky B_p (the indirect Chandrasekhar–Fermi method used by Lai et al., 2001 needs to be tested). Without direct measurement of the full field strengths, the role magnetic fields play in star formation can only be assessed using some statistical arguments (Crutcher, 1999).

Recently, Houde et al. (2000a, 2000b, also see Houde's review in this journal) proposed an interesting new approach to probe the presence of magnetic fields. From basic MHD equations, Houde et al., show that even under weak fields ($\sim$10 μG) the cyclotron interaction between the ions and the field can reduce the linewidth and suppress the high velocity wings in the ion spectra when compared to coexistent

Astrophysics and Space Science **292:** 247–254, 2004.

neutral spectra. Their HCN and HCO^+ observations with the Caltech Submillimeter Observatory (CSO) show evidence of these predicted features. The ion-to-neutral linewidth ratio depends largely on the angle between the flow direction and the magnetic fields. Houde et al. (2001) use this relation to illustrate that the outflows in NGC2071 have good alignment with the magnetic fields.

The magnitude and the orientation of magnetic fields can be determined by combining the observations of Zeeman splitting, polarimetry of dust continuum and/or CO emission, and the ion-to-neutral linewidth ratio. The ion-to-neutral linewidth ratio helps infer the inclination angle of the magnetic fields to the line-of-sight as demonstrated by Houde et al. (2002) for M17. However, the differences in the velocity dispersions of the ion/neutral spectra may not unambiguously represent the cyclotron interaction, especially, if the beam is large enough to include the spectral line emission from physically and chemically different components. Furthermore, the presence of chemical differentiation between the ion and the neutral species across the beam will reduce the applicability of the ion/neutral spectra comparison. For example, if a velocity gradient exists in a large region with HCN, but HCO^+ only coexists with HCN in a small part of this region, the HCN linewidth is naturally larger than the HCO^+ linewidth. Therefore, it is important to obtain high-spatial resolution spectral line maps with interferometers to separate regions with widely different physical and chemical characteristics.

In this paper, we present the comparison between $H^{13}CN$ and $H^{13}CO^+$ in DR21(OH) observed with the millimeter array of the Owens Valley Radio Observatory (OVRO). DR21(OH) is an excellent object for a detailed study of the magnetic field structure, because it is in the early stage of massive star formation. In addition, DR21(OH) is also one of the two sources with CN Zeeman detections (Crutcher et al., 1999). Unlike CN (which represents high densities as traced by HCN and HCO^+), most of the other Zeeman detections were done with HI or OH lines which trace much lower density regions than HCN and HCO^+. Furthermore, its field directions in the plane of sky have been mapped with the Berkeley–Illinois–Maryland–Association (BIMA) interferometer using both dust and CO polarimetry (Lai et al., 2003), which makes it possible to construct the 3D direction of the magnetic fields in DR21(OH), if the cyclotron effect is proved to be applicable in this region.

2. Observations and Results

The observations were made with the OVRO Millimeter Array at 3 mm in the Compact, Low, and High resolution array configurations between October 2002 and May 2003. We obtained simultaneously spectral line images of HCN & HCO^+ and $H^{13}CN$ & $H^{13}CO^+$; ^{13}CS and N_2H^+ were observed in the opposite sideband. The synthesis beam is $6.4'' \times 6.9''$ and spectral resolutions are 0.434 km s^{-1}. Among these images, the HCN and HCO^+ spectra show features of optically thick emission

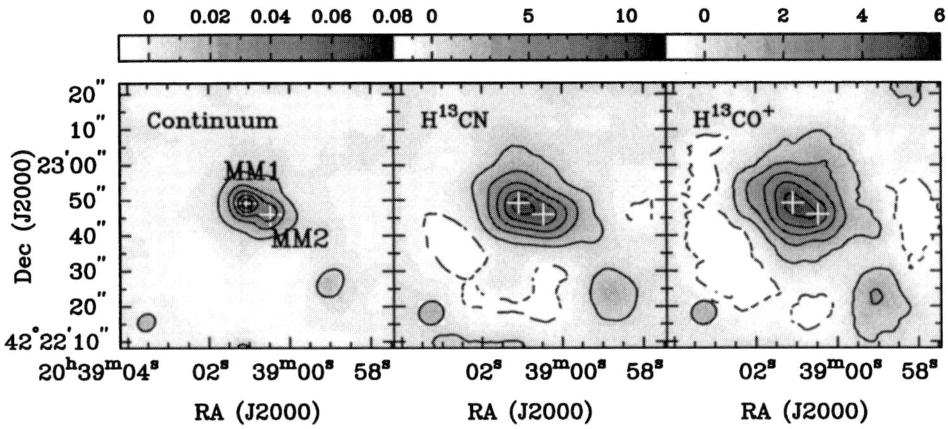

Figure 1. The 3 mm dust continuum and the velocity integrated intensity maps of $H^{13}CN$ and $H^{13}CO^+$. The lowest contour and the interval are at 10% and 20%, respectively, of the peak intensities.

with strong self-absorption; N_2H^+ has a very different spatial distribution compared to other species; $H^{13}CN$, $H^{13}CO^+$, and ^{13}CS have a similar spatial distribution, but ^{13}CS has a different velocity structure from the other two. Therefore, $H^{13}CN$ and $H^{13}CO^+$ are the only neutral/ion pair that can be used to examine the cyclotron effect in our data.

Figure 1 shows the 3 mm continuum and the integrated flux of $H^{13}CN$ and $H^{13}CO^+$. The two compact cores in DR21(OH), MM1 and MM2, are resolved. $H^{13}CN$ and $H^{13}CO^+$ both trace the dense cores very well. $H^{13}CO^+$ is slightly more extended than $H^{13}CN$, which is probably due to the lower critical density of $H^{13}CO^+$. Therefore, $H^{13}CN$ and $H^{13}CO^+$ seem to occupy the same volume.

Figure 2 shows the normalized $H^{13}CN$ and $H^{13}CO^+$ spectra overlaid on their flux ratio map. Note that $H^{13}CN$ has three hyperfine lines with intensity ratios of 1:5:3 for the optically thin case. The spectra show that DR21(OH) contains two dominant velocity components—one at ~-5 km s^{-1} and the other at ~-1 km s^{-1}, which are likely to be associated with MM1 and MM2, respectively. At the positions between MM1 and MM2, the spectra are more complex and may contain an absorption feature at ~-2.5 km s^{-1}. Overall, the spectra of $H^{13}CN$ and $H^{13}CO^+$ trace each other pretty well. A similar velocity structure is strong evidence for the coexistence of $H^{13}CN$ and $H^{13}CO^+$ as also suggested by their spatial intensity correlation.

In Figure 2 is shown the distribution of the $H^{13}CN$-to-$H^{13}CO^+$ flux ratio (in greyscale) which is a measure of the chemical differentiation in the core. Although we see a large variation in their flux ratio (by factors of 0.5 to 4), it appears to be gradual except near the southwest parts of the core. Especially in the eastern parts of the core the variation is small and there may not be any significant variation on angular sizes smaller than the OVRO beam. Thus, the use of the high angular resolution in the OVRO spectral data minimizes any contribution to the ion/neutral linewidths differences from chemical inhomogeneity.

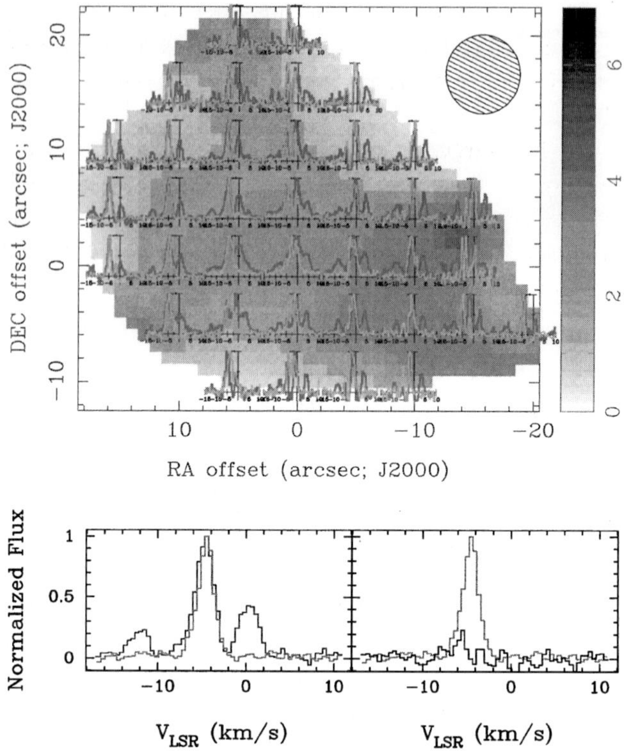

Figure 2. Top: the $H^{13}CN$-to-$H^{13}CO^+$ flux ratio (greyscale), $H^{13}CN$ spectra (red), and $H^{13}CO^+$ spectra (green). The beam size is indicated. Bottom Left: $H^{13}CN$ (red) and $H^{13}CO^+$ (green) spectra at offset position (10″, 5″). Bottom Right: $H^{13}CO^+$ spectrum (green) and the difference spectrum between observed $H^{13}CN$ and simulated $H^{13}CN$ (using $H^{13}CO^+$ as template – see Section 2) shown in blue.

In the eastern part of MM1 where the $-5\,\mathrm{km\ s^{-1}}$ component dominates and the spectral line shapes are relatively more well defined, we can see that $H^{13}CO^+$ is slightly narrower than $H^{13}CN$. To eliminate the complication of the $H^{13}CN$ hyperfine lines, we simulate $H^{13}CN$ spectra with all three hyperfine lines using the $H^{13}CO^+$ as "template". The difference spectrum between the observed $H^{13}CN$ spectrum and that simulated from $H^{13}CO^+$ is a useful measure of the cyclotron interaction. The difference spectrum is largely free from the confusion caused by the hyperfine components. The only contribution from the hyperfine lines to the difference spectrum is that corresponding to the ion/neutral linewidth differences in the other velocity component, weighted by its intensity and the strength of the hyperfine line. In the eastern parts of the core the difference spectrum under the $-5\,\mathrm{km\ s^{-1}}$ component is relatively "clean", because any contribution from the $-1\,\mathrm{km\ s^{-1}}$ component is only through the weakest (20% level) hyperfine component, whereas in the central and western parts of the core the difference spectrum under $-1\,\mathrm{km\ s^{-1}}$ is contaminated by the contribution from $-5\,\mathrm{km\ s^{-1}}$ component through the stronger (60% level) hyperfine component. Therefore, in this paper we limit our analysis of cyclotron interaction to the eastern parts (MM1) of the core as observed in the $-5\,\mathrm{km\ s^{-1}}$ velocity feature.

If indeed $H^{13}CN$ has larger linewidths than $H^{13}CO^{+}$, the difference spectra will have two positive peaks near the wings of the line profile, which is demonstrated in the bottom panel of Figure 2. If the line shapes of $H^{13}CN$ and $H^{13}CO^{+}$ are Gaussians, integrating the difference spectra over the extent of a single velocity component gives a measure of the difference of the dispersions in the spectra.

$$\int (I_n - I_i)\, dv = \sqrt{2\pi}(\sigma_n - \sigma_i), \tag{1}$$

where I_n and I_i are the normalized $H^{13}CN$ and $H^{13}CO^{+}$ spectra and σ_n and σ_i are the dispersions of $H^{13}CN$ and $H^{13}CO^{+}$ spectra, respectively. Figure 3 shows the map of $(\sigma_n - \sigma_i)$ in the left panel, which is calculated by integratiing the difference spectra over the $-5\,\mathrm{km\ s^{-1}}$ component ($v_{\mathrm{LSR}} = -9$ to $-2.5\,\mathrm{km\ s^{-1}}$). The signal-to-noise ratio (S/N) of $(\sigma_n - \sigma_i)$ is presented in the central panel of Figure 3. We use $(\sigma_n - \sigma_i)$ and σ_i measured from $H^{13}CO^{+}$ spectra to derive (σ_i/σ_n) (the right panel of Figure 3). It is evident from the results shown in Figure 3 that in the eastern side of MM1 the ion linewidth is smaller than the neutral linewidth. We do not present similar analysis for the $-1\,\mathrm{km\ s^{-1}}$ component, because of confusion from the $-5\,\mathrm{km\ s^{-1}}$ component as discussed above.

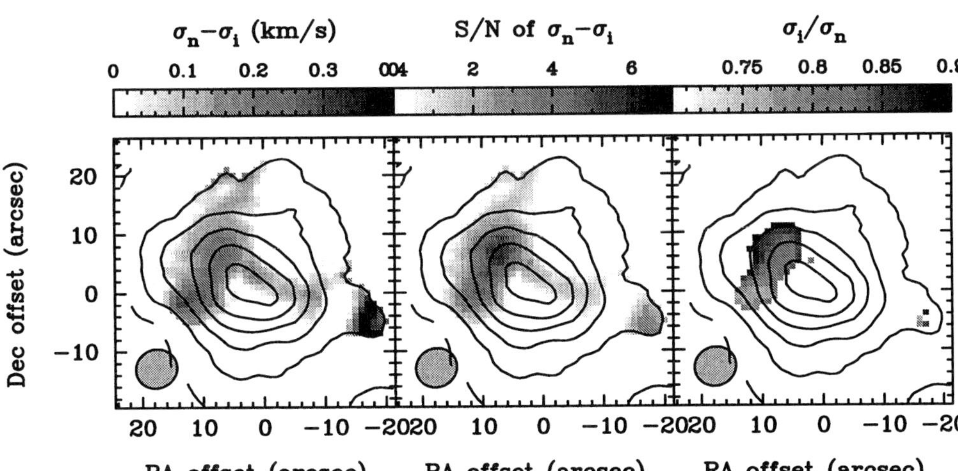

Figure 3. Left: the difference of $H^{13}CN$ and $H^{13}CO^{+}$ line dispersion $(\sigma_n - \sigma_i)$ in $\mathrm{km\ s^{-1}}$. Center: signal-to-noise ratio of $(\sigma_n - \sigma_i)$. Right: (σ_i/σ_n). The contours shows the −10%, 10%, 30%, 50%, 70%, and 90% of the $H^{13}CO^{+}$ integrated flux.

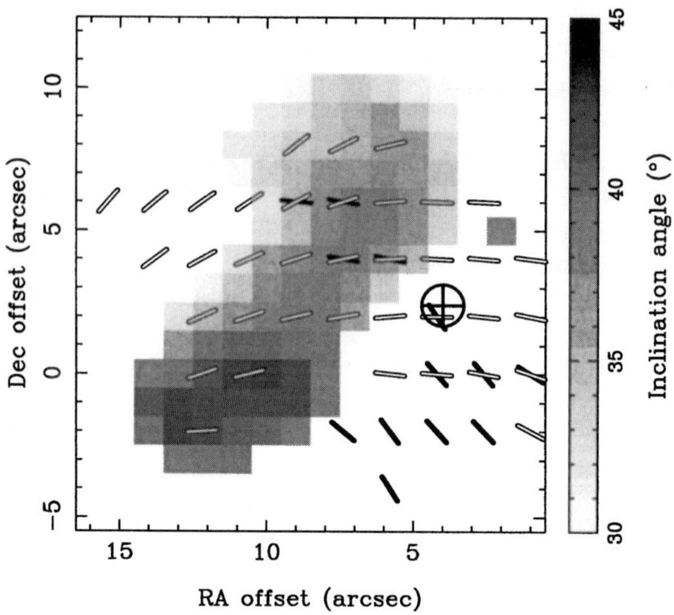

Figure 4. The greyscale shows the inclination angle of magnetic fields (α) to the line of sight varying from 30° to 42°. The black and white vectors are B_p vectors inferred from dust and CO polarization, respectively. The line-of-sight field component obtained from the CN Zeeman measurement at MM1 is indicated by ⊕ and it is pointing toward the observer with a strength of 0.36 mG.

In Figure 4, we use the ion/neutral linewidth ratio (σ_i/σ_n) to infer the inclination angle of the magnetic field to the line of sight (α) according to Equation (11) of Houde et al. (2002) and assuming the neutrals and ions are fully turbulent (Lai et al., 2003b).

3. Discussion

Figure 4 shows that the inclination angle α ranges from 30° in the northern part of MM1 to 42° in the south, and the uncertainty in α is ~4–8°. The variation in α is small (12°) and is comparable to the variation of the overlaid B_p vectors derived from the dust polarization (black) and CO polarization (white). Therefore, the magnetic field directions seem to be approximately uniform across the MM1 region. The average line-of-sight magnetic field strength in MM1 $\sim -0.36 \pm 0.10$ mG as measured by CN Zeeman observations (Crutcher et al., 1999). With the average $\alpha = 36°$, the field strength in the plane of sky (B_p) is 0.25 ± 0.07 mG, and the full magnetic field strength is 0.44 ± 0.12 mG. The direction of the field is out of the plane of the sky toward the observer at an angle of 36° to the line of sight, and its projection on the sky is at a position angle −75°.

Lai et al. (2003) estimate B_p ~0.9 mG for MM1 from the observed dispersion in the CO polarization position angles using the Chandrasekhar–Fermi formula.

This value is a factor of 3 larger than what is derived from the cyclotron interaction data presented here. This discrepancy could be partly due to the low values for the dispersion caused by the smoothing of polarization by the finite beam size of the BIMA observations. On the other hand, in our analysis of the cyclotron interaction we have assumed isotropic turbulence in the core MM1. If collimated flows of the ions and neutrals are present we could be underestimating α and hence a low value for B_p (as seen in Figure 4 in Houde et al. (2002). However, although MM1 does power high velocity CO outflow (Lai et al., 2003), it is unlikely that the spectra of ions and neutrals observed at the rest velocity of the core is contaminated by the outflow.

With the knowledge of the full magnetic field strength, we can evaluate the role that magnetic fields play for star formation in MM1. First, we estimate the mass to magnetic flux ratio M/Φ_B to be 4.5. Therefore, MM1 is clearly magnetically supercritical. We can also evaluate the relative importance of magnetic and turbulent support. We find that in MM1 the magnetic-to-gravitational energy ratio is 0.05 and the turbulent-to-gravitational energy ratio is 0.66. Therefore, turbulence provides the dominant support in MM1.

4. Conclusion

With high-angular resolution interferometric data, we derive the 3D structure of the magnetic fields in the MM1 core of DR21(OH) by comparing the ion and neutral linewidths and using the model of Houde's et al. The inferred inclination angles of magnetic fields to the line of sight ranges from 30° to 42°across region and full field strength B $= 0.44 \pm 0.12$ mG.

Acknowledgements

This research was conducted at JPL (Caltech) with support from NASA, while SPL held a National Research Council Resident Research Associateship at JPL. The OVRO mm array is supported by NSF Grant AST-99-81546.

References

Crutcher, R.M.: 1999, *ApJ* **520**, 706 .
Crutcher, R.M., Troland, T.H., Lazareff, B., Paubert, G. and Kazès, I.: 1999, *ApJL* **514**, L121.
Dotson, J.L., Davidson, J., Dowell, C.D., Schleuning, D.A. and Hildebrand, R.H.: 2000, *ApJS* **128**, 335.
Girart, J.M., Crutcher, R.M. and Rao, R.: 1999, *ApJL* **525**, L109.
Houde, M., Bastien, P., Peng, R., Phillips, T.G. and Yoshida, H.: 2000a, *ApJ* **536**, 857.
Houde, M., Peng, R., Phillips, T.G., Bastien, P. and Yoshida, H.: 2000b, *ApJ* **537**, 245.
Houde, M., Phillips, T.G., Bastien, P., Peng, R. and Yoshida, H.: 2001, *ApJ* **547**, 311.

Houde, M. et al.: 2002, *ApJ* **569**, 803.
Lai, S.-P., Crutcher, R.M., Girart, J.M. and Rao, R.: 2001, *ApJ* **561**, 864.
Lai, S.-P., Crutcher, R.M., Girart, J.M. and Rao, R.: 2002, *ApJ* **566**, 925 .
Lai, S.-P., Crutcher, R.M. and Girart, J.M.: 2003a, *ApJ* in press.
Lai, S.-P., Velusamy, T. and Langer, W.D.: 2003b, *ApJ* **596**, L239.

FROM CORES TO STARS

FROM CORES TO STARS, BROWN DWARFS AND PLANETS

RAFAEL REBOLO
Instituto de Astrofísica de Canarias, IAC and Consejo Superior de Investigaciones Cientficas, CSIC, Spain; E-mail: rrl@ll.iac.es

Keywords: molecular cores, stars, substellar objects

1. Introductory Remarks

Multiple observations support the association of dense cores of molecular clouds with young stellar objects, groups and clusters. It is generally accepted that star formation takes place in these relatively high density regions ($\sim 10^4$ cm^{-3}) of the interstellar medium. However, in spite of the large effort devoted to explain the origin of stars, the processes that lead to star formation in the dense cores of molecular clouds are not well understood yet. A severe limitation is imposed by the scarce knowledge of the initial conditions for protostellar collapse and for the lack of information on the role of magnetic fields and supersonic turbulence in molecular clouds (Andrè et al., Tafalla et al., this volume). How the initial density perturbations are originated in any of these contexts and to what extent these phenomena determine the characteristics of a major observable as the stellar initial mass function (IMF) are fundamental questions awaiting a definitive answer.

Understanding the origin of stars requires a complete picture of the formation, collapse and evolution of the dense cores of molecular clouds. There is a remarkable similarity between the mass spectrum of pre-stellar cores and the IMF (Motte et al., 1998) at least down to $\sim 0.2\ M_{\odot}$. This may suggest that the characteristics of the IMF are set already during the fragmentation process, an issue that clearly demands further observational effort. New facilities like ALMA will be crucial in providing data on pre-stellar cores at much lower masses, possibly extending more than a factor 10 the mass range available for comparison. Besides, a proper description of the initial conditions of pre-stellar cores and an adequated treatment of fragmentation are essential. The dependence of fragmentation of collapsing molecular cores on the initial density perturbations and ultimately, on the density structure and geometry of the clouds has been subject of study for decades (see, e.g., Boss and Bodenheimer, 1979). More recently significant efforts have been devoted to understand how fragmentation proceeds in dense molecular core structures (discs and rings) that are significantly supported by rotation (Burkert et al., 1997; Matsumoto and Hanawa, 2004) and/or by magnetic fields. The density and temperature profiles

Astrophysics and Space Science **292:** 293–295, 2004.

of protostellar cores and their time evolution are extremely sensitive to rotation, turbulence and magnetic fields. Empirical determination of the effects of these variables at such early evolutionary stages is very challenging given the large extinction of these regions and the relatively low spatial resolution associated with observations at long wavelengths. Fortunately, the final products of star formation are much better determined both in the last stages of pre-Main Sequence evolution and in Main Sequence Stars. Observations indicate that a large fraction of field stars (about 60%) are actually found in binaries or multiple systems (Abt and Levy, 1976; Duquennoy and Mayor, 1991) and that many stars also host giant planets (more than 7%). It is likely that stars result from a combination of the various types of fragmentation mechanisms. Theories have to develop further to provide detail explanations of the frequency of binary and multiple systems and the separation and mass ratio of components.

What is the minimum mass of a fragment? The discovery of brown dwarfs and giant planets in the past decade requires an extension of the proposed paradigms of star formation into the substellar domain. Objects below the minimum mass for hydrogen burning (about 75 $M_{\rm Jup}$ for solar metallicity) have been found free floating in the field, in stellar clusters and orbiting stars. The first determinations of the substellar initial mass function (see e.g., Béjar et al., 2001) suggest that brown dwarfs may even outnumber stars in the Galaxy. Thus, these substellar objects have to be considered in any scenario of star formation. Theoretical models and numerical simulations of star formation from the collapse of molecular cloud cores not only have to explain the distribution of stellar masses, and the formation of accretion and protoplanetary discs but also the existence, and mass distribution of substellar objects. There is observational evidence for a smooth extrapolation of the stellar initial mass function to masses as low as $\sim 5\ M_{\rm Jup}$. A substellar mass function described by a power law $dN/dM \sim M^{-\alpha}$ with an index $\alpha = 0.8 \pm 0.4$ appears to fit the observations (Béjar et al., this volume). Remarkably, the existence of free-floating objects with masses a few times that of Jupiter (Zapatero–Osorio et al., 2002, 2004) challenges the classical opacity limit for fragmentation (Larson, 1969). Are these objects formed as a direct result of cloud fragmentation or do they form via gravitational instability in accretion disks and are later ejected in dynamical interactions?

The formation of stars and brown dwarfs has been addressed using hydrodynamical calculations (see, e.g., Bate et al., 2002, Bate this volume). It is suggested that brown dwarfs may form via fragmentation of dense molecular gas in unstable multiple systems and are dynamically ejected before they have been able to accrete to stellar masses. In these simulations, the resulting number of brown dwarfs is comparable to that of stars and the resulting substellar initial mass functions agree reasonably well with observations.

According to Bate et al., about 75% of the brown dwarfs would form in gravitationally unstable circumstellar discs and the rest in collapsing filaments of molecular gas. An interesting prediction is that the number of binary brown dwarfs will be small ($\leq 5\%$) and the typical separations between binary components smaller than 10 au. Both results are consistent with the observations, although the number of known binary brown dwarfs is still rather small for a statistically significant comparison. The simulations also point out that many of the discs originally formed around stars and brown dwarfs that are ejected result truncated dynamically. Only a small fraction of these ejected objects would retain discs. Again, the observations hint toward a low fraction of brown dwarfs with discs, but the sample investigated so far is not large.

It appears that the essential characteristics of the stellar/substellar mass function, the binarity/multiplicity rates and the frequency of discs around stars and brown dwarfs can be essentially understood without including the effects of magnetic fields in hydrodynamical simulations. However, magnetic pressure may be required if fragmentation has to be enhanced and/or extended to even lower masses. Magnetic fields may also be required to support clouds and certainly to drive the evolution of angular momentum leading to the formation of binary systems and discs (Tomisaka and Matsumoto, this volume). Discerning the role of magnetic fields and rotation in star, brown dwarf and protoplanetary disc formation is a major task that will demand a large effort, the development of new observational techniques (see, e.g., Desmurs, this volume), better physical descriptions (e.g., Wardle, this volume) and more complex numerical simulations.

References

Abt, H.A. and Levy, S.: 1976, *ApJ Supp Ser* **30**, 273.
Bate, M.R., Bonnell, I.A. and Bromm, V.: 2002, *MNRAS* **332**, L65.
Béjar, V.J.S. et al.: 2001, *ApJ* **556**, 830.
Boss, A.P. and Bodenheimer, P.: 1979, *ApJ* **234**, 289.
Burkert, A., Bate, M.R. and Bodenheimer, P.: 1997, *MNRAS* **289**, 497.
Duquennoy, A. and Mayor, M.: 1991, *A&A* **248**, 485.
Larson, R.B.: 1969, *MNRAS* **145**, 271.
Motte, F., Andrè, P. and Neri, R.: 1998, *A&A* **136**, 150.
Zapatero–Osorio, M.R. et al.: 2002, *ApJ* **578**, 536.

FROM MOLECULAR CORES TO STARS AND BROWN DWARFS

MATTHEW R. BATE
School of Physics, University of Exeter, UK; E-mail: mbate@astro.ex.ac.uk

Abstract. In this paper, we review the results from numerical simulations that investigate the evolution of molecular cores into stars and brown dwarfs. We begin by discussing the four main processes involved in this evolution, fragmentation, accretion, interactions with discs, and stellar dynamical interactions. We then discuss large-scale calculations in which all of these processes are model simultaneously. Finally, we review results that have been obtained from past magnetohydrodynamical (MHD) calculations of fragmentation and look towards future MHD calculations.

Keywords: accretion, accretion discs, binaries: general, brown dwarfs, hydrodynamics, stars: formation

1. Processes Involved in the Evolution from Cores to Stars

1.1. Fragmentation

The first process involved in the evolution from molecular cores to stars is that of collapse and gravitational fragmentation. We assume here that gravitationally-unstable dense molecular cores have already been formed, perhaps by turbulent fragmentation or ambipolar diffusion (see other reviews in this book).

1.1.1. *Types of Fragmentation*

Although there are many sub-classes of gravitational fragmentation that can be defined (e.g., Matsumoto, 2004), there are two main classes: 'prompt' (Pringle, 1989) fragmentation, and rotational fragmentation. In the former, a gravitationally unstable dense molecular core collapses nonhomologously and fragments due to initial density perturbations (e.g., Boss and Bodenheimer, 1979; Boss, 1986; Bonnell et al., 1991). Such fragmentation has been studied numerically for more than 20 years, typically by adding $m = 2$ density perturbations to initial dense cores. In this case, the fragmentation depends primarily on the initial density structure or geometry of the cloud and the initial ratio of thermal to gravitational energies.

Rotational fragmentation may occur to an object that is in quasi-equilibrium but which derives a large fraction of its support from rotation. The two structures that undergo such fragmentation are rotationally supported discs (Bonnell, 1994; Whitworth et al., 1995; Burkert et al., 1997) or rings (Norman and Wilson, 1978) that may form during the collapse of a dense molecular core. If the ratio of rotational energy to the magnitude of the gravitational energy of the object is large, nonaxisymmetric structure can grow from small initial fluctuations on the

Astrophysics and Space Science **292:** 297–307, 2004.

object's dynamical timescale, leading to fragmentation. An object that is dynamically unstable must be assembled on its dynamical timescale otherwise it would adjust to a more stable state as it was being assembled without undergoing an instability violent enough to lead to fragmentation. In practice, this means that the disc or ring must have a high accretion rate (Bonnell, 1994), accreting a significant fraction of its mass and angular momentum on its dynamical timescale to drive it into a highly unstable equilibrium.

1.1.2. *The Opacity Limit for Fragmentation*

The combination of the prompt and rotational fragmentation mechanisms can produce most observed binary and multiple systems. However, there is a problem in forming the closest binary systems (separations $\lesssim$10 AU) by fragmentation directly due to the opacity limit for fragmentation. A molecular core collapses roughly isothermally because the compressional heating rate of the gas is much lower than the rate at which the gas can cool. However, as the collapse proceeds, it accelerates and eventually the compressional heating rate exceeds the cooling rate. The temperature of the densest gas increases forming a pressure-supported core with a typical radius of $\approx$5 AU and an initial mass of a few Jupiter masses (Larson, 1969). Thus, there is a minimum initial binary separation of $\approx$10 AU. Fragmentation within a pressure-supported core (i.e., on length scales $\lesssim$5 AU) is inhibited (Boss, 1988; Bate, 1998). Thus, it seems that the closest binaries must form by another mechanism.

1.2. Accretion

Fragmentation is only the first step in the formation of a stellar system. When fragmentation occurs, the resulting fragments may only contain a few Jupiter masses of material. In order to become stars, they may increase their mass by 4–5 orders of magnitude through accretion. This accretion alters the masses and orbits of a multiple system and is primarily responsible for the formation of circumstellar discs.

1.2.1. *Binaries*

The effect of accretion on a binary system has been studied in detail by Artymowicz (1983), Bate (1997), Bate and Bonnell (1997), and Bate (2000). The amount of gas that must be accreted by a binary system in order to obtain some final system mass depends on the initial separation of the fragments (Boss, 1986; Bate, 2000). This is because, in order to fragment into two objects, the Jeans length at the time of fragmentation must be less than the separation of the two fragments that will form and, for a given temperature, the Jeans mass (i.e., the initial mass of the fragments) is proportional to the Jeans length. Thus, the initial mass of a binary depends on its initial separation. Bate and Bonnell (1997) used hydrodynamical simulations to study the effect of accretion from an infalling envelope on the mass ratio, separation and discs of an embedded binary. Bate (2000) used these results to study the evolution of binaries from their initial to their final masses. He found that the long-term effect of accretion from most types of molecular cloud core was to drive the stars towards equal masses. The greater the factor by which a binary increased its mass, the more equal the masses became. To obtain the same total mass, close systems must increase their mass by a larger factor than wide systems because the initial mass of the binary depends on its initial separation, as described above. Thus, closer systems are predicted to have mass ratios biased towards unity when compared with wider systems. When applied to brown dwarf companions to solar-type stars, the implication is that brown dwarf companions should be rare in close orbits because the binary has had to increase its initial mass by a large factor to produce the solar-mass star and yet maintain very unequal-mass components. Because the initial fragment masses of wider systems can be larger the binary does not have to accrete as much material to reach its final mass and, thus, brown dwarf companions should be more common at larger separations. This prediction is in good agreement with observations. For circumstellar discs, Bate (2000) found that the circumprimary disc should be more massive than the circumsecondary disc and circumbinary discs are more likely around closer systems than wide systems.

1.2.2. *Triples*

Accretion can have even more dramatic effects on triple systems (Smith et al., 1997). Because the stability of a triple system depends on the relative orbital separations and masses, changes in these quantities can lead to a stable triple becoming unstable and breaking up into a binary and a single star. Conversely, because the break up of a triple is a statistical process and may take many crossing times to occur, accretion onto an unstable system can stablise the system before it breaks up. Whether a specific system is stabilised or destabilise depends on its orbital separations, masses, and the angular momentum of the gas being accreted. However, we note that triples comprised of a binary orbiting a star whose mass exceeds that of the binary are destabilised for most types of accretion and, therefore, are unlikely to survive significant accretion.

1.2.3. *Competitive Accretion*

In clusters of stars, the individual accretion rates vary widely. This "competitive accretion" (Zinnecker, 1982) can lead to a distribution of masses. Bonnell et al. (1997, 2001a,b) and Bonnell and Bate (2002) have studied this competitive accretion process in clusters of stars ranging from ~10–1000 stars. They find that even if each protostar begins with the same mass, in the long-term, competitive accretion results in a mass distribution that has a Salpeter-type slope at the high-mass end. Thus, competitive accretion between protostars embedded in a gas cloud is a possible mechanism for the origin of the stellar initial mass function.

1.3. Disc interactions

Angular momentum conservation during the collapse of rotating or turbulent molecular clouds inevitably leads to the formation of discs around young stars. These discs can play an important role in the formation of multiple systems.

1.3.1. *Circumbinary Discs*

If a binary system is surrounded by a circumbinary disc, the binary will generate waves in the disc. These waves transfer angular momentum from the orbit of the binary to the disc material. In the long term, a large amount of orbital angular momentum can be removed from the binary by the disc, decreasing the separation of the binary and driving eccentricity (Artymowicz et al., 1991). Such interactions may therefore help us to solve the problem that close binaries cannot form by fragmentation directly (Section 1.1.2) by allowing the orbits of a wide systems to decay to produce close systems (e.g., Bate et al., 2002b). Furthermore, the accretion of gas with high angular momentum from a circumbinary disc by the binary tends to equalise the masses of the components (Bate, 2000). An example of a system where all three of these effects may be seen is the close binary system DQ Tau which appears to be accreting from a circumbinary disc, and has a mass ratio of $q = 0.97$ and an eccentricity of $e = 0.56$ (Mathieu et al., 1997). This mechanism is likely to be responsible for the apparent excess of "twins" (Tokovinin, 2000), close binaries with mass ratios $q > 0.95$.

Similar to the effects of accretion on triple systems, a triple system can be dramatically affected by its interaction with a circumtriple disc. For example, in a hierarchical triple system, the orbital separation of the distant component may decay due to its interaction with the circumtriple disc to the point that the system becomes unstable (Bate et al., 2002b).

1.3.2. *Star-Disc Encounters*

Although fragmentation is capable of forming most multiple systems directly, it is not the only way to form binaries. If a single star passes through the disc of another star enough kinetic energy may be dissipated in the encounter to form a bound binary system (Larson, 1990; Clarke and Pringle, 1991; Hall et al., 1996). This mechanism may be efficient enough to form a significant number of wide binaries in small-N clusters. However, in large-N clusters the stellar velocity dispersion is likely to be too high to allow enough kinetic energy to be dissipated and the disc would simply be truncated without forming a bound system (Clarke and Pringle, 1991).

1.4. Dynamical interactions

Even neglecting the effects of discs, dynamical interactions between stars are likely to play an important role in star formation. An important ratio is the binding energy of the binary compared to the typical kinetic energy of the stars in a cluster. In general, wide binaries with small binding energies will be broken up by encounters with other stars, while close binaries will be hardened or undergo chaotic interactions (Hut and Bahcall, 1983). Chaotic interactions may lead to ejection of the incoming star, the formation of a triple, or an exchange where the incoming star becomes a member of the binary and one of the original components is ejected. Usually, the lowest mass object is ejected from such chaotic interactions. Thus, successive encounters tend to result in the mass ratio of the binary being driven towards unity and an increase in the primary's mass.

The effect of dynamical biasing on the properties of multiple systems, assuming stars form in small-N groups, has been studied extensively by McDonald and Clarke (1993, 1995), Sterzik and Durisen (1998, 2003), and Durisen et al. (2001). Durisen et al. (2001) find that dynamical effects can reproduce many of the observed properties of binary and multiple systems if the stellar groups are assembled in a two-step process where the mass of a group is selected from a clump-mass distribution and the stellar masses are then selected from the stellar initial mass function (IMF), constrained so as to give the correct cluster mass.

In larger clusters, the main effect of stellar dynamical interactions is to break up wide binaries. This has been studied extensively by Kroupa (1995, 1998). He found that starting with a cluster consisting entirely of binaries with a broad range of separations, a large fraction of the binaries could be destroyed in the first few Myrs after its formation and that this destruction sets the form of the separation distribution of binaries at large separations. The fraction of binaries destroyed and the characteristic 'maximum' separation depend on the stellar density of the cluster. However, dynamical interactions also tend to harden close binaries, prompting Kroupa and Burkert (2001) to inquire whether or not dynamical interactions in cluster could broaden an initially peaked distribution of separations. For example, they performed simulations of clusters with initial binary periods between 10^4 and 10^6 days. However, they found that very few binaries with periods less than 10^4 days were produced. Thus, dynamical interactions alone do not seem a viable way to form close binaries.

2. Large-Scale Calculations of Cluster Formation

Although all of the above processes play a part in star formation, we would like to determine which processes are primarily responsible for determining stellar properties and why. One way we can tackle this question is to perform calculations of star cluster formation which include all of the above effects and form enough stars to obtain their statistical properties.

2.1. Past Calculations

Hydrodynamical calculations of cluster formation are very computationally intensive and have only recently become possible. Chapman et al. (1992) performed the first hydrodynamical calculations to follow the formation of large numbers of protostars. However, they could not follow the evolution until these objects had reached their final states. Klessen et al. (1998) and Klessen and Burkert (2000, 2001), followed the formation of $\sim$100 objects and their evolution until most of the gas had been used up in the star formation process. They found that the mass function of their objects was log-normal, reminiscent of the stellar IMF, with a mean mass of order the mean Jeans mass of the cloud. However, these calculations did not have the resolution to follow small-scale fragmentation, most binaries, or circumstellar discs.

2.2. Fully-Resolved Calculations

Bate et al. (2002a,b; 2003) performed the first fully-resolved calculation of star cluster formation. The calculation was fully-resolved in the sense that fragmentation was followed down to the opacity limit for fragmentation (mass of a few Jupiter masses). Thus, all stars and brown dwarfs that should have formed in the cloud were modelled. The calculation allowed binaries to become as close as 1 AU, and resolved circumstellar discs with radii greater than $\approx$20 AU.

2.2.1. *The Initial Mass Function*

In the left panel of Figure 1, weplot the IMF obtained from the calculation. It is consistent with $dN/d\log M = M^{\Gamma}$ where $\Gamma = -1.35$ for $M > 0.5\ M_{\odot}$, $\Gamma = 0.0$ for $0.006 < M < 0.5\ M_{\odot}$, and there are no objects below the opacity limit for fragmentation ($\approx 0.005\ M_{\odot}$). The Salpeter slope is $\Gamma = -1.35$. This mass function is consistent with recent determinations of the IMF in young stellar clusters and star-forming regions (e.g., Luhman et al., 2000). Furthermore, there are roughly equal numbers of stars and brown dwarfs, a result also supported by observations (Reid et al., 1999).

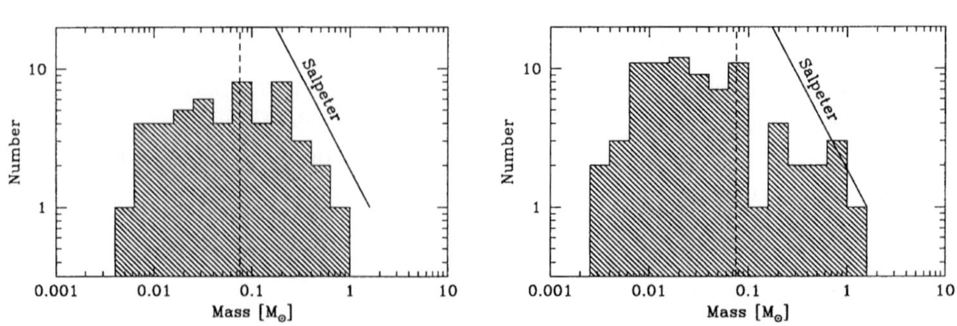

Figure 1. The IMFs of star clusters whose progenitor clouds had mean thermal Jeans masses of 1 $M_{\odot}$ (left) and 1/3 $M_{\odot}$ (right).

2.2.2. *The Formation Mechanism of the Brown Dwarfs*
Although the mean thermal Jeans mass in the cloud is initially 1 $M_\odot$, the calculation produces many brown dwarfs. The formation mechanism of the brown dwarfs is discussed in detail by Bate et al. (2002a). They find that the brown dwarfs form in dense gas where the local Jeans mass is lower than that in the cloud as a whole. Roughly three quarters form via the fragmentation of massive gravitationally-unstable circumstellar discs, while the remainder form in dense filaments of molecular gas. However, in either case, the objects must avoid accreting to stellar masses. This is accomplished by the objects forming in, or quickly falling into, unstable multiple systems where they are dynamically ejected from the cloud before they have been able to accrete to stellar masses. This formation mechanism has been discussed by Reipurth and Clarke (2001), although they could only speculate on its efficiency and the ways in which the multiple systems might form. The hydrodynamical calculation shows that this mechanism is capable of producing the observed numbers of brown dwarfs.

2.2.3. *The Properties of the Brown Dwarfs*
The close dynamical interactions that occur during the ejection process have two important implications for the properties of brown dwarfs. First, binary brown dwarf systems should be rare, $\approx$5%. Current observations suggest a frequency of $\approx$15% (Close et al., 2003). With only $\approx$20 brown dwarfs formed in the hydrodynamical calculation and a similar number observed, both the theoretical and observational fractions are currently limited by small number statistics.

Second, the frequency of young brown dwarfs with large circumstellar discs (greater than $\approx$20 AU in radius) should also be low at $\approx$5%. There are two main reasons for this. First, to avoid becoming stars the brown dwarfs must be ejected from the cloud soon after their formation and, thus, many do not have time to accrete the high angular momentum gas required to form large discs. Second, the majority of the dynamical encounters that eject the brown dwarfs occur at separations $<$20 AU so that any existing large disc is truncated.

2.2.4. *The Formation of Close Binaries*
As discussed in Section 1.1.2, the opacity limit for fragmentation sets a minimum initial binary separation of $\approx$10 AU. However, Bate et al. (2002b) found that their calculation produced 7 close binary systems (separations $<$10 AU). Rather than forming directly by fragmentation, these systems formed from initially wider multiple systems through a combination of accretion (reducing the separation of a system), the interaction of binaries and triples with circumbinary and circumtriple discs (Section 1.3.1), and dynamical interactions (which can harden tight binaries, Section 1.4). Together, these mechanisms can lead to efficient decay of wide multiple systems to form close systems. The frequency of close binaries at the end of the calculation was $7/43 \approx 16\%$. This is in agreement with the observed frequency of

close ($<$10 AU) binaries of $\approx$20% (Duquennoy and Mayor, 1991), demonstrating that close binaries need not be created by fragmentation in situ.

These formation mechanisms lead to several consequences for the properties of close binaries (Bate et al., 2002b). First, there is a preference for mass ratios near unity due to the mass-equalising effect of accretion (Section 1.2.1) and dynamical exchange interactions (Section 1.4). This gives a natural explanation for the observation that close binaries (periods $\lesssim$10 years) tend to have higher mass ratios than wider binaries (Mazeh et al., 1992; Halbwachs et al. 1998). Second, the close binary frequency increases with primary mass due to successive dynamical exchanges (Section 1.4). Finally, many of the close binaries are members of multiple systems. Although the true frequency of wide companions to close binaries is not yet well known, many close binaries do have wider components (e.g., Mayor and Mazeh, 1987; Tokovinin, 2000).

2.2.5. *Discs*

The calculation resolves circumstellar discs with radii greater than $\approx$20 AU. Although many of the stars and brown dwarfs form with large discs initially, dynamical interactions truncate many of the discs. At the end of the calculation, of the stars that have been ejected from the cloud (and thus have reached their final states) only 2 out of 10 systems (20%) have resolved discs. For the brown dwarfs, the fraction is even lower with only 1 out of 18 ($\approx$5%). Even if objects that are still evolving are included, these fractions only increase to $\approx$40% for stars and $\approx$17% for brown dwarfs. Thus, the majority of the stars and brown dwarfs, in this relatively density star-forming region ($\approx 10^3$ stars/pc^{-3}), do not retain discs large enough to form our solar system.

2.3. Variation with Initial Conditions

The calculation of Bate et al. (2002a,b; 2003) began with a reasonable estimate for the conditions in molecular clouds and obtains stellar properties that are in reasonable agreement with observations. However, the question arises as to how these stellar properties depend on the conditions in the molecular cloud. One might imagine that varying the mean thermal Jeans mass, the gas opacity, and/or the properties of the turbulence leads to differences in stellar properties.

Recently, Bate and Bonnell, in preparation, performed a calculation identical to that of Bate et al. (2002a,b; 2003), but reduced the cloud radius so that the mean thermal Jeans mass was reduced by a factor of 3. The IMFs obtain from the original calculation and the new calculation are compared in Figure 1. It can be seen that the calculation with the lower thermal Jeans mass produces a larger fraction of brown dwarfs. A Kolmogorov–Smirnov test shows that this result is significant at the 98.1% level (i.e., 2.4 σ). However, despite a reduction in the mean thermal Jeans mass by a factor of 3 between the first and second calculations, the mean object mass is only 15% lower in the second calculation. Thus, although the mean 'stellar' mass does seem to depend on the mean thermal Jeans mass in the progenitor cloud, it is only a very weak dependence. Further calculations are now being performed to confirm this result and test the dependence of stellar properties on other properties of the progenitor clouds.

3. The Effect of Magnetic Fields

The calculations described above do not include magnetic fields. As described in some of the other contributions to this proceedings, magnetic fields may play an important role in the evolution from cores to stars and brown dwarfs. Processes such as ambipolar diffusion and magnetic braking may modify the above results. On the other hand, it has also been suggested (Lubow and Pringle 1996), that the dense cores from which stars form may be assembled from material that has been freed from the magnetic field that permeates molecular clouds on larger scales through magnetic reconnection.

A few three-dimensional magnetohydrodynamic (MHD) calculations following the collapse of dense cores to form stars have been performed in the past. Phillips (1986a,b) performed calculations following the collapse of dense molecular cores with ideal MHD. None of these cores fragmented. This is potentially a major problem since most stars are members of binary or multiple systems. Boss (2000, 2002), performed pseudo-MHD calculations of the collapse of dense cores. He found that magnetic pressure effects could actually enhance fragmentation. However, his calculations did not include the effects of magnetic tension such as magnetic braking and angular momentum transport. Hosking (2002) performed two-fluid (ions and neutrals) MHD calculations of the collapse of dense molecular cores, thus allowing ambipolar diffusion and potentially decreasing the effects of magnetic tension. However, he found that using standard ion-neutral coupling coefficients resulted in so much angular momentum transport that fragmentation was inhibited – initial conditions that fragmented in the absence of magnetic fields did not fragment when magnetic fields were added due to the angular momentum transport provided by magnetic tension. He had to reduce the coupling coefficients by a factor of 25 over the standard values before any sign of fragmentation was observed. This result is in qualitative agreement with the ideal MHD calculations presented by Tomisaka in this paper who finds that very high initial rotation rates are required to permit fragmentation.

The number of calculations following the evolution from cores to stars including MHD effects is likely to increase rapidly in the next few years. In particular, it now seems possible to incorporate magnetic fields into Smoothed Particle Hydrodynamics (SPH), one of the main techniques for simulating star formation (Price, this proceedings). The new formulation seems to have overcome the past problems with numerical instabilities and maintaining $\nabla \cdot B = 0$. SPH and MHD nested grid or adaptive mesh refinement codes are likely to be the primary methods of simulating these problems.

4. Conclusions

In this review, I have summarised the main processes involved in dense molecular cores producing stars and brown dwarfs. These processes have been studied primarily in the absence of magnetic fields, but can give results that are in good agreement with observed stellar properties. Magnetic fields may be important in earlier phases (e.g., the formation of dense cores) or later processes (e.g., protostellar jet formation), but there is no obvious need to invoke MHD effects to explain disagreements between theory and observation in the evolution from dense cores to stars at this time.

If magnetic fields are an important element in this evolution, it is likely to be mainly due to their ability to transport angular momentum. Calculations with ideal MHD seem to give angular momentum transport that is too strong, inhibiting fragmentation. In order to calculate MHD effects correctly, future calculations will have to include non-ideal MHD effects such as ambipolar diffusion and possibly the Hall effect and magnetic resistivity at high densities. Even including these effects, there are indications from Hosking's work that magnetic fields may still be very efficient at inhibiting fragmentation, forcing us to consider the possibility that either magnetic fields are weaker in dense cores than generally thought, or that the coupling between the field and the gas is weaker than generally thought.

References

Artymowicz, P.: 1983, *Acta Astron.* **33**, 223.
Artymowicz, P., Clarke, C.J., Lubow, S.H. and Pringle, J.E.: 1991, *ApJ* **370**, L35.
Bate, M.R.: 1997, *MNRAS* **285**, 16.
Bate, M.R.: 1998, *ApJ* **508**, L95.
Bate, M.R.: 2000, *MNRAS* **314**, 33.
Bate, M.R. and Bonnell, I.A.: 1997, *MNRAS* **285**, 33.
Bate, M.R., Bonnell, I.A. and Bromm, V.: 2002a, *MNRAS* **332**, L65.
Bate, M.R., Bonnell, I.A. and Bromm, V.: 2002b, *MNRAS* **336**, 705.
Bate, M.R., Bonnell, I.A. and Bromm, V.: 2003, *MNRAS* **339**, 577.
Bonnell, I.A.: 1994, *MNRAS* **269**, 837.
Bonnell, I.A., Bate, M.R.: 2002, *MNRAS* **336**, 659.
Bonnell, I.A., Bate, M.R., Clarke, C.J. and Pringle, J.E.: 1997, *MNRAS* **285**, 201.
Bonnell, I.A., Bate, M.R., Clarke, C.J. and Pringle, J.E.: 2001a, *MNRAS* **323**, 785.
Bonnell, I.A., Clarke, C.J., Bate, M.R. and Pringle, J.E.: 2001b, *MNRAS* **324**, 573.
Bonnell, I., Martel, H., Bastien, P., Arcoragi, J.-P. and Benz, W.: 1991, *ApJ* **377**, 553.
Boss, A.P.: 1986, *ApJS* **62**, 519.
Boss, A.P.: 1988, *ApJ* **331**, 370.
Boss, A.P.: 2000, *ApJ* **545**, L64.
Boss, A.P.: 2002, *ApJ* **568**, 743.

Boss, A.P. and Bodenheimer, P.: 1979, *ApJ* **234**, 289.
Burkert, A., Bate, M.R. and Bodenheimer, P.: 1997, *MNRAS* **289**, 497.
Chapman, S., Pongracic, H., Disney, M., Nelson, A., Turner, J. and Whitworth, A.: 1992, *Nature* **359**, 207.
Clarke, C.J. and Pringle, J.E.: 1991, *MNRAS* **249**, 584.
Close, L.M., Siegler, N., Freed, M. and Biller, B.: 2003, *ApJ* **587**, 407.
Duquennoy, A. and Mayor, M.: 1991, *A&A* **248**, 485.
Durisen, R.H., Sterzik, M.F. and Pickett, B.K.: 2001, *A&A* **371**, 952.
Halbwachs, J.L., Mayor, M. and Udry, S. in R. Rebolo, E.L. Martin and M.R. Zapatero Osorio (eds.), *Brown Dwarfs and Extrasolar Planets*, ASP Conference Series 134, Brigham Young University, Provo, p. 308.
Hall, S.M., Clarke, C.J. and Pringle, J.E.: 1996, *MNRAS* **278**, 303.
Hosking, J.G.: 2002, PhD thesis, University of Wales, Cardiff.
Hut, P. and Bahcall, J.N.: 1983, *ApJ* **268**, 319.
Klessen, R.S. and Burkert, A.: 2000, *ApJS* **128**, 287.
Klessen, R.S. and Burkert, A.: 2001, *ApJ* **549**, 386.
Klessen, R.S., Burkert, A. and Bate, M.R.: 1998, *ApJ* **501**, L205.
Kroupa, P.: 1995, *MNRAS* **277**, 1522.
Kroupa, P.: 1998, *MNRAS* **298**, 231.
Kroupa, P. and Burkert, A.: 2001, *ApJ* **555**, 945.
Larson, R.B.: 1969, *MNRAS* **145**, 271.
Larson, R.B. in R. Capuzzo-Dolcetta, C. Chiosi, A. di Fazio (eds.), *Physical Processes in Fragmentation and Star Formation*, Kluwer Academic Publishers, Dordrecht, The Netherlands. p. 389.
Lubow, S.H. and Pringle, J.E.: 1996, *MNRAS* **279**, 1251.
Luhman, K.L., Rieke, G.H., Young, E.T., Cotera, A.S., Chen, H., Rieke, M.J., Schneider, G. and Thompson, R.I.: 2000, *ApJ* **540**, 1016.
Mathieu, R.D., Stassun, K., Basri, G., Jensen, E.L.N., Johns-Krull, C.M., Valenti, J.A. and Hartmann, L.W.: 1997, *AJ* **113**, 1841.
Mayor, M. and Mazeh, T.: 1987, *A&A* **171**, 157.
Mazeh, T., Goldberg, D., Duquennoy, A. and Mayor, M.: 1992, *ApJ* **401**, 265.
McDonald, J.M. and Clarke, C.J.: 1993, *MNRAS* **262**, 800.
McDonald, J.M. and Clarke, C.J.: 1995, *MNRAS* **275**, 671.
Norman, M.L. and Wilson, J.R.: 1978, *ApJ* **224**, 497.
Phillips, G.J.: 1986a, *MNRAS* **221**, 571.
Phillips, G.J.: 1986b, *MNRAS* **222**, 111.
Pringle, J.E.: 1989, *MNRAS* **239**, 361.
Reipurth, B. and Clarke, C.J.: 2001, *AJ* **122**, 432.
Reid, I.N., Kirkpatrick, J.D., Liebert, J., Burrows, A., Gizis, J.E., Burgasser, A., Dahn, C.C., Monet, D., Cutri, R., Beichman, C.A. and Skrutskie, M.: 1999, *ApJ* **521**, 613.
Smith, K.W., Bonnell, I.A. and Bate, M.R.: 1997, *MNRAS* **288**, 1041.
Sterzik, M.F. and Durisen, R.H.: 1998, *A&A* **339**, 95.
Sterzik, M.F. and Durisen, R.H.: 2003, *A&A* **400**, 1031.
Tokovinin, A.A.: 2000, *A&A* **360**, 997.
Whitworth, A.P., Chapman, S.J., Bhattal, A.S., Disney, M.J., Pongracic, H. and Turner, J.A.: 1995, *MNRAS* **277**, 727.
Zinnecker, H.: 1982, *Ann. NY Acad. Sci.* **395**, 226.

CONTRACTION AND FRAGMENTATION OF MAGNETIZED ROTATING CLOUDS AND FORMATION OF BINARY SYSTEMS

KOHJI TOMISAKA[1], MASAHIRO N. MACHIDA[1] and TOMOAKI MATSUMOTO[2]

[1]*Theoretical Astrophysics, National Astronomical Observatory, Mitaka, Tokyo, Japan;*
E-mail: tomisaka@th.nao.ac.jp

[2]*Faculty of Humanity and Environment, Hosei University, Fujimi, Chiyoda-ku, Tokyo, Japan*

Abstract. Using three-dimensional (3D) magnetohydrodynamical (MHD) nested-grid simulations, the fragmentation of a rotating magnetized molecular cloud core is studied. An isothermal rotating magnetized cylindrical cloud in hydrostatic balance is considered. We studied non-axisymmetric evolution of the cloud. It is found that non-axisymmetry hardly evolves in the early phase, but it begins to grow after the gas contracts and forms a thin disk. The disk formation and thus growth of non-axisymmetric perturbation are strongly promoted by rotation and magnetic field strength. We found two types of fragmentations: fragmentation from a ring and that from a bar. These two types of fragmentations occur in thin adiabatic cores with the thickness being smaller than 1/4 of the radial size. For the fragments to survive, they should be formed in a heavily elongated barred core or a flat round disk. In the models showing fragmentation, outflows from respective fragments are found as well as that driven by the rotating bar or the disk.

Keywords: ISM: magnetic fields, stars: formation, ISM: jets and outflow, stars: binary

1. Introduction

Since a major part of stars are found as binaries or multiple system rather than single stars, it is important to understand how these multiple systems form from the molecular gas cloud (Abt, 1983; Duquennoy and Mayor, 1991). It is difficult for a single star to capture another one and to form a binary system in the main sequence phase (Kroupa and Burkert, 2001). Therefore, most stars seem to be born as multiples rather than single stars in the gas cloud. In this paper, we study fragmentation in the first H_2 core, which leads to multiple star formation.

Fragmentation of isothermal cloud has been studied by many authors with three-dimensional hydro codes (Miyama et al., 1984; Boss, 1993; Boss and Myhill, 1995; Tsuribe and Inutsuka, 1999). The criterion for an initially uniform spherical cloud to fragment is given by the condition that the thermal-to-gravitational energy ratio $\alpha \simeq 0.2$–0.5, if we restrict ourselves to the isothermal gas. The cloud with larger $\alpha \gtrsim 0.5$ experiences the run-away collapse (Larson, 1969; Norman et al., 1980). However, on the fragmentation of magnetized clouds there have been published only a few papers. Boss (2002 and references therein) studied the fragmentation of the molecular cloud core by three-dimensional (3D) numerical simulations taken the magnetic fields into account by a simple approximation. In his studies, the angular momentum transfer by the magnetic field is neglected. However, the angular momentum is largely removed by the magnetic breaking and the molecular outflow in the course of cloud contraction (Tomisaka, 2000). Therefore, to explore the fragmentation of the magnetized cloud, full 3D MHD simulations are essential. Here, we employ the nested grid MHD code, which always keeps sufficient spatial resolution in the central region and has a large dynamic range necessary for the cloud collapse simulations. In this study, we calculate the non-axisymmetric dynamical contraction of the cloud from $4 \times 10^2\ \mathrm{cm}^{-3}$ to $1 \times 10^{17}\ \mathrm{cm}^{-3}$ in the density, and investigate how the fragmentation proceeds and the outflow occurs in the course of the binary star formation.

2. Model and Numerical Method

To study the nonaxisymetric evolution, we consider a cylindrical isothermal cloud in hydrostatic balance as the initial condition. The rotation axis and the direction of magnetic fields are assumed parallel to the central axis of the cylinder. We assumed an infinitely long cylinder and added a perturbation whose wavelength L is equal to that of the most unstable Jeans mode acquired by the linear analysis (Matsumoto et al., 1994). Using the cylindrical coordinate, the density, azimuthal velocity, and magnetic flux density distributions in the radial direction are taken as follows (Stodółkiewicz, 1963): $\rho_0(r) = \rho_c[1 + (r^2/8H^2)]^{-2}$, $v_{\phi 0}(r) = r\Omega_c[1 + (r^2/8H^2)]^{-1/2}$, and $B_{z0}(r) = B_c[1 + (r^2/8H^2)]^{-1}$, where ρ_c, Ω_c and B_c represent the density, angular rotation speed, and magnetic flux density at the center of the cylindrical cloud respectively, and H is the scale-height as $H^2 = (c_s^2 + B_c^2/8\pi\rho_c)/(4\pi G\rho_c - 2\Omega_c^2)$, where c_s denotes the isothermal sound speed. Molecular gas obeys the isothermal equation of state below $\rho_{cri} \approx 10^{10}\ \mathrm{cm}^{-3}$ but it becomes adiabatic if $\rho \gtrsim \rho_{cri}$ (Tohline, 1982). To mimic this, we adopt two-component equation of state as follows: $p = c_s^2\rho + c_s^2\rho_{cri}(\rho/\rho_{cri})^\gamma$, with $\gamma = 7/5$.

The initial central density and the critical density are chosen as $\rho_{c0} = 4 \times 10^2\ \mathrm{cm}^{-3}$ and $\rho_{cri} = 10^{10}\ \mathrm{cm}^{-3}$. To begin the contraction, we added axisymmetric and non-axisymmetric perturbations to the above magnetohydrostatic equilibrium as $\rho(r, z, \phi) = \rho_0(r)[1 + \delta\rho_z(z)][1 + \delta\rho_\phi(r, \phi)]$, $B_z(r, z, \phi) = B_{z0}(r, \phi)[1 + \delta B_\phi(r, \phi)]$, with $\delta\rho_z(z) = A_z \cos(2\pi z/L)$, $\delta\rho_\phi(r, \phi), \delta B_\phi(r, \phi) = A_\phi(r/H)^2 \cos(m\phi)$, for $r \le H$, and $= A_\phi \cos(m\phi)$, for $r > H$. In this paper, we restrict ourselves to the relative amplitude of the axisymmetric perturbation $A_z = 0.1$. As the non-axisymmetric perturbation, only $m = 2$ mode is included. Models are parameterized with three non-dimensional parameters: the amplitude of the non-axisymmetric perturbation as A_ϕ, the magnetic-to-thermal pressure ratio as $\alpha (\equiv B_c^2/4\pi\rho_c c_s^2)$ and the angular rotation speed normalized by the free-fall timescale as $\omega \equiv \Omega_c/(4\pi G\rho_c)^{1/2}$. We calculated 51 different models with a combination of $A_\phi = (0, 0.01, 0.1, 0.2)$, $\alpha = (0, 0.01, 0.1, 1, 5)$ and $\omega = (0, 0.1, 0.5, 0.7)$.

In order to solve the central high-density region with higher spatial resolutions, we adopt an MHD nested grid method based on the Cartesian coordinate (for detail, see Machida, 2003). In this method, a number of grids with different spacings are prepared, in which finer grids cover the central high-density portion and the coarser ones cover the cloud as a whole. Each grid has successively by a factor two different cell widths. The MHD code uses the numerical fluxes proposed by Fukuda and Hanawa (1999) but modified to solve the barotropic gas. We adopt the mirror symmetry with respect to the $z = 0$ plane and set the fixed boundary condition on the surface of $r = H$. And the periodic boundary condition is applied to $z = L/2$. We use $128 \times 128 \times 32$ cubic cells in the x-, y- and z-directions. We added a new finer grid to keep the Jeans condition, $\lambda_J/4 > h$, with ample margin (Truelove et al., 1997), where λ_J and h are the Jeans length and the cell width, respectively. Whenever one-eighth of the minimum Jeans length becomes smaller than the cell width of the finest grid, a new finer grid is added to the nested grid. Only when a high density fragment escapes from the region covered by the finest grid, the Jeans condition is violated in our simulations.

3. Results

The cylindrical gas cloud contracts to become spherical shape in early collapse phase. If the initial gas cloud has neither magnetic field nor rotation, the cloud continues to collapse spherically and it forms a small massive core in the center of cloud (Larson, 1969) even in the cylindrical cloud (Tomisaka, 1995). In the case of the gas cloud with magnetic fields or rotation, it collapses to form a pseudo-disk due to the magnetic pressure or centrifugal force (Tomisaka, 1995; Nakamura and Hanawa, 1997; Matsumoto et al., 1997). If only the axisymmetric perturbation exists, a disk is formed but it does not fragment as long as the axisymmetry is held. When the non-axisymmetric perturbation is added, the cloud evolves to form a non-axisymmetric shape, and fragments later for some range of the parameters. In any case, the central density (ρ_c) increases with time and the gas becomes adiabatic when $\rho_c > \rho_{cri}$. For convenience, we divide the evolution into two phases: isothermal ($\rho_c < \rho_{cri}$) and adiabatic phase ($\rho_c > \rho_{cri}$). The outflow is driven by the twisted magnetic fields (Tomisaka, 1998, 2002) after $\sim 10^3$ yr has passed from the core formation epoch $\rho_c = \rho_{cri}$.

Figure 1 shows the pole-on view of the adiabatic core (upper panels) and the side-view of the outflow region (lower panels) for three models in the adiabatic phase. The thick line denotes the boundary of the adiabatic core ($\rho \geq \rho_{cri}$) for the upper panels. In the lower panels, the thick lines, which represent the isovelocity lines of $v_z = 0.1$, 1 and 4 km s^{-1}, indicate an outflow region. In Figure 1a, a model without non-axisymmetric perturbations ($A_\phi = 0$) is plotted. Model parameters $(\alpha, \omega) = (0.1, 0.1)$ correspond to the cloud with a weak magnetic field and slow rotation. This cloud collapses almost spherically and forms a small spherical core at the center of the cloud. We have no evidence of fragmentation in

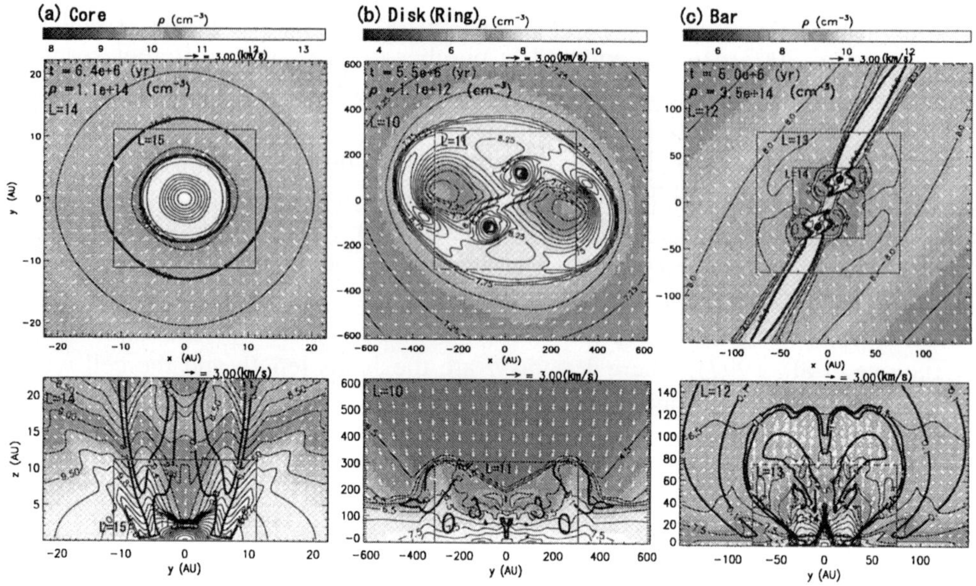

Figure 1. Three typical examples of the adiabatic core are plotted (core-type: left, disk- or ring-type: middle, and bar-type: right). Model parameters are $(A_\phi, \alpha, \omega) =$ (a): (0.0, 0.1, 0.1), (b): (0.01, 0.01, 0.5) and (c): (0.2, 1.0, 0.5). Density (contour and false color) and velocity vectors (arrows) are plotted on a plane including the major axis of the core and the z-axis (lower panels) and on the $z = 0$ plane (upper panels). The elapsed time, maximum density, grid level and scale length are also plotted. Thick lines in the upper panels denote the contour line of $\rho = 10^{10}$ cm^{-3} showing the outline of the adiabatic core, while those in the lower panels mean the isovelocity curves representing the outflow region ($v_z = 0.1$, 1 and 4 km s^{-1}).

this model. We found a strong outflow is driven near the adiabatic core after the central density reaches at $\sim 10^{13}$ cm^{-3} (lower panel). A model with small non-axisymmetric perturbation ($A_\phi = 0.01$) is shown in the middle panel (b). In this model, a pseudo-disk is formed in the isothermal phase, because the gas cloud is rotating rapidly ($\omega = 0.5$). The formed disk evolves into a ring in the adiabatic phase, and the fragmentation occurs in the inner side of the ring. Two contracting cores at $(x, y) =$ (75AU, 125AU) and (−75AU, −125AU) are formed. The right panel (c) shows the model with a large amplitude of the non-axisymmetric perturbations ($A_\phi = 0.2$). In this case, non-axisymmetric patterns grow enough in the isothermal phase and a bar is formed at the beginning of the adiabatic phase. In the adiabatic phase, the bar grows longer and thinner. Finally, the bar fragments and forms two contacting cores [$(x, y) =$ (−10AU, −25AU), (10AU, 25AU)]. Figure 1 shows that there are at least two kinds of fragmentation processes: fragmentation from a ring (b) and that from a bar (c).

Figure 1c shows that two lobes of outflow are ejected from respective contracting cores ($y \approx \pm 25$AU), whose velocities reach ≈ 8 km s^{-1}. As for the outflow, low-velocity outflows are seen also in Figure 1b, each of which is connected to the contracting core. In both the bar and ring fragmentations, two types of outflows co-exist: the inner one is connected to the cores formed by the fragmentation and driven by their spin motions, and the other is driven by the rotation motion of the ring or the bar. However, the strength of the outflow is different for respective fragmentation patterns. Fast outflows are appeared in the bar fragmentation (c), while slow ones appear in the ring fragmentation case (b). This difference comes from the angular momentum distribution between the spin motion of the cores and their orbital one. In the ring fragmentation, most angular momentum has a form of the orbital one. On the other hand, in the bar fragmentation, the angular momentum is almost evenly shared between the spin and the orbital ones. In model of Figure 1c, the fragmented cores are expected to merge each other due to their small orbital angular momentum.

4. Discussion

As mentioned in the third section, patterns of fragmentation and shapes of the outflow depend on the growth of the non-axisymmetry. This seems to show us that the cloud evolution is classified by the growth of non-axisymmetry. In order to characterize the shape of the central region, we define oblateness ($\varepsilon_{\rm ob}$) and axis ratio($\varepsilon_{\rm ar}$) as follows: $\varepsilon_{\rm ob} \equiv \sqrt{h_{\rm l} \cdot h_{\rm s}}/h_{\rm z}$, $\varepsilon_{\rm ar} \equiv h_{\rm l}/h_{\rm s}$, where $h_{\rm l}$, $h_{\rm s}$ and $h_{\rm z}$ mean the length of the long-, short-, and z-axis for the gas with $\rho > (1/10)\rho_{\rm max}$, respectively.

In Figure 2, the oblateness and the axis ratio of the central region are plotted for all models at the beginning of the adiabatic phase ($\rho_{\rm c} = \rho_{\rm cri} = 10^{10}\ {\rm cm}^{-3}$). Curves in this figure illustrate the evolutionary tracks for four different models $(A_\phi, \alpha, \omega) = (0.01, 0.01, 0.5)$, $(0.2, 0.1, 0.5)$, $(0.2, 1, 0.5)$, and $(0.2, 5, 0.5)$. The cloud evolves from the lower-left corner, which means the early shape is almost spherical, to the upper-right region via the upper-left region. First, the cloud evolves vertically upwardly. This indicates that the oblateness increases. Then, the curves change their directions to the right for three models with $A_\phi = 0.2$. Evolutionary tracks of the high-density region in Figure 2 show very clearly that the non-axisymmetric perturbation grows only after the disk is formed. That is, the axis ratio increases after the oblateness grows sufficiently large ($\varepsilon_{\rm ob} \geq 4$). Because the gas is partially supported in the radial direction by the magnetic pressure and the centrifugal force, the radial contraction is delayed than that of the z-direction. As a result, the disk is formed ($\varepsilon_{\rm ob} > 4$) earlier for models with larger α and ω. In such models, the non-axisymmetry grows sufficiently and forms a large bar in the isothermal phase. On the contrary, the non-axisymmetry hardly evolves in the model with small α and ω, because it takes a long time for disk formation.

Adiabatic cores are classified into three types: *core* (Figure 1a: $\varepsilon_{\rm ob} < 4$ and $\varepsilon_{\rm ar} < 4$), *disk* (Figure 1b: $\varepsilon_{\rm ob} > 4$ and $\varepsilon_{\rm ar} < 4$), and *bar* (Figure 1c: $\varepsilon_{\rm ob} > 4$, $\varepsilon_{\rm ar} > 4$). Some *disks* evolve into rings, in which the central density is lower than the periphery. The model with a cross ($\times$) experiences fragmentation. In the models

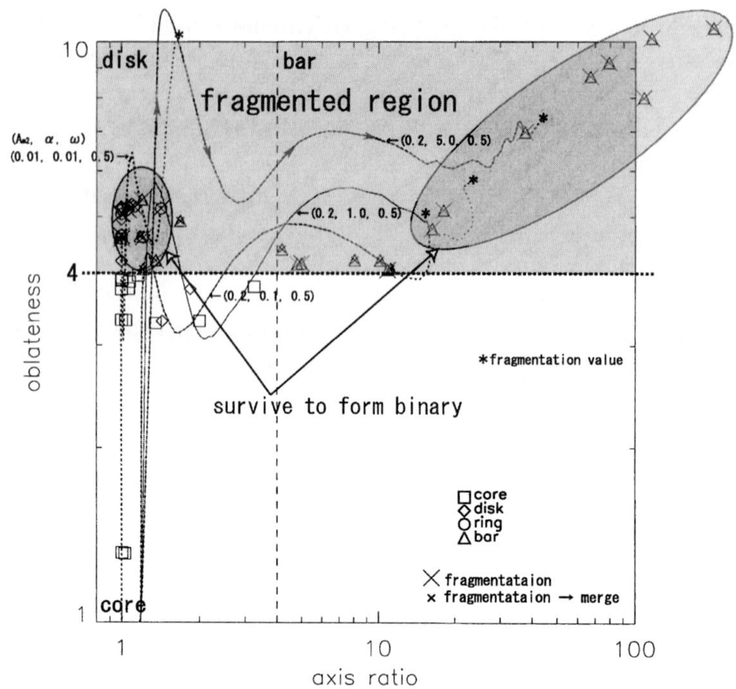

Figure 2. The axis ratio and oblateness are plotted when the gas at the cloud center becomes adiabatic $\rho_c = \rho_{cri}$. The abscissa indicate the axis-ratio and the ordinate is the oblateness. Both are calculated for the gas with $\rho > (1/10)\rho_{max}$. The symbols indicate the shape of the adiabatic core when the fragment occurs or calculation ends. Cross represents a core which experiences the fragmentation. Fragmentation occurs above the horizontal dotted line ($\varepsilon_{ob} > 4$). The domains in large circles mean that fragmented cores survive to form binary or multiple stars. The solid lines represent the evolutional tracks for some typical parameters. Dotted lines represent the evolution after fragmentation. Asterisk shows the point of fragmentation. The symbols □ (core), ◇ (disk), ○ (ring) and △ (bar) denote the shape of the adiabatic core when the fragmentation occurs or the calculation ends.

with small crosses, the fragments finally merge each other before the calculation ends. On the other hand, the large cross means that the fragments survive without merging at the end of the calculation. Since the models with crosses distribute only in the region $\varepsilon_{ob} > 4$, it is concluded that the necessary condition for fragmentation is $\varepsilon_{ob} > 4$, which means that a thin disk or bar is necessary for fragmentation. There are exceptional three models which do not fulfill the criterion of the fragmentation. In these three models the adiabatic core evolves to form an almost axisymmetric thin disk or ring because non-axisymmetry is counteracted in the isothermal phase. It is concluded that for the adiabatic core to fragment, the central gas needs to form a sufficiently thin disk or bar in the isothermal phase. The large crosses are distributed in two regions enclosed by circles. The fragmented cores in the circles seem to evolve to the binary stars because the length of the semi-major axis of them expands or vibrates. Models in which the fragments merge with each other are distributed outside the domain. Outside of the domain, a few fragments survive at the end of the calculation but the separation between two cores shrinks with time. These will not grow into the binary stars.

In conclusion, if the following criteria are fulfilled at the end of the isothermal phase, fragmentation occurs and fragments survive to form binary or multiple stars.

(1) $\varepsilon_{\rm ob} > 4$: fragmentation condition.
(2) $0 < \varepsilon_{\rm ar} < 2$, or $\varepsilon_{\rm ar} > 10$: survival condition against merger.

The symbol in this figure represents the shape of the adiabatic core when the fragmentation occurs or the calculation ends. Figure 2 confirms the idea that fragmentation occurs from a ring (○) or bar (△) not a core (□).

We showed the domain which may evolve into binary stars in Figure 2. Outside the binary-forming domain, the core forming model (□) results in a compact massive core at the center. The bar-fragmentation model (△) in which a mutual merger of the fragments occurs leads a compact massive core at the center. These configurations seem to result in a single star formation.

Acknowledgements

We are greatly benefited from discussion with Dr T. Hanawa and Dr H. Koyama. Numerical calculations were carried out by Fujitsu VPP5000 at the Astronomical Data Analysis Center, the National Astronomical Observatory of Japan. This work is supported partially by the Grants-in-Aid from MEXT (11640231, 14540233 [KT], and 14740134[TM]).

References

Abt, H.A.: 1983, *ARA&A* **21**, 343.
Boss, A.P.: 1993, *ApJ* **410**, 157.
Boss, A.P.: 2002, *ApJ* **568**, 743.
Boss, A.P. and Myhill, E.A.: 1995, *ApJ* **451**, 218.
Duquennoy, A. and Mayor, M.: 1991, *A&A* **248**, 485.
Fukuda, F. and Hanawa, T.: 1999, *ApJ* **517**, 226.
Kroupa, P. and Burkert, A.: 2001, *ApJ* **555**, 945.
Larson, R.B.: 1969, *MNRAS* **145**, 271.
Machida, M.N.: 2003, *Binary Star Formation and Mass Outflow in the Molecular Cloud Core*, PhD thesis, Hokkaido U.
Matsumoto, T., Hanawa, T. and Nakamura, F.: 1997, *ApJ* **478**, 569.
Matsumoto, T., Nakamura, F. and Hanawa, T.: 1994, *PASJ* **46**, 243.
Miyama, S.M., Hayashi, C. and Narita, S.: 1984, *ApJ* **279**, 621.
Nakamura, F. and Hanawa, T.: 1997, *ApJ* **480**, 701.
Norman, M.L., Wilson, J.R. and Barton, R.T.: 1980, *ApJ* **239**, 968.
Stodółkiewicz, J.S.: 1963, *Acta Astron.* **13**, 30.
Tohline, J.E.: 1982, *Fundam. Cosm. Phys.* **8**, 1.

Tomisaka, K.: 1995, *ApJ* **438**, 226.
Tomisaka, K.: 1998, *ApJ* **502**, L163.
Tomisaka, K.: 2000, *ApJ* **528**, L41.
Tomisaka, K.: 2002, *ApJ* **575**, 306.
Truelove, J.K., Klein, R.I., Mckee, C.F., Holliman, J.H., Howell, L.H. and Greenough, J.A.: 1997, *ApJ* **489**, L179.
Tsuribe, T. and Inutsuka, S.: 1999, *ApJ* **523**, L155.

STAR FORMATION AND THE HALL EFFECT

MARK WARDLE

Physics Department, Macquarie University, NSW, Australia; E-mail: wardle@physics.mg.edu.au

Abstract. The breakdown of flux-freezing in molecular clouds and protostellar discs is usually approximated by ambipolar diffusion at low densities or by resistive diffusion at high densities. Here an intermediate regime is discussed in which the Hall term in the conductivity tensor is significant, and the vector evolution of the magnetic field, and therefore the evolution of the system under consideration is dramatically altered. Calculations of charged particle abundances in dense gas in molecular clouds and protostellar discs demonstrate that Hall diffusion is important over a surprisingly broad range of conditions.

Keywords: molecular clouds, star formation, accretion discs, instabilities, magnetohydrodynamics

1. Introduction

The magnetic field in molecular clouds provides pressure support against gravity and carries away angular momentum prior to and during the collapse of cloud cores to form stars. Magnetic fields may also drive the evolution of protostellar discs through the magnetorotational instability, dynamo activity, or by magnetic launching and collimation of jets from their surfaces or inner edges.

The diffusion of magnetic field through the weakly ionised molecular gas plays a crucial role, allowing gas to slip through the magnetic field, or vice-versa. It is usually considered using one of the two limits: (i) ambipolar diffusion, in which the magnetic field is frozen into the charged species and drifts along with them through the neutrals; and (ii) resistive diffusion, in which charged particles are completely decoupled from the field by collisions with neutrals. Both limits neglect Hall diffusion, which qualitatively changes the evolution of the magnetic field from any given initial configuration and has profound implications for the magnetically-mediated processes associated with star formation.

Here some calculations illustrating the relevance of Hall diffusion to molecular clouds and protostellar discs are presented.

2. Hall Diffusion

The diffusion of a magnetic field in weakly ionised gas is determined by the drift of charged particles through the dominant neutral component in response to the electric field in the neutral rest frame, $\mathbf{E}'$. The drift speed parallel to the magnetic field is set by the drag associated with neutral collisions. In the plane perpendicular

Astrophysics and Space Science **292:** 317–323, 2004.

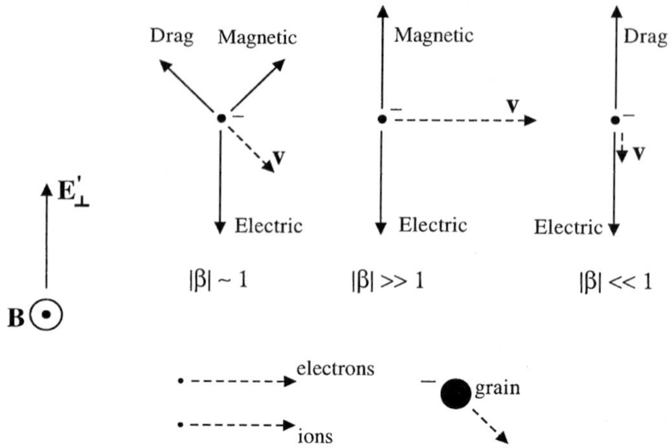

Figure 1. Charged particle drifts in a weakly ionised gas. The upper panel shows the drift speed and force balance for negatively charged particles with different hall parameters (see text). At typical molecular cloud densities ions and electron drifts are set by a balance between the electric and magnetic forces and they drift perpendicular to both the electric and magnetic fields (lower panel). The grain drift is affected by collisions with the neutral particles, the resultant drag causes them to drift at an oblique angle to $\mathbf{E}'_\perp$

to the magnetic field, the electric force $Ze\mathbf{E}'_\perp$ on a particle of charge Ze and mass m is balanced by the magnetic and drag forces (see Figure 1). The relative importance of these two forces is determined by the Hall parameter

$$\beta = \frac{ZeB}{mc}\frac{1}{\gamma\rho}. \tag{1}$$

In the limit $|\beta| \gg 1$, the drag force is negligible, and the drift speed $\mathbf{v}$ satisfies $\mathbf{E}'_\perp + \mathrm{v} \times \mathbf{B}/c = 0$. In this case the charged particle is tied to the magnetic field line. In the other limit $|\beta| \ll 1$, the magnetic force is negligible and the drag force $\gamma m\rho\mathbf{v} = Ze\mathbf{E}'_\perp$.

At typical molecular cloud densities, ions and electrons have $|\beta| \gg 1$ and drift perpendicular to both $\mathbf{B}$ and $\mathbf{E}'$. Charged grains, however, have a large collisional cross-section, so that drag is much more important and $|\beta| \lesssim 1$. Grains therefore drift obliquely to $\mathbf{E}'_\perp$. The grain population acquires a significant charge through the sticking of electrons, and the ensemble of drifting grains modifies the vector relationship between the current density $\mathbf{J}$ and $\mathbf{E}'$. This relationship can be expressed in terms of a tensor conductivity $\mathbf{J} = \sigma\mathbf{E}'$:

$$\mathbf{J} = \sigma \cdot \mathbf{E}' = \sigma_\parallel \mathbf{E}'_\parallel + \sigma_\mathrm{H}\hat{\mathbf{B}} \times \mathbf{E}'_\perp + \sigma_\mathrm{P}\mathbf{E}'_\perp\,, \tag{2}$$

where $\mathbf{E}'_\parallel$ and $\mathbf{E}'_\perp$ are the components of $\mathbf{E}'$ parallel and perpendicular to $\mathbf{B}$, and $\sigma_\parallel$, σ_H and σ_P are the field-parallel, Hall, and Pedersen conductivities, respectively.

The magnetic field evolves according to

$$\frac{\partial \mathbf{B}}{\partial t} = \nabla \times (\mathbf{v} \times \mathbf{B}) - c\nabla \times \mathbf{E}' . \quad (3)$$

The departure from flux-freezing magnetic diffusion arises from the second term on the right-hand side and is determined by the magnetic field configuration via the conductivity tensor: $\mathbf{E}' = \sigma^{-1}\nabla \times \mathbf{B}/4\pi$. There are three distinct diffusion regimes:

1. ambipolar diffusion: $|\sigma_{\mathrm{H}}| \ll \sigma_{\mathrm{P}} \ll \sigma_{\|}$;
2. hall diffusion: $\sigma_{\mathrm{P}} \lesssim |\sigma_{\mathrm{H}}| \ll \sigma_{\|}$; and
3. resistive diffusion: $|\sigma_{\mathrm{H}}| \ll \sigma_{\mathrm{P}} \approx \sigma_{\|}$.

The vector evolution of the magnetic field in the presence of Hall diffusion is quite different from the other two regimes. Many of the magnetic field geometries adopted in studies of ambipolar or resistive diffusion, in which the magnetic field lies within a plane of symmetry, break down when Hall drifts are present. The evolution is not invariant under global reversal of the magnetic field and plane-polarised damped Alfvén waves do not exist (Wardle and Ng, 1999). The consequences of Hall diffusion during star formation and the subsequent evolution of protostellar discs are profound, as found for example in the magnetorotational instability (Wardle, 1999; Balbus and Terquem, 2001; Salmeron and Wardle, 2004).

3. Charged Species and Conductivity in Molecular Clouds

The degree of Hall diffusion in molecular clouds depends on charged particle abundances; especially of grains. Figure 2 shows the abundances of charged species in molecular clouds for two standard grain models: 0.1 μm grains, and grains with an MRN size distribution. These ionisation equilibrium calculations follow Umebayashi and Nakano (1990) and Nishi et al. (1991) in following the abundance of H^+, H_3^+, He^+, C^+, representative molecular (m^+) and metal (M^+) ions, electrons, and grains. The calculations have been improved by allowing for the higher grain charge states that appear at temperatures in excess of 10 K. Here I have adopted an ionisation rate $\zeta = 10^{-17}\,\mathrm{s}^{-1}\,\mathrm{H}^{-1}$ and gas temperature 30 K. At this temperature, 0.1 μm grains acquire one or two electrons via sticking of electrons. At low densities this has negligible effect on the gas phase ionisation levels because of the relatively small numbers of grains ($n_{\mathrm{g}}/n_{\mathrm{H}} \sim 10^{-12}$). Once the density exceeds $10^{11}\,\mathrm{cm}^{-3}$, the sticking of electrons and ions to grains becoming increasingly important, with recombinations occurring predominantly on grain surfaces rather than in the gas phase. An MRN grain size distribution contains many more small grains which typically acquire a single electron. These more numerous grains dominate the recombination process at lower densities ($n_{\mathrm{H}} \gtrsim 10^7\,\mathrm{cm}^{-3}$).

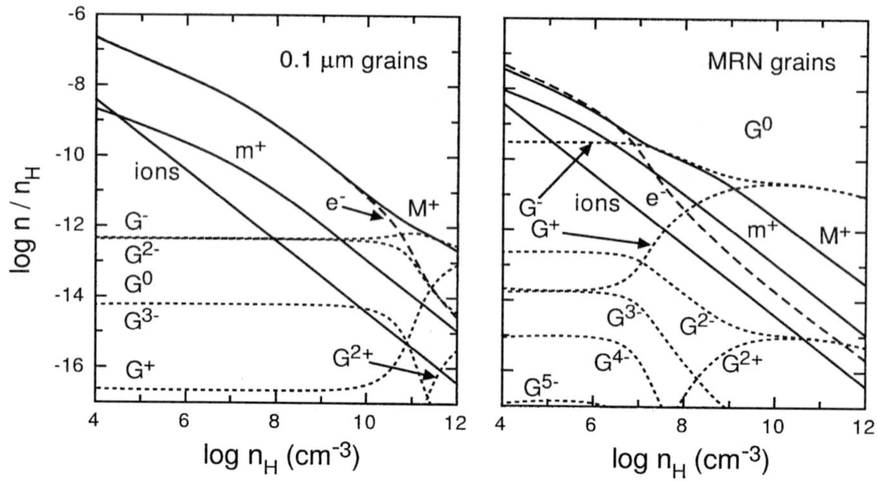

Figure 2. Abundances of charged species in molecular clouds for (*left*) 0.1 μm and (*right*) MRN grain models.

The conductivity tensor for the two grain models can be calculated once the magnetic field is specified. It is instructive to consider the ratio of the Hall and Pedersen conductivities (σ_H and σ_P as a measure of the significance of the Hall effect in magnetic diffusion. Contours of $|\sigma_H|/\sigma_P$ are plotted in the B–n_H plane for the 0.1 μm and MRN grain models in Figures 3 and 4 respectively. In light of the qualitative changes to magnetic field evolution, I recommend feeling nervous about neglecting Hall diffusion when $|\sigma_H|/\sigma_P$ is above 0.1, the second contour level in these plots.

For the 0.1 μm grain model (Figure 3), two islands of Hall diffusion are apparent. The left-hand island follows the locus where the grain Hall parameter is of order

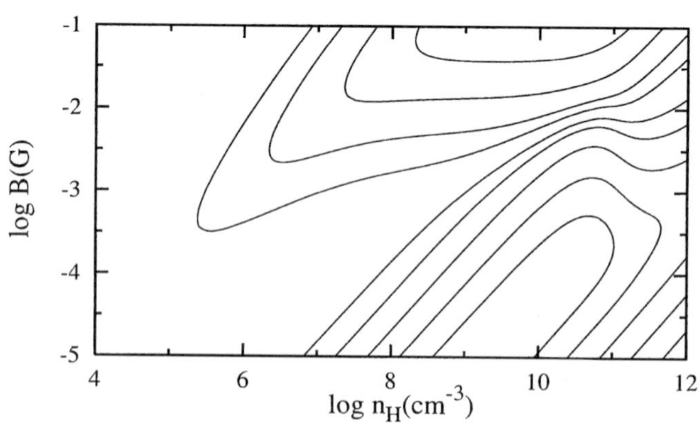

Figure 3. Ratio of the Hall and Pedersen conductivities for the 0.1 μm grain model. The contour levels are set at $|\sigma_H|/\sigma_P = 0.01, 0.1, 1, 10$ and 100.

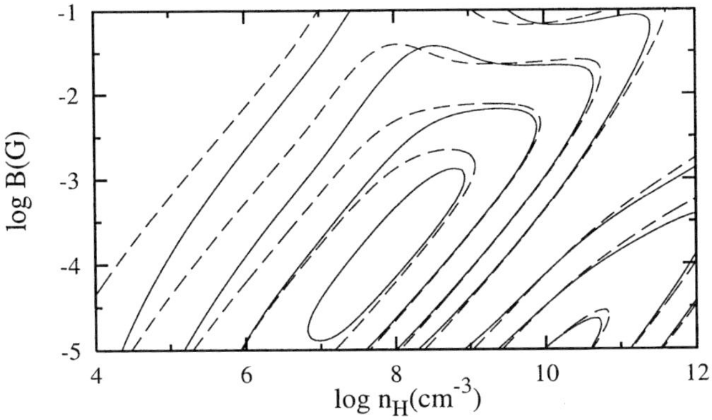

Figure 4. Ratio of the magnitudes of the Hall and Pedersen conductivities for an MRN grain model (*solid contours*) and for MRN grain distribution with ice mantles (*dashed contours*). The contour levels are set at $|\sigma_H|/\sigma_P = 0.01, 0.1, 1$ and 10.

unity and grains partially decouple from the magnetic field. The island is truncated at low densities where the relative abundance of charged grains becomes too small to influence the conductivity significantly (see Figure 2). The right-hand island occurs where the ion Hall parameter is of order unity. A zero-line occurs between these two islands where the contributions of grains and ions to σ_H cancel because they carry charge of opposite signs.

The location and size of the grain island depends on the grain size distribution, as size determines both the typical charge acquired by sticking of thermal electrons and ions and the drag coefficient with the neutrals. In Figure 4, the solid contours show $|\sigma_H|/\sigma_P$ for an MRN size distribution. This has many small grains that acquire a single negative charge (see Figure 2), so the grain island extends to lower densities but falls off at the highest densities when the abundances of positive and negatively charged grains are almost equal and their Hall contributions cancel. The addition of ice mantles to the grains (dashed contours in Figure 4) tends to extend the island to the left, as grains have a higher drag coefficient and decouple at lower densities for a given magnetic field strength.

Clearly, Hall diffusion is significant for both the 0.1 μm and MRN grain models at densities $\gtrsim 10^6$ cm^{-3} and the 0.1–10 mG magnetic field strengths in molecular clouds. However, the effect is reduced if there are many more small grains (such as PAHs, or an MRN distribution that extends down to a few Angstrom), or if the characteristic grain size is much larger than 0.1 μm. In the former case, small grains are the dominant charged species at all densities (see Nishi et al., 1991) and the near-equality in abundances of positively- and negatively-charged grains supresses the Hall effect. In the latter case, grains are sufficiently rare that the grain island is shifted to $\gtrsim 0.1$ G, in excess of molecular cloud magnetic field strengths.

4. Conductivity in Protostellar Discs

Magnetic coupling in protostellar discs is thought to be poor because of two contributing factors. First, the magnetic field must couple to the gas on timescales orders of magnitude shorter than those in molecular clouds. Second, because the gas density is high, recombination occurs more rapidly and the fractional ionisation is low. Third, within a few AU of the central protostar the disc column density shields the gas from cosmic-rays or X-rays. This has led to a picture of magnetic activity occurring only in the surface layers over much of the disc (Gammie, 1996; Wardle, 1997).

By way of illustration, let us examine the conductivity tensor in a minimum-mass solar nebula model at 1 AU from the central star. The temperature and midplane density are $n_{\rm H} = 6 \times 10^{14}\,{\rm cm}^{-3}$ and $T = 280\,{\rm K}$ respectively. The disc is assumed to be ionised from above and below by cosmic rays at a rate $10^{-17} \exp(-\Sigma(z)/96\,{\rm g\,cm}^{-2})\,{\rm s}^{-1}\,{\rm H}^{-1}$ where $\Sigma(z)$ is the surface density from height z to the surface, by X-rays from the protostar according to Figure 3 of Igea and Glassgold (1999) and by natural radioactivity at a uniform rate $10^{-21}\,{\rm s}^{-1}\,{\rm H}^{-1}$.

In principle, the grain population is determined by competition between growth or sublimation of ice mantles, sticking, shattering, gravitational settling to the disc midplane, and stirring by convection or turbulence (e.g., Weidenschilling and Cuzzi, 1993). For the sake of definiteness, I consider two simple models: (i) 0.1 μm grains, and (ii) no grains (assuming that they have settled to the midplane). In the latter case, metal atoms play a key role in determining the ionisation fraction in the absence of grains and are conservatively assumed to be depleted by a factor of 10^3 over the interstellar value.

The resulting vertical profile of the conductivity tensor in these two models is plotted in the top panel of Figure 5 for a uniform 100 mG magnetic field. The much larger conductivities in the no-grain model within three scale heights of the midplane reflects the lack of grains—the charge is carried by more mobile free ions and electrons. The decoupling of ions from the magnetic field for densities $\gtrsim 10^9\,{\rm cm}^{-3}$ means that Hall component is larger than the Pedersen component beween 1.5 and 4 scale heights, and is 80% of $\sigma_{\rm P}$ even at the midplane. In the 0.1 μm grain model, the Hall conductivity dominates the Pedersen conductivity within 4 scale heights of the midplane.

One measure of whether the conductivity is sufficient for the magnetic field to interact with the disc material, is whether the field is unstable to the magnetorotational instability. This is determined by comparing the coupling parameter $\chi = \omega_{\rm c}/\Omega$ to the ratio of Alfvén speed to sound speed, $v_{\rm A}/c_{\rm s}$. Here $\omega_{\rm c}$ is the frequency above which ideal MHD breaks down and Ω is the Keplerian frequency. If Hall diffusion is unimportant the criterion is $\chi \gtrsim v_{\rm A}/c_{\rm s}$, if it is dominant (and the magnetic field has the correct orientation), then $\chi \gtrsim (v_{\rm A}/c_{\rm s})^2$ is necessary.

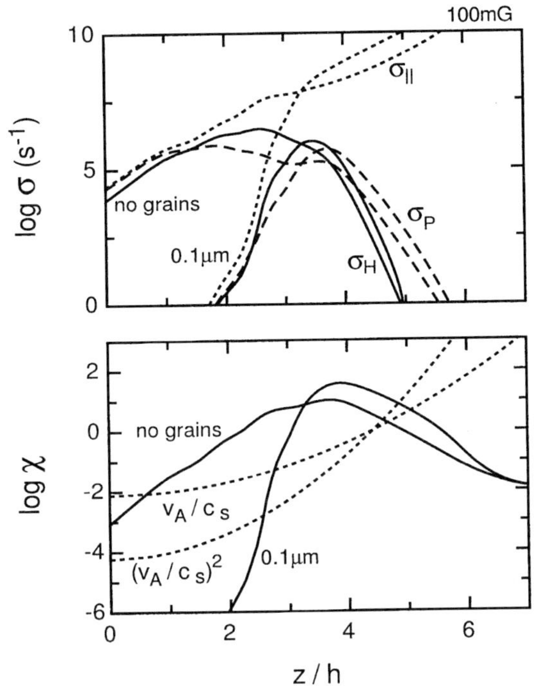

Figure 5. Upper panel: conductivity tensor components as a function of height above the midplane at 1AU in a minimum solar nebula for a dusty disc with 0.1 μm grains and a model in which grains have settled to the midplane. *Lower panel:* The vertical profile of the coupling parameter $\chi = \omega_c/\Omega$.

The coupling parameter is plotted as a function of height in the lower panel of Figure 5. The entire disc is magnetically active in the settled-grain case, whereas in the single-size grain model the layers above 2.5 scale heights are active. In both cases, Hall diffusion is important throughout the active regions.

References

Balbus, S.A. and Terquem, C.: 2001, *ApJ* **552**, 235.
Gammie, C.F.: 1996, *ApJ* **457**, 355.
Hawley, J.F. and Balbus, S.A.: 1991, *ApJ* **376**, 223.
Igea, J. and Glassgold, A.E.: 1999, *ApJ* **518**, 848.
Nishi, R., Nakano, T. and Umebayashi, T.: 1991, *ApJ* **368**, 181.
Umebayashi, T. and Nakano, T.: 1990, *MNRAS* **243**, 103.
Wardle, M. and Salmeron, R.: 2004, *Ap&SS* **292**, 449.
Wardle, M.: 1997, in: D. Wickramasinghe, L. Ferrario and G. Bicknell (eds.), IAU Colloq. 163, *Accretion Phenomena and Related Outflows*, ASP: San Francisco, p. 561.
Wardle, M.: 1999, *MNRAS* **307**, 849 (W99).
Wardle, M. and Ng, C.: 1999, *MNRAS* **303**, 239.
Weidenschilling, S.J. and Cuzzi, J.N.: 1993, in: E.H. Levy and J.I. Lunine (eds.), *Protostars and Planets III*, University Arizona Press, Tucson, p. 1031.

SUBMILLIMETER STUDIES OF PRESTELLAR CORES AND PROTOSTARS: PROBING THE INITIAL CONDITIONS FOR PROTOSTELLAR COLLAPSE

PHILIPPE ANDRÉ[1], JEROEN BOUWMAN[1], ARNAUD BELLOCHE[2] and PATRICK HENNEBELLE[2]

[1]*CEA Saclay, Service d'Astrophysique, Orme des Merisiers, Bât. 709, F-91191 Gif-sur-Yvette Cedex; E-mail: pandre@cea.fr*

[2]*Laboratoire de Radioastronomie, ENS, Paris Cedex France*

Abstract. Improving our understanding of the earliest stages of star formation is crucial to gain insight into the origin of stellar masses, multiple systems, and protoplanetary disks. We discuss recent advances made in this area, thanks to submillimeter mapping observations with large single-dish telescopes and interferometers. Although ambipolar diffusion appears to be too slow cores, there is nevertheless good evidence that the gravitational collapse of isolated protostellar cores is strongly magnetically controlled. We also argue that the beginning of protostellar collapse is much more violent in cluster-forming clouds than in regions of distributed star formation.

Keywords: ISM: structure, ISM: gravitational collapse, stars: protostellar accretion rates

1. Introduction

One of the main limitations in our present understanding of the star formation process is that we poorly know the initial conditions for protostellar collapse. In particular, there is a major controversy at the moment between two schools of thought for the formation and evolution of dense cores within molecular clouds: The classical picture based on magnetic support and ambipolar diffusion (e.g., Shu et al., 1987; Mouschovias and Ciolek, 1999) has been seriously challenged by a new, more dynamical picture, which emphasizes the role of supersonic turbulence in supporting clouds on large scales and generating density fluctuations on small scales (e.g., Vázquez–Semadeni et al., 2000; Klessen et al., 2000; Ballesteros–Paredes, 2004). Improving our knowledge of the earliest stages of star formation is thus of prime importance, especially since there is now good evidence that these stages control the origin of the stellar IMF (e.g., Motte et al., 1998).

Observationally, there are two complementary approaches to constraining the initial conditions for collapse: (1) studying the structure and kinematics of gravitationally-bound starless cores observed prior to protostar formation (i.e., *prestellar* cores such as L1544, cf Figure 1a and Ward–Thompson et al., 1994), and (2) studying the structure of young accreting protostars observed sufficiently early after point mass formation that they still retain detailed memory of their

Astrophysics and Space Science **292:** 325–337, 2004.

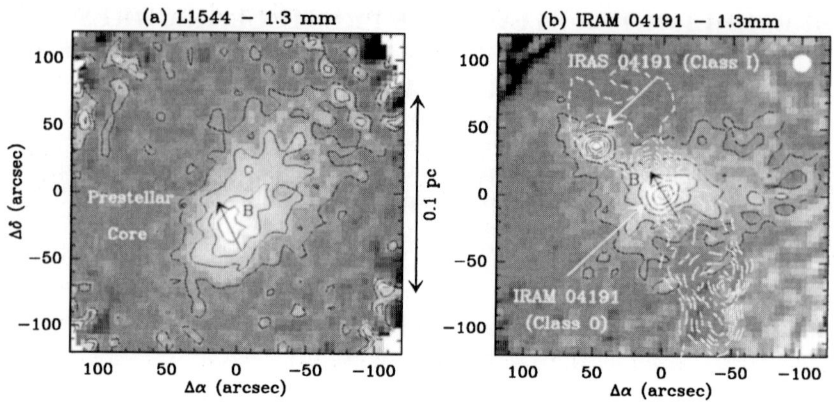

Figure 1. Dust continuum maps of L1544 (a) and IRAM 04191 (b) at 1.3 mm taken with the IRAM 30 m telescope and the MPIfR bolometer array (from Ward-Thompson et al., 1999 and André et al., 1999, respectively). Base contour and contour step: 20 mJy/13″-beam. The directions of the magnetic field measured in L1544 and IRAM 04191 by submillimeter polarimetry with SCUBA on JCMT (Ward-Thompson et al., 2000, André et al., in prep.), as well as the collimated CO(2-1) bipolar flow emanating from IRAM 04191, are shown.

initial conditions (e.g., Class 0 objects such as IRAM 04191, cf Figure1b and André et al., 1993).

Here, after a brief introduction on collapse models (Section 1.1), we review recent observational advances concerning the density and velocity structure of cloud cores in Sections 2 and 3, respectively. We conclude in Section 4 with a comparison between observations and theoretical models.

1.1. Collapse Initial Conditions: Theory

The inside-out collapse model of Shu (1977), starting from a singular isothermal sphere (SIS) or toroid (cf Li and Shu, 1997), is well known and underlies the 'standard' picture of isolated, low-mass star formation (e.g., Shu et al., 1987).

Other collapse models exist, however, which adopt different initial conditions. In particular, Whitworth and Summers (1985) have shown that there is a two-parameter continuum of similarity solutions to the problem of isothermal spherical collapse. One of the parameters measures how close to hydrostatic equilibrium the system is initially, while the other parameter reflects how important external compression is in initiating the collapse. In this continuum, the solutions proposed by Shu (1977) and Larson (1969)–Penston (1969) represent two extreme limits. All of the similarity solutions share a universal evolutionary pattern. At early times ($t < 0$), a compression wave (initiated by, e.g., an external disturbance) propagates inward leaving behind it a $\rho(r) \propto r^{-2}$ density profile. At $t = 0$, the compression wave reaches the center and a point mass forms which subsequently grows by accretion. At later times ($t > 0$), this wave is reflected into a rarefaction or expansion wave, propagating outward through the infalling gas, and leaving behind it a free-fall $\rho(r) \propto r^{-1.5}$ density distribution. Several well-known features of the Shu model (such as the expansion wave) are thus in fact common to all solutions. The various solutions can be distinguished by the *absolute* values of the density and velocity at $t \sim 0$. In particular, the Shu (1977) solution has $\rho(r) = (a_s^2/2\pi G)\, r^{-2}$ (where a_s is the isothermal sound speed) and is static ($v = 0$) at $t = 0$, while the Larson–Penston (1969) solution is ~4.4 times denser and far from equilibrium ($v \approx -3.3\ a_s$). During the accretion phase ($t > 0$), the infall envelope is a factor ~7 denser in the Larson–Penston solution. Accordingly, the mass infall rate is also much larger in the Larson–Penston case ($\sim 47\, a_s^3/G$) than in the Shu case ($\sim a_s^3/G$).

In practice, however, protostellar collapse is unlikely to be strictly self-similar, and the above similarity solutions can only be taken as plausible asymptotes. More realistic initial conditions than the SIS are provided by the so-called 'Bonnor-Ebert' spheres (e.g., Bonnor, 1956), which represent the equilibrium states for self-gravitating isothermal spheres and have a flat density profile in their central region. Such spheres are stable for a center-to-edge density contrast <14 and unstable for a density contrast >14 (e.g., Bonnor, 1956). Numerical hydrodynamic simulations of cloud collapse starting from such initial conditions (e.g., Foster and Chevalier, 1993; Hennebelle et al., 2003) find that the Larson–Penston similarity solution is generally a good approximation near point-mass formation ($t = 0$) *at small radii*, but that the Shu solution is more adequate at intermediate $t \geq 0$ times, before the expansion wave reaches the edge of the initial, precollapse dense core. In general, the mass accretion rate is thus expected to be time-dependent.

2. Density Structure

2.1. Prestellar Cores

Two main approaches have been used to trace the density structure of cloud cores (see Figure 2a):(1) mapping the optically thin (sub)millimeter continuum *emission* from the cold dust contained in the cores, and (2) mapping the same cold core dust in *absorption* against the background infrared emission (originating from warm cloud dust or remote stars).

Ward–Thompson et al. (1994, 1999) and André et al. (1996) employed the first approach to probe the structure of prestellar cores (see also Shirley et al., 2000). Under the simplifying assumption of spatially uniform dust temperature and emissivity properties, they concluded that the radial density profiles of isolated prestellar cores were *not* consistent with the single $\rho(r) \propto r^{-2}$ power law of the SIS but were flatter than $\rho(r) \propto r^{-1}$ in their inner regions (for $r \leq R_{\rm flat}$), and approached $\rho(r) \propto r^{-2}$ only beyond a typical radius $R_{\rm flat} \sim$ 2500–5000 AU.

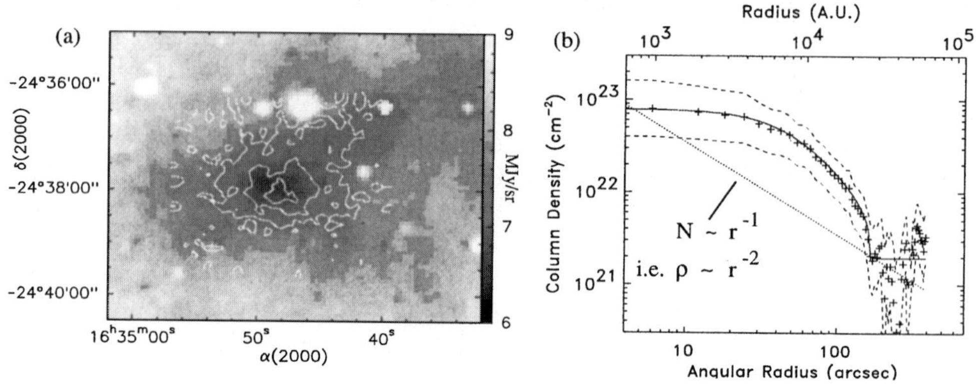

Figure 2. (a) ISOCAM 6.75 μm absorption image of the prestellar core L1689B in Ophiuchus (scale on the right in MJy/sr). The 1.3 mm continuum emission map of Andrè et al. (1996) with the IRAM 30 m telescope is superposed as contours (levels: 10, 30, 50 mJy/13″-beam). (b) Column density profile of L1689B (crosses) derived from the absorption map shown in (a) by averaging the intensity over elliptical annuli for a 40° sector in the southern part of the core. The dashed curves show the most extreme profiles compatible with the data given the uncertainties affecting the absorption analysis. The solid curve is the best fit of a Bonnor-Ebert sphere model (embedded in a medium of uniform column density), obtained with the following parameters: $\rho_c/\rho_{\rm out} = 40 \pm 15$ (i.e., well into the unstable regime), $T_{\rm eff} = 50 \pm 20$ K, $P_{\rm ext}/k_B = 5 \pm 3 \times 10^5$ K cm^{-3}. For comparison, the dotted line shows the $N_{\rm H2} \propto \bar{r}^{-1}$ profile of a SIS at $T = 10$ K. (Adapted from Bacmann et al., 2000.)

More recently, the use of the *absorption* approach, both in the mid-IR from space (e.g., Bacmann et al., 2000; Siebenmorgen and Krügel, 2000) and in the near-IR from the ground (e.g., Alves et al., 2001), made it possible to confirm and extend the (sub)millimeter emission results, essentially independently of any assumption about the dust temperature distribution.

The typical column density profile found by these emission and absorption studies of prestellar cores has the following characteristics (see, e.g., Figure 2b): (a) a flat inner region (of radius $R_{\rm flat} = 5000 \pm 1000$ AU for L1689B according to the fitting analysis shown in Figure 2(b), (b) a region roughly consistent with $N_{\rm H2} \propto \bar{r}^{-1}$ (corresponding to $\rho \propto r^{-2}$ for a spheroidal core), (c) a sharp edge where the column density falls off more rapidly than $N_{\rm H2} \propto \bar{r}^{-2}$ with projected radius defining the core outer radius $R_{\rm out}$ ($R_{\rm out} = 28000 \pm 1000$ AU for L1689B).

2.2. Comparison with Models of Core Structure

The results summarized above set strong constraints on the density structure at the onset of protostellar collapse. The circularly-averaged column density profiles can often be fitted remarkably well with models of pressure-bounded Bonnor-Ebert spheres, as first demonstrated by Alves et al. (2001) for B68. This is also the case of L1689B as illustrated in Figure 2b. The quality of such fits shows that Bonnor-Ebert spheroids provide a good, first order model for the structure of isolated prestellar cores. In detail, however, there are several problems with these Bonnor-Ebert models. First, the inferred density contrasts (from center to edge) are generally larger (i.e., $\gtrsim$ 20–80, see Figure 2b) than the maximum contrast of $\sim$14 for stable Bonnor-Ebert spheres (cf Section 1). Second, the effective core temperature needed in these models (for thermal pressure gradients to balance self-gravity) is often significantly larger than both the average dust temperature measured with ISOPHOT (e.g., Ward–Thompson et al., 2002) and the gas temperature measured in NH_3 (e.g., Lai et al., 2003). In the case of L1689B, for instance, the effective temperature of the Bonnor-Ebert fit shown in Figure 2b is $T_{\rm eff} \sim 50$ K, while the dust temperature observed with ISOPHOT is only $T_{\rm d} \sim 11$ K (Ward–Thompson et al., 2002). Third, the physical process responsible for bounding the cores at some external pressure is unclear. These arguments suggest that prestellar cores cannot simply be described as isothermal hydrostatic structures and are either already contracting (see Section 3.1 below) or experiencing extra support from static or turbulent magnetic fields (e.g., Curry and McKee, 2000). It has also recently been pointed out that good Bonnor–Ebert fits can often be found for non-equilibrium, transient "cores" produced by turbulence (Ballesteros–Paredes et al., 2003).

One way to account for large density contrasts and high effective temperatures is to consider models of cores initially supported by a static magnetic field and evolving through ambipolar diffusion (e.g., Ciolek and Mouschovias, 1994; Basu and Mouschovias, 1995). In these models, at any time prior to protostar formation, the cores are expected to feature a uniform-density central region whose size corresponds to the instantaneous Jeans length. This agrees well with the characteristics of the flat inner regions in starless cores. Furthermore, the observed sharp edges (cf Figure 2b) are consistent with the model predictions (shortly) after the formation of a magnetically supercritical core. Physically, this is because when a supercritical core forms, it collapses dynamically inward, while the outer, subcritical envelope is still efficiently supported by the magnetic field and remains essentially "held in place". As a result, a steep density profile develops at the outer boundary of the supercritical core (see Figure 8 of Basu and Mouschovias, 1995).

A serious problem, however, with this explanation of the sharp edges is that it requires a strong field *in the low-density ambient cloud* ($\sim$30–100 μG, see Bacmann et al., 2000), which seems inconsistent with existing Zeeman measurements (e.g., Crutcher, 1999, Crutcher and Troland, 2000). Furthermore, when the ambient field is strong (highly subcritical case), the model cores evolve on the ambipolar diffusion timescale $t_{AD} \sim 10\, t_{\rm ff}$ (Ciolek and Basu, 2001; $t_{\rm ff}$ is the free-fall time), while observed cores have significantly shorter lifetimes $\sim 4\, t_{\rm ff}$ at $n_{\rm H_2} \sim 10^4$ cm^{-3} (cf Jessop and Ward–Thompson, 2000). (Note that isolated starless cores nevertheless appear to live $\gtrsim$ 3 times *longer* than typical density fluctuations generated by turbulence –cf Ballesteros–Paredes et al., 2003.) It is possible that more elaborate ambipolar diffusion models, incorporating the effects of a non-static, turbulent magnetic field in the outer parts of the cores and in the ambient cloud, would be more satisfactory and could also account for the filamentary shapes often seen on large ($\gtrsim$0.25 pc) scales (cf Jones and Basu, 2002).

2.3. Temperature distribution in starless cores

The models discussed above assume that prestellar cores are isothermal, which is quite a good first approximation (e.g., Larson, 1969; Tohline, 1982). In actual fact, however, the central regions of starless cores must be somewhat cooler than their outer regions. Indeed, starless cores are heated only from outside by the local interstellar radiation field (ISRF), with no evidence for any central heating source (e.g., Ward–Thompson et al., 2002). In such a situation, dust radiative transfer calculations (e.g., Evans et al., 2001; Zucconi et al., 2001) predict that there should be a positive temperature gradient from the cores' centers (with $T_{\rm d}$ as low as $\sim$ 5–7 K) to their edges (at T_d $\sim$ 15 K). Bouwman et al. (2003) have recently carried out similar calculations in which the levels of the diffuse mid-IR and far-IR backgrounds observed toward the cores are used to make realistic estimates of the effective radiation field directly impinging on their surfaces. Figure 3 gives an illustration for the prestellar core L1689B discussed earlier (see Figure 2).

The Bouwman et al. (2003) study confirms that the dust temperature generally reaches a minimum $\leq$10 K in the centers of prestellar cores and that the central temperature depends primarily on the central optical depth (directly related to the degree of shielding from the external ISRF). However, it appears that the minimum temperature may not be as low as initially reported by Evans et al. (2001) and Zucconi et al. (2001) and that the shape of the temperature profile may differ from that calculated by these authors: It can be seen in Figure 3 that the major drop in

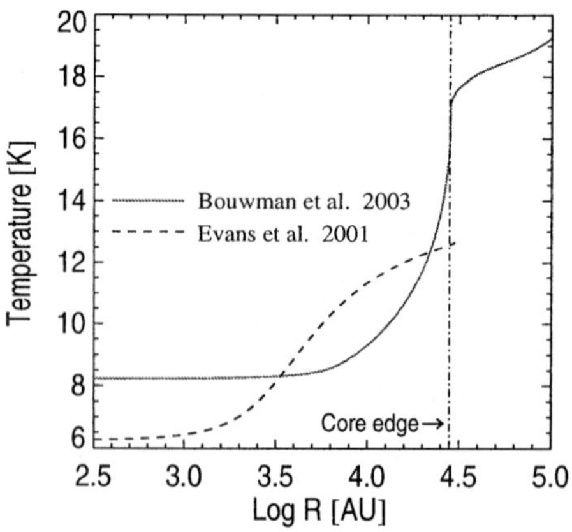

Figure 3. Comparison of two dust temperature profiles calculated for the prestellar core L1689B: the solid curve shows the model of Bouwman et al. (2003) based on the ISO observations of Bacmann et al. (2000) and Ward–Thompson et al. (2002); the dashed curve corresponds to the model favored by Evans et al. (2001).

temperature occurs in the outer parts of the core in the Bouwman et al. (2003) model of L1689B (solid curve), while it occurs closer to the center in the Evans et al. model (dashed curve). Interestingly, all models of L1689B have a relatively high mass-averaged dust temperature, comparable to the temperature of ~11 K estimated by fitting a greybody to the global SED (cf Figure 13 of Ward–Thompson et al., 2002).

The presence of a temperature gradient changes slightly the density profile expected in hydrostatic equilibrium and modifies the stability properties of prestellar cores compared to strictly isothermal Bonnor-Ebert spheres, allowing stable equilibria with density contrasts up to $\lesssim 40$ (Galli et al., 2002). The latter effect remains small, however, and is restricted to a narrow range of core masses, so that it is unlikely to account for the large density contrasts observed in real prestellar cores (see Figure 2b above).

2.4. Structure of Protostellar Envelopes

The density structure of YSO envelopes has been primarily studied using the (sub)millimeter emission approach. In contrast to prestellar cores, protostellar envelopes are always found to be strongly centrally condensed (see, e.g., Figure 1b) and do *not* exhibit any marked inner flattening in their radial (column) density profiles. As an example, Figure 4a showsthat the circularly-averaged radial intensity

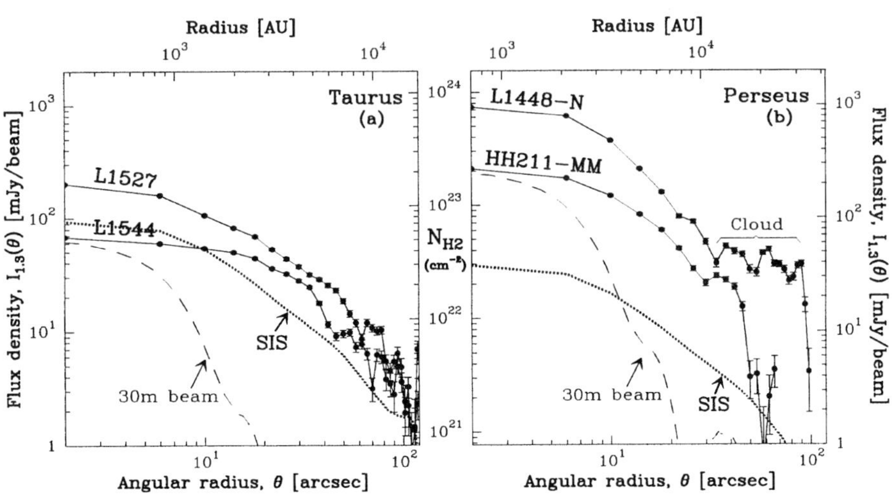

Figure 4. Averaged 1.3 mm radial intensity profiles of (a) L1544 (prestellar) and L1527 (Class 0) in Taurus (d = 140 pc), and (b) HH211–MM and L1448-N (Class 0s) in Perseus (d = 350 pc), compared to a synthetic profile, simulated for a singular isothermal sphere (SIS) model at T = 10 K (see Appendix of Motte and André, 2001, for details about such simulations). The profile of the 30 m beam is shown. An approximate column density scale, calculated assuming representative dust properties (T_d = 15 K, $\kappa_{1.3}$ = 0.0075 cm^2 g^{-1}) is also indicated. (Adapted from Ward–Thompson et al., 1999 and Motte and André, 2001.)

profile measured at 1.3 mm is significantly steeper at small radii for the envelope of the Class 0 object L1527 than for the prestellar core L1544.

In *regions of isolated star formation* such as Taurus, protostellar envelopes have radial density gradients consistent with $\rho(r) \propto r^{-p}$ with $p \sim 1.5$–2 over more than ~10000–15000 AU in radius (e.g., Chandler and Richer, 2000; Hogerheijde and Sandell, 2000; Shirley et al., 2000; Motte and Andrè, 2001). The density gradient estimated for (Class 0 and Class I) protostars (i.e., $p \sim 1.5$–2) agrees with most collapse models which predict a value of p between 2 and 1.5 during the protostellar accretion phase (before and after the passage of the collapse expansion wave, respectively, see Section 1.1 above). Furthermore, the absolute level of the density distributions observed toward Taurus Class I sources is roughly consistent with the predictions of the inside-out collapse model of Shu (1977) for ~10^5 yr-old protostars (Motte and André, 2001).

The situation is markedly different in regions where stars form in tight groups or *clusters*, such as Serpens, Perseus, and the ρ Oph main cloud. In this case, the observed envelopes are clearly *not scale-free*: they merge with dense cores, other envelopes, and/or the diffuse ambient cloud at a *finite* radius $R_{out} \lesssim 5\,000$ AU (Motte et al., 1998; Motte and André, 2001; Looney et al., 2003). Moreover, as can be seen in Figure 4b for HH211-MM and L1448-N, the Class 0 envelopes mapped in cluster-forming regions are found to be *3 to 12 times denser* than the SIS at $T = 10$ K (which is the typical gas kinetic temperature expected in these clouds prior to massive star formation, see Evans, 1999; Goldsmith, 2001 and Figure 3).

3. Velocity Structure

3.1. Internal Motions in Starless Cores

Both isolated starless cores and prestellar condensations in cluster-forming regions are characterized by very *narrow line widths* (e.g., Myers, 1983; Belloche et al., 2001). In the ρ Oph protocluster, for instance, the nonthermal velocity dispersion is about half the thermal velocity dispersion of H_2 ($\sigma_{NT}/\sigma_T \sim 0.7$) toward the starless condensations of the dense cores Oph B1, C, E, F (Belloche et al., 2001). This indicates that, *even in cluster-forming clouds*, the initial conditions for individual protostellar collapse are "thermally-dominated" (cf Caselli et al., 2002) and largely free of turbulence. The dissipation, on small (<0.1 pc) scales, of a significant fraction of the turbulent motions observed on larger scales (e.g., Falgarone and Phillips, 1990) is thus a prerequesite for the formation of prestellar condensations (e.g., Nakano, 1998).

Apart from low levels of internal turbulence, low-mass prestellar cores are also characterized by small rotational velocity gradients in general ($\lesssim$1 km s^{-1} pc^{-1}—e.g., Goodman et al., 1993), and subsonic, extended infall motions (e.g., Tafalla et al., 1998; Lee et al., 2001).

3.2. Rotation and infall in class 0 objects

There have been very few quantitative studies of the velocity structure of Class 0 objects so far. Here, we describe two recent, contrasted examples.

The first object, IRAM 04191+1522 (IRAM 04191 for short), is relatively *isolated*. It was found at 1.3 mm by André et al. (1999) in the southern part of the Taurus cloud ($d = 140$ pc). IRAM 04191 is associated with a prominent, flattened protostellar envelope, seen in the (sub)millimeter dust continuum and in dense gas tracers such as N_2H^+, C_3H_2, $H^{13}CO^+$, and DCO^+ (Belloche et al., 2002). All of the maps taken at the IRAM 30 m telescope in small optical depth lines show a clear rotational velocity gradient across the envelope of $\sim$9 km s^{-1} pc^{-1} (after deprojection), perpendicular to the outflow axis (seen in Figure 1b). This gradient is one order of magnitude larger than those typically observed in starless cores (see Section 3.1 above). Furthermore, the rotation of the protostellar envelope does not occur in a rigid-body, but differential, fashion: the inner $\sim$3500 AU-radius region rotates significantly faster than the outer parts of the envelope, as indicated by the characteristic "S" shape of the position-velocity diagrams obtained by Belloche et al. (see, e.g., Figure 5). Fast, differential rotation is expected in protostellar envelopes because of conservation of angular momentum during dynamical collapse. In the present case, however, the dramatic drop in rotational velocity observed at $r \geq$ 3500 AU(Figure 6a), combined with the flat infall velocity profile (see below), points to *losses of angular momentum* in the outer envelope (see discussion in Section 4).

Direct evidence for infall motions over a large portion of the envelope of IRAM 04191 is observed in optically thick lines such as CS(2–1), CS(3–2), $H_2CO(2_{12} - 1_{11})$, and $H_2CO(3_{12} - 2_{11})$. These lines are double-peaked and skewed

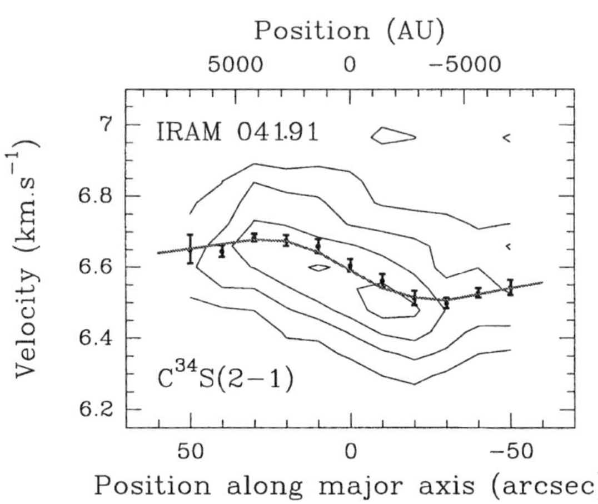

Figure 5. Position-velocity diagram along the major axis of the IRAM 04191 envelope (i.e., perpendicular to the flow) based on a $C^{34}S(2–1)$ map taken at the IRAM 30 m telescope (Belloche et al., 2002). Contours: 0.2 to 0.8 by 0.2 K. The dots with error bars mark the observed velocity centroids. The solid curve shows the profile of a model with differential rotation [$V_{rot}(r > 3500\,\mathrm{AU}) \propto r^{-1.5}$].

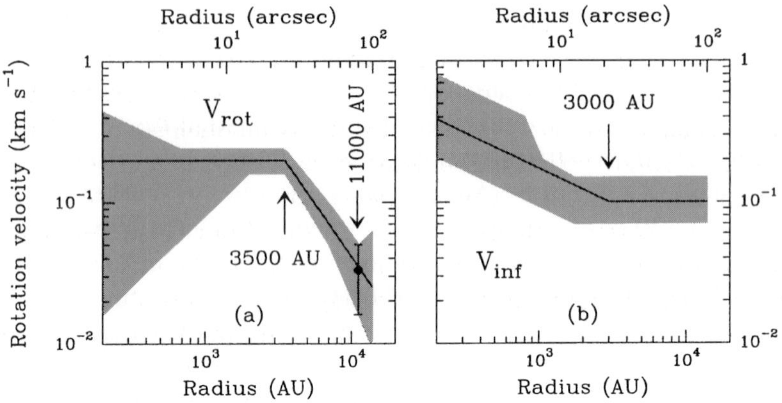

Figure 6. Rotational velocity (a) and infall velocity (b) inferred in the IRAM 04191 envelope based on radiative transfer modeling of multi-transition CS and $C^{34}S$ observations with the IRAM 30 m telescope (Belloche et al., 2002). The shaded areas show the estimated domains where the models match the observations.

to the blue up to $\gtrsim 40''$ from source center, which is indicative of infall motions up to a radius $R_{\rm inf} \gtrsim$ 5000 AU (cf Evans, 1999, 2003). Radiative transfer modeling confirms this view, suggesting a flat, subsonic infall velocity profile ($V_{\rm inf} \sim 0.1$ km s^{-1}) for $3000 \lesssim r \lesssim 11000$ AU and larger infall velocities scaling as $V_{\rm inf} \propto r^{-0.5}$ for $r \lesssim 3000$ AU (see Figure 6b and Belloche et al., 2002 for details). The mass infall rate is estimated to be $\dot{M}_{\rm inf} \sim 2 - 3 \times a_s^3/G \sim 3 \times 10^{-6}\ M_\odot$ yr^{-1} (with $a_s \sim 0.15 - 0.2$ km s^{-1} for $T \sim 6 - 10$ K), roughly independent of radius.

Another Class 0 object whose kinematics has been quantified in detail is IRAS 4A in the NGC 1333 *protocluster* (Di Francesco et al., 2001). Using the IRAM Plateau de Bure interferometer, Di Francesco et al. observed inverse P-Cygni profiles in $H_2CO(3_{12} - 2_{11})$ toward IRAS 4A, from which they derived a very large mass infall rate of $\sim 1.1 \times 10^{-4} M_\odot$ yr^{-1} at r ~2000 AU. Even if a warmer initial gas temperature (~20 K) than in IRAM 04191 and some initial level of turbulence are accounted for (see Di Francesco et al., 2001), this value of $\dot{M}_{\rm inf}$ corresponds to more than ~15 times the canonical $a_{\rm eff}^3/G$ value (where $a_{\rm eff} \lesssim 0.3$ km s^{-1} is the effective sound speed). This very high infall rate results both from a very dense envelope (a factor ~12 denser than a SIS at 10 K – see Motte and André 2001) and a large, supersonic infall velocity (~0.68 km s^{-1} at ~2000 AU, Di Francesco et al., 2001). Evidence for fast rotation in the IRAS4A envelope, producing a velocity gradient as high as ~40 km s^{-1} pc^{-1}, was also reported by Di Francesco et al. (2001).

4. Conclusions: Comparison with Collapse Models

In the case of *isolated dense cores* such as those of Taurus, the SIS model of Shu (1977) describes global features of the collapse reasonably well (e.g., the mass infall rate within a factor ~2) and thus remains a useful, approximate guide. In detail, however, the extended infall velocity profiles observed in prestellar cores (see Section 3.1) and in the very young Class 0 object IRAM 04191 (Section 3.2) are inconsistent with a pure inside-out collapse picture. The *shape* of the density profiles observed in prestellar cores are well fitted by purely thermal Bonnor–Ebert sphere models, but the *absolute* values of the densities are suggestive of some additional magnetic support (Section 2.2). The observed infall velocities are also marginally consistent with isothermal collapse models starting from Bonnor–Ebert spheres (e.g., Foster and Chevalier, 1993, Hennebelle et al., 2003), as such models tend to produce somewhat faster velocities. This suggests that the collapse of 'isolated' cores is essentially *spontaneous* and somehow moderated by magnetic effects in (mildly) magnetized, non-isothermal versions of Bonnor–Ebert cloudlets. Indeed, the contrast seen in Figure 6 between the steeply declining rotation velocity profile and the flat infall velocity profile of the IRAM 04191 envelope beyond ~3500 AU is very difficult to account for in the context of non-magnetic collapse models. In the presence of magnetic fields, on the other hand, the outer envelope can be coupled to, and spun down by, the (large moment of inertia of the) ambient cloud (e.g., Basu and Mouschovias, 1994).

Based on a qualitative comparison with the ambipolar diffusion models of Basu and Mouschovias (1994, 1995), Belloche et al. (2002) propose that the rapidly rotating inner envelope of IRAM 04191 corresponds to a magnetically (slightly) supercritical core decoupling from an environment still supported by magnetic fields and strongly affected by magnetic braking. A magnetic field $\sim 60\ \mu$G is required at 3500 AU where $n_{\mathrm{H2}} \sim 1 - 2 \times 10^5$ cm^{-3}, which is comparable to the field strengths recently estimated at such densities by Crutcher et al. (2003) in three prestellar cores (see Crutcher, this volume). In this picture, the inner ~3500 AU radius envelope of IRAM 04191 would correspond to the effective mass reservoir ($\sim 0.5\ M_{\odot}$) from which the central star is being built. Moreover, comparison of these results with the rotational characteristics of other objects in Taurus (Ohashi et al., 1997) indicates that IRAM 04191 behaves in a typical manner and is simply observed particularly soon after point mass formation (i.e., at $t \gtrsim 0$). The IRAM 04191 example therefore suggests that *the masses of stars forming in clouds such as Taurus are largely determined by magnetic decoupling effects.*

In *protoclusters* such as NGC 1333, by contrast, the large overdensity factors measured for Class 0 envelopes compared to hydrostatic isothermal structures (Section 2.4 and Figure 4b), as well as the fast supersonic infall velocities and very large infall rates observed in some cases (Section 3.2), are inconsistent with self-initiated forms of collapse and require a *strong external influence*. This point is supported by the results of recent SPH simulations by Hennebelle et al. (2003). These simulations follow the evolution of a Bonnor–Ebert sphere whose collapse has been induced by an increase in external pressure P_{ext}. Large overdensity factors (compared to a SIS), together with supersonic infall velocities, and large infall rates ($\gtrsim 10\, a_s^3/G$) are found near $t = 0$ when (and only when) the increase in P_{ext} is strong and very rapid (e.g. Figure 7), resulting in a violent compression wave.

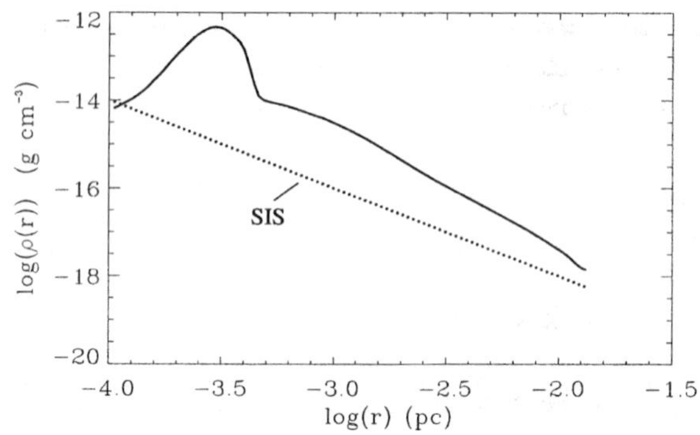

Figure 7. Density profile (solid curve) obtained near point mass formation ($t \approx 0$) in SPH simulations of the collapse of an initially stable, rotating ($\beta = E_{rot}/E_{grav} = 2\%$) Bonnor-Ebert sphere ($T = 10$ K) induced by a very rapid increase in external pressure (with $P_{ext}/\dot{P}_{ext} = 0.03 \times$ the initial sound crossing time) (Hennebelle et al., 2003; Hennebelle et al., this volume). Note the large overdensity factor compared to the $\rho \propto r^{-2}$ profile of a SIS at 10 K (dotted line).

Such a violent collapse in protoclusters may be conducive to the formation of both massive stars (through higher accretion rates) and multiple systems (when realistic, non-isotropic compressions are considered). Future high-resolution studies with the next generation of (sub)millimeter instruments (e.g., ALMA) will greatly help test this view and shed further light on the physics of collapse in cluster-forming regions.

References

Alves, J.F., Lada, C.J. and Lada, E.A.: 2001, *Nature* **409**, 159.
André, P., Motte, F. and Bacmann, A.: 1999, *ApJ* **513**, L57.
André, P., Ward–Thompson, D. and Barsony, M.: 1993, *ApJ* **406**, 122.
André, P., Ward–Thompson, D. and Motte, F.: 1996, *A&A* **314**, 625.
Bacmann, A., André, P., Puget, J.-L., Abergel, A., Bontemps, S. and Ward-Thompson, D.: 2000, *A&A* **361**, 555.
Ballesteros-Paredes, J., Klessen, R.S. and Vázquez–Semadeni, E.: 2003, *ApJ* **592**, 188.
Basu, S. and Mouschovias, T.C.: 1994, *ApJ* **432**, 720.
Basu, S. and Mouschovias, T.C.: 1995, *ApJ* **453**, 271.
Belloche, A., André, P. and Motte, F.: 2001, in: T. Montmerle and P. André (eds.), *From Darkness to Light*. ASP Conf. Ser., San Francisco, **243**, p. 313.
Belloche, A., André, P., Despois, D. and Blinder, S.: 2002, *A&A* **393**, 927.
Bonnor, W.B.: 1956, *MNRAS* **116**, 351.
Bouwman, J., André, P. and Galli, D.: 2003, in preparation.
Caselli, P., Benson, P.J., Myers, P.C. and Tafalla, M.: 2002, *ApJ* **572**, 238.
Chandler, C.J. and Richer, J.S.: 2000, *ApJ* **530**, 851.
Ciolek, G.E. and Basu, S.: 2001, *ApJ* **547**, 272.
Ciolek, G.E. and Mouschovias, T.C.: 1994, *ApJ* **425**, 142.
Crutcher, R.M.: 1999, *ApJ* **520**, 706.
Crutcher, R.M., Nutter, D.J., Ward-Thompson, D. and Kirk, J.M.: 2004, *ApJ* **600**, 279.
Crutcher, R.M. and Troland, T.H.: 2000, *ApJ* **537**, L139.

Curry, C.L. and McKee, C.F.: 2000, *ApJ* **528**, 734.
Di Francesco, J., Myers, P.C., Wilner, D.J., Ohashi, N. and Mardones, D.: 2001, *ApJ* **562**, 770.
Evans, N.J. II.: 1999, *ARA&A* **37**, 311.
Evans, N.J. II.: 2003, in: C.L. Curry and M. Fich (eds.), Chemistry as a Diagnostic of Star Formation, NRC Research Press, Ottawa, 157.
Evans, N.J. II, Rawlings, J.M.C., Shirley, Y.L. and Mundy, L.G.: 2001, *ApJ* **557**, 193.
Falgarone, E. and Phillips, T.: 1990, *ApJ* **359**, 344.
Foster, P.N. and Chevalier, R.A.: 1993, *ApJ* **416**, 303.
Galli, D., Walmsley, M. and Goncalves, J.: 2002, *A&A* **394**, 275.
Goldsmith, P.F.: 2001, *ApJ* **557**, 736.
Goodman, A.A., Benson, P.J., Fuller, G.A. and Myers, P.C.: 1993, *ApJ* **406**, 528.
Hennebelle, P., Whitworth, A.P., Gladwin, P.P. and André, P.: 2003, *MNRAS* **340**, 870.
Hogerheijde, M.R. and Sandell, G.: 2000, *ApJ* **534**, 880.
Jessop, N.E. and Ward–Thompson, D.: 2000, *MNRAS* **311**, 63.
Jones, C.E. and Basu, S.: 2002, *ApJ* **569**, 280.
Klessen, R.S., Heitsch, F. and Mac Low, M.-M.: 2000, *ApJ* **535**, 887.
Lai, S.-P., Velusamy, T., Langer, W.D. and Kuiper, T.B.H.: 2003, AJ **126**, 311.
Larson, R.B.: 1969, *MNRAS* **145**, 271.
Lee, C.W., Myers, P.C. and Tafalla, M.: 2001, *ApJS* **136**, 703.
Li, Z.-Y. and Shu, F.H.: 1997, *ApJ* **475**, 237.
Looney, L.W., Mundy, L.G. and Welch, W.J.: 2003, *ApJ* **592**, 255.
Motte, F. and André, P.: 2001, *A&A* **365**, 440.
Motte, F., André, P. and Neri, R.: 1998, *A&A* **336**, 150.
Mouschovias, T.C. and Ciolek, G.E.: 1999, in: C.J. Lada and N.D. Kylafis (eds.), *The Origin of Stars and Planetary Systems*, Kluwer, Dordrecht, p. 305.
Myers, P.C.: 1983, *ApJ* **270**, 105.
Nakano, T.: 1998, *ApJ* **494**, 587.
Ohashi, N., Hayashi, M., Ho, P.T.P., Momose, M., Tamura, M., Hirano, N. and Sargent, A.I.: 1997, *ApJ* **488**, 317.
Penston, M.V.: 1969, *MNRAS* **144**, 425.
Shirley, Y., Evans II, N.J., Rawlings, J.M.C. and Gregersen, E.M.: 2000, *ApJS* **131**, 249.
Shu, F.: 1977, *ApJ* **214**, 488.
Shu, F.H., Adams, F.C. and Lizano, S.: 1987, *ARA&A* **25**, 23.
Siebenmorgen, R. and Krügel, E.: 2000, *A&A* **364**, 625.
Tafalla, M., Mardones, D., Myers, P.C., Caselli, P., Bachiller and Benson, P.J.: 1998, *ApJ* **504**, 900.
Tohline, J. E.: 1982, *Fund. of Cos. Phys.* **8**, 1.
Vázquez–Semadeni, E., Ostriker, E.C., Passot, T., Gammie, C.F. and Stone, J.M.: 2000, in: V. Mannings, A.P. Boss and S.S. Russell (eds.) *Protostars and Planets IV*, The University of Arizona Press: Tucson, p. 3.
Ward-Thompson, D., Kirk, J.M., Crutcher, R.M., Greaves, J.S., Holland, W.S. and André, P.: 2000, *A&A* **537**, L135.
Ward–Thompson, D., André, P. and Kirk, J.M.: 2002, *MNRAS* **329**, 257.
Ward–Thompson, D., Motte, F. and André, P.: 1999, *MNRAS* **305**, 143.
Ward–Thompson, D., Scott, P.F., Hills, R.E. and André, P.: 1994, *MNRAS*, **268**, 276.
Whitworth, A. and Summers, D.: 1985, *MNRAS* **214**, 1.
Zucconi, A., Walmsley, C.M. and Galli, D.: 2001, *A&A* **376**, 650.

THE SUBSTELLAR POPULATION IN THE YOUNG σ ORIONIS CLUSTER, SPATIAL DISTRIBUTION

V.J.S. BÉJAR[1], J.A. CABALLERO[1], R. REBOLO[1], M.R. ZAPATERO OSORIO[2] and D. BARRADO Y NAVASCUÉS[2]

[1]*Instituto de Astrofísica de Canarias, Spain; E-mail: vbejar@ll.iac.es*
[2]*Laboratorio de Astrofísica Espacial y Física Fundamental, Spain*

Abstract. We review different surveys, in the optical and infrared, conducted in the very young (age 1–8 Myr), nearby (d $\sim$ 350 pc) σ Orionis cluster aimed to characterize the substellar population. We describe spectral characteristics of very low mass stars, brown dwarfs and planetary mass objects in the cluster with spectral types from K7 to T6. We study the spatial distribution of the substellar population detected in a IZJ survey covering an area of 1.12 deg.^{2} We find that the radial distribution of substellar objects can be well fitted by an exponential law ($\rho = \rho_0 e^{-r/r_0}$), with a central density (ρ_0) of 0.26 $\pm$ 0.03 objects/arcmin^2 and a characteristic radius (r_0) of 8.8 arcmin $\pm$ 0.6 (equivalent to 0.90 $\pm$ 0.06 pc at the distance of the cluster). We discuss the presence of possible inhomogeneities in this distribution due to the existence of subclustering. We also compare the spatial distribution of the substellar population with previously known stars in the cluster. We report the initial mass spectrum in the substellar domain.

Keywords: brown dwarfs, stellar clusters, stars: σ Orionis

1. Brown Dwarf and Planets

Brown dwarfs are astronomical bodies unable to sustain stable hydrogen burning in their interiors. For solar metallicity they have masses lower than $\sim$75 M_{Jup}[1] (Baraffe et al., 1998). Recently, several authors have suggested to establish the frontier between brown dwarfs and giant planets according to the intrinsic properties of these objects, independently of their origin and on wether they are orbiting a star or not. This frontier is proposed at the minimum mass which can sustain the deuterium fusion, so, for solar metallicity, planets are objects with masses below 12–13 M_{Jup} (Saumon et al., 1996). Here we adopt this criterium.

2. The σ Orionis Cluster

The σ Orionis cluster is located in the south-west of the Orion belt and belongs to the Orion OB1b association. The star that gives the name to the cluster is a multiple system composed by an O9.5V, two B0.5V and two A0. It is a very interesting site to

[1] $1\ M_\odot = 1047\ M_{Jup}$.

Astrophysics and Space Science **292:** 339–346, 2004.

investigate substellar objects because of its youth, proximity and low extinction. The *Hipparcos* satellite has measured a distance of 350 pc to the central star (Perryman et al., 1997). The age of σ Orionis has been restricted using observations of Li in stars later than K7. According to this, the cluster is younger than 8 Myr, with a most probable age of 2–4 Myr (Zapatero Osorio et al., 2002a).

3. Searches for the Brown Dwarf Population in the Cluster

We have performed several surveys in order to detect very low-mass stars and brown dwarfs in the cluster. The first (WFC97) was a RIZ survey covering a central area of 870 arcmin^2, using the wide field camera (WFC) mosaic, mounted at the Isaac Newton telescope (INT, Observatorio del Roque de los Muchachos). We extended this survey covering another 430 arcmin^2 in the RI filters with the CCD-camera on the IAC-80 telescope (Observatorio del Teide). The limiting magnitude in the I-band was 21 in both surveys. From colour-magnitude diagrams, we selected a total of 100 very low-mass stars and brown dwarf candidates, with some objects in common to both surveys, which, according to theoretical models, span a mass range from 0.1 down to 0.02 $M_\odot$.

In order to detect less massive brown dwarfs, we performed a deeper survey in a more extended area (WFC98 survey). We used the WFC on the INT and filters IZ. We covered an area about 1 deg.^{2} reaching a I-band limiting magnitude of 23. We selected 171 candidates between I magnitude 16–24 , which, according to models, cover the whole brown dwarf domain down to planetary-mass objects with 5 M_{Jup}. Figure 1 shows the I vs. I–Z colour-magnitude (CM) diagrams, where the selected candidates are indicated in solid circles, and with asterisks those objects which were spectroscopically confirmed. Completeness and limiting magnitudes are also shown.

We have performed follow-up infrared photometry of our candidates above completeness in order to discriminate between bona fide cluster members and reddened field objects. Figure 2 shows a I, I–J colour-magnitude diagram. We indicate with asterisks candidates from the WFC97 and IAC-80 surveys, in solid circles bona fide candidates from the WFC98 which belong to the cluster sequence and in open circles probable non cluster members older than 10Myr. We also show 1, 5 and 10 Myr isochrones from the Lyon group (Baraffe et al., 1998, Chabrier et al., 2000).

We have obtained low-resolution optical spectra for a total of 30 brown dwarfs candidates. In Figure 3 we show several of them, observed with the ISIS spectrograph, mounted on the William Herschel Telescope (WHT), and the LRIS spectrograph on the KeckI telescope, which provides a total spectral coverage of 635–920 nm and a resolution of around 20 Å . Spectral types range from M6 to M8.5, showing the typical TiO and VO absorption bands. These objects also show spectral features suggestive of youth, like strong emission of Hα and weak alkaline lines (Béjar et al., 1999). Several of the candidates ($\sim$6%) show near infrared

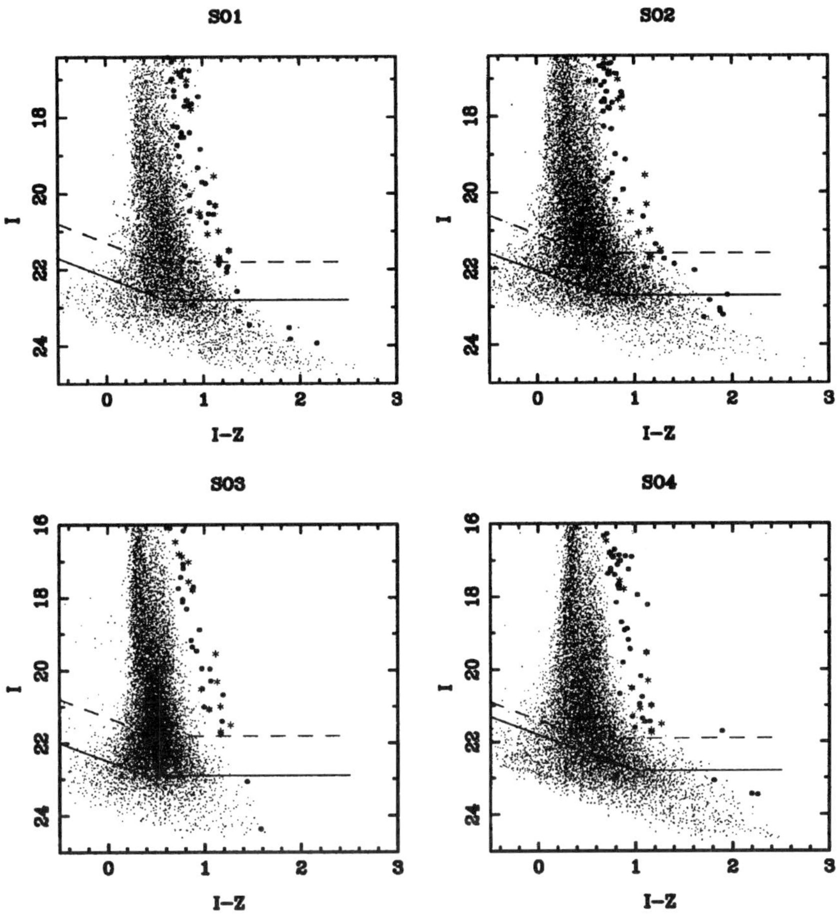

Figure 1. I versus I–Z colour-magnitude diagrams: The selected candidates are indicated with solid circles, while those previously confirmed members are represented with asterisks. The completeness and limiting magnitude are also shown in dashed and solid lines, respectively.

excesses that could be associated to the presence of disks (Barrado et al., 2003). For a few of them we have obtained higher resolution spectra, e. g. SOri 45, which also shows forbidden lines, thought to be associated with outflow events (Zapatero Osorio et al., 2002a).

4. Isolated Planetary-Mass Objects

To detect the free-floating planetary-mass population of σ Orionis, we performed a J-band survey with the Omega-Cass instrument, mounted on the 3.5 m Calar Alto telescope. This near-infrared study overlaps with previous optical surveys in a sky region of 847 arcmin^2, with limiting I_{Cousin} and J_{UKIRT} magnitudes of 23 and 21.2. From a correlation of both data sets, we have constructed the I versus I– J

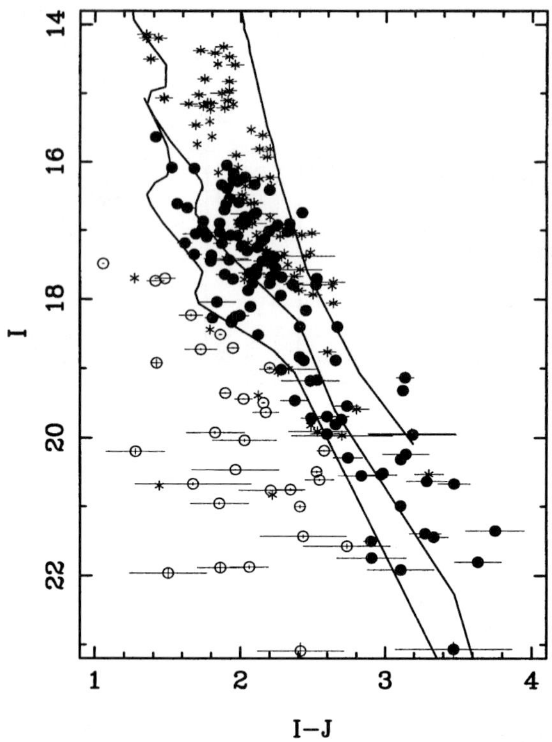

Figure 2. I versus I–J colour-magnitude diagram. Asterisks represent objects from previous surveys. Solid circles indicate candidates which follow the photometric cluster sequence, while open circles denote likely non-cluster members. The 1, 5 and 10 Myr Lyon group isochrones are also indicated.

colour–magnitude diagram and we selected 17 candidates, between I-band magnitudes 20.5 and 24, that smoothly extrapolate the previous cluster sequence, and fit the location predicted by theoretical isochrones for cluster members with masses below 15 M_{Jup}. We have obtained low-resolution spectra for all of these candidates but one, and found that all, except two, are very cool objects which belong to the expected spectral sequence of the cluster (Zapatero Osorio et al., 2000; Barrado et al., 2001, Martín et al., 2001). Figure 3 shows near-infrared spectra of some of the planetary mass objects, having spectral types from L1 to L8 ($T_{eff} \sim 2200$–1500 K). In the infrared, these objects are characterized by the presence of strong water absorption bands which become stronger in later spectral types.

We have performed a very deep JH-band survey with the Ingrid camera on the WHT (J-band completeness magnitude of 20), covering a small area of ~55 arcmin^2 in the center of the cluster. We have found a T6-class object ($T_{eff} \sim 1000$ K), SOri 70. Its near-infrared spectrum shows the typical methane absorption bands of T spectral type dwarfs, not present in L-class objects. SOri 70 shows spectral features characteristic of a young object, like the presence of water and methane absorption bands and alkaline lines stronger than those present in older field T

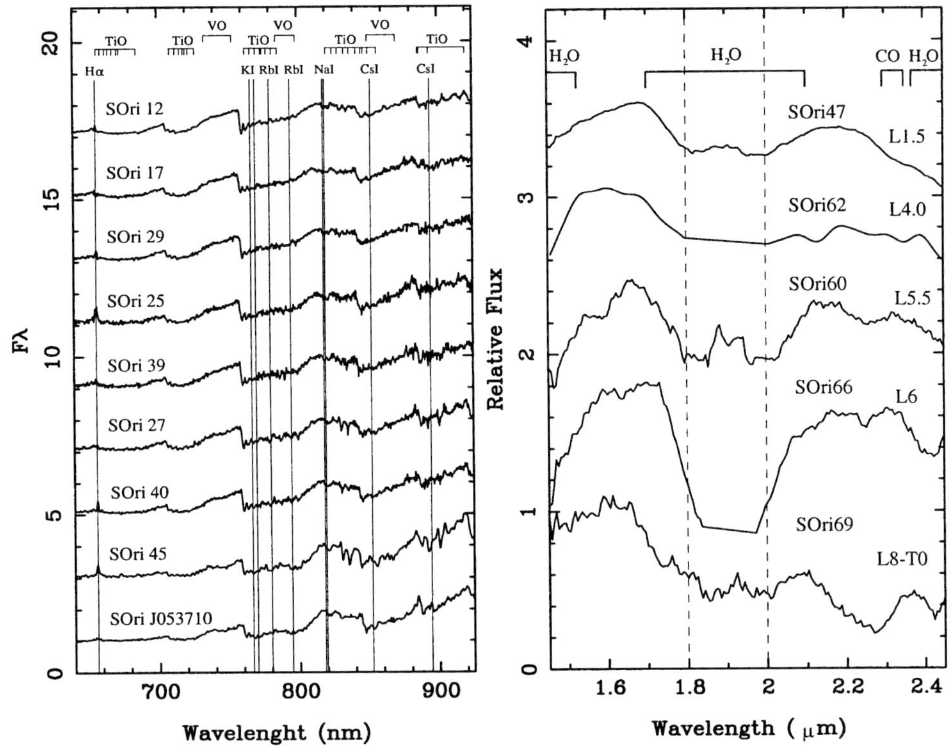

Figure 3. Left pannel: Optical spectroscopy of late M-type brown dwarfs. *Right pannel:* Near-infrared spectroscopy of L-type planetary-mass objects. Main spectroscopic features are indicated in both figures.

dwarfs (see Zapatero Osorio et al., 2003 contribution in these proceedings), which confirm it as a good cluster member. SOri 70, with an estimated mass of only 2–8 M_{Jup}, is the least massive object ever imaged out of the Solar System (Zapatero Osorio et al., 2002b).

We have investigated the environment of these isolated planets, in order to see if they are really free-floating. Although the knowledge of the stellar population of the cluster is not complete, we can see that typical distances between these objects and other cluster members are in the range 10^4 to 10^5 AU, so they are not expected to be gravitationally bounded to them for times longer than a few million years.

5. Spatial Distribution of the Substellar Population

In the extended WFC98 survey, we have found (see Figure 4) that the projected surface density can be well reproduced by an exponential law, $\rho = \rho_o e^{-r/r_o}$, similar to other open clusters like Pleiades or Praesepe, with a central density ρ_o of 0.26 ± 0.03 obj/arcmin^2 and effective radius r_o of 0.90 ± 0.06 pc. The increase

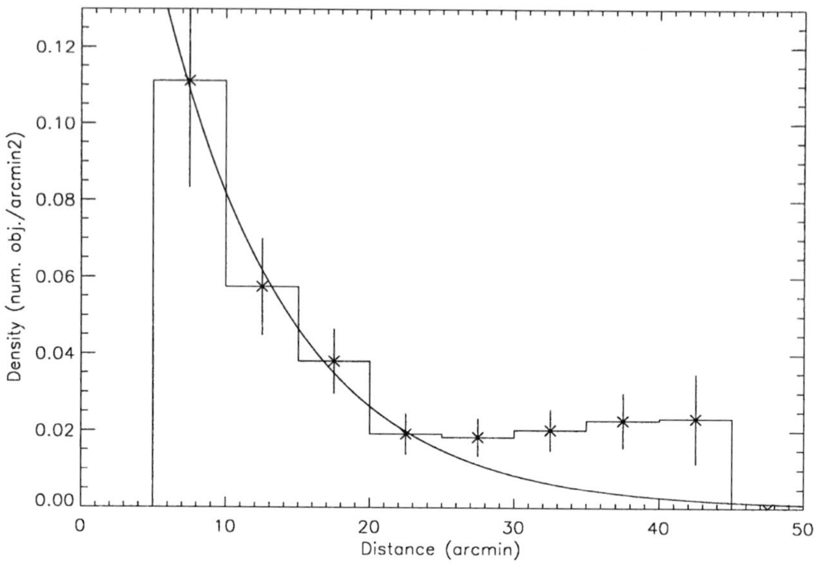

Figure 4. Projected spatial distribution of the substellar population. The best exponential law fit $\rho = 0.26\, e^{-r/0.9(pc)}$ obj/arcmin^2 is also indicated.

for distances larger than 30 arcmin is due to the contribution of brown dwarfs in the north region, near ζ Orionis, which is possibly related to the existence of a cluster around this star. According to the derived spatial distribution, our WFC98 survey comprises more than 80% of the total number of brown dwarfs in the cluster.

We have also checked if the distribution of the brown dwarfs are Poissonian for a given distance to the center, in order to study the existence of subclustering. We have compared the distribution of the distances to the nearest neighbourhood with Monte Carlo simulations of identical number of objects and the same spatial distribution. We can see that derived distances are in agreement with those obtained from simulations, implying that the distribution seems to be Poissonian and that there is no evidence of subclustering.

6. The Substellar Mass Function

From the combination of the deep optical IZ and near-infrared J-band data we have obtained a census of the substellar population in the central region of the cluster. Using evolutionary models, we derive a mass function from the low-mass stars (0.2 $M_{\odot}$), across the complete brown dwarf domain (0.075 $M_{\odot}$ – 0.013 $M_{\odot}$) We find that the mass spectrum (see Figure 5) can be represented by a power-law, $dN/dM \sim M^{-\alpha}$ and increases toward lower masses with an exponent $\alpha = 0.8 \pm 0.4$. An interesting feature is that the planetary domain, although is not complete, is well populated and smoothly extrapolates the results in the brown dwarf domain. These results indicate that substellar objects can be very common, and both brown

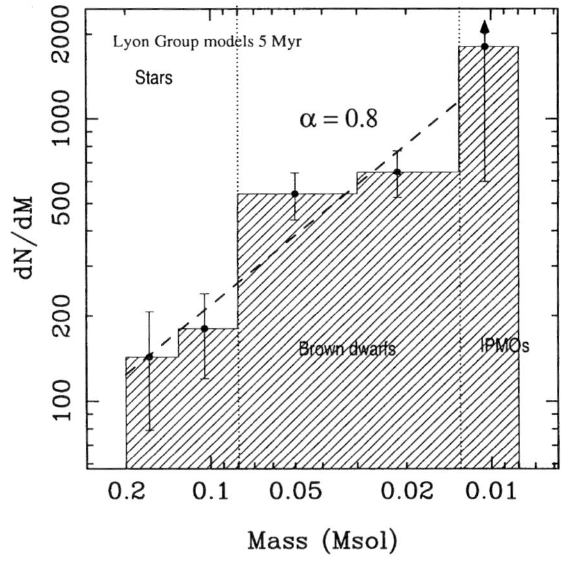

Figure 5. The substellar mass spectrum of the σ Orionis cluster for a 5 Myr age. The best power-law fit ($dN/dM \sim M^{-0.8}$) in the brown dwarf domain is also indicated in dashed line. Error bars correspond to Poissonian uncertainties, where the arrow denotes the incompleteness in the isolated planetary-mass objects (IPMOs).

dwarfs and isolated planetary mass objects could outnumber stars, although their contribution to the total mass of the cluster is not significant (Béjar et al., 2001).

7. Conclusions

- We have found in the σ Orionis cluster a numerous population of isolated brown dwarfs and planetary mass objects down to 2–5 M_{Jup}.
- The projected spatial distribution of brown dwarfs can be reproduced by a exponential law with a characteristic radius of 0.9 pc.
- The mass spectrum is an increasing function in the substellar domain ($dN/dM \sim M^{-0.8}$), but the contribution of substellar objects to the total mass of the cluster is less than 10%.

References

Baraffe, I., Chabrier, G., Allard, F. and Hauschildt, P.: 1998, *A&A* **337**, 403.
Barrado y Navascués, D., Béjar, V. J. S., Mundt, R., Martín, E. L., Rebolo, R., Zapatero Osorio, M. R. and Bailer-Jones, C.A.L.: 2003, *A&A* **404**, 171.
Barrado y Navascués, D., Zapatero Osorio, M. R., Béjar, V. J. S., Rebolo, R., Martín, E. L., Mundt, R. and Bailer-Jones, C.A.L.: 2001. *A&A* **377**, L9.
Béjar, V. J. S., Zapatero Osorio, M. R. and Rebolo, R.: 1999, *ApJ* **521**, 671.
Béjar, V. J. S. et al.: 2001, *ApJ* **556**, 830.

Chabrier, G., Baraffe, I., Allard, F. and Hauschildt, P. H.: 2000, *ApJ* **542**, 464.

Martín, E. L., Zapatero Osorio, M. R., Barrado y Navascués, D., Béjar, V. J. S. and Rebolo, R.: 2001, *ApJ* **558**, L117.

Perryman, M.A.C., Lindegren, L., Kovalevsky, J., Hoeg, E., Bastian, U., Bernacca, P.L., Crézé, M., Donati, F., Grenon, M., van Leeuwen, F., van der Marel, H., Mignard, F., Murray, C.A., Le Poole, R.S., Schrijver, H., Turon, C., Arenou, F., Froeschlé, M. and Petersen, C.S.: 1997, *A&A* **323**, L49.

Saumon D., Hubbard, W. B., Burrows, A., Guillot, T., Lunine, J. I. and Chabrier, G.: 1996, *ApJ* **460**, 993.

Zapatero Osorio, M. R., Béjar, V.J.S., Martín, E.L., Rebolo, R., Barrado y Navascués, D., Bailer-Jones, C.A.L. and Mundt, R.: 2000, *Science* **290**, 103.

Zapatero Osorio, M.R., Béjar, V.J.S., Pavlenko, Ya., Rebolo, R., Allende Prieto, C., Martín, E.L. and García López, R.J., 2002a, *A&A* **384**, 937 .

Zapatero Osorio, M.R., Béjar, V.J.S., Martín, E.L., Rebolo, R., Barrado y Navascués, D., Mundt, R., Eislöffel, J. and Caballero, J.A.: 2002b, *ApJ* **578**, 536.

Zapatero Osorio, M.R., Caballero, J.A., Martín, E.L., Béjar, V.J.S and Rebolo, R., 2004, these proceedings.

ON THE INTERNAL STRUCTURE OF STARLESS CORES. PHYSICAL AND CHEMICAL PROPERTIES OF L1498 AND L1517B

M. TAFALLA[1], P.C. MYERS[2], P. CASELLI[3] and C.M. WALMSLEY[3]

[1]*Observatorio Astronómico Nacional, Spain; E-mail: m.tafalla@oan.es*

[2]*Harvard-Smithsonian Center for Astrophysics, USA*

[3]*Osservatorio Astrofisico di Arcetri, Italy*

Abstract. We have characterized the internal structure of two close-to-round starless cores in Taurus, L1498 and L1517B, setting constraints on the initial conditions of star formation and on models of core condensation. Our analysis is based on high angular resolution observations in at least two transitions of NH_3, N_2H^+, CS, $C^{34}S$, $C^{18}O$, and $C^{17}O$, together with maps of the 1.2 mm continuum. For both cores, we derive radial profiles of constant temperature and constant turbulence, and density distributions close to those of non-singular isothermal spheres. Using a Monte Carlo radiative transfer model, we derive abundance profiles for all species and find a pattern of strong chemical differentiation. NH_3 has a higher abundance toward the core centers while N_2H^+ has a constant abundance over most of the cores. Both $C^{18}O$ and CS (and isotopomers) are strongly depleted in the core interiors, most likely due to their freeze out onto cold dust grains. Concerning the kinematics of the dense gas, we find (in addition to constant turbulence) a pattern of internal motions at the level of 0.05 km s^{-1}. These motions seem correlated with asymmetries in the pattern of molecular depletion, and we interpret them as residuals of core contraction. Their distribution and size suggest that core formation happens in rather irregular manner. A comparison with supersonic turbulence models of core formation shows that our observed cores are much more quiescent than allowed by these models.

Keywords: dense cores, molecular gas abundances, molecular gas kinematics

1. Introduction

Dense cores represent the simplest environments where stars form and their study holds the promise of revealing the basic physics of star formation. Despite their apparent simplicity, however, dense cores remain mysterious in their internal structure, formation and evolution, and in the manner in which they collapse under gravity. Work over the last few years has shown that previous inconsistencies in core studies arise in part from cores not being chemically homogeneous, as initially assumed, but having strongly differentiated interiors (Kuiper et al., 1996; Caselli et al., 1999; Tafalla et al., 2002). Understanding and characterizing the chemical structure of dense cores has therefore become a top priority for star formation research, as solving this problem may finally produce a coherent picture of dense cores from which their star-forming evolution may be inferred.

In this contribution, we present the results of a systematic study and modeling of two starless dense cores in the Taurus complex, L1498 and L1517B. These two cores were chosen because of their starless nature and regular geometry, and

Astrophysics and Space Science **292:** 347–354, 2004.

they were subject to systematic mapping in different tracers of molecular gas. These observations have been modeled simultaneously with a Monte Carlo radiative transfer code in order to derive a self consistent description of the core physical and chemical properties. A full account of this work is presented in Tafalla et al. 2004 (A&A submitted, TMCW04 hereafter).

2. Observations and Overview of the Results

We observed L1498 and L1517B in the 1.2 mm continuum with the MAMBO array at the IRAM 30 m telescope, and in $C^{18}O(1–0)$, $C^{18}O(2–1)$, $C^{17}O(2–1)$, CS(2–1), CS(3–2), $C^{34}S(2–1)$, $C^{34}S(3–2)$, $N_2H^+(1–0)$, and $N_2H^+(3–2)$ with heterodyne receivers also at the IRAM 30 m telescope. We also observed the two cores in $NH_3(1,1)$ and (2,2) with the 100 m Effelsberg telescope and in $C^{17}O(1–0)$ with the FCRAO antenna. The continuum data were obtained in on-the-fly mode, and the line data were obtained using frequency switching. All the data were reduced with the GILDAS software and the line data were converted into the main beam brightness scale using the appropriate efficiencies. The angular resolution of the observations ranges between 26″ at 3 mm and 11″ at 1 mm (the NH_3 and $C^{17}O(1–0)$ data have resolutions of 40″ and 50″, respectively).

Figure 1 presents a summary of our L1517B data in the form of integrated intensity maps for the line observations and an emission map for the 1.2 mm continuum (for equivalent maps of L1498, see TMCW04). As it can be seen, the top three panels (1.2 mm continuum, N_2H^+, and NH_3) are similar to each other and centrally concentrated, while the maps of the CO and CS isotopomers present either a central depression or have offset maxima. This difference in morphologies is a first indication of a pattern of chemical differentiation, which is confirmed below by detailed modeling.

In order to describe self consistently the internal structure of L1498 and L1517B, we have modeled simultaneously the emission distributions shown in Figure 1 plus the central emergent spectra. This has been done using a Monte Carlo radiative transfer code based on the one presented by Bernes (1979). Both cores have been assumed spherical and the model predictions have been compared with azimuthal averages of the observations. The following sections describe the main results of the modeling.

3. Density Distributions

Molecules are not in general reliable tracers of the mass distribution in cores given their abundance variations. The dust continuum, on the other hand, seems more adequate, especially in starless cores, where the dust emissivity is expected to be close to constant (Ward-Thompson et al., 1994; André et al., 1996). For this reason, we have used the 1.2 mm continuum maps to derive the density distributions of

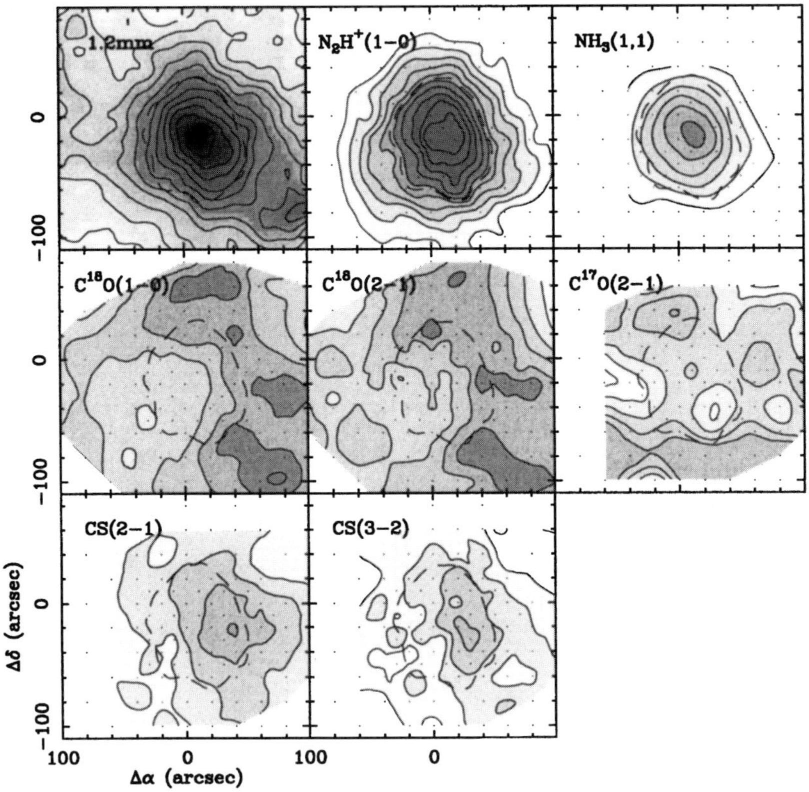

Figure 1. Integrated intensity maps of continuum and lines for L1517B. For reference, the N_2H^+ half maximum contour is shown with a dashed line.

L1498 and L1517B. We have assumed a constant dust temperature of 10 K (section 4) and a 1.2 mm emissivity of 0.005 cm^2 g^{-1}. Two types of models we have used for describing the density distributions: analytical expressions of the form $n(r) = n_0/(1 + (r/r_0)^\alpha)$, and isothermal (Bonnor–Ebert) spheres. As Figure 2 shows,both models can fit the data, with the analytical profile doing slightly better in L1498.

4. Temperature Distributions

The NH_3(1,1) and (2,2) lines provide a sensitive indication of the gas kinetic temperature, and we have used both standard LTE and Monte Carlo analysis to derive radial distributions of this important parameter. As Figure 3 shows, constant temperature models provide excellent fits to the data, suggesting that the gas kinetic temperature is very close to 10 K with an rms of about 1 K. The higher NH_3 abundance toward the core centers (Section 5) guarantees that this tracer is sensitive to the

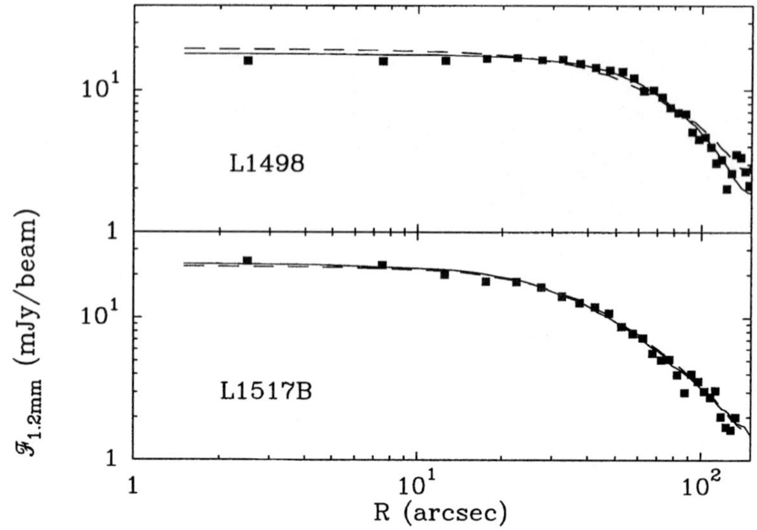

Figure 2. Radial profiles of 1.2 mm continuum emission (from the maps in Figure 1). The squares represent observed data and the lines are the predictions from two density models: analytic model (solid lines) and isothermal model (dashed line).

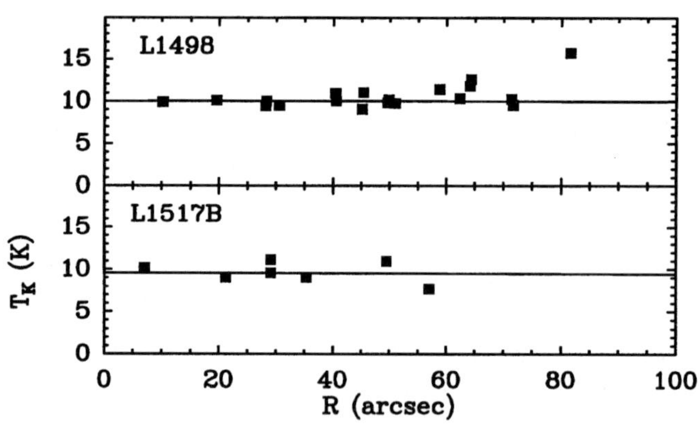

Figure 3. Radial profiles of gas kinetic temperature for L1498 and L1517B derived from NH_3. The squares represent data and the solid lines represent the predictions from Monte Carlo models having constant gas temperature (10 K for L1498 and 9.5 K for L1517B).

innermost regions of the cores even despite the relatively large beam size of the NH_3 observations (40″ FWHM).

5. Molecular Abundance Distributions

The above density and temperature profiles are the basis of our radiative transfer model. Together with the velocity profiles (Section 6), they fully determine the

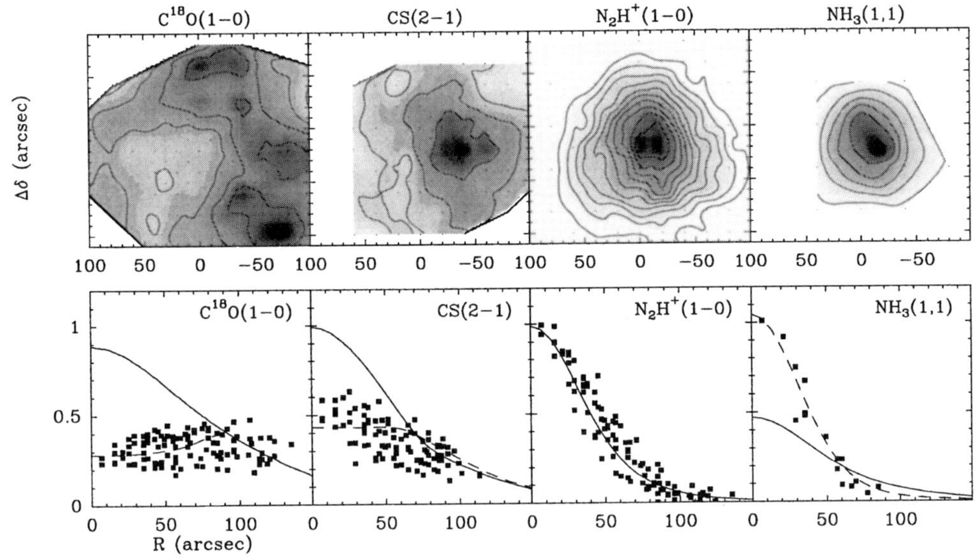

Figure 4. *Top*: Integrated intensity maps of different molecular species toward L1517B as in Figure 1. *Bottom*: radial profiles of emission for the same molecules. The squares represent data and the lines are the result of non-LTE Monte Carlo models. The solid lines are models with spatially constant abundance, and only provide a good fit for the N_2H^+ emission. To fit the $C^{18}O$ and CS radial profiles, a sharp abundance drop toward the center is needed (dashed lines), while to fit the NH_3 emission, an abundance that increases inwards is required.

physical parameters of the core, leaving the molecular abundance as the only free parameter to fit the molecular line observations. As each molecular species has been observed in two transitions, the fitting is over-determined, and this provides a self consistency check of our models. Given the limitations of this article, we only present (in Figure 4) the results for one transition toward L1517B (see TMCW04 for the full data set).

As expected from the maps of Figure 1, the $C^{18}O$ and CS emissions cannot be fitted with constant abundance models (solid lines in Figure 4), and sharp drops of abundance are needed to reproduce the data. These drops are so sudden that the models in Figure 4 have step-function abundances with no $C^{18}O$ and CS molecules in the inner core (dashed lines). In contrast, the N_2H^+ emission is fit at all radii by a model with constant abundance. This model, which fits both the (1–0) and (3–2) transitions, shows that N_2H^+ is the most reliable tracer of the molecular gas in the core. Finally, the NH_3 emission can only be fitted if the abundance of this species is higher at the center than at intermediate radii. As the figure shows, a constant abundance model is too flat to fit the NH_3 emission both at small and large radii, and only models with a central abundance increase of a factor of about 5 can go through all points in the intensity radial profile.

The observed behavior of the different molecular species can be understood as the result of the selective freeze out of molecules onto cold dust grains as the gas becomes centrally concentrated during core formation. CO and CS are particularly vulnerable to this process, while the lower binding energy of N_2 to grain surfaces makes this species relatively over abundant at high densities, and this in turn favors the presence of molecules composed of N and H (like NH_3 and N_2H^+) at the core centers (Bergin and Langer, 1997; Charnley, 1997). A somewhat surprising result is that NH_3 and N_2H^+ do not follow the same abundance trend, while they were expected to do so because they have N_2 as a common parent species. This discrepancy, not well understood yet, may result from a reduction of the abundance of N_2 over that of N via interactions with H_3^+ and He^+ at high levels of CO depletion. This process is expected to favor the formation of NH_3 over N_2H^+.

Our Monte Carlo analysis assumes spherical symmetry so it is only sensitive to radial gradients in the molecular abundances. These gradients for CO and CS have order-of-magnitude size, and therefore dominate the chemical behavior of these molecules. The CO and CS emissions, however, are strongly asymmetric, in contrast with that of 1.2 mm continuum, N_2H^+, and NH_3 (Figure 1), and this suggests that superposed upon the radial gradients there are (factor-of-2) non axisymmetric abundance variations of CO and CS (e.g., higher abundances toward the west in L1517B). These variations most likely represent differences in the amount of molecular freeze out, suggesting that the bright CO/CS positions are somehow chemically younger. As we will see below, this interpretation is supported by the core kinematics.

6. Kinematics

The fit of the line profiles toward the core centers (not shown here but see TMCW04) requires constant turbulent linewidth and small velocity gradients along the central lines of sight. Using the spectra of the depletion-resistant NH_3 and N_2H^+, we now study the distribution of these parameters in the plane of the sky. Figure 5 shows

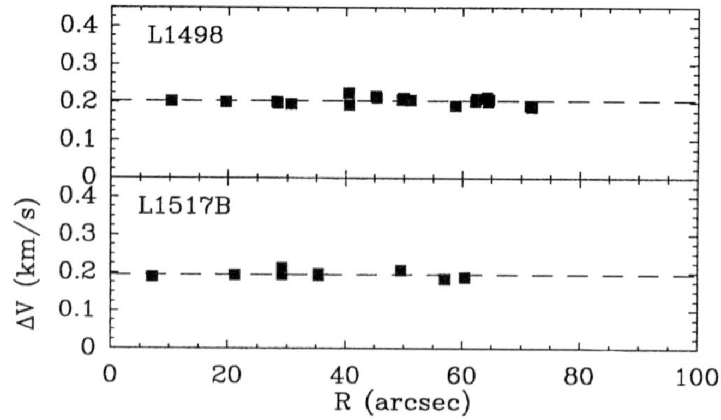

Figure 5. Radial profiles of intrinsic linewidth for L1498 and L1517B from NH_3 hyperfine-structure fits. Note the constant value and the low dispersion.

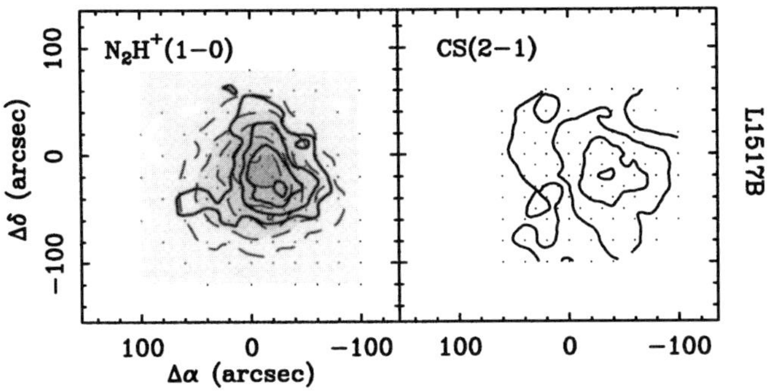

Figure 6. Comparison of line-wing red N_2H^+ emission (0.2 km s^{-1} from line center) and full integrated CS emission showing their similar distribution (solid lines). The dashed lines and grey scale in the left panel represent the full integrated N_2H^+ map.

the radial distributions of linewidth derived from NH_3 for both L1498 and L1517B (the N_2H^+ data are similar). This parameter has a constant value over the central core ($\approx$0.2 km s^{-1}) and a very small dispersion ($\leq$0.01 km s^{-1}, consistent with random errors). Subtracting the contribution of thermal motions for gas at 10 K, we derive a constant non-thermal velocity dispersion σ_{NT} component of about 0.05 km s^{-1} in both cores, which is subsonic by a large margin ($\sigma_T = 0.19$ km s^{-1}).

Radial profiles of the line center velocity for both NH_3 and N_2H^+ show a significant degree of dispersion ($\approx$0.02 km s^{-1}, which is 7 times what is expected from random errors), and this indicates the presence of internal motions inside both L1498 and L1517B. These bulk motions do not arise from simple patterns like rotation, given their spatial distributions, and imply an irregular velocity structure inside the cores (again, at the subsonic level). To determine the origin of these motions, we comparein Figure 6 the emission from the N_2H^+ line wings in L1517B with the integrated maps of CS. As mentioned before, the regions brighter in CS are expected to represent chemically younger positions because freeze out has not had time to act. The fact that these regions correspond with positions of slightly accelerated gas is consistent with the interpretation that the residual kinematics is a remnant of core formation, and that the 'young' gas is still settling onto the core (the same effect is seen in L1498).

7. Comparison with Models of Core Formation

Our observations show that L1498 and L1517B have density distributions close to those expected for gas with constant temperature and turbulence, which corresponds to the conditions measured with NH_3. Whether the cores are in a true state of physical equilibrium, however, is unclear, as the density contrast between center and edge is very close to the stability limit for Bonnor–Ebert spheres, and the measured thermal pressure seems insufficient to balance self gravity (although magnetic support may not be negligible). The presence of internal motions, in addition, suggest that part of the cores is contracting, and in a rather irregular manner (Figure 6). Although these properties may seem to support a more dynamical picture of core formation than that provided by the ambipolar diffusion model, the very small velocity gradients and the low level of turbulence seem inconsistent with models of core formation via supersonic turbulence (e.g., Klessen et al., 2003).

References

André, P., Ward-Thompson, D. and Motte, F.: 1996, *A&A* **314**, 625.
Bergin, E.A. and Langer, W.D.: 1997, *ApJ* **486**, 316.
Bernes, C.: 1979, *A&A* **73**, 67.
Caselli, P., Walmsley, C.M., Tafalla, M., Dore, L. and Myers, P.C.: 1999, *ApJ* **523**, L165.
Charnley, S.B.: 1997, *MNRAS* **291**, 455.
Klessen, R.S., Ballesteros-Paredes, J., Várquer-Semadeni, E. and Durján-Rojas, C.: 2003, astro-ph/0306055.
Kuiper, T.B.H., Langer, W.D. and Velusamy, T.: 1996, *ApJ* **468**, 761.
Tafalla, M., Myers, P.C., Caselli, P. and Walmsley, C.M.: 2004, *A&A* **416**, 191.
Tafalla, M., Myers, P.C., Caselli, P., Walmsley, C.M. and Comito, C.: 2002, *ApJ* **569**, 815.
Ward-Thompson, D., Scott, P.F., Hills, R.E. and Andre, P.: 1994, *MNRAS* **268**, 276.

ACCRETION DISKS

ACCRETION AND EJECTIONS IN MAGNETIZED DISKS

MICHEL TAGGER

Service d'Astrophysique (CNRS FRE 2591), CEA Saclay, France; E-mail: tagger@cea.fr

Abstract. In this contribution I review the main challenges for theory and numerical simulation of accretion turbulence in disks. I then present briefly a solution we have elaborated in recent years to a part of these questions: we have found an MHD instability which occurs in the inner region of a disk in the configuration (poloidal field of the order of equipartition with the gas pressure) used for MHD jet models. This instability has the unique characteristic that it re-emits toward the corona a fraction of the energy and angular momentum it extracts from the disk. It is a good candidate to explain the low-frequency Quasi-Periodic Oscillation observed in X-ray binaries, and we believe that it might occur also in YSO.

Keywords: accretion, disks, jets, MHD

It is now twelve years ago that our view of accretion in astrophysical disks has radically changed, as Balbus and Hawley (1991) showed that a magnetized disk is strongly unstable to what has since been called the Magneto-Rotational Instability (MRI). This gave at last a clear path to explaining the turbulence believed to allow accretion in these disks, whether they are formed in a Young Stellar Object, an X-ray binary, an AGN or even, as we now believe, in γ-ray bursts.

Much work has followed since then. As discussed by Hawley (this session), theory and numerical simulations have made steady progress concerning the essential properties of the instability, its ability to generate turbulence and accretion, its relations with hydrodynamical shear turbulence, etc. As discussed in Dutrey's review, and in other contributions in this session, much has also been learned in the observation of protostellar and protoplanetary disks, clarifying the physical conditions in these disks and the relevance of the MRI to explain accretion and other processes.

Rather than review this progress, which is well represented in the contributions in this session, I will thus try and look beyond it, and explore the difficulties which still lie ahead and briefly present what we consider as possible way forward. Indeed, before the matter of accretion can be considered as solved, theory and simulation still have to answer important questions. These have to do with the limits of the domain in which the MRI exists, with the present limitations of numerical simulations, and with the question of jets which are, in all these types of sources, observed to be directly connected with the accretion process.

One of these limits is at low ionization, leading to the 'dead zone' problem. Clearly progress is still necessary there, to understand whether the MRI can apply

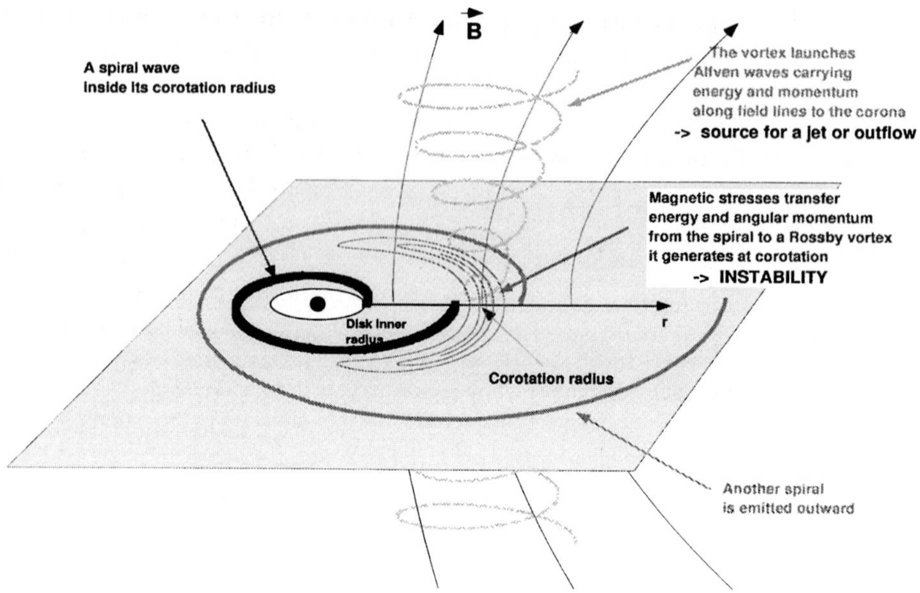

Figure 1. The AEI is formed of a *spiral wave*, propagating between the inner disk edge and corotation (the radius where the wave corotates with the gas). There the spiral couples to a crescent-shaped *Rossby vortex*, to which it transfers the accretion energy and angular momentum it has extracted from the disk. If the disk has a low-density corona, the twisting motion of the vortex can then propagate upward to the corona, as *Alfvén waves*, a substantial fraction of this energy and momentum.

in any form (e.g. Hall MHD), or whether we still need to find a different form of turbulence in unmagnetized disks.

Other limits are in fact connected:

- the MRI stops acting when the plasma β (the ratio of thermal to magnetic pressure) in the disk is below one.
- it seems to be more active, but less easy to follow numerically when the disk is threaded by a net poloidal magnetic flux.
- although numerical simulations have reached the point where global 3D, geometrically realistic models can be built, they have not yet allowed (in good part because of these limits) a detailed characterization of its non-linear properties. In particular, we need to understand the statistics of the MRI turbulence, which defines the manner in which it transports matter, angular momentum, and magnetic flux.
- neither simulations nor theory have found a clue of how the MRI could connect accretion the jet.

Characterization of the statistical properties of the MRI turbulence is critical to understand how it transports matter, angular momentum, energy, and even dust or chemical species. These are questions where the α model cannot be of any help.

That this model is dangerously simple has always been clear to anybody who has had to deal with plasma turbulence, but we need to analyze this in more details. For instance, our main observable is bound to be the radial temperature profile in the disk. But this does NOT rely on the α model, only one one of its assumptions that the turbulence deposits locally the accretion energy (rather than, as e.g. discussed by Hawley and farther in this paper, carry it away by waves). Different mathematical modelizations of turbulence which preserve this will yield the same temperature profile, but different density profile and temporal spectra, which will be much harder to observe, but could matter a lot for the long-term evolution of the disk.

Among the quantities whose evolution and radial transport need to be understood is the magnetic flux. We already know (see Blackman, these proceedings) that the MRI can generate magnetic field by a turbulent dynamo. We still have to get, from the simulations, a full description of the resulting average field. In particular, we need to know what happens to the global magnetic flux in the disk. For simplicity let us divide this into 'horizontal' and 'vertical' flux: horizontal magnetic flux can always vary, because it can be expelled from the disk by buoyancy (this is the basic physics of the Parker instability), and regenerated by a dynamo. On the other hand the disk cannot get rid of vertical flux unless it transports it radially: outward, preventing it from entering the disk, or inward, accreting it with the gas. The first possibility, with the magnetic flux transported against the accretion flow, would require a rather strong magnetic diffusivity. Advection with the gas, on the other hand, is certainly possible: we must remember that linearly the MRI is an ideal MHD instability, so that the flux is frozen in the gas. Although magnetic diffusivity acts non-linearly, when the instability has a high enough amplitude, it is likely that at least a fraction of whatever vertical flux the gas carries as it enters the disk will be advected with it, so that the amplitude of the vertical field threading the disk has to grow with time. As is well known in galactic dynamo theory, dynamo modes can also generate fields with a strong vertical component in its inner region.

This is good because a common feature of MHD jet models (Blandford and Payne, 1982; Lovelace et al., 1986; Pelletier and Pudritz, 1992; for a recent review see Casse and Ferreira, 2000) is that they find necessary to use that geometry: a disk threaded by a vertical (poloidal) magnetic field (leading to the familiar hourglass configuration), at equipartition with the gas pressure ($\beta \sim 1$). And these are precisely the conditions where our knowledge of disk turbulence fails, because of the limits of the MRI and because the simulations have a hard time getting there.

Indeed the simulations do show β decreasing from high values toward 1, but have to stop before that because the description of a $\beta \sim 1$ plasma is a tough numerical challenge. It is just as important to notice that simulations have a hard time dealing with a disk threaded by a net vertical flux, because the boundary conditions at the top of the numerical box become very hard to manage: the numerical setup must allow the boundaries to be crossed freely by magnetic field lines, but prevent any reflection of waves back toward the disk. The condition at the inner edge of the disk is also critical, since it involves the interaction of the disk with the magnetic flux anchored in the star or trapped around it. For the study of disks around black holes, the inclusion of general relativistic effects, mimicked by a pseudo-newtonian potential creating an innermost stable circular orbit, has been very helpful at this inner boundary. For young stars or neutron stars, the task will certainly be much more difficult.

All this—concentration of magnetic field reaching equipartition with the gas pressure, jet formation, final steps of the gas flow to the central object—is expected to occur in the innermost region of the disk. And this is where comparison with another type of accreting sources, X-ray binaries, might prove useful: there, since the launch in 1995 of the dedicated satellite RXTE, observables have been found which may contain clues to these final steps of accretion and ejections, in the form of sudden relativistic ejections or as a continuous jet. These observables are Quasi-Periodic Oscillations (QPO), which occur at different frequencies but are believed to originate in the innermost region of the disk. Two of these QPO occur at frequencies of the order of a kHz in neutron-star binaries, and of 10^2 Hz in microquasars. They are believed to be very close to the rotation frequency at the inner edge of the disk.

The third QPO, on which I will concentrate, occurs at a frequency roughly ten times lower (Psaltis et al., 1999), corresponding to the keplerian frequency at ~4–5 times the inner radius of the disk. Its amplitude (rms) can be up to 40% of the luminosity of the source, so that it must mark an extremely strong phenomenon occurring in the inner region of the disk, and controlling the luminosity associated with accretion. Many models have been presented, but none has gained a wide acceptance until now. I will conclude by presenting an explanation we have presented in the last few years, which seems to correspond well to the observed properties of the QPO and which, we believe, should also occur in other sources such as YSO.

Our model is based on what we (Tagger and Pellat, 1999) have called an accretion–ejection instability (AEI), which occurs in the conditions I have outlined above, and which are used in MHD jet models: a disk threaded by a poloidal magnetic field of the order of equipartition ($\beta \sim 1$). We have found that the inner region of such a disk is subject to a spiral instability, similar to the one of disk galaxies but driven by magnetic stresses rather than self-gravity (with only pressure stresses, this becomes the Papaloizou–Pringle instability). As in galaxies it forms normal modes, i.e. standing wave patterns between the galaxy center (or here the inner disk edge) and the radius where the wave corotates with the gas.

Magnetic stresses and differential rotation allow the spiral to grow by exchanging energy and angular momentum with a Rossby vortex it generates at its corotation radius. In this process the spiral extracts energy and angular momentum from the inner region of the disk (thus causing accretion) and storing them in the Rossby vortex. The structure of the spiral and the vortex is nearly constant across the disk thickness (technically it corresponds to modes with $n_z = 0$ nodes in their vertical structure, while the MRI corresponds to $n_z \geq 1$), so that we can study the AEI with 2D numerical simulations, as was done in the early simulations of galactic spirals. These simulations (Caunt and Tagger, 2001) show, in agreement with the theory, that the one-armed spiral is the dominant mode, and that it forms as expected a quasi-steady pattern (known in the galactic context as a quasi-stationary spiral structure) which has the right properties to explain the low-frequency QPO.

But the process doesn't stop there, and this explains the name we have given to this instability: the vortex twists the footpoints of the field lines threading the disk. If the disk has a low-density corona, this twist will propagate along the field lines as Alfvén waves. We have shown (Varnière and Tagger, 2003) that this can be very efficient, and thus re-emit toward the corona a substantial fraction of the energy and momentum extracted from the disk. This theoretical result can however not yet be checked in numerical simulations, which only represent an infinitely thin disk in vacuum.

Much remains to be done, on theory (e.g. finding how the Alfvén waves are damped in the corona to energize a jet), on simulation (going to 3D models of the AEI) and on the detailed comparison with observations, in the nice laboratory provided by the time variability of X-ray binaries. We can even hope that this can tell us about the evolution of the disks on a longer time scale (Tagger et al., submitted to ApJ). And of course we will need to know whether this applies to other accreting sources, including YSOs.

Although periodicities have been seen, no clear QPO has yet been seen in these sources. This may have to do with the time scales days or weeks, much harder to follow than fractions of seconds); with the geometry of the disk (for instance AGN disks are most often hidden by the torus if their inclination angle is not high enough); with the details of the emission, irradiation and absorption processes in the disk and the corona; or, if no QPO exists, it might have to do with the different origin and evolution of the magnetic flux in the disk: for instance in the case of YSO, the dead zone might prevent the penetration of magnetic flux in the inner disk region. But then we would face a tough challenge to explain their jets.

References

Balbus, S.A. and Hawley, J.F.: 1991, *ApJ* **376**, 214.
Blandford, R.D. and Payne, D.G.: 1982, *MNRAS* **199**, 883.
Casse, F. and Ferreira, J.: 2000, *A&A* **361**, 1178.
Caunt, S.E. and Tagger, M.: 2001, *A&A* **367**, 1095.
Lovelace, R.V.E., Mehanian, C., Mobarry, C.M. and Sulkanen, M.E.: 1986, *ApJ* (supplements) **62**, 1.
Pelletier, G. and Pudritz, R.E.: 1992, *ApJ* **394**, 117.
Psaltis, D., Belloni, T. and van der Klis, M.: 1999, *ApJ* **520**, 262.
Tagger, M. and Pellat, R.: 1999, *A&A* **109**, 301.
Varnière, P. and Tagger, M.: 2002, *A&A* **394**, 329.

TURBULENCE IN ACCRETION DISKS

JOHN F. HAWLEY

Department of Astronomy, University of Virginia, U.S.A. E-mail:jh8h@virginia.edu

Abstract. An accretion disk is an inevitable part of the star forming process. Recent years have witnessed dramatic progress in our understanding of how turbulence arises and transports angular momentum in astrophysical accretion disks. The key conceptual point is that the combination of a subthermal magnetic field and outwardly decreasing differential rotation is subject to the magnetorotational instability. This rapidly generates magnetohydrodynamical (MHD) turbulence, leading to greatly enhanced angular momentum transport. Purely hydrodynamic disks, on the other hand, are stable. Disks that are too cool to couple effectively to the magnetic field will not be turbulent. Fully global three dimensional MHD simulations are now beginning to probe the properties of accretion disks from first principles.

Keywords: instabilities, MHD, protostellar disks

1. Introduction

When considering the physics of the star formation process, from the interstellar medium down to the zero age main sequence star itself, one must eventually confront the properties of an accretion disk. Disk structures are the inevitable consequence of the dynamical combination of gravity and rotation, both of which are certain to be present in the star-forming gas. In fact, a disk-like structure has been a primary feature of star formation theories from the very beginning. In 1796 the French mathematician and astronomer Simon-Pierre de Laplace revised an earlier disk-like picture of the nascent solar system originally envisioned by Rene Descartes, and adapted it into a new nebular theory of solar system formation. He demonstrated mathematically that the conservation of angular momentum necessarily leads to a rotating disk. In Laplace's nebular theory, the rotating disk then contracts toward the center, where it forms the Sun. As it contracts, the disk leaves behind a series of rings that form the individual planets. The plane of the ecliptic reflects the disk's ancestral angular momentum.

Laplace's nebular theory is, at least in broad terms, incorporated into the current leading theory of solar system formation. The concept of a protostellar disk has not always been highly regarded, however. For a long time the nebular theory was considered fatally flawed for several reasons, the most important of which was known as the angular momentum problem. Since angular momentum is conserved in the absence of appropriately applied torques, most of the nebula's angular momentum should reside with the bulk of the mass. In the solar system most of the mass is

Astrophysics and Space Science **292:** 383–394, 2004.

located, of course, in the Sun. But the Sun rotates slowly and contains little of the total angular momentum of the solar system. More to the point, if the angular momentum of the gas in the initial disk had been retained, the collapse to the center to form the Sun could not have happened.

The angular momentum problem is no longer considered fatal to protostellar disk theory. Instead, it is now appreciated that angular momentum transport is the central dynamical feature of disk structure and evolution. There are enough examples of disk structure in the universe that we know that Nature solves the angular momentum problem, regardless of whether or not we fully appreciate how. Traditionally, our ignorance has been cordoned off into the Shakura–Sunyaev α parameter (Shakura and Sunyaev, 1973). In the original formulation of this parameter it was assumed that turbulence within the disk accounted for the transport, but the α formulation did not require it. Matters have become somewhat clearer now. As shall review below, it has been shown that magnetohydrodynamic (MHD) disks are quite generally turbulent, and transport angular momentum outward quite effectively. Purely hydrodynamic disks are another matter, however, and this makes the subject of protostellar disks particularly interesting. The answer to the question "Are protostellar disks turbulent?" is itself determined by the question "Are protostellar disks magnetized or not?"

2. Disk Stability and Turbulence

The transport of angular momentum through a disk has historically been considered a problem because the molecular viscosity is far too small given the size of the system. More precisely, the Reynolds number, that is, the ratio of the characteristic size and velocity to the viscosity, $Re = VL/\nu$, is huge. However, high Re systems are known in terrestrial contexts to become turbulent, and turbulent eddies can, in principle, provide a substantial angular momentum transport. The onset of turbulence requires an instability, however. A hydrodynamic disk is linearly stable to the Rayleigh criterion, which requires only that angular momentum increase with radius, a condition easily satisfied by a Keplerian rotation law. Thus the conundrum.

Things became clearer with the elucidation of the magnetorotational instability (MRI) in accretion disks (Balbus and Hawley, 1991). Weak magnetic fields render differential rotation highly unstable. The physics of the MRI has been previously reviewed (Balbus and Hawley, 1998), and a quick summary of the basic ideas are presented here. A magnetic field provides a tension force through the Maxwell stress component $-B_R B_\phi/4\pi$ that allows two orbiting fluid elements to exchange angular momentum. The transfer of angular momentum by the Maxwell stress from one element to the other causes them to move to lower and higher orbits, respectively. The relative angular velocity between the two fluid elements *increases*, thereby increasing the separation. This, in turn, increases the magnetic tension and the angular momentum transfer. The result is an instability.

The characteristic wavelength of the ideal MHD MRI is approximately the distance that an Alfvén wave travels in one orbital period, i.e., $\lambda_{MRI} = 2\pi v_A/\Omega$; the growth rate is in order Ω^{-1}. All longer wavelengths are also unstable with reduced growth rates comparable to the Alfvén crossing time. The presence of the MRI and the peak growth rate are independent of the strength of the magnetic field which only sets the characteristic wavelength. The fact that the existence of the MRI is independent of the strength of the magnetic field (for ideal MHD) is often regarded as one of its more surprising aspects. The concept of a 'weak' magnetic field is generally expressed in the ratio of the total field energy to other energies of the system, e.g. internal or kinetic. But such a concept misses the point: what is important is the existence of a tangential Maxwell stress force, a force that has no hydrodynamic analog. The magnetic field enables degrees of freedom in the plasma that are simply unavailable with hydrodynamics alone. The magnetic field strength sets the wavelength of the most unstable mode of the MRI, but does not determine whether or not the system is unstable. The transition to hydrodynamic stability occurs when the field is so weak, or the magnetic diffusivity so large, that the characteristic MRI wavelength is comparable to the diffusion lengthscale. This limit is discussed in Section 3 below.

An extensive body of local, three dimensional, numerical simulations (Hawley, et al., 1995; Brandenburg et al., 1995; Matsumoto and Tajima, 1995; Hawley et al., 1996; Stone et al., 1996) have demonstrated that the MRI quite generally results in turbulence. Turbulence is dissipative, and requires some source of energy to be maintained. The only source of available energy in an accretion disk is the gravitational energy of accretion itself, available from the differential rotation. This is the standard accretion scenario: gravitational energy is tapped through the transport of specific angular momentum outward through the disk. This requires a net stress; specifically the r, ϕ component of the stress tensor, $T_{r\phi}$, must be positive, that is, the turbulence must produce a positive stress to be self-sustaining. As the magnetorotational instability is driven precisely by the transfer of angular momentum, the turbulence it produces automatically has this property.

This is not the case, however, for purely hydrodynamic disks, and this has the main reason for the difficulty in finding the source of the 'α viscosity' throughout much of the history of disk theory. Because a hydrodynamical Keplerian disk is linearly stable to the Rayleigh criterion, the search for hydrodynamic turbulence had to proceed along less direct lines. One approach has been to seek some other hydrodynamical instability and attempt to apply it to the task. An example of this in the protostellar context has been to propose that the disks might become convectively unstable in the vertical direction, leading to turbulence, and, hopefully, net angular momentum transport (Lin and Papaloizou, 1980). Another approach has been to assert that while the disk may be linearly stable, it is nevertheless subject to nonlinear instabilities (e.g., Zahn, 1991).

It was the consideration of the first possibility (convectively-driven turbulence) that led to a clearer understanding why the second possibility (nonlinear instabilities) is actually an impossibility. Stone and Balbus (1996; see also Cabot 1996) simulated a local section of a disk with a vertical convective instability. They found that the turbulence died out promptly once the vertical entropy gradient was stabilized, with no significant angular momentum transport. Next, they maintained the gradient with an applied heat source at the equator. This led to with the remarkable result that a small amount of angular momentum was transferred *inward*. Far from extracting energy from the rotating flow, the convective instability was putting energy into it. This was possible only because the equator of the disk was maintained at a high temperature by the imposed boundary condition.

As it happens, there is nothing special about convection in this result. Rather, it is the manner in which the fluid fluctuations couple to the background flow, as demonstrated by a formal analytic development of the moment equations (Balbus et al., 1996; Balbus and Hawley, 1998). The linear Rayleigh criterion derives from the conservation of angular momentum. If a fluid element is displaced radially, the coriolis term acts as a restoring force, causing the fluid element to execute epicyclic oscillations around its original orbital radius. The coriolis force, the source of the stability in the linear equations, is stabilizing in the nonlinear equations as well. Because angular momentum is a conserved quantity for hydrodynamic flows, turbulent mixing tends to even out the angular momentum gradient, not the angular velocity gradient (Balbus, 1995; Balbus, 2000). But it is the angular velocity gradient that is the source of free energy. These things work at cross purposes, and self-sustaining hydrodynamic turbulence is not possible.

These analytic arguments are supported by a series of hydrodynamic simulations of differentially rotating flows that were carried out using a variety of grid resolutions and different numerical techniques (Balbus et al., 1996; Hawley et al., 1999). These simulations were done in the local limit, obtained by centering a locally Cartesian coordinate system at a radius R_o, and rotating at angular velocity Ω. Using coordinates $x = R - R_o$, $y = R_o\phi$, $z = z$, the momentum equation becomes

$$\rho\frac{\partial \mathbf{v}}{\partial t} + (\rho\mathbf{v}\cdot\nabla)\mathbf{v} = -\nabla P - 2\Omega\times\mathbf{v} + 2q\Omega^2 x\hat{\mathbf{x}} \qquad (1)$$

where $q \equiv -d\ln\Omega/d\ln R$. The equilibrium velocity profile is $\mathbf{v} = -qx\hat{\mathbf{y}}$. Flows with Keplerian angular momentum distributions ($q = 1.5$) prove completely stable. Even when strong fluctuations are added to the initial conditions, the resulting turbulence quickly died out. On the other hand, simple shear flows, which lack the coriolis and tidal force terms, proved to be unstable, and the instability was observed even with surprisingly crude resolution.

Interestingly, a constant angular momentum flow ($q = 2$) also proves to be unstable, and this illustrates an important point. The equations for a shear layer are identical to those for a constant angular momentum disk. For such a disk the coriolis plus tidal force term vanishes and the epicyclic frequency is zero. This is the boundary between Rayleigh stable and unstable angular momentum distributions; there are no epicyclic oscillations. This means that the nonlinear modes that are destabilized in a shear layer correspond to these zero frequency inertial modes. In other words, the sum of all linear amplitude forces is precisely zero for such disturbances so the smallest contributing forces evoking a dynamical response are nonlinear. Clearly, then, nonlinear instabilities are a very special case and not a generic expectation for accretion disks. Simulations easily uncover nonlinear instabilities in the $q = 2$ disk and shear layer, but find rapid decline of imposed turbulence when q is even slightly below 2 (Hawley et al., 1999).

It has been suggested that the numerical simulations are too diffusive to capture the nonlinear instability (e.g. Longaretti, 2002). But this cannot be the case. In Eq. (1) there are no special scales; the box can represent as small or as large a section of a differentially rotating flow as one wishes. The coriolis and tidal forces are set by local constants. There is nothing for a viscous diffusion length to be tiny with respect to.

The stark difference between a hydrodynamic and an MHD disk was made clear by Balbus (1995), who derived the general stability criteria for a magnetized differentially rotating medium. In the case of a hydrodynamic gas the stability criteria are the the classic Høiland criteria,

$$N^2 + \kappa^2 > 0, \tag{2}$$

where N is the Brunt–Väisälä frequency and κ is the epicyclic frequency, and

$$-\left(\frac{\partial P}{\partial Z}\right)\left(\frac{\partial l^2}{\partial R}\frac{\partial \ln P\rho^{-5/3}}{\partial Z} - \frac{\partial l^2}{\partial Z}\frac{\partial \ln P\rho^{-5/3}}{\partial R}\right) > 0. \tag{3}$$

Briefly, these show the stability consequences of entropy gradients and angular momentum gradients. In a magnetized fluid, however, these become

$$N^2 + \frac{\partial \Omega^2}{\partial \ln R} > 0, \tag{4}$$

and

$$-\left(\frac{\partial P}{\partial Z}\right)\left(\frac{\partial \Omega^2}{\partial R}\frac{\partial \ln P\rho^{-5/3}}{\partial Z} - \frac{\partial \Omega^2}{\partial Z}\frac{\partial \ln P\rho^{-5/3}}{\partial R}\right) > 0. \tag{5}$$

What happens is that in the presence of a magnetic field the angular momentum term, l, is replaced by the angular *velocity* Ω. The difference is profound. Whereas the Rayleigh criterion, $\partial l^2/\partial R > 0$ is always satisfied, the Balbus stability criterion, $\partial \Omega^2/\partial R > 0$ is never satisfied. This, in turn, is the difference between laminar and turbulent flow.

3. Nonideal MHD

The presence or absence of turbulence comes down to whether or not the disk is MHD or hydrodynamic. For accretion disks in compact binary systems or active galactic nuclei, one can be fairly confident that the ionization levels are high and that the ideal MHD limit holds. Things are far less certain in the protostellar context, where the gas may be too cool and too dense in many regions to be well coupled to magnetic fields.

Non-ideal MHD represents the transition between magnetic and hydrodynamic disks, and, by inference, between turbulent and nonturbulent disks. There are at least three important effects to consider: ohmic resistivity, which corresponds to collisions between electrons and neutrals, ambipolar diffusion, which is due to collisions between ions and neutrals, and the Hall effect, which arises from differences in the ion and electron velocities. For each of these we must consider both the linear stability properties, and, if unstable, the *nonlinear* outcome of the instability. Although in a protostellar disk an ionization fraction as small as 10^{-13} may be sufficient to allow the linear MRI (Blaes and Balbus, 1994), there remain a number of uncertainties. One is whether or not nonlinear turbulence can be sustained.

A physically motivated way to gauge the relative importance of these nonideal effects is to compare their associated lengthscales and frequencies with those of the MRI. The effect of ohmic resistivity on the linear MRI has been examined by a number of authors (Jin, 1996; Balbus and Hawley, 1998; Sano and Miyama, 1999). The relative importance of ohmic resistivity is measured by the magnetic Reynolds number which is the ratio of a characteristic velocity times a characteristic length to the magnetic diffusivity $Re_M = VL/\eta$. If we take the Alfvén speed to be the characteristic velocity, and the MRI wavelength to be the length, then $Re_M \propto v_A^2/\eta\Omega$. The MRI is seriously compromised when Re_M is on order or less than unity. Physically, the field slips through the gas due to diffusion faster than the rate of MRI growth. Nonlinear simulations (Sano et al., 1998; Fleming et al., 2000) reveal an even more interesting behavior. Although the initial configuration may be linearly unstable and develop turbulence, resistive dissipation can cause the turbulence to die away. If there is a net field through the computational domain (which therefore cannot be eliminated by ohmic diffusion), a subsequent round of linear growth, turbulence, and decay can result (Sano et al., 1998). Without a net field, however, the decay can be permanent. The critical value of Re_M to sustain nonlinear turbulence is roughly 100 times larger than that required to eliminate the linear instability (Fleming et al., 2000).

In the ambipolar diffusion limit (Blaes and Balbus, 1994) the MRI is effectively unchanged if the ion-neutral collision frequency greatly exceeds the orbital frequency. In the limit of low collision frequencies the MRI operates only in the ion fluid and the neutrals are unaffected. In this case since the ion density is very low the Alfvén speed is correspondingly high. In the intermediate regime the MRI is substantially reduced by drag with the neutrals. Simulations of the MRI with ambipolar diffusion have been done by Hawley and Stone (1998) and Mac Low et al. (1995). Again, the conditions for significant nonlinear turbulence are more stringent than those for linear instability. Hawley and Stone found that significant turbulence in both the ions and neutrals was produced only when the collision frequency was greater than about 100Ω.

Wardle (1999; see also this volume) pointed out that in protostellar disks Hall electromotive forces (EMFs) are likely to be important. This conclusion was further supported by the analysis of Balbus and Terquem (2001), who showed that, for the physical conditions in a protostellar disk, the Hall effect is significant whenever resistivity is as well. However, in contrast to resistivity, the effects of Hall currents are more varied and interesting. The Hall term can be destabilizing as well as stabilizing; indeed, it can produce instability under conditions that are stable to the ideal MRI (Balbus and Terquem, 2001). It also has a dependence on the angle between the rotation vector $\mathbf{\Omega}$ and the magnetic field vector $\mathbf{B}$ (alignment increases growth rates). The relative importance of the Hall terms can be estimated by the ratio of the Hall EMF term to the usual induction term (Balbus and Terquem, 2001; Sano and Stone, 2002a):

$$\frac{\mathbf{J} \times \mathbf{B}}{en_e(\mathbf{v} \times \mathbf{B})} = \frac{cB}{4\pi en_e VL} = \frac{cB\Omega}{4\pi en_e v_A^2}. \tag{6}$$

When this quantity is greater than 1, the Hall EMFs are important. Simulations of rotating Hall plasmas have been carried out by Sano and Stone (2002a,b). Of particular interest was whether the destabilizing properties of Hall EMFs could decrease the critical Reynolds number for the suppression of turbulence. They conclude that while the Hall term can reduce the critical Re_M, the effect is not great and the stability of protostellar disks is determined primarily from the effective ohmic dissipation.

Perhaps the greatest uncertainty at this point is the true state of the disk itself. Many issues pertain to the effective coupling within the disk. The inner regions of disks (<1 AU; Stone et al., 2000) seem likely to be sufficiently ionized, but the outer regions are more problematic. Here things may well be complicated. When the precise level of free electrons is crucial, questions such as the amount and type of dust present and the abundance of alkali elements must be answered with precision. The sources of ionization must also be considered with care. Gammie (1996) has drawn attention to the role of cosmic rays in ionizing an outer layer of the disk, thereby introducing the idea of a *dead zone* inside the disk, surrounded by a magnetically active region. Such layered disks might also be produced by ionizing X-rays from the central star (Glassgold et al., 2000; Fromang et al., 2002). Recent simulations by Stone and Fleming (2003) suggest that turbulence in the active zone can drive velocity fluctuations in the dead zone and produce a moderate net stress there. If a net stress is present the resulting turbulent heating may be sufficient to increase the ionization level and further enhance the turbulence. Clearly, a great deal of detailed work remains to be done on this interesting and complex problem.

Finally, it is also worth noting that if the magnetic field is sufficiently decoupled from the disk that the MRI can be ignored, it is highly unlikely that the magnetic field will be sufficiently well-coupled for some other dynamical role. At the very least, asserting some dynamical importance to magnetic fields while simultaneously claiming immunity to the MRI due to nonideal effects would require careful justification.

4. Are α-Disk Models Appropriate?

Formal treatments of linear stability analyses, expositions on Rayleigh decompositions of turbulent flows, and extensive details of numerical simulations can leave the phenomenologist at a loss. What from all of these complex results can one use to understand real systems and the observations of them? When, if ever, is a standard α model an appropriate description of a protostellar system?

The original Shakura and Sunyaev (1973) disk model was based on the general idea that turbulence produced the stress responsible for the outward transport of angular momentum and the local release of gravitational energy. On dimensional and energetic grounds Shakura and Sunyaev argued that the turbulent stress would be roughly proportional to the pressure, $T_{r\phi} = \alpha P$. The limit $\alpha < 1$ follows from the expectation that the turbulent velocities should be subsonic; larger velocities would rapidly dissipate in shocks and, presumably, could not be sustained. Other critical assumptions of the model are that the disk is steady state, vertically thin, i.e., $c_s \ll R\Omega$, and that the energy released by the stress is promptly thermalized and locally radiated.

Balbus and Papaloizou (1999) examined α-disk theory in light of the properties of MRI-driven MHD turbulence. They found that because of the local character of MHD turbulence, an α-type formalism remains appropriate so long as the other assumptions that go into α disks remain valid, particularly that the disk is thin so that only the leading terms in an expansion in the turbulent velocities need be retained. However, they point out that the α formalism is not appropriate for nonlocal angular momentum transport mechanisms such as waves. This is an important point. Although purely hydrodynamic disks will not be turbulent, there are nevertheless purely hydrodynamic angular momentum transport mechanisms. These depend, however, on the nonlocal stress of spiral waves and include shock waves, and waves generated by self-gravity. An α description for hydrodynamic disks is not appropriate.

Even within the context of MHD an α model presumes a local relationship between the stress and the local heating. In fact, the immediate product of the MRI is not heat but kinetic and magnetic fluctuations. If there is an efficient local turbulent cascade to resistive and viscous scales, then the stress will produce local heat in keeping with the α model. But much of this is uncertain. How efficient is the turbulent dissipation? Is the disk optically thick or thin, and what radiative processes dominate? What fraction of the magnetic field rises buoyantly from the disk to remove energy in a wind or a jet?

There is also a question of time-scale. One expects that there should be some time interval over which one can average to recover the relationships between disk quantities consistent with α disks. Numerical simulations show large temporal and spatial fluctuations in all values, including the stress. The averaging interval is at least several orbits (Papaloizou and Nelson, 2003), and may be much longer because the turbulence exhibits chaotic behavior (Winters et al., 2003a). This means that while there may be circumstances in which an α disk provides a reasonable overall steady state model, it cannot be used to address detailed issues of evolution on dynamical timescales.

Another caution is worth reiterating. In accretion disk modeling it is sometimes fashionable to convert the α stress into a shear viscosity. While there are certainly occasions in which this will do no harm (mainly when it doesn't make any difference to regard the stress in this way), one must avoid the temptation to treat the stress as a real viscosity, either mathematically or physically. A viscous laminar flow differs in too many essential ways from a turbulent flow for this to be very useful.

An example is provided by the recent simulations of protoplanets in MHD turbulent disks. The processes of planet formation and migration depend intimately on the interaction between planetesimals and the gaseous disks in which they form. For example, it is known that nonlinear gravitational coupling between the disk gas and the planet clear out a gap in the disk, and that this, in turn, affects the rate of planet growth and inward migration. The presence of turbulent stresses alters the process and drives accretion across the gap. Traditionally, this turbulence has been simulated using a viscosity. Now it is possible to model the stress directly using MHD simulations (Nelson and Papaloizou 2003; Winters et al., 2003b). While the results from these simulations show some qualitative agreement with previous viscous and inviscid simulations of planetary gaps, they also find significant differences. In the presence of MHD turbulence, the gaps produced are shallower and asymmetrically wider than those produced with inviscid hydrodynamics (Winters et al., 2003). Nelson and Papaloizou (2003b) found that gap in an MHD turbulent disk model was deeper than the viscous model with equivalent α value, and that the magnetic field could transfer angular momentum directly across the gap. The MHD turbulence does not provide a constant source of friction, unlike a viscosity. Most importantly, viscous hydrodynamics cannot describe the interdependence between the planet and the stress. That is, not only does the turbulence affect the planet and its gap, but planets capable of producing gaps also proved capable of affecting the MHD turbulence itself. The spiral waves generated by the planet reduced the MHD stresses particularly near the gap where the waves were strong.

5. How Do Disks Create Jets?

One of the most striking signatures of star formation is the collimated outflow, or jet. It has long been argued that jets are launched from the inner surface of a disk, and images by HST provide direct evidence for that scenario. There have been numerous numerical simulations of MHD winds that demonstrate that magnetic fields can provide the necessary collimation and acceleration. But these simulations generally begin with an assumed large scale magnetic field, and treat the disk as a boundary condition. If the disk is evolved, it generally begins threaded with a large-scale field which dominates the subsequent evolution. What we would really like to see are jets forming from a fully self-consistent global disk simulation that generated its own magnetic field through a dynamo process.

We are a long way as yet from that goal. However, global simulations of disks in the context of black hole accretion (Hawley and Balbus, 2002; Hawley and Krolik, 2002; De Villiers and Hawley, 2003) have found a few promising features. First, it is clear that the MRI does act like a dynamo; it amplifies field and sustains it in the face of losses. Within the disk the field remains subthermal, and in rough equipartition with the kinetic energy of the turbulence. Field is lost from the disk due to buoyancy, leading to a magnetized corona. In the corona the magnetic and gas pressures are comparable, with the magnetic pressure having a slight edge. Much of the material in the corona is outflowing, although at modest velocities and in an uncollimated fashion. There is no obvious structure to the coronal field. There is an unbound hollow jet seen in some of these simulations. It is confined to near the centrifugal funnel wall, appears to be launched by pressure forces at the inner boundary of the disk, and is collimated by the surrounding magnetic corona.

Several intriguing questions are suggested by the results to date. First, if a large-scale ordered field is truly required for the sort of jet that is observed, can that field be created by the MRI-driven turbulence in the disk alone? It is possible that fields of certain polarity are preferentially ejected out one side of the disk, in which case the coronal outflow could eventually ‘comb out’ the ejected field into (say) a dipole configuration. But perhaps something more needed, like a net vertical field. This could be externally supplied and drawn in with the inflow from large radius. Simulations should soon be able to address these questions directly.

6. Conclusion

The past decade has seen several important developments. Linear stability analyses have clarified when disks are unstable and become turbulent, and also, significantly, when they are stable and do not produce turbulence. We now know that turbulence and turbulent transport are the inevitable consequence of magnetism in disks. Hydrodynamic disks, in contrast, have no way to tap locally into the free energy of the flow and will be laminar. Waves, global instabilities, and effects from self-gravity can transport angular momentum within hydrodynamic disks, but these do not lead to turbulence and they are not well modeled by the usual α formalism. Numerical simulations of disk physics have increased in size and complexity in keeping with the substantial growth in computer power. Simulations have gone from local two dimensional boxes to fully three dimensional global domains. By moving beyond the α-viscosity picture, we can start to address questions of protostellar disk physics in greater detail. Turbulence is a flow state with its own distinct properties, and these can be investigated. There are many areas for further ongoing work. Non-ideal MHD and the chemical and ionization properties of protostellar disks are of particular importance for the presence of turbulence. Understanding the generation of large scale magnetic fields and the launching of jets is another area where large scale simulations should soon provide new insights.

These and other developments have led us to a more complex picture of protostellar disks, and numerical simulations will be increasingly required to make progress. To those accustomed to analytic closed models this is not necessarily welcome news. While we may hope that protostellar disks may ultimately be understood in terms of a relatively simple model, we cannot let our desire for simplicity, nor our reluctance to pursue more complicated treatments, outweigh the realities of the physics.

Acknowledgements

The author wish to thank the organizers of the conference for their invitation to present this review, and their hospitality during the conference. This work is supported in part under NSF grant AST-0070979, and NASA grants NAG5-9266 and NAG5-13288.

References

Balbus, S.A.: 1995, *ApJ* **453**, 380.
Balbus, S.A.: 2000, *ApJ* **534**, 420.
Balbus, S.A. and Hawley, J.F.: 1991, *ApJ* **376**, 214.
Balbus, S.A. and Hawley, J.F.: 1998, *Rev. Mod. Phys.* **70**, 1.
Balbus, S.A., Hawley, J.F. and Stone, J.M.: 1996, *ApJ* **467**, 76.
Balbus, S.A. and Papaloizou, J.C.B.: 1999, *ApJ* **521**, 650.
Balbus, S.A. and Terquem, C.: 2001, *ApJ* **552**, 235.
Blaes, O.M. and Balbus, S.A.: 1994, *ApJ* **421**, 163.
Brandenburg, A., Norlund, Å., Stein, R.F. and Torkelsson, U.: 1995, *ApJ* **446**, 741.
Cabot, W.: 1996, *ApJ* **465**, 874.
De Villiers, J.-P. and Hawley, J.F.: 2003, *ApJ* **592**, 1060.
Fleming, T.P., Stone, J.M. and Hawley, J.F.: 2000, *ApJ* **530**, 464.
Fromang, S., Terquem, C. and Balbus, S.A.: 2002, *MNRAS* **329**, 18.

Gammie, C.F.: 1996, *ApJ* **457**, 355.
Glassgold, A.E., Feigelson, E.D. and Montmerle, T.: 2000, in: V. Mannings, A.P. Boss and S.S. Russell (eds.), *Protostars and Planets IV*, Univ. Arizona, Tucson, 429.
Hawley, J.F. and Balbus, S.A.: 2002, *ApJ* **573**, 738.
Hawley, J.F., Balbus, S.A. and Winters, W.F.: 1999, *ApJ* **518**, 394.
Hawley, J.F., Gammie, C.F. and Balbus, S.A.: 1995, *ApJ* **440**, 742.
Hawley, J.F., Gammie, C.F. and Balbus, S.A.: 1996, *ApJ* **464**, 690.
Hawley, J.F. and Krolik, J.H.: 2002, *ApJ* **566**, 164.
Hawley, J.F. and Stone, J.M.: 1998, *ApJ* **501**, 758.
Jin, L.: 1996, *ApJ* **457**, 798.
Lin, D.N.C. and Papaloizou, J.C.B.: 1980, *MNRAS* **191**, 37.
Longaretti, P.-Y.: 2002, *ApJ* **576**, 587.
Mac Low, M.-M., Norman, M.L., Königl, A. and Wardle, M.: 1995, *ApJ* **442**, 726.
Matsumoto, R. and Tajima, T.: 1995, *ApJ* **445**, 767.
Nelson, R. and Papaloizou, J.C.B.: 2002a, *MNRAS* **339**, 983.
Nelson, R. and Papaloizou, J.C.B.: 2002b, *MNRAS* **339**, 993.
Sano, T., Inutsuka, S. and Miyama, S.M.: 1998, *ApJ* **506**, L57.
Sano, T. and Miyama, S.M.: 1999, *ApJ* **515**, 776.
Sano, T. and Stone, J.M.: 2002a, *ApJ* **570**, 314.
Sano, T. and Stone, J.M.: 2002b, *ApJ* **577**, 534.
Shakura, N.I. and Sunyaev, R.A.: 1973, *A&A* **24**, 337.
Stone, J.M. and Balbus, S.A.: 1996, *ApJ* **464**, 364.
Stone, J.M. and Fleming, T.: 2003, *ApJ* **585**, 908.
Stone, J. M., Gammie, C. F., Balbus, S. A. and Hawley, J.F.: 2000, Transport Processes in Protostellar Disks, in: V. Manning, A. Boss and S. Russell (eds.), *Protostars and Planets IV*, Tuscon, Univ. Arizona Press, 589.
Stone, J.M., Hawley, J.F., Gammie, C.F. and Balbus, S.A.: 1995, *ApJ* **463**, 656.
Wardle, M.: 1999, *MNRAS* **307**, 849.
Winters, W.F., Balbus, S.A. and Hawley, J.F.: 2003a, *MNRAS* **340**, 519.
Winters, W.F., Balbus, S.A. and Hawley, J.F.: 2003b, *ApJ* **589**, 543.
Zahn, J.-P.: 1991, in: C. Bertout et al. (eds.), *Structure and Emission Properties of Accretion Disks*, Editions Frontières, Gif-sur-Yvette, 87.

CORONAE & OUTFLOWS FROM HELICAL DYNAMOS, COMPATIBILITY WITH THE MRI, AND APPLICATION TO PROTOSTELLAR DISKS

ERIC G. BLACKMAN[1] and JONATHAN C. TAN[2]

[1]*Department of Physics & Astronomy, University of Rochester, Rochester, U.S.A.; E-mail: blackman@pas.rochester.edu*

[2]*Princeton University Observatory, Peyton Hall, Princeton, U.S.A.*

Abstract. Magnetically mediated disk outflows are a leading paradigm to explain winds and jets in a variety of astrophysical sources, but where do the fields come from? Since accretion of mean magnetic flux may be disfavored in a thin turbulent disk, and only fields generated with sufficiently large scale can escape before being shredded by turbulence, *in situ* field production is desirable. Nonlinear helical inverse dynamo theory can provide the desired fields for coronae and outflows. We discuss the implications for contemporary protostellar disks, where the (magneto-rotational instability (MRI)) can drive turbulence in the inner regions, and primordial protostellar disks, where gravitational instability drives the turbulence. We emphasize that helical dynamos are compatible with the magneto-rotational instability, and clarify the relationship between the two.

Keywords: outflows, dynamos, accretion disk, magnetic fields, primordial protstars

1. Introduction: The Dynamo-Corona-Outflow Paradigm

Bipolar jet-like outflows are commonly observed in protostellar systems (Richer et al., 2000), as is enhanced x-ray activity (Feigelson and Montmerle, 1999). Though the correlation between outflows and x-ray activity is not entirely systematic, magnetic fields probably play an important role for both phenomena. Large scale outflows are likely magneto-centrifugally ‘fling’ driven (Blandford and Payne, 1982) or magnetically ‘spring’ driven (e.g. Lynden-Bell, 1996) from large scale open field lines, while coronal dissipation is likely the result of reconnection from closed loops, perhaps as they open (relax) to form larger scale structures. Both jet and coronal fields are plausibly the result of large scale fields produced inside protostellar disks because (1) magnetic flux is difficult to accrete in a thin turbulent disk (2) only large scale fields are able to escape from the disk, before being shredded by turbulence therein. A working paradigm is this: helical dynamo $\rightarrow$ large scale fields $\rightarrow$ coronal fields $\rightarrow$ field lines that relax and open up $\rightarrow$ magnctocentrifugally launched outflow with a significant supersonic vertical velocity component $\rightarrow$ asymptotic outflow opening angle, whose tangent is the ratio of the expansion speed to the vertical wind speed. The angle may be further reduced by magnetic collimation.

We focus here on understanding the principles that govern the strength of large scale disk fields produced by dynamos, and on the mechanical luminosities of

Astrophysics and Space Science **292:** 395–406, 2004.

the resulting winds. We do not focus on the possible role of stellar dynamos or the accretion ejection instability (Tagger and Pellat, 1999; Varnière and Tagger, 2002).

2. Need For *In Situ* Generation of Large Scale Fields

2.1. Small scale fields shred before they escape

For magnetic fields to launch jets or power coronae, the field buoyancy time t_b, must be < its diffusion time in the disk. Since $t_b \gtrsim h/\overline{U}_b$, where h is the disk 1/2 thickness and $\overline{U}_b$ is the buoyancy speed ($\leq$ Alfvén speed $\overline{U}_A$) associated with a structure whose smallest gradient scale is L, we require $h/\overline{U}_b < L^2/\beta$ for escape, where $\beta \sim u_2 l$ is the turbulent diffusion coefficient for the magnetic field, and u_2 and l are a measure of the dominant turbulent speed and scale, where (e.g. $l \sim$cube root of the turbulent cell volume $(l_x l_y l_z)^{1/3}$). The escape condition can be rewritten $L^2 > lh(u_2/\overline{U}_b)$ and since the ratio on the right (more on this later) is $\gtrsim u_2/\overline{U}_A \sim 1$, the condition $L^2 > lh$ is required for escape.

Now consider a Shakura–Sunyaev (Shakura and Sunyaev, 1973) disk viscosity $\nu_{ss} \sim \beta = u_2 l \sim \alpha_{ss} c_s h \sim u_2^2/\Omega \sim u_{A,2}^2/\Omega$, where $u_{A,2}$ is the Alfvén speed associated with the turbulent B-field and $u_2 \sim u_{A,2}$ follows from simulations (Hawley et al., 1995; Brandenburg et al., 1995). Then $\alpha_{ss} \sim u_{A,2}^2/c_s^2$ or $u_{A,2} \sim \alpha_{ss}^{1/2} c_s$, so that $l = \alpha_{ss}^{1/2} h$. The escape condition $L^2 > lh$ then gives $L > \alpha_{ss}^{1/4} h$. Note that the minimum L applies to the *smallest* dimension of a given structure since this determines the diffusion time. This escape condition is not easily satisfied in the nonhelical MRI dynamo (see minimum k_y, k_z in Figure 3a (from Hawley et al., 1995)). More on this point later.

2.2. Mean flux is not frozen: diffusion may beat accretion

Can the large scale fields required by jets simply be accreted? Though 3-D MHD simulations are needed, it may be difficult for thin disks. The reason is that turbulent magnetic diffusivity $\beta \sim \nu_{ss}$ may keep the large scale field from being frozen on an advection time scale. (In general, the extent of flux freezing even for the total field depends on the turbulent spectrum. Turbulence produces many small scale strucures, which thus enhances the importance of the diffusion term in the induction equation.) To see the potential problem, we compare the ratio of the rate of flux advection with that of diffusion. Call d the dominant vertical variation scale of the field strength. The dominant variation scale of the radial velocity is r. The above mentioned ratio is then $\nabla \times (\overline{\mathbf{v}} \times \overline{\mathbf{B}})/(\beta \nabla^2 \overline{\mathbf{B}}) \sim (d^2/r^2)(\alpha_{ss} h c_s/\beta)$. The diffusion term thus dominates when the plausble (though not proven) relations $d < r$ and $\alpha_{ss} h c_s \sim \beta$ apply. That diffusion can be faster than advection is supported by numerical 2-D calculation in (Lubow et al., 1994), but future 3-D numerical testing

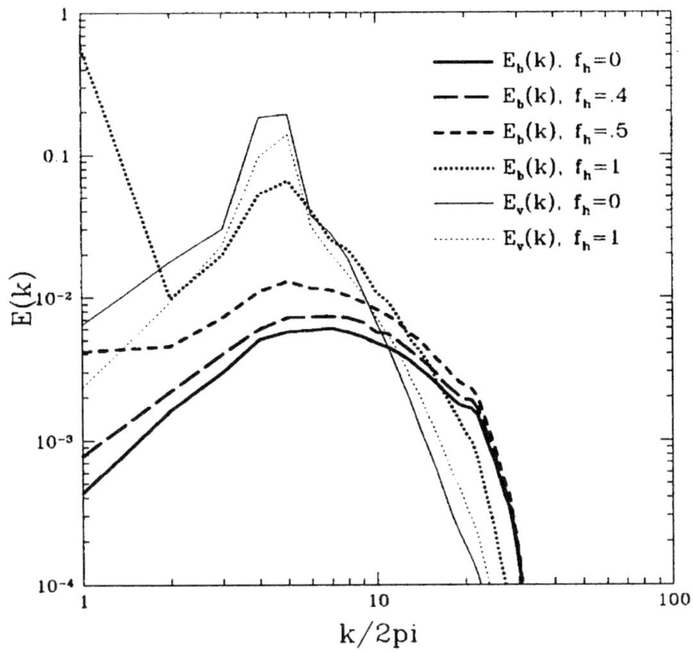

Figure 1. Saturated dynamo energy spectra (Maron and Blackman, 2002).

is required. This motivates an *in situ* helical dynamo, whose growth rate can exceed the diffusion rate.

3. MHD *In Situ* Amplification: Direct vs Inverse Dynamos

We classify dynamos into two types: **direct** and **inverse**. Direct dynamo action sustains magnetic fields on scales at or below the dominant turbulent scale. This does not require helical turbulence, although helicity does influence the magnetic spectrum. In contrast, inverse dynamo action describes amplification on scales larger than that of the dominant scale of the turbulence. The label 'inverse' is used to suggest an inverse cascade. The turbulence must be helical to generate and sustain large scale flux over times longer than an eddy turnover time, which can in turn escape to coronae and drive outflows. The distinction is illustrated in Figure 1 (Maron and Blackman, 2002) for forced turbulence. The magnetic and kinetic energy spectra are plotted for different values of the fractional helicity $f_h \equiv |\mathbf{u}_2 \cdot \nabla \times \mathbf{u}_2| / |u_2^2 k_2|$. The nonhelical forcing at $k_2 = 4.5$ produces no magnetic energy at wavenumbers $k < k_2$, whereas the $f_h = 1$ case produces a large peak at $k = 1$.

3.1. Direct Dynamo

For nonhelical direct dynamo action, the field grows by random walk, field line stretching (Kazantsev, 1968; Parker, 1979). The turbulence can be externally driven by isotropic forcing (e.g. Maron and Cowley, 2002; Haugen et al., 2003) , or self-generated by an angular velocity gradient (e.g. Balbus and Hawley, 1998). In either case, turbulent stretching compensates for exponential decay from turbulent diffusion. The latter operates because the random motions of the gas also mix the field lines, inducing a cascade to small scales where dissipation occurs. In 3-D, a steady state balance can be achieved. In the saturated state, the magnetic energy $\sim$ turbulent kinetic energy integrated below the dominant turbulent scale. How far the magnetic energy peak is from the forcing scale and the overall nonhelical dynamo spectrum is an active area of research (e.g. Maron and Cowley, 2002; Schekochihin et al., 2002; Haugen et al., 2003), which we do not discuss further here.

3.2. Helical inverse dynamo

The helical inverse dynamo amplifies field on scales larger than that of the turbulence. This is most desirable for coronae and outflows. Figure 2a. shows a traditional $\alpha\Omega$ kinematic dynamo diagram (Parker, 1979). Consider an initially weak toroidal (=encircling the rotation axis) loop of the magnetic field embedded in the rotator. With a vertically decreasing density gradient, a rising turbulent eddy threaded by a magnetic field will twist oppositely to the underlying global rotation to conserve angular momentum. Statistically, northern (southern) eddies twist field clockwise (counterclockwise). This is the 'α_d' effect and the result is a large scale poloidal field loop. Differential rotation shears this loop (the 'Ω'-effect). The bottom part reinforces the initial toroidal field and the top part diffuses away. The result is exponential growth. The α_d effect can be supplied by any source of 3-D turbulence in a stratified rotator, including gravitational instability or the MRI.

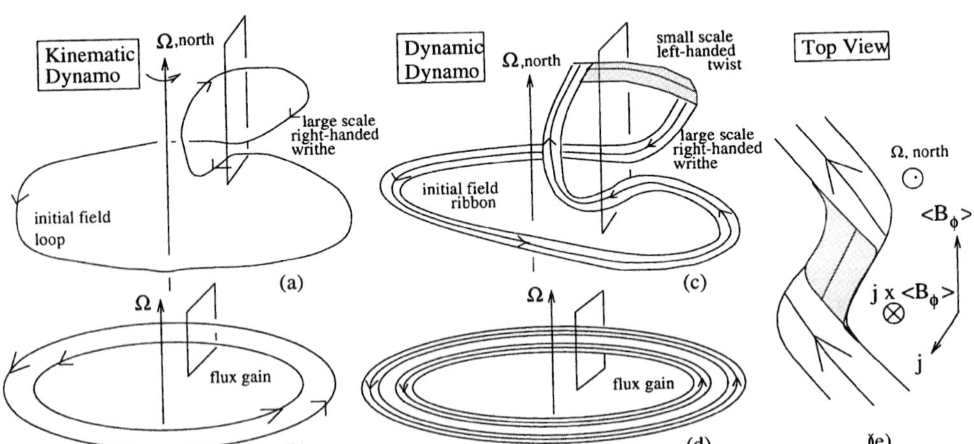

Figure 2. Kinematic helical dynamo diagram (left) ignores magnetic helicity conservation. In dynamic theory field should be replaced by ribbon, showing the conservation (Blackman and Brandenburg, 2003).

This dynamo is revealed by averaging the magnetic induction equation over a local volume and breaking all quantities (velocity $\mathbf{U}$, magnetic field $\mathbf{B}$ in Alfvén velocity units, and normalized current density $\mathbf{J} \equiv \nabla\times\mathbf{B}$) into their mean (indicated by an overbar) and fluctuating (lower case) components. The result is

$$\partial_t\overline{\mathbf{B}} = \nabla\times(\alpha_d\overline{\mathbf{B}} + \overline{\mathbf{U}} \times \overline{\mathbf{B}}) + (\beta + \lambda)\nabla^2\overline{\mathbf{B}}, \tag{1}$$

where λ is the magnetic diffusivity. The $\overline{\mathbf{U}}$ term incorporates the Ω-effect, the β term incorporates the turbulent diffusion (we assume constant β) and the first term on the right incorporates the α_d-effect. Accretion disk $\alpha - \Omega$ dynamo models have been proposed (Pudritz, 1981; Reyes-Ruiz and Stepinksi, 1997, 1999; Rüdiger et al., 1995; Brandenburg and Donner, 1997; Campbell, 1999; Rüdiger and Pipin, 2000) but without a fully dynamic theory of the field saturation value. We now discuss this for forced turbulence, and expand further for disks in Section 4.2.

For forced helical turbulence, α_d is (Blackman and Field, 2002) $\alpha_d = \alpha_{d0} + \alpha_m$ where $\alpha_{d0} = -(\tau/3)\overline{\mathbf{u}\cdot\nabla\times\mathbf{u}}$, and $\alpha_m = (\tau/3)\overline{\mathbf{b}\cdot\nabla\times\mathbf{b}}$, where τ is a turbulent damping time. The α_m correction term arises from including the magnetic forces that backreact on the velocity (Pouquet et al., 1976; Blackman and Field, 2002). From the form of α_d, it is evident that growth of α_m can offset α_{d0} and quench the dynamo (Field and Blackman, 2002; Blackman and Brandenburg, 2002; Blackman and Field, 2002). Magnetic helicity conservation determines the dynamical suppression. The magnetic helicity, a volume integral $H \equiv \int \mathbf{A}\cdot\mathbf{B}dV \equiv \langle\mathbf{A}\cdot\mathbf{B}\rangle V$ satisfies (Woltjer, 1958; Berger and Field, 1984)

$$\partial_t H = -2\lambda C - \text{surface terms}, \tag{2}$$

where $\nabla\times\mathbf{A} \equiv \mathbf{B}$, and $C \equiv \langle\mathbf{J}\cdot\mathbf{B}\rangle V$. Without non-diffusive surface terms H is typically well conserved in astrophysics. Since H is a measure of field line 'linkage' and 'twist,' its conservation implies that the α_d-effect cannot produce a net magnetic twist, just positive and negative magnetic twists on different scales. The conservation equations like (2) for H_1 and H_2 (where 1(2) indicates large (small) scales) are

$$\partial_t H_1 = 2S - 2\lambda k_1^2 H_1 - \text{surface terms} \tag{3}$$

$$\partial_t H_2 = -2S - 2\lambda k_2^2 H_2 - \text{surface terms}, \tag{4}$$

where $S = (\alpha_d k_1/\epsilon - \beta k_1^2)H_1$ and $\alpha_d = \alpha_0 + \alpha_m$, where $\epsilon \leq 1$ is the fraction of the large scale field energy which is force-free. The solution without surface terms and with $\epsilon = 1$ provides an estimate for the fully helical field large scale field component, which can be further augmented by differential rotation. Because $\alpha_m \propto k_2^2 H_2$, the solution (Field and Blackman, 2002; Blackman and Brandenburg, 2002; Blackman and Field, 2002) shows that for initially small H_2 but large α_0,

H_1 grows. Growth of H_1 implies the oppositely signed growth of H_2. This H_2 backreacts on α_0, ultimately quenching α_d and the dynamo.

Figure 2a does not capture magnetic helicity conservation because a net helicity is generated. The resolution of this problem is Figure 2b (Blackman and Brandenburg, 2003), where the field is treated as a ribbon. The large scale writhe corresponds to to the large scale magnetic helicity and to mean field dynamo action, while the small scale twist of opposite sense along the ribbon is the result of magnetic helicity conservation. The latter twist produces a force which resists the bending of the field, and thus encapsulates the backreaction and growth of α_m. Before discussing the saturation, more on the relation between the MRI and the direct versus inverse helical dynamos is addressed.

4. MRI is Compatible with Direct and Inverse Dynamos

4.1. MRI IN NONSTRATIFIED DISK: DIRECT DYNAMO ONLY

In a sufficiently ionized disk with a weak magnetic field (note: field must be strong enough for growth to beat diffusion at critical wavenumber (Balbus and Papaloizou, 1999)) the MRI results from an outward decreasing angular velocity gradient. Without vertical stratification, and thus without any pseudoscalar helicity, the MRI does not produce magnetic fields on scales larger than that of the developed turbulence. The marginally unstable (and dominant) growth mode of the MRI satisfies $ku_A \sim \Omega$ (Balbus and Hawley, 1998) so that as the magnetic field is amplified from the instability, larger and larger scales become unstable to growth. The largest scale that can grow is that associated with the scale height h. Without stratification or helicity, the largest scale of the turbulent magnetic field matches the largest turbulent velocity scale (Figure 3a (Hawley et al., 1995); k_y is the toroidal wavenumber).

The flux associated with the field on the largest scale in the nonstratified MRI would change every turnover time, as the field is shredded. The field is large scale only in the sense that the box height limits the scale of the turbulence NOT because there is any sustained ordered flux on scales larger than the turbulence. Furthermore, for a periodic box, the net flux cannot grow, as it is conserved. Note also from Figure 3a, that the largest field structure in the vertical (k_z)and radial (k_x) directions are smaller than in the toroidal direction (k_y). This is important because the constraints of section 2.1 apply to the smallest dimension of the largest scale structures. In sum, a nonstratified nonhelical MRI does *not* produce the fields required by jets/coronae.

Nonstratifed, nonhelical MRI driven turbulence and nonhelically forced turbulence behave qualitatively similar with respect to the absence of fields on scales larger than the dominant scale of the turbulent motions. Note however, that the nonhelical MRI, *unlike* the nonhelical forced case, more nearly matches the turbulent kinetic energy spectrum on larger scales (compare the first two panels in Figure 3

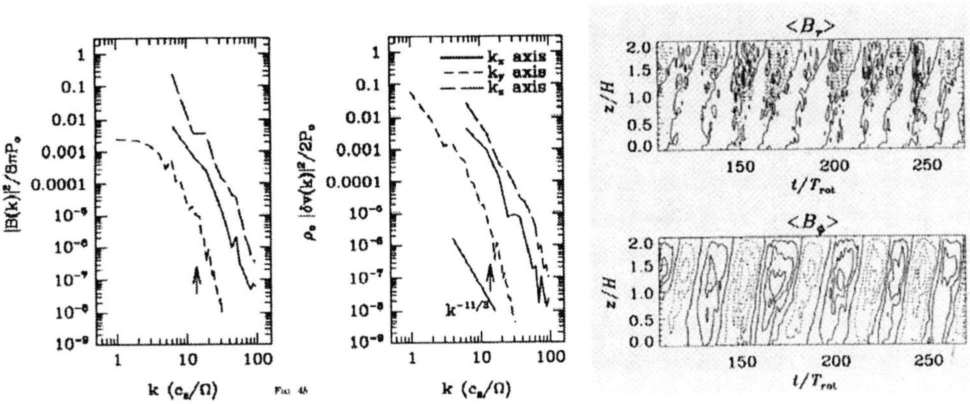

Figure 3. (a) Nonhelical MRI spectra from Hawley et al. (1995) and (b) MRI driven helical dynamo cycle in stratified non-periodic box (Brandenburg and Donner, 1997).

with the curves in Figure 1 with $f_h = 0$). This is likely due to the fact that the global shear growth time is the same on all scales in a shearing box, and thus more able to beat diffusion on the larger scales, where it can provide additional magnetic energy compared with the case without global shear.

4.2. MRI IN STRATIFIED DISK: BOTH DIRECT & INVERSE DYNAMOS

Unlike the nonstratified case, when the MRI operates in a vertically stratified medium, it can have a built in inverse helical dynamo: the stratification means that turbulence produced by the MRI can be helical, in the manner discussed in section 3.2 and also from magnetic buoyancy (Brandenburg, 1998; Rüdiger and Pipin, 2000). One can then use a nonlinear mean field theory to model the growth and saturation of the sustained large scale flux. This is independent of the fact that the MRI will always also produce direct dynamo action that amplifies fields as described. That is, when the MRI operates in a stratified medium, one gets BOTH direct and inverse helical dynamo action.

Mean field dynamo growth is seen in magnetorotational instability simulations (Brandenburg et al., 1995; Brandenburg and Donner, 1997). Figure 3b shows a dynamo cycle in which mean flux is maintained over 30 orbits. In that simulation, large and small scale magnetic helicities are seen to have opposite sign (as expected from discussion of section 3) but $\langle \mathbf{u}\cdot\nabla\times\mathbf{u}\rangle$ has the opposite sign to $\langle \mathbf{j}\cdot\mathbf{b}\rangle$ which is not expected if $\alpha_d = \alpha_{d0}+\alpha_m$. There is an additional buoyancy term $\alpha_b \sim -\frac{\tau}{3}|g_z\langle\frac{\delta\rho}{\rho}b_x\rangle|$ (g_z is vertical gravity constant) which can exceed α_{d0} and carry the opposite sign (Brandenburg, 1998; Rüdiger and Pipin, 2000). Then, for example, in the northern hemisphere $\alpha_d \sim -\tau|g_z\langle(\delta\rho/\rho)b_x\rangle| + |\tau\langle\mathbf{v}\cdot\nabla\times\mathbf{v}\rangle|/3 + \tau\langle\mathbf{j}\cdot\mathbf{b}\rangle/3$. The 1st and 2nd terms on the right have opposite signs and the sign of the large scale magnetic helicity is determined by whichever of those two initially dominates (since α_d is the growth coefficient). This then predicts the sign of the last term on the right, which grows with opposite sign to the large scale magnetic helicity, eventually quenching the larger of the 1st two terms on the right.

The fact that the stratfied MRI leads to both helical dynamo action and direct dynamo action, whereas the nonstratified MRI leads only to direct dynamo action is the key to clarifying previous discussions of the MRI and the tradititional mean field dynamo: It is sometimes stated that even a nonstratified MRI generates 'large scale' fields without the need for helical turbulence. But this statement relates directly to section 4.1. The fields referred to as 'large scale' in this nonhelical context are just the result of direct (not helical) dynamo action: (1) they only reach the maximum scale of the turbulence, not larger, and (2) they change the sign of their flux every eddy turnover time (every rotation period). That such fields are generated without helicity, does not make the MRI inconsistent with helical dynamos: as discussed above, when stratified turbulence is present both direct dynamo and inverse helical dynamos operate together, and large scale fields on scales larger than that of the dominant turbulence arise.

Like Brandenburg et al. (1995), Stone et al. (1996), studied the stratified MRI. But the latter used vertically periodic boxes, so no net flux could grow. Flux could have appeared in each $\frac{1}{2}$ of the (Stone et al., 1996) box with opposite sign, but this was not a large region, given the resolution: Measuring finite pseudoscalars in a periodic stratified box requires averaging only over $\frac{1}{2}$ of the vertical thickness, since the sign changes across the midplane. For a periodic stratified box, the pseudoscalars must vanish at the midplane and at the top and bottom.

5. Helical Dynamo Saturation in Disks & Outflow Power

The helical dynamo grows helical large scale field first 'kinematically' (independent of dissipation and with growth rate not too different from α_{d0}) until $|\alpha_m| \sim |\alpha_b - \alpha_{d0}|$, after which the field growth is resistively limited. Setting $|\alpha_m| \sim |\alpha_b - \alpha_{d0}|$, and using magnetic helicity conservation, then provides a reasonable estimate of the helical field by the end of the kinematic phase. Growth beyond that value is often too slow to be of interest. The large scale helical field at the end of the kinematic phase is (Blackman and Field, 2002; Tan and Blackman, 2003)

$$\overline{B}_H \sim (4\pi\rho f_h)^{1/2} u_2 (l/h)^{1/2} \simeq (4\pi\rho)^{1/2} \alpha_{ss} c_s, \tag{5}$$

where f_h is the fractional kinetic helicity, l is the dominant turbulent scale, h is the disk height (=the vertical scale of the growing mean field), and the latter similarity follows for disks, since $l/h \sim \alpha_{ss}^{1/2}$ and $u_2 \sim \alpha_{ss}^{1/2} c_s$, which in turn follow from the scalings of Section 2.1.

Equation (5) is an estimate of the helical field but the nonhelical toroidal field is further amplified by shear. The toroidal field is linearly stretched above $\overline{B}_r$ during a buoyancy time scale. The buoyancy loss time is $\tau_b \sim h(4\pi\rho)^{1/2}/\overline{B}_\phi$, so linear growth of $\overline{B}_\phi$ above (5) in a time τ_b gives $\overline{B}_\phi = \overline{B}_r\Omega\tau_b = \alpha_{ss}^{1/2}c_s(4\pi\rho)^{1/2}$ so $\overline{B}_r/\overline{B}_\phi = \alpha_{ss}^{1/2}$.

The above mean fields have their smallest variation scale $\sim h$ and thus satisfy the conditions of section 2.1 and can rise to the corona, open up, and launch outflows. By using $\nabla \cdot \overline{\mathbf{B}}$ one can use the disk field to estimate the field at the surface and the Poynting flux there (Tan and Blackman, 2003). This provides an estimate for the magnetic luminosity available for magneto-centrifugal (Blandford and Payne, 1982) or magnetic 'spring' (Lynden-Bell, 1996) driven outflows

$$\begin{aligned} L_{\text{mag}} &\sim \frac{c}{4\pi}\int(\overline{\mathbf{E}}\times\overline{\mathbf{B}})\cdot d\mathbf{S} \sim \frac{\alpha_{ss}^{1/2}Gm_*\dot{m}_*}{r} \\ &\simeq 220\,L_\odot\left(\frac{\alpha_{ss}}{10^{-2}}\right)^{\frac{1}{2}}\frac{m_*}{M_\odot}\frac{\dot{m}_*}{10^{-2}M_\odot/\text{yr}}\left(\frac{10^{13}\text{cm}}{r}\right), \end{aligned} \tag{6}$$

using normalizations for primordial protostars (Tan and McKee, 2003).

6. Outflows in Contemporary and Primordial Protostars

The time scale for which the helical dynamo operates at a given radius is not the viscous time, but the accretion disk's lifetime ($\sim$ star formation time). This is because the mean field is NOT frozen into the flow. The amplification is statistical, lasting as long as turbulence is supplied. For contemporary star formation, we expect that growth to the kinematic saturation value occurs in a small fraction of the star formation time, since the initial fields after collapse from molecular clouds are already quite strong, and the star formation time scales are relatively long. Contemporary protostellar disks should be able to easily sustain outflows powered by fields from a helical dynamo, with the helical turbulence in the inner regions of the disk supplied by the MRI.

For primordial protostellar disks, although magnetic Reynolds numbers are high, the seed fields are weaker and the MRI may be unfavorable for the maximally unstable mode. Instead, gravitational instability (Gammie, 2001, Wada et al., 2002; though these are 2-D simulations) may supply the 3-D helical turbulence for much of the dynamo evolution. In addition, we expect the star formation time scales to be relatively short since the accretion rates are $\sim 10^{-2}M_\odot/\text{yr}$ (Tan and McKee, 2003). Gravitational instability in the disk may maintain the Toomre parameter $Q \lesssim 1$ (Gammie, 2001; Tan and Blackman, 2003): Where the disk is initially stable (such that viscosity is small), mass accumulates, triggering the instability. This can induce turbulent accretion rather than global angular momentum transport (e.g. Gammie, 2001). Typical accretion rates, while large by contemporary standards, are small enough that fragmentation is not expected (Tan and Blackman, 2003).

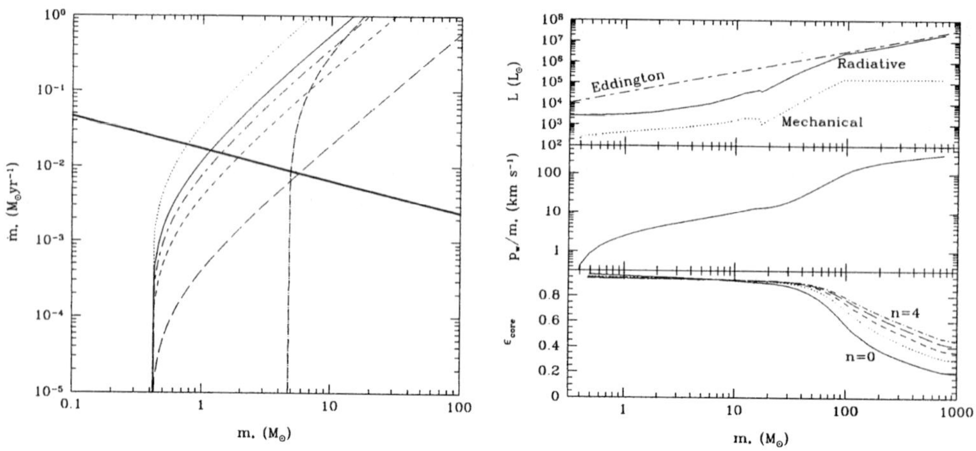

Figure 4. (a) Dynamo saturation regime for primordial protostellar disks (b) Evolution of radiative and wind mechanical luminosities from protostar and disk; outflow momentum; effect of wind on star formation efficiency (Tan and Blackman, 2003).

Tan and Blackman (2003) used a primordial protostellar model (Tan and McKee, 2003) to determine the m_* versus $\dot{m}_*$ regime for which a helical dynamo reaches the end of the kinematic regime (discussed above). In Figure 4a, the rising curves mark the condition where the helical field has been able to saturate, as the disk is old enough to have allowed sufficient dynamo amplification. The disk dynamo is assumed to start operating once the centrifugal radius of the accretion flow is $\sim$ twice the stellar radius, which happens when $m_{*,\rm init} = 0.4 M_\odot$ in the fiducial case. The dashed, solid, and dotted lines show the saturation condition for $\alpha_{\rm ss} = 10^{-3}, 10^{-2}, 0.1$, all with radius $r = 10^{13}$ cm and seed field of 10^{-16} G. With $\alpha_{\rm ss} = 0.01$, the long-dashed line shows the condition at $r = 10^{14}$ cm, the dashed-long-dashed line shows it for a seed field of 10^{-26} G, and the dot-long-dashed line shows it evaluated with $m_{*,\rm init} = 4.7 M_\odot$, corresponding to a more slowly rotating initial gas core.

Protostellar growth starts from low masses with accretion rates that decline with time (and so also with protostellar mass), as shown by the straight thick solid line. In the fiducial case, the field in the inner disk saturates when $m_* \sim 1 M_\odot$. This does not depend sensitively on $\alpha_{\rm ss}$. In lower angular momentum cores the disk emerges later and the dynamo then rapidly saturates. The main message of Figure 4a is that dynamo saturation and thus dynamo-driven winds may occur at relatively low protostellar masses, even in the primordial case.

Figure 4b (Tan and Blackman, 2003) shows the evolution of protostellar luminosities (radiative and mechanical), the cumulative specific outflow momentum relative to stellar mass (assuming that the wind speed is the escape speed from the protostellar surface and that the flow kinetic energy dominates far from the star), and the resulting star formation efficiency due to erosion of the core by the outflow for the fiducial case of Figure 4a, with $\alpha_{\rm ss} = 0.01$. The efficiency calculation is based on Matzner and McKee (1999, 2000). The initial core's axisymmetric angular density profile is specified by $dM/d\Omega = (1/4\pi)Q(\mu)M$, with $\mu = \cos\theta$ and $Q(\mu) = (1-\mu^2)^n / \int_0^1 (1-\mu^2)^n \, d\mu$. The solid line is $n = 0$ (isotropic core), dotted is $n = 1$, dashed is $n = 2$, long-dashed is $n = 3$, dot-dashed is $n = 4$. The message from Figure 4b is that dynamo driven winds can strongly affect the star formation efficiency for masses $m_* \gtrsim 100 M_\odot$, at which the protostar has contracted to the main sequence radius, and the contraction has enhanced the accretion and wind luminosities.

References

Balbus, S.A. and Hawley, J.F.: 1998, *Rev. Mod. Phys.* **70**, 1.
Balbus, S.A. and Papaloizou, J.C.B.: 1999, *ApJ* **521**, 650.
Berger, M. and Field, G.B. 1984: *J. Fluid Mech.* **147**, 133.
Blackman, E.G.: 2003, in: R. Ouyed, J. Hjorth and A. Nordlund (eds.), *Proceedings of the 1st Niels Bohr Summer Institute*, Beaming and Jets in Gamma-Ray Bursts, Copenhagen, Aug 2002, astro-ph/0211187.
Blackman, E.G. and Brandenburg, A.: 2002, *ApJ* **579**, 359.
Blackman, E.G. and Brandenburg, A.: 2003, *ApJL* **584**, L99.
Blackman, E.G. and Field, G.B.: 2002, *Phys. Rev. Lett.*, **89**, 265007.
Blandford, R.D. and Payne, D.G.: 1982, *MNRAS* **199**, 883.
Brandenburg, A: 1998, in: M.A. Abramowicz, G. Bjornsson and J.E. Pringle (eds.), *Theory of Black Hole Accretion Disks*, Cambridge University Press, Cambridge, p. 61.
Brandenburg, A.: 2001, *ApJ* **550**, 824.
Brandenburg, A., Nordlund, A., Stein, R.F. and Torkelsson, U.: 1995, *ApJ* **446**, 741.
Brandenburg, A. and Donner, K.: 1997, *MNRAS* **288**, L29.
Campbell, C.G.: 1999, *MNRAS* **310**, 1175.
Feigelson, E.D. and Montmerle, T.: 1999, *Ann. Rev. A&A* **37**, 363.
Field, G.B. and Blackman, E.G.: 2002, *ApJ* **572**, 685.
Gammie, C.F.: 2001, *ApJ* **553**, 174.
Haugen, N.E.L, Branendburg, A., Dobler, W.: 2003, sub. to PRL, astro-ph/0303372.
Hawley, J.F., Gammie, C.F. and Balbus, S.A.: 1995, *ApJ* **440**, 742.
Kazantsev, A.P.: 1968, *Sov. Phys., JETP* **26** 1031.
Konigl, A. and Pudritz, R.E.: 2000, *Protostars and Planets IV*, 759.
Lubow, S.H., Papaloizou, J.C.B. and Pringle, J.E.: 1994, *MNRAS* **267**, 235.
Lynden-Bell, D.: 1996, *MNRAS* **279**, 389.
Maron, J. and Cowley, S.: 2002, sub. to *ApJ* astro-ph/0111008.
Maron, J. and Blackman, E.G.: 2002, *ApJL* **556**, L41.
Matzner, C.D. and McKee, C.F.: 1999, *ApJ* **526**, 109.
Matzner, C.D. and McKee, C.F.: 2000, *ApJ* **545**, 364.
Moffatt, H.K.: 1978, *Magnetic Field Generation in Electronically Conducting Fluids* Cambridge University Press, Cambridge, UK.
Ouyed, R., Clarke, D.A. and Pudritz, R.E.: 2003, *ApJ* **582**, 292.
Parker, E.N.: 1979, *Cosmical Magnetic Fields* Clarendon Press, Oxford.
Pouquet, A., Frisch, U. and Léorat, J.: 1976, *J. Fl. Mech.* **77**, 2, 321 (PFL).
Pudritz, R.E.: 1981, *MNRAS* **195**, 897.
Reyes-Ruiz, M. and Stepinski, T.F.: 1997, *MNRAS* **285**, 501.
Reyes-Ruiz, M. and Stepinski, T.F.: 1999, *A&Ap* **342**, 892.

Richer, J., Shepherd, D., Cabrit, S., Bachiller, R. and Churchwell, E.: 2000, in: V. Mannings, A.P. Boss and S.S. Russell (eds.), *Protostars & Planets IV*, The University of Arizona Press, Tucson, 867.
Rüdiger, G., Elstner, D. and Stepinski, T.F.: 1995, *A&Ap* **298**, 934.
Rüdiger, G. and Pipin, V.V.: 2000, *A&Ap* **362**, 756.
Schekochihin, A.A., Cowley, S.C., Hammett, G.W., Maron, J.L. and McWilliams, J.C.: 2002, *New Journal of Physics* **4**, 84.
Shakura, N.I. and Sunyaev, R.A.: 1973, *A& A.* **24**, 337.
Stone, J.M., Hawley, J.F., Gammie, C.F. and Balbus, S.A.: 1996, *ApJ* **463**, 656.
Tagger, M. and Pellat, R.: 1999, *A. & Ap* **349**, 1003.
Tan, J.C. and Blackman, E.G.: 2004, *ApJ* **603**, 401.
Tan, J.C. and McKee, C.F.: 2004, *ApJ* **603**, 383.
Tout, C.A. and Pringle, J.E.: 1996, *MNRAS* **281**, 219.
Varnière, P. and Tagger, M.: 2002, *A&Ap* **394**, 329.
Wada, K., Meurer, G. and Norman, C.A.: 2002, *ApJ* **577**, 197.
von Rekowski, B., Brandenburg, A., Dobler, W. and Shukurov, A.: 2003, *A&Ap* **398**, 825.
Woltjer, L.: 1958, *Proc. Nat. Acad. Sci.* **44** 489.

OBSERVATIONS OF CIRCUMSTELLAR DISKS

ANNE DUTREY and STÉPHANE GUILLOTEAU

L3AB, UMR 5804, Observatoire de Bordeaux, 2 rue de l'Observatoire, F-33270 Floirac, France; E-mail: anne.dutrey@obs.u-bordeaux1.fr

Abstract. This review presents recent results on protoplanetary disks obtained from angularly resolved observations. Observations with mm arrays show that disks are in Keplerian rotation, with radius as large as 1000 AU. Optical images show disks to be flared. Both type of observations imply the dust in disk has evolved and grown from interstellar dust. Measurement of the gas temperature from CO isotopes indicate temperature gradient, consistent with the disk flaring and heating by the central star. Disks which appear to have started to dissipate their initial gas content have also been discovered, but their very diverse aspects leaves the dissipation process unclear. Current data mostly concern the outer disk (>50 AU), although near-IR interferometry has started to unveil the innermost regions (<1 AU). The next generation of instruments (MIDI on VLTI, ALMA) will allow to probe the intermediate regime, where planet formation is expected to occur.

Keywords: star formation, planetary formation, proto-planetary disks, observations

1. Introduction

Keplerian circumstellar disks are now observed not only around pre-main-sequence (PMS) low-mass stars but also intermediate-mass Herbig Ae/Be stars. Continuum surveys, performed in the mm/submm domain, suggest a tendency for the disk mass to increase with the stellar mass (e.g. Natta et al., 2000). In this review, we will mainly discuss the observed properties of protoplanetary disks found around class II PMS low and intermediate-mass objects, namely T Tauri and Herbig Ae stars of ages around 1–5 Myrs and masses ranging between 0.5 to 2.5 $M_\odot$. The dust emission of such disks is optically thick in the near-infrared (NIR) but the main component of these disks remains H_2, as a remnant of the parent cloud. One usually assumes a gas-to-dust ratio of $\sim$100, as in molecular clouds.

We will illustrate our current knowledge by presenting recent results from angularly resolved observations. Disk features are routinely revealed by observations performed from optical to the mm/cm wavelengths. Ground-based optical telescopes and the Hubble space telescope (HST) are able to image the small dust particles at the disk surface, which are scattering the stellar light and directly show, for edge-on disks, that the disk is flaring (e.g. HH 30, see Figure 2). Thanks to their high spectroscopic resolution, large millimeter arrays, such as the OVRO and IRAM interferometers, have clearly demonstrated by mapping the CO J $= 1 - 0$ and J $= 2 - 1$ emission lines from the gas that disks are in Keplerian rotation.

Astrophysics and Space Science **292:** 407–418, 2004.

Interferometric continuum observations of disks in the mm and cm (VLA data) allow us to image the thermal emission of the dust. Since this emission is mostly optically thin at such wavelengths, these data provide estimate of the dust surface density radial distribution and emissivity spectral indexes in disks. Optical and near-infrared interferometers provide now quantitative information within the structure of the very inner disk, close to the star at ~0.1–5 AU. The possibilities offered by current and future observational techniques are summarized in Figure 1.

Since most stars form in binary or multiple system, we will discuss in Section 2, the properties of the circumstellar material around binary and multiple young systems through examples such as GG Tau and HK Tau. Then, we will describe the dust and gas content of disks found around single PMS stars. We will present in Section 3 their outer disk properties as deduced from optical and mm/submm observations before discussing the inner properties of disks in Section 4, as seen by optical to Mid-IR interferometry. In Section 5, we summarize the properties of peculiar objects or *transition* disks, objects which are thought to be in the phase of dissipating their primary gas and dust. We then conclude.

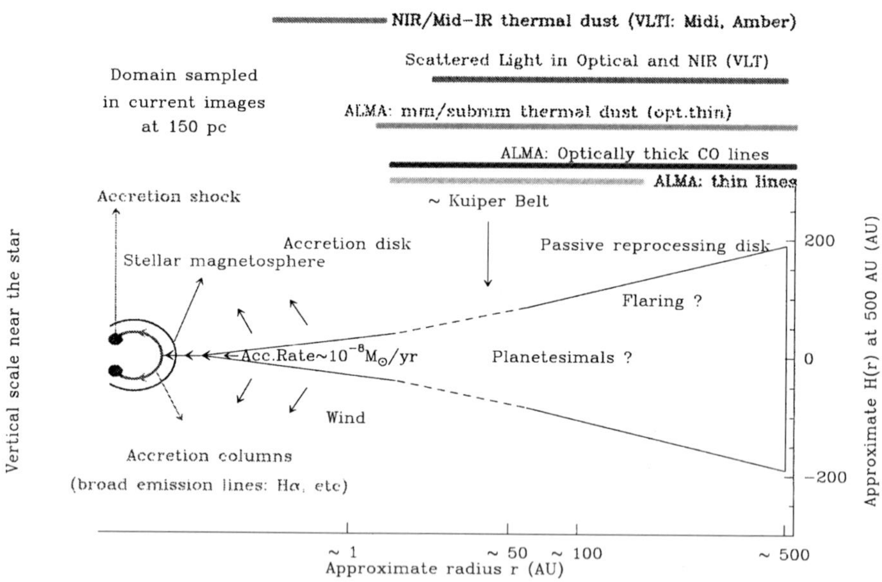

Figure 1. This ideal view of a protoplanetary disk, located around a single T Tauri star at 150 pc from us, shows which part of the disk is sampled depending on the observational technique and wavelength. The approximate location of the Kuiper Belt in our Solar System is given for comparison. This simple scheme allows us to distinguish between the very inner disk ($r \leq$ 0.1–10 AU), mainly traced by optical to Mid-IR interferometers and the outer disk ($r >$ 30–50 AU), traced by current mm arrays and observations in scattered light. Contrary to current mm arrays which are sensitivity limited, ALMA will also provide information from the inner disk.

2. Young Multiple Systems

Most T Tauri stars form in binary or multiple systems (Mathieu et al., 2000). Many observational results show that binarity strongly affects the dust and gas distribution as a result of tidal truncations. Binarity affects the surrounding material by creating an unstable zone outside the Lagragian points L2 and L3. Inside the Roche lobes, two inner disks can survive, feeded by a circumbinary disk (which location depends on the eccentricity of the system) through streamers. Hence, the material can be in a circumbinary ring as in the GG Tau disk (Dutrey et al., 1994) or in UY Auriga (Close et al., 1998; Duvert et al., 2000) or confined in small, truncated, circumstellar disks as in the case of HK Tau (Stapelfeldt et al., 1998).

2.1. The circumbinary disk of GG Tau

GG Tau is a T Tauri binary star of separation 0.26″ in the plane of sky. Observed first at mm wavelengths (Dutrey et al., 1994) in ^{13}CO and in continuum, then in the NIR (Roddier et al., 1996), the circumbinary disk of GG Tau is a Keplerian disk of mass $\sim 0.14\ M_{\odot}$. The cavity created by tidal truncation was clearly revealed both by mm interferometry and NIR images. More recently, Guilloteau and Dutrey (2000) observed the disk in ^{12}CO $J = 2-1$. More optically thick than the ^{13}CO $J = 2-1$ line, the ^{12}CO line reveals transient material flowing from the circumbinary ring to the inner disks. A preliminary analysis of the data by comparison with the ^{13}CO line suggests that the inner disks are feeded with an accretion rate of a few $\sim 10^{-6} M_{\odot}$/yr and that the quantity of flowing gas is $\sim 10^{-4} M_{\odot}$.

The existence of the inner disks was demonstrated by the 1.3 mm interferometric data (Guilloteau et al., 1999). However, the lack of angular resolution precludes to conclude on the geometry of the material distribution (one or two disks?). So far, all observations are consistent with the GG Tau ring being nearly coplanar with the orbit of the binary.

2.2. The inner disks of HK Tau

HK Tau is a T Tauri binary star of separation $\sim 2.4''$. Optical observations performed with the HST have revealed the existence of an optically thick edge-on inner dust disk around the companion while no disk was detected around the primary star (Stapelfeldt et al., 1998). However, images at 1.3 mm from the IRAM array reveal inner dust disks around both stars. Duchêne et al. (2003) show that these results can be reconciled if the disk around the primary is seen close to pole-on, the less favorable case to detect a disk at optical wavelength because this minimizes the opacity along the line of sight. HK Tau is the first direct evidence for non-coplanar inner disks in a binary system, revealing a scenario not as simple as the ideal scheme presented above.

3. Tracing the Outer Disk

Current telescopes and performing methods of data analysis allow observers to retrieve in some cases quantitative information on the gas and dust content of the outer disk (typically $r \geq 30 - 50$ AU).

3.1. HH 30: A CASE STUDY

So far, the HH 30 observations, shown in Figure 2, give one of the most complete picture of the material surrounding a PMS star because the disk is studied at several wavelengths. Optical observations performed with the HST (shown in false color in Figure 2) by Burrows et al. (1996) have revealed an almost edge-on disk which appears as a dark lane, only the disk surface or atmosphere being bright. Figure 2 also shows the jet, perpendicular to the disk plane. Since the disk is seen edge-on, the vertical distribution can be estimated from the optical observations of the dust (Burrows et al., 1996). The disk appears in hydrostatic equilibrium, and the authors estimate the surface density law to be $\Sigma(r) \propto 1/r$ and the disk mass $\sim 6.10^{-3} M_{\odot}$.

More recently, HH 30 has been observed with the IRAM array by Pety et al. (2004). In Figure 2, Pety et al., have superimposed in contours to the HST image the thermal dust emission at 1.3mm (left panel) and the blueshifted and redshifted integrated emission (with respect to the systemic velocity) of the ^{13}CO J $= 2 - 1$ line (right panel). The central panel shows the ^{12}CO J $= 2 - 1$ emission which traces the outflow cavities of the jet seen in optical.

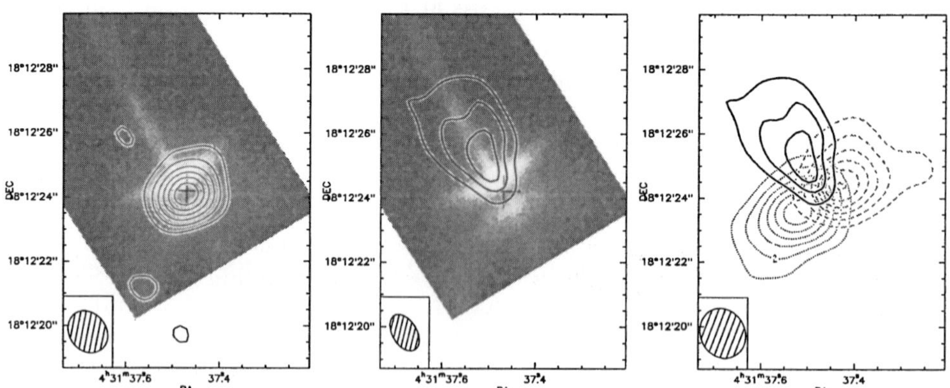

Figure 2. Background in false color: a montage of the dust disk emission from the HST (from Burrows et al., 1996) and its perpendicular jet. Contours present ^{12}CO J $= 2 - 1$ emission associated to the jet (black and white) corresponding to the extreme velocities and the redshifted and blueshifted integrated emission with respect to the systemic velocity of ^{13}CO J $= 2 - 1$ line coming from the disk (from Pety et al., 2004). Note that the velocity gradient of the ^{13}CO J $= 2 - 1$ emission is along the major disk axis, as expected for rotation.

The analysis reveals that the disk is in Keplerian rotation around a central star of mass 0.5 $M_\odot$ and has an outer radius of $R_{out} = 440$ AU. Interestingly, this is in agreement with the best fit of the optical data which gives $R_{out} \sim 400$ AU (Burrows et al., 1996). While ^{12}CO J $= 2-1$ clearly traces the outflow at extreme velocities, the high opacity of the ^{12}CO J $= 2-1$ emission near the systemic velocity does not allow Pety et al., to properly disentangle the individual contributions of the outflow, the disk and the cloud. Even in ^{13}CO, some confusion remains. This is a common problem which cannot be fully solved out by current interferometric observations. Selecting sources which are located in a region devoid of CO emission strongly minimizes the confusion.

However, both mm and optical images are tracing two different aspects of the same physical object and contribute to give a coherent understanding of its physical properties.

3.2. Dust disks as seen by optical/NIR telescopes

Optical and/or NIR images are particularly useful for edge-on disks, where they allow to retrieve scale heights and some of the dust properties.

The Flying Saucer: Using the VLT, Grosso et al. (2002) have recently detected a new disk in ρ Oph.: the 'Flying Saucer', observed in J,H and K bands. The disk, close to edge-on, is flaring and has an outer radius of ~300 AU. The data are correctly modeled by a viscous disk model of scale height $H \simeq 15$ AU at $r = 100$ AU. As in the case of HH 30, the size of the dust particles scattering the stellar light appear larger than in the interstellar medium with size of order ~1 μm (Wood et al., 2002). The observations also reveal the existence of a tenuous envelope of smaller dust particles.

The Butterfly Nebula: Similar radius and scale height have also been found by Wolf et al. (2003) for the Butterfly Nebula, another edge-on disk imaged by the HST. However, from 1.3 mm and 2.7 mm images obtained at OVRO, Wolf et al., also demonstrate that dust particles with sizes up to ~0.15 mm must exist in the disk. The data also show the existence of a tenuous surrounding envelope of dust particles with similar properties than those found in the interstellar medium.

Since the dust is optically thick in the optical and in NIR, analysis of images in the scattering regime preclude detailed studies of the disk interior and dust properties. Observations at a given frequency are mainly sensitive to grains of size $a \leq$ a few λ because absorption or diffusion cross-sections cannot significantly exceed the geometrical cross-section, even for more complex grain features and aggregates (Pollack et al., 1994; Krügel and Siebenmorgen, 1994). Only multi-wavelength analysis of resolved images, from the optical to the cm domains, can allow to conclude about the dust size distribution. This is clearly illustrated by the examples given above.

3.3. Dust disks observed with mm/submm arrays

The thermal dust emission of disks becomes mostly optically thin in the mm and cm domain. Therefore, detailed studies of resolved images provided by mm and cm arrays allows direct estimates of the surface density distribution and dust properties. Dutrey et al. (1996) found from a high angular resolution 2.7 mm continuum survey that the mean dust emissivity index $\beta \simeq 1$ in T Tauri disks, and that the exponent of the surface density law $\Sigma(r) \propto r^{-p}$ is $p \simeq 1$. Using the VLA, Testi et al. (2003) have resolved the CQ Tau disk at 7 mm. The 7 mm image and integrated flux densities from 3 to 0.8 mm, are analyzed with a model of superheated disk (e.g. Chiang and Goldreich, 1997). They find very large grains in the disk, up to ~cm sizes and estimate that the dust emissivity varies as $\beta \simeq 0.5$–0.7, another clear evidence of grain growth and evolution. Best models give a surface density law with $p = 0.5$–1.5, depending on the disk size which is ranging between $R_{\mathrm{out}} \simeq 100$–300 AU for various models. Such studies are a promising step towards a better understanding of the mass distribution in dust disks.

3.4. Disk structure inferred from CO lines

Carbon monoxide is the most abundant molecule after H_2. Its first rotation lines are observable with current mm arrays allowing to trace the outer gas disk properties. Since the density in disks is very high ($n(H_2) > 10^6$ cm^{-3}), the $J = 1 - 0$ and $J = 2 - 1$ CO lines are thermalized by collision with H_2 in almost the whole disk. Hence, a simple model of Keplerian disk assuming LTE conditions is sufficient to retrieve the CO disk properties (Dutrey et al., 1994).

CO maps not only reveal that disks are in Keplerian rotation around T Tauri (Koerner et al., 1993) and Herbig Ae stars (Piétu et al., 2003), they also allow us to retrieve the physical conditions in outer disks. Comparing resolved CO maps to a disk model by performing a χ^2 minimization of the disk parameters provides quantitative information about the density and temperature distributions, provided the analysis properly takes into account the transfer function of the interferometer (Guilloteau and Dutrey, 1998). The radial profiles of the temperature deduced from ^{12}CO images are consistent with stellar heating in flared disks and the turbulence appears to be small, less than 0.1 km·s^{-1} (Dutrey et al., 2004, in prep.).

Since the ^{12}CO and ^{13}CO $J = 1 - 0$ and $J = 2 - 1$ lines have different opacities, they samples different disk layers and allow to probe the existence of a vertical temperature gradient (Dartois et al., 2003). Dartois et al. (2003) have shown that in the DM Tau disk, the 'CO disk surface', traced by ^{12}CO, is located around 3 scale heights above the disk mid-plane, the ^{13}CO $J = 2 - 1$ transition samples material at 1 scale height, while the $J = 1 - 0$ line is representative of the disk

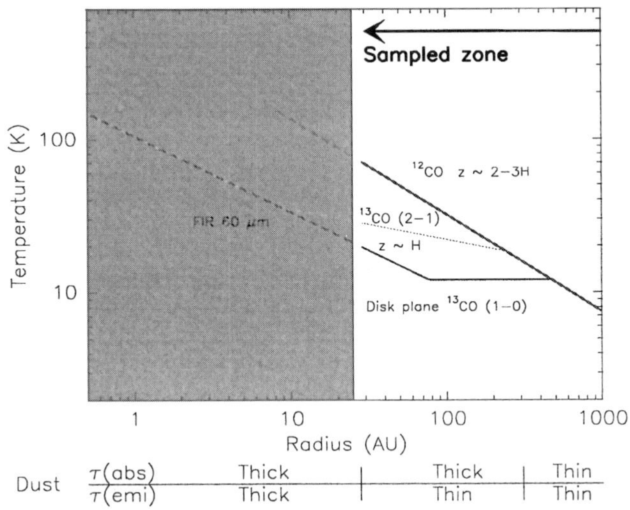

Figure 3. Temperature gradient in the DM Tau disk as deduced from a global analysis of the ^{13}CO and ^{12}CO J $= 1 - 0$ and J $= 2 - 1$ lines. From Dartois et al. (2003).

mid-plane. Figure 3 presents their results: the mid-plane is cooler ($\sim$13 K) than the CO disk surface ($\sim$30 K at 100 AU). This appears in the region of the disk where the dust is still optically thick in absorption to the stellar radiation while it is already optically thin to its own emission, around $r \sim$ 50–200 AU in the DM Tau case. Beyond $r \geq 200$ AU where the dust becomes optically thin to both processes, the temperature profile appears vertically isothermal. Such a vertical kinetic gradient is in agreement with disk models (e.g. D'Alessio et al., 1999).

The study by Dartois et al. also reveals that the outer radius of the disk is larger in ^{12}CO, than in ^{13}CO. The difference between both outer radii is compatible with selective photo-dissociation of CO. A significant fraction of the DM Tau disk has a temperature below the CO freeze out point (17 K) but there remains enough CO in the gas phase to allow the J $=$ 2–1 line of the main isotope to be optically thick.

So far, there are only a few attempts to survey a large set of molecules in protoplanetary disks (Dutrey et al., 1997; Kastner et al., 1997; Zadelhoff et al., 2001). Today, in addition to ^{13}CO and C^{18}O, only the more abundant species after the carbon monoxyde are detectable, like HCO^+, CS, HCN, CN, HNC, H_2CO or C_2H.

4. Tracing the Inner Disk by Optical to Mid-IR Interferometry

Optical to Mid-IR interferometry is a promising tool to trace the physical properties of the very inner disks. This can be illustrated through the recent study of Monnier and Millan-Gabet (2002). Using existing observations of disks surrounding T Tauri and Herbig Ae/Be stars performed with IOTA and PTI, Monnier and Millan-Gabet (2002) have analyzed these data with simple disks models. They found, in particular,

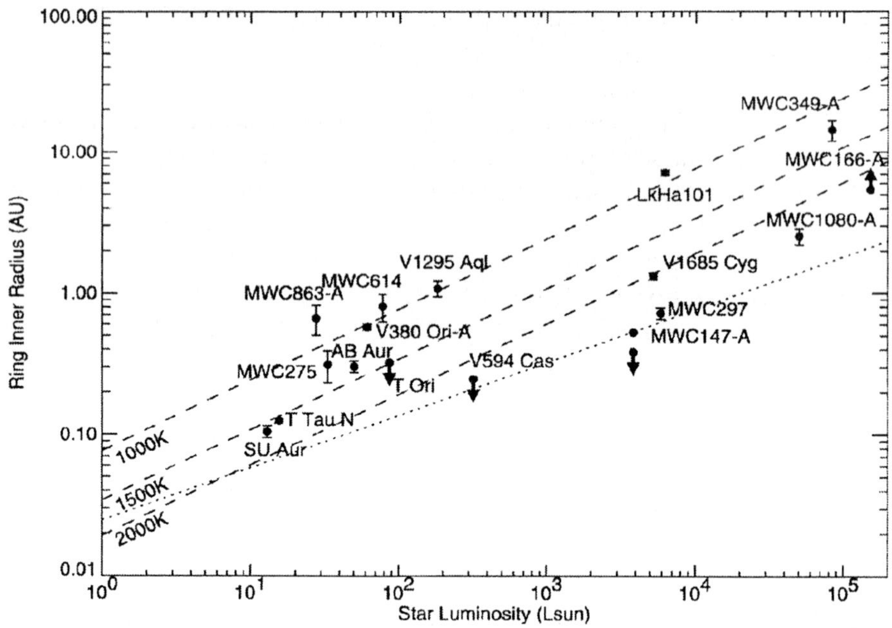

Figure 4. From Monnier and Millan-Gabet (2002). Inner radii of disks around Herbig Ae/Be and T Tauri stars estimated by NIR and optical interferometry. This compilation shows that the larger is the stellar luminosity, the larger is the inner radii, as expected if the inner radius is truncated by dust grains sublimation.

that the observed inner disk sizes are increasing with the stellar luminosity, as shown in Figure 4. T Tauri stars have smaller inner radii ($r_{\rm in} \leq 0.1$ AU) than Herbig Ae/Be stars ($r_{\rm in}$ up to 10 AU, for Herbig Be objects). The observed inner radii appear consistent with the presence of an optically thin cavity for a NIR emission arising from Silicate grains of sizes $a \geq 0.5 - 1\ \mu$m and which are heated close to their temperature of sublimation.

5. From Class II to Class III Objects

Depending on the wavelengths of observations, the observed properties of a given disk can be very different. This point can be illustrated through two recent examples of very well-known disks in the optical domain, which present apparent contradictory properties in the mm regime. For different reasons, these two object can be considered as transition disks which have started to dissipate their remnant gas and dust.

5.1. The surprising BP Tau case

BP Tau is often mentioned as the prototype of the CTTs. It has a high accretion rate of $\sim 3 \cdot 10^{-8} M_\odot$/yr from its circumstellar disk which produces its strong excess emission in the ultraviolet, visible and NIR (Gullbring et al., 1998). This object is also very young ($6 \cdot 10^5$yr, Gullbring et al., 1998).

Recent CO J $= 2-1$ and continuum at 1.3 mm images have revealed a weak and small CO and dust disk (Simon et al., 2000). With a radius of about $\simeq$120 AU, the disk is small and is in Keplerian rotation around a $(1.3 \pm 0.2)(D/140\text{pc})M_{\odot}$ mass star. The analysis of the CO data shows that the J $= 2-1$ transition is marginally optically thin, contrary to what is observed in other T Tauri disks (Dutrey et al., 2003). The disk mass, estimated from the mm continuum emission by assuming a gas-to-dust ratio of 100, is very small $1.2 \cdot 10^{-3} M_{\odot}$. By reference to this mass, the CO depletion factor is estimated to $\sim$150 with respect to H_2. Even taking into account possible uncertainties such as a lower gas-to-dust ratio or a higher value for the dust absorption coefficient, this CO depletion remains unique compared with other T Tauri CO disks. Finally, the kinetic temperature derived from the CO data is also relatively high, about $\sim$50 K at 100 AU.

Both the relatively high temperature and the low disk mass suggest that a significant fraction of the disk might be superheated (above the black body temperature) similarly to a disk atmosphere (see model from Chiang and Goldreich, 1997). Dutrey et al. (2003) have estimated the fraction of small grains ($a \leq 0.1\mu$m) still present in the disk to reach in the visible $\tau_V = 1$ at the disk mid-plane. The mass of small grains is about 10% of the total mass of dust ($1.2 \cdot 10^{-5} M_{\odot}$) derived from the mm continuum data. This should be confirmed by optical and NIR observations.

Although CO is underabundant, the CO content of BP Tau is far too high to result from evaporation of proto-comets: a few times 10^{11} large comets similar to Hale–Bopp would be simultaneously required to explain the amount of CO, well above the number of FEBs falling on β Pic per year (a few hundred).

All these unusual mm properties suggest that BP Tau may be a transient object in the phase of clearing its outer disk.

5.2. The HD141569 case

Located at 100 pc, HD 141569 is classified as a Herbig Ae/Be stars of spectral type B9 and age of 5–10 Myr (van den Ancker et al., 1998). The presence of circumstellar gas was revealed by CO observations (Zuckerman et al., 1995) and by the identification of emission bands tentatively attributed to PAH (Sylvester and Skinner, 1996) which are frequently observed around HAeBe. But the lack of IR excess in the spectral energy distribution, characteristic of an optically thin dust disk, and the low disk to star luminosity ratio (8.4×10^{-3}) rather correspond to the description of a Vega-like star. The detection of a substantial amount of cold gas associated with an optically thin dust disk is yet unusual.

HD 141569 exhibits a spatially resolved optically thin dust disk in the NIR, tilted by about $\sim 45^{\circ}$ along the line of sight. The HST observations reveal a complex dust structure in scattered light with a size $\sim$500 AU (Augereau et al., 1999). ^{12}CO J $= 2-1$ image obtained with the IRAM array by Augereau et al. (2004) resolves

Observations of Disks

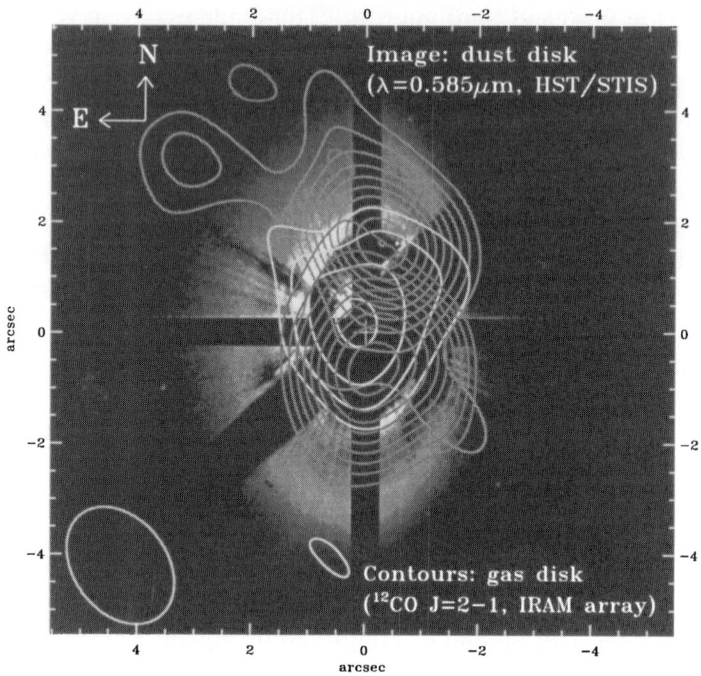

Figure 5. From Augereau et al. (2004). The optically thin dust disk, as seen with STIS at 0.5 μm, is shown in false colors. Three channels (corresponding to blueshifted, systemic and redshifted velocities) of the ^{12}CO J $= 2 - 1$ map from the IRAM array have been superimposed. Note that the velocity gradient is along the major dust disk axis, as expected for rotation.

the gaseous counterpart of the optically thin dust disk. The CO gas is in Keplerian rotation and perfectly matches the extended dust disk resolved in scattered light. Figure 5 is a montage of the HST observations (dust disk) and the CO map from the IRAM array.

The amount of cold CO gas is too large to be characteristic of a debris disk. This suggests that HD 141569 is less evolved than the β Pic disk, with a significant fraction of its mass still in the form of cold molecular gas (Augereau et al., 2004).

6. Summary

Resolved observations of disks from the cm to the optical now allow quantitative studies of the dust and gas physical properties:

- Many disks are large with outer radii $R_{\rm out} \geq 300$ AU
- Disks are in Keplerian rotation
- Both mm and optical data reveal that disks are flaring

– Vertical temperature gradients in outer disks seem to be compatible with heating by the central star
– Optical images reveal in several cases the existence of a tenuous envelope of small grains
– Both mm and optical data show that grains in disks are more evolved (and larger) than in interstellar clouds
– There exist now a few cases of disks in a transition phase which are dissipating their primary gas and dust.

Single wavelength analysis does not allow a global understanding of physical properties of disks and can lead in some cases to a misclassification. As a consequence, it is time to include mm/submm properties of gas disks in the standard classification of stellar formation. This is motivated by the fact that the mm/submm data remain the best tracers of the primary gas originating from the parent molecular cloud.

Because of the limit in angular resolution and sensitivity of current mm arrays, the data mostly concerns the outer disk (>50 AU). A similar restriction applies to scattered light, which cannot be traced closer to the star. Optical and near-IR interferometry has started to unveil the innermost regions (1 AU). However, the intermediate regime, where the bulk of planet formation is expected to occur, will only be revealed by the next generation of instruments. Mid-IR interferometers will play a role, but dust will often be optically thick even at 10 microns. With its capability of reaching AU-scale angular resolution at wavelengths where dust is much more optically thin, ALMA is expected to be the premium instrument to probe these regions. ALMA also has the potential of providing much higher angular resolution observations of molecular lines than currently possible, and, by studying polarization, to start probing the magnetic field in such regions.

Acknowledgements

We would like to thank N. Grosso, J. Monnier and L. Testi who provided material for this review. J.-Ch. Augereau and J. Pety are also acknowledged for providing us prior to publications images of HD 141569 and HH 30, respectively.

References

Augereau, J.C., Lagrange, A.M., Mouillet, D. and Ménard, F.: 1999, *A&A* **350**, L51.
Augereau, J.C., Dutrey, A., Lagrange, A.M., Mouillet, D. and Forveille, T.: 2004, *A&A*, in preparation.
Boccaletti, A., Augereau, J.C., Marchis, F. and Hahn, J.: 2003, *ApJ* **585**, 494.
Burrows, C.J., et al.: 1996, *ApJ* **473**, 437.
Chiang, E.I. and Goldreich, P.: 1997, *ApJ* **490**, 368.
Close, L., et al.: 1998, *ApJ* **499**, 883.
Dartois, E., Dutrey, A. and Guilloteau, S.: 2003, *A&A* **399**, 778.
Duchêne, G., et al.: 2003, *A&A* **400**, 559.

Dutrey, A., Guilloteau, S. and Simon, M.: 1994, *A&A* **286**, 149.
Dutrey, A., et al.: 1996, *A&A* **309**, 493.
Dutrey, A., Guilloteau, S. and Guélin, M.: 1997, *A&A* **317**, L55.
Dutrey, A., Guilloteau, S. and Simon, M.: 2003, *A&A* **402**, 1003.
Dutrey, A., Guilloteau, S. and Simon, M.S.: 2004, *A&A*, in preparation.
Duvert, G., et al.: 2000, *A&A* **355**, 165.
Grosso, et al.: 2003, *ApJ* **586**, 296.
Guilloteau, S. and Dutrey, A.: 1998, *A&A* **339**, 467.
Guilloteau, S. and Dutrey, A.: 2000, in *Proceedings of the IAU 200 ASP Conf. Series*, p. 229.
Guilloteau, S., Dutrey, A. and Simon, M.: 1999, *A&A* **348**, 570.
Gullbring, E., et al.: 1998, *ApJ* **492**, 323.
Kastner, J., et al.: 1997, *Science* **277**, 61.
Koerner, D.W., Sargent, A.I. and Beckwith, S.W.W.: 1993, *Icarus* **106**, 2.
Krügel, E. and Siebenmorgen, R.: 1994, *A&A* **288**, 929.
Mathieu, R.D., Ghez, A.M., Jensen, E.L.N. and Simon, M.: 2000 in: *Protostars and Planets IV*, p. 703.
Monnier, J. and Millan-Gabet, R.: 2002, *ApJ* **579**, 694.
Mouillet, D., Lagrange, A.M., Augereau, J.C. and Ménard, F.: 2001, *A&A* **372**, L61.
Natta, A., Grinin, V. and Mannings, V.: 2000, in: *Protostars and Planets IV*, p. 559.
Pety, J., Gueth, F., Guilloteau, S. and Dutrey, A.: 2004, *A&A*, in preparation.
Pietu, V., Dutrey, A. and Kahane, C.: 2003, *A&A* **398**, 568.
Pollack, J., et al.: 1994, *ApJ* **421**, 615.
Roddier, C., et al.: 1996, *ApJ* **463**, 326.
Simon, M., Dutrey, A. and Guilloteau, S.: 2000, *ApJ* **545**, 1034.
Stapelfeldt, Krist, J.E., Menard, F., Bouvier, J., Padgett, D.L. and Burrows, C.J.: 1998, *ApJ* **502**, L65.
Sylvester, R.J. and Skinner, C.J.: 1996, *MNRAS* **283**, 457.
Testi, et al.: 2003, *A&A* **403**, 323.
Wood, K.S., Wolff, M.J., Bjorkman, J.E. and Whitney, B.: 2002, *ApJ* **564**, 887.
van Zadelhoff, G.-J., et al.: 2001, *A&A* **377**, 566.
van den Ancker, M.E., de Winter, D. and Tjin A Djie, H.R.E.: 1998, *A&A* **330**, 145.
Wolf, S., Padgett, D. and Stapelfeldt, K.: 2003, *ApJ* **588**, 373.
Zuckerman, B., Forveille, T. and Kastner, J.H.: 1995, *Nature* **373**, 494.

ON THE ALIGNMENT OF T TAURI STARS WITH THE LOCAL MAGNETIC FIELD IN THE TAURUS MOLECULAR CLOUD COMPLEX

FRANCOIS MÉNARD[1] and GASPARD DUCHÊNE[2]

[1]*Laboratoire d'Astrophysique de Grenoble, Grenoble Cedex 9, France;*
E-mail: menard@obs.ujf-grenoble.fr
[2]*Department of Physics & Astronomy, UCLA, Los Angeles, U.S.A.*

Abstract. Magnetic field is believed to play an important role in the collapse of a molecular cloud. In particular, due to the properties of magnetic forces, collapse should be easier along magnetic field lines. This is supported by the large-scale sheet-like structures observed in the Taurus giant molecular cloud for instance. Here we investigate whether such a preferred orientation for collapse is present at a much smaller scale, that of individual objects, i.e., about 100AU. We use recent high-angular resolution images of T Tauri stars located in the Taurus star-forming region to find the orientation of the symmetry axis of each star+jet+disk system and compare it with that of the local magnetic field. We find that (i) T Tauri stars that are associated to a jet or an outflow are generally oriented parallel to the magnetic field, as previously demonstrated. More surprising, given our current knowledge of these objects, we also find that (ii) T Tauri stars that are not at present believed to be associated to a jet or an outflow are oriented very differently, i.e., mostly perpendicular to the magnetic field. We present some implications of this puzzling new result.

Keywords: T Tauri stars, magnetic field, orientation, disks, jets

1. Introduction

Stars form as a consequence of the gravitational collapse of dense molecular cores. This process is, however, sensitive to other phenomena than gravity, such as rotation or magnetic field. For instance, neutral species are only affected by gravity while ions are tightly bound to the magnetic field. Friction between ions and neutrals, i.e., ambipolar diffusion, modifies the kinematics of the neutrals, leading to protostars that are surrounded by pseudo-disks, even in the absence of rotation (Galli and Shu, 1993). In the presence of straight magnetic field lines threading the cloud, Galli and Shu (1993) showed that the major axis of the pseudo-disks are perpendicular to the direction of the original magnetic field, as expected if the collapse occurs preferentially along the field lines. The orientation of young stellar objects (YSOs) can therefore inform us directly on the importance of magnetic field during the collapse process. In the event of a leading role for magnetic forces, we expect circumstellar disks to be oriented with their symmetry/rotation axis (roughly) parallel to the local magnetic field in the cloud. On the other hand, if magnetic field does not influence much the collapse, disks might well be oriented at random.

Astrophysics and Space Science **292:** 419–425, 2004.

In the Taurus–Auriga giant molecular cloud, both the dense gas clouds and the YSOs shows a sheet-like structure fully consistent with a large scale collapse along the magnetic field lines (Goodman et al., 1990; Hartmann, 2002). At the scale of individual objects, Lee and Myers (1999) showed that pre-stellar cores are elongated, with an average aspect ratio of $\sim$2.4. If one attributes this elongation to rotation, it can be shown that their major axis are also preferentially found perpendicular to the magnetic field (Hartmann, 2002). These findings support the case for a causal link between the direction of the large scale magnetic field and the orientation of young stars.

With the advent of high-angular resolution imaging devices, it is now possible to trace the orientation of already-formed YSOs. First, collimated jets and outflows from Myr-old T Tauri stars are commonly observed and can be used to trace their orientation. Strom et al., (1986) noted, during a deep imaging survey of Taurus sources driving Herbig-Haro objects, that jets and outflows have a tendency to align parallel with the local magnetic field. MHD models predict that the ionised atomic jets of young stars are expelled perpendicular to the accretion disk (e.g., Shu et al., 1995; Ferreira and Pelletier, 1995), in agreement with spectacular high resolution observations of objects like HH 30 (Burrows et al., 1996). The preferred jet orientation therefore suggests that the rotation axis of T Tauri circumstellar disks are parallel to the local magnetic field. A similar conclusion was reached by Tamura and Sato (1989) who found linear polarisation vectors of YSOs to be preferentially parallel to the local magnetic field. Although the interpretation of this result is affected by a 90° ambiguity, it has reinforced the general belief that the orientation of young stars is in some way related to the direction of the local magnetic field.

Since early surveys for optical jets from T Tauri stars, many new jets and outflows have been identified. Furthermore, it is now also possible to spatially resolve circumstellar disks and accurately estimate their orientation through high angular resolution imaging. It is therefore time to readdress this issue and study the link between the orientation of the local magnetic field and the orientation of YSOs in more detail. We will focus here on the well-studied Taurus star-forming regions.

2. Direction of the Magnetic Field in Taurus

In the presence of a magnetic field, elongated interstellar grains tend to spin along a preferred direction, with their axis of smallest moment of inertia parallel to the magnetic field (e.g., Lazarian, 2002, and references therein). Dichroism is induced as the grains act like a picket-fence to absorb the light of background stars more efficiently along the direction perpendicular to the grains' rotation axis, i.e., generally perpendicular to the field. A net linear polarisation results in the other direction, i.e., parallel to the projected direction of the magnetic field.

This technique has been used by numerous authors to infer the structure of the magnetic field in molecular clouds. For our analysis, we have compiled over 400 linear polarisation measurements of background stars in Taurus (left panel in

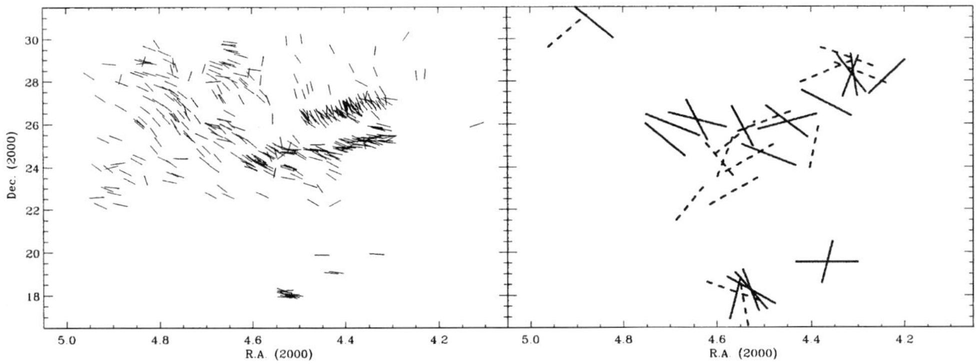

Figure 1. Left: Polarization measurements for background stars seen through the Taurus molecular cloud, tracing the direction of the magnetic field in the cloud. Only well-defined measurements ($P/\sigma_P > 3$) are considered here. The length of all vectors is uniform and not proportional to the polarisation level. *Right*: Orientation of T Tauri stars in Taurus as defined by an optical jet or outflow (solid vectors) or by a spatially resolved circumstellar or circumbinary disk in the absence of an outflow (dashed vectors). In both cases, the symmetry axis of the system is shown.

Figure 1) from Vrba et al. (1976), Heyer et al. (1987), Moneti et al. (1984), Tamura et al. (1987), Tamura and Sato (1989) and Goodman et al. (1990, 1992). To estimate the direction of the magnetic field at the location of each star in our sample (see below), we searched all the interstellar polarisation data in increasingly larger circles of radii 0.5–2.0°. We selected the smallest zone containing at least four different measurements of the magnetic field and retained the median of the measured position angles as the direction of the local magnetic field. As can be seen in Figure 1, the polarisation vectors form a smooth structure throughout the region and we estimate that the orientation of the local magnetic field can be estimated to within 5° for most objects.

3. Orientation of T Tauri Stars in Taurus

We have first compiled a complete list of T Tauri stars in the Taurus–Auriga star forming region from the Herbig and Bell (1988, hereafter HBC) catalogue. We restrict our study to the zone $4h00 < \alpha < 5h00$ in right ascension and $+17^\circ < \delta < +30^\circ$ in declination. This yields a sample of over 100 pre-main sequence objects, to which we added HH 30 and IRAS 04158 + 2805 (Ménard et al., 2003), which we classify as normal T Tauri stars, once their edge-on orientation is taken into account. They are not present in thc HBC catalogue because of their extreme faintness induced by the occulting presence of their opaque disks. There might be a limited number of additional objects that we did not include in our study, but we believe that they would not statistically affect our conclusions.

For each objects, we searched the literature for the presence of a spatially resolved jet and/or disk. Only morphological evidences were used in this process.

Collimated jets are usually identified in deep narrow-band images obtained at the wavelength of optical forbidden lines (e.g., Mundt et al., 1991; Dougados et al., 2000) or through long slit spectroscopy observations, which locate the redshifted and blueshifted parts of the jet on opposite sides of the star (e.g., Hirth et al., 1997). In most cases, jets are clearly resolved and their position angle is known within 10° or better. Disks around young stars can be identified in two main ways: thermal imaging in the submillimeter and millimeter ranges and scattered light imaging in the optical and near-infrared. Because, current instruments have a limited dynamic range, the latter technique is usually more sensitive to edge-on disks (e.g., Burrows et al., 1996) and to circumbinary rings (e.g., Roddier et al., 1996). In the millimeter continuum, disks appear as sources of thermal emission that are spatially resolved when observed with long baseline interferometers (e.g., Dutrey et al., 1996; Kitamura et al., 2002). Furthermore, observations of some disks in CO lines at millimeter wavelength reveal velocity profiles that are consistent with Keplerian rotation (e.g., Simon et al., 2000). When available, we used these resolved CO maps to define the orientation of the disk's semi-major axis. Disk orientations are known to within 5° or better when a Keplerian velocity gradient is detected and within 5–15° otherwise. More details on the individual sources as well as data tables are presented elsewhere (Ménard and Duchêne, 2004).

4. Relative Orientations of T Tauri Stars

Among our targets, we identified 34 objects with a resolved jet and/or disk, including 12 that possess both of them. Among those 12 sources, all but one dubious case (DO Tau, see Ménard and Duchêne, 2004) have perpendicular jets and disks to within 20° or better, as illustrated by the prototypical object HH 30 (Burrows et al., 1996). The symmetry axis of the system is determined by the orientation of the jet if it is present. Otherwise, we assume that the symmetry axis lies perpendicular to the major axis of the disk. As for the magnetic field direction, we only find the projection in the plane of the sky of the symmetry axis of the object. Despite possible projection effects, which we discuss below, it is worthwhile searching for possible correlations, which would trace an underlying physical link between the orientation of YSOs and the magnetic field.

The orientation of the symmetry axis for the 34 T Tauri stars in our sample is plotted in the right panel of Figure 1 with different symbols for objects with and without jets. *We plot the same axis, i.e. the symmetry axis of the star+jet+disk system, for both categories of objects.* Although the direction of the symmetry axes shows significant scatter, two general trends can be identified from Figure 1, especially in the central area of the figure (R.A. ~4h30, Dec. ~25°). First, T Tauri stars for which a jet or outflow has been spatially resolved (solid vectors) are mostly oriented along the same direction as the local magnetic field. Second, T Tauri stars without a jet or an outflow but with a spatially resolved circumstellar disk are oriented roughly perpendicularly to the local magnetic field. While the former conclusion is similar to those reached in past studies, the latter is revealed here for the first time. We now quantify it in more detail, before exploring its implication for star formation studies.

For each object with a known orientation in the plane of the sky, we measured the "misalignment angle" between the axis of symmetry of the T Tauri star and that of the local magnetic field. A value of 0° indicate that both are parallel while it is 90°, if they are perpendicular. The distributions of these angles for both sub-samples (objects with and without a spatially resolved jet) are shown in Figure 2 (left panel). The two histograms are statistically different at the $\sim 3\,\sigma$ level: *there is a significant difference in the respective orientations of T Tauri stars with and without jets. Systems with a jet have their symmetry axis roughly parallel to the local magnetic field* (*median angle* $= 25^\circ$) *while systems without a jet or an outflow are preferentially perpendicular to it (median angle* $= 75^\circ$*)*. Interestingly, the distribution of orientations for our complete sample of 34 objects is consistent with random orientation, as illustrated in the right panel of Figure 2.

As pointed out earlier, our sample is unlikely to be 100% complete and an unknown selection effect may be responsible for this apparent difference. For example, much deeper images of objects without jets (as of today) may actually reveal the presence of a faint outflow, calling for a revision of our samples.

In the meantime, the imaging studies used to build our database of stellar orientations are equally sensitive to structures (jets, outflows, disks) in any direction in the plane of the sky. *It is difficult to think of a selection effect that would preferentially sample stars oriented in a specific direction.* Although there are strong selection

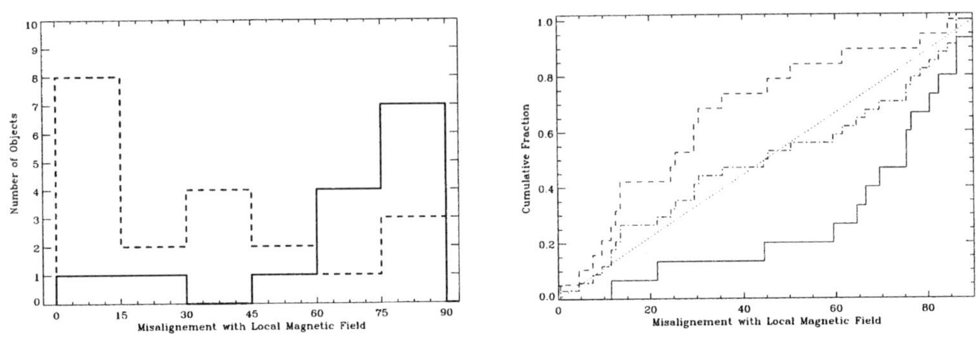

Figure 2. Left: Distribution of the angles between the orientation of the symmetry axis of T Tauri stars and the local magnetic field. From 0° to 90°, these two directions go from parallel to perpendicular. The dashed histogram represents T Tauri stars with a jet or an outflow while the solid histogram is for stars where only a disk could be spatially resolved. *Right*: Cumulative distributions of angles between the symmetry axis of T Tauri stars and their local magnetic field for objects with a jet/outflow (dashed), with only a disk (solid) and for the whole sample of objects with known orientation (dot-dashed). The dotted line represents the expected distribution if YSOs were randomly oriented with respect to the local magnetic field.

effects regarding inclination, in particular against pole-on systems, we do not find any significant difference between the distributions of inclinations of objects with and without a jet. Therefore, we do not believe that the distinction between the two subsamples compared here can be due to specific geometric configurations that lead to undetectable jets for some YSOs.

5. Open Issues

T Tauri stars, as a group, seem to orient randomly with respect to the local magnetic field. However, the presence of a powerful bipolar jet appears linked to the specific orientation of a given object with respect to that field. The origin of the differential orientation trend discovered here is not clear yet and only suggestions can be put forward for the moment.

A first important point to note is that, while jets are most likely launched through a specific magnetic configuration linking the star, its accretion disk and the jet, this involves the *stellar magnetic field*. On the other hand, our study of orientations concerns the *molecular cloud magnetic field*, which may have a different topology and orientation.

It is unclear either whether T Tauri stars retain the same orientation throughout their evolution or not. It could be that the collapse is strongly driven by the magnetic field, with all systems having their symmetry axis roughly parallel to the local magnetic field, and that some of them later become misaligned through an unidentified process, e.g., close dynamical encounters. It is worth noting that the orientation of bipolar outflows from more embedded (i.e., younger) Class I sources in Taurus show the same tendency to align parallel to the magnetic field as T Tauri stars that possess a jet (see Ménard and Duchêne, 2004). If all YSOs retain their original orientation, where are the precursors of those T Tauri stars that now only have a disk? Despite this suggestive trend for protostars, we fail to identify any mechanism that could provide enough torque to rotate by 90° a star+disk system.

Another possibility is that YSOs are formed with random orientations, as suggested by the distribution of our complete sample. The subsequent evolution would then lead to the formation of a powerful jet/outflow only in some conditions. For instance, Ferreira (1997) showed that a quadrupolar magnetic field configuration in the disk leads to a much weaker disk-wind than a dipolar configuration does. We can then propose that the configuration of the *stellar* magnetic field would be dipolar if its axis of symmetry is more or less aligned with the local field in the molecular cloud, while only quadrupolar (or high order configurations) would be created if the relative orientation is close to perpendicular. This could be the consequence of a feedback of the cloud magnetic field on the (currently unknown) growth mechanism of the T Tauri star field. The fact that Class I sources do not appear to be randomly oriented is somewhat problematic in this picture.

A third possibility to explain the observed trend is a projection effect, in which T Tauri stars that have a jet are located in a different part of the cloud than those who do not have one. The latter could be seen in projection over the Taurus cloud, while being located well in front, where the magnetic field orientation could be different. In that case however, the field at the periphery of the cloud would need to be twisted by $\sim 90^\circ$ with respect to the field found inside the cloud where T Tauri stars with jets would presumably be located. It is unclear why this would be the case. Furthermore, one would have to explain why the objects located in the outer parts of the cloud are systematically devoid of jets and outflows.

Only tentative interpretations of the observational result presented here can be given for now; none of them is fully satisfying. Conducting similar studies in other star-forming regions, such as ρ Ophiuchus or Orion, and measuring submillimeter polarisation vectors from much younger prestellar cores will help to understanding whether this trend is a general one, and possibly identify other factors that play a role in this preferred alignment of objects with the magnetic field. Also, a much deeper search for outflows and jets is needed to fully confirm the reality of the sample of objects believed to lack jets for now.

References

Burrows, C.J. et al.: 1996, *ApJ* **473**, 437.
Dougados, C., Cabrit, S., Lavalley, C. and Ménard, F.: 2000, *A&A* **356**, L41.
Dutrey, A., Guilloteau, S., Duvert, G., Prato, L., Simon, M., Schuster, K. and Ménard, F.: 1996, *A&A* **309**, 493.
Ferreira, J. and Pelletier, G.: 1995, *A&A* **295**, 807.
Ferreira, J.: 1997, *A&A* **319**, 340.
Galli, D. and Shu, F.H.: 1993, *ApJ* **417**, 243.
Goodman, A.A., Bastien, P., Ménard, F. and Myers, P.C.: 1990, *ApJ* **359**, 363.
Goodman, A.A., Jones, T.J., Lada, E.A. and Myers, P.C.: 1992, *ApJ* **399**, 108.
Hartmann, L.: 2002, *ApJ* **578**, 914.
Herbig, G.H. and Bell, K.R.: 1988, *Lick Obs. Bull.*, No. 1111.
Heyer, M.H., Vrba, F.J., Snel, R.L., Schloerb, F.P., Strom, S.E., Goldsmith, P.F. and Strom, K.M.: 1987, *ApJ* **321**, 855.
Hirth, G.A., Mundt, R. and Solf, J.: 1997, *A&AS* **126**, 437.
Kitamura, Y., Momose, M., Yokogawa, S., Kawabe, R., Tamura, M. and Ida, S.: 2002, *ApJ* **581**, 357.
Lazarian, A.: 2002, Elsevier preprint (astro-ph/0208487).
Lee, C.W. and Myers, P.C.: 1999, *ApJS* **123**, 233.
Ménard, F., Dougados, C., Magnier, E., Duchêne, G., Cuillandre, J.C., Martín, E.L., Fahlman, G., Forveille, T., Lai, O., Manset, N., Martin, P., Veillet, C. and Magazzu, A.: 2004, *ApJ*, in press.
Ménard, F. and Duchêne, G.: 2004, *A&A* submitted.
Moneti, A., Pipher, J.L., Helfer, H.L., McMillan, R.S. and Perry, M.L.: 1984, *ApJ* **282**, 508.
Mundt, R., Ray, T.P. and Raga, A.C.: 1991, *A&A* **252**, 740.
Roddier, C., Roddier, F., Northcott, M.J., Graves, J.E. and Jim, K.: 1996, *ApJ* **463**, 326.
Shu, F.H., Najita, J., Ostriker, E.C. and Shang, H.: 1995, *ApJ* **455**, L155.
Simon, M., Dutrey, A. and Guilloteau, S.: 2000, *ApJ* **545**, 1034.
Strom, K.M., Strom, S.E., Wolff, S.C., Morgan, J. and Wenz, M.: 1986, *ApJS* **62**, 39.
Tamura, M., Nagata, T., Sato, S. and Tanaka, M.: M. 1987, *MNRAS* **224**, 413.
Tamura, M. and Sato, S.: 1989, *AJ* **98**, 1368.
Vrba, F.J., Strom, S.E. and Strom, K.M.: 1976, *AJ* **81**, 958.

POLARIZATION IN THE YOUNG CLUSTER NGC 6611: CIRCUMSTELLAR, INTERSTELLAR, OR ... BOTH?*

PIERRE BASTIEN[1], FRANÇOIS MÉNARD[2,3], PATRICE CORPORON[1], NADINE MANSET[1,3], FRÉDÉRICK POIDEVIN[1], GASPARD DUCHÊNE[4] and JEAN-LOUIS MONIN[2]

[1]*Département de physique and Observatoire du mont Mégantic, Université de Montréal*
[2]*Laboratoire d'Astrophysique de l'Observatoire de Grenoble*
[3]*Canada-France-Hawaii Telescope Corporation*
[4]*Department of Physics and Astronomy, University of California, Los Angeles, U.S.A.*

Abstract. We obtained linear polarization observations of 82 A/B-type stars in the young cluster NGC 6611, in order to probe the circumstellar material and to search for any evidence of intracluster or interstellar material that could also contribute to the polarization. We found linear polarization values that reach up to 14%. We consider the distribution of the polarization, its position angle, correlations with extinction and membership probability, polarization variability and wavelength distribution to identify the origin of the polarization toward NGC 6611. The polarization is found to be dominated by interstellar polarization, although some stars also have some circumstellar polarization. There is no evidence for intracluster dust. Rather, the dust must be located in a low density cloud toward the general line of sight to NGC 6611 and in front of it. The depth of that cloud along the line of sight increases slowly from the south–east to the north–west. The cloud is threaded by a very uniform magnetic field.

Keywords: polarization, stars: pre-main-sequence

1. Introduction

As part of a large survey of the linear polarization of young stars, T Tauri and Herbig Ae/Be stars, and also of stars in young clusters, we carried out observations of NGC 6611, located at 2.1 kpc in the Sagittarius arm (Belikov et al., 1999). Its hottest stars ionize the H II region M16 (the "Eagle Nebula"). It is a young cluster, as its oldest stars are about 6 Myr old (Hillenbrand et al., 1993). There is still ongoing star formation as many embedded stars are found in deep IR images. Its stellar population is relatively well known (see de Winter et al., 1997), hence its interest for this study. The extinction is known from published photometric studies to be anomalous and patchy.

*Based on observations obtained on the 1.6 m telescope at the mont Mégantic Observatory (Canada) and on the 2m Bernard-Lyot telescope (TBL) operated by INSU/CNRS and Pic du Midi observatory (CNRS USR 5026). Financial support for the observations at Pic du Midi was provided by the *Programme National de Physique Stellaire* (PNPS) of CNRS/INSU, France.

2. Observations

The sample includes almost all known A- and B-type stars in NGC 6611, which are brighter than $V = 14$ mag, according to various papers and catalogues, in particular the compilation by Belikov et al (1999). There are only a few exceptions, a few fainter stars included and two brighter ones, which are missing. This sample also includes stars, which are members of NGC 6611 and some which are not, according to two criteria: proper motion and the presence of the lithium line at 6707 Å.

The observations were carried out at the mont Mégantic Observatory in Québec, Canada in May, June, and July 1998. The 1.6 m Ritchey-Chrétien telescope was equiped with the polarimeter Beauty and the Beast (Manset and Bastien, 1995) which has a Pockels cell as a modulator and a Wollaston prism as an analyser. The detectors are two RCA photomultipliers. Additional observations were performed at the Pic du Midi Observatory in France in June 1999, August 2000, and August 2001. The polarimeter STERREN was mounted on the 2.0 m Bernard Lyot telescope. The modulator is a half-wave plate rotating at 20 Hz and the analyser is a Wollaston prism. The detectors used are two avalanche photodiodes.

3. Results

For the 82 stars included in our sample, 115 measurements were obtained, usually in one broad bandpass red filter. For some stars, a few measurements were taken to check about possible variability. For two stars, which were found to be particularly interesting, the wavelength distribution was obtained using four different filters.

For stars with multiple measurements in the red filter, a weighted average was performed. The resulting polarization map is presented in Figure 1. One can easily see that the alignment of the polarization vectors over the whole area is very good. Nearly, all vectors are close to being parallel to each other. In addition, one can notice a systematic increase of the polarization going from the south–east to the north–west.

Histograms of the polarization and its position angle are presented in Figures 2 and 3 respectively. In addition to the rather large average polarization, 2.81%, there is a significant high polarization 'tail', which is typical of polarization distributions of the classical T Tauri stars in the Taurus complex, but not of the weak-line T Tauri stars in the same region (e.g., Ménard et al., 2004).

The distribution of position angles (Figure 3) is strongly peaked, as expected from an examination of Figure 1. Such a narrow distribution, $\sigma(\theta) = 9.3°$, is quite remarkable. It makes NGC 6611, the cluster with the smallest dispersion in position angle known (see Myers and Goodman, 1991). These authors identified three different structures, in order of increasing position angle dispersion, dark clouds, clusters, and complexes (e.g., Taurus is considered a complex). When compared to

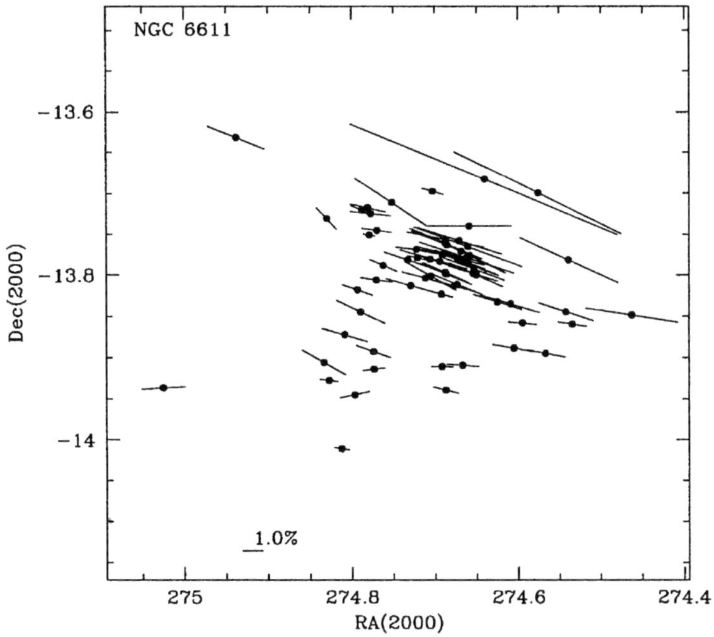

Figure 1. Polarization map of NGC 6611. The vectors are proportional to the magnitude of the polarization. The scale is indicated in the figure.

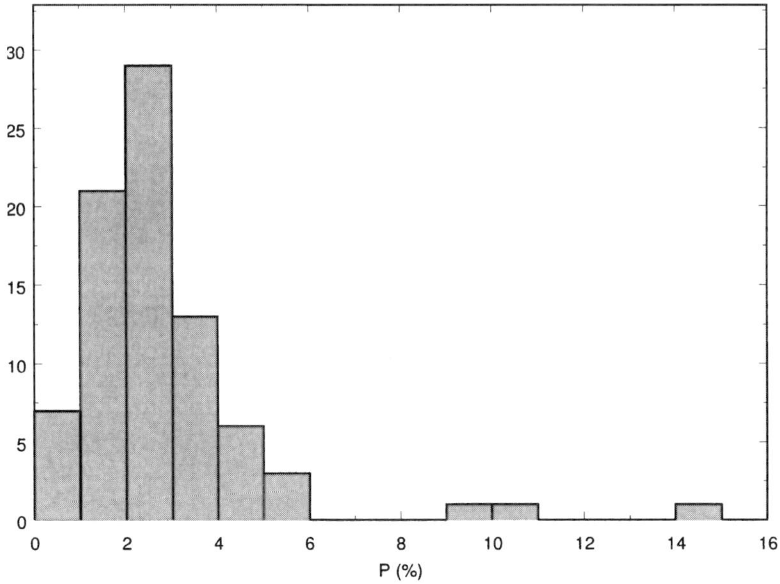

Figure 2. Histogram of the linear polarization of NGC 6611. The number of stars plotted is 82. The average polarization $\langle P \rangle$ is 2.81%, and the standard deviation $\sigma(P) = 2.10\%$.

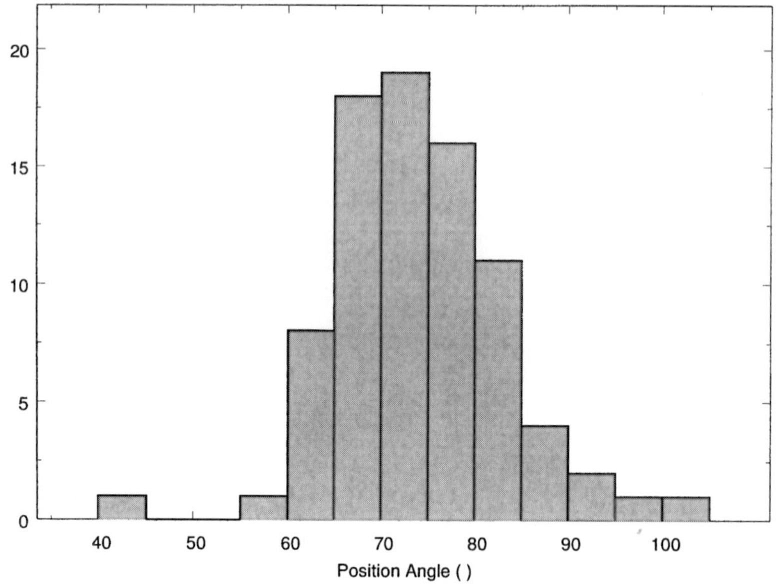

Figure 3. Histogram of the position angles of the linear polarization in NGC 6611. 82 values are plotted. $\langle\theta\rangle = 74.2^\circ$, and $\sigma(\theta) = 9.3^\circ$.

dark clouds, NGC 6611 is comparable to those with the smallest distribution, e.g., Lup 4, Cha 2, and GF 7. This is a clear indication that the magnetic field in the direction of NGC 6611 is very well organized and very uniform.

The polarization is plotted in Figure 4 as a function of extinction for all those stars which have extinction measurements available. There is a clear correlation between these quantities. The slope of the best fit line is 3.9 ± 0.2. The maximum slope observed, which defines the upper envelope of the distribution for all interstellar polarization measurements towards other regions in the Galaxy, including molecular clouds, is 9.0 (Serkowki et al., 1975). This could mean that either the magnetic field toward NGC 6611 is not perpendicular to our line of sight, or that the alignment efficiency is not as high as observed in other regions in the Galaxy. However, we note that one star, with the largest polarization in our sample, W188, visible in the upper right portion of Figure 4, would just fit on the line with the maximum slope. We will rediscuss this star below.

Arce et al. (1998) showed in the molecular clouds of Taurus that the polarization keeps increasing with extinction only for those stars, which are located in the more diffuse regions, or in the cloud envelopes. For stars located in the denser dark clouds, the polarization reaches a plateau at an A_V value of $\approx$1.3 or E_{B-V} of $\approx$0.4. Clearly in NGC 6611, we are not in the dark cloud regime, but rather in the less dense regime. The interpretation given by Arce et al. is that above a certain density threshold, the additional grains do not contribute to the polarization anymore, even

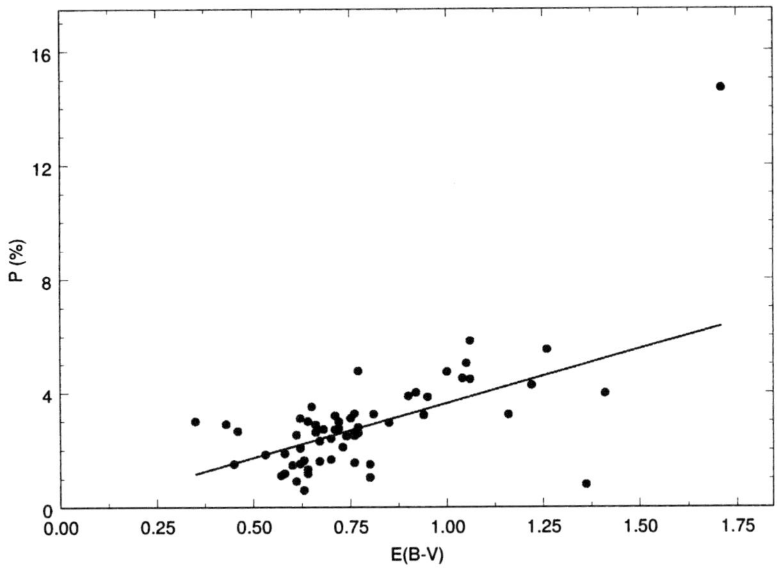

Figure 4. Plot of the polarization as a function of the extinction $E(B-V)$ for NGC 6611. The slope of the fitted line is 3.9 ± 0.2. The star at the upper right is W188.

though they still contribute to the extinction. However, even though the new grains do not contribute to the polarization, we still measure the correct position angle in that line of sight.

Figure 5 shows the polarization plotted as a function of the probability of membership in the NGC 6611 cluster as determined from the intensity of the lithium line at 6707 Å and from proper motions. It is obvious that there is no dependence of the polarization on the membership probability. This means that stars which do not belong to the cluster are polarized in a similar way as those that belong to the cluster. Therefore, we conclude that the polarization must arise in a cloud located in front of the cluster.

Is there any evidence for circumstellar polarization? For that purpose, we took a few more measurements for some stars, as already mentioned above. A variability analysis reveals that 10 out 19 stars with two or more measurements in the same filter do show polarization variability; of course, this variability has to be circumstellar.

In addition, two stars with a large polarization were measured in four different bandpasses. The wavelength distribution of the polarization for these two stars is clearly not compatible with the standard interstellar polarization law. One star, W188, in fact displays a nice rotation of position angle with wavelength, an indication that there must be (at least) two different contributions to the polarization.

Even though some stars definitely have some intrinsic, circumstellar, polarization, this intrinsic polarization is usually such that the total polarization still gives a good alignment with all the other interstellar polarization vectors. This means

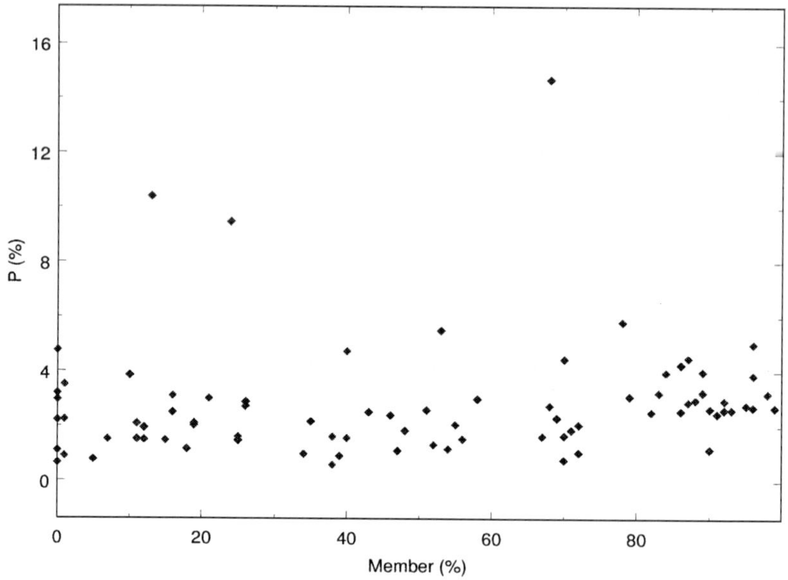

Figure 5. Polarization of the program stars as a function of the probability of being a member in the NGC 6611 cluster.

that either the interstellar component is the dominant one, or that the position angle of the intrinsic component is not too different from the direction of the general interstellar polarization.

The observations are compatible with the following scenario. There is a cloud of dust in front of the general area which covers NGC 6611. This cloud is *not* associated with the cluster itself and is in front of it. Its density is low, i.e., smaller than that of dark molecular clouds ($\approx 10^6$ cm^{-3}). The thickness of the cloud (but not its density) increases regularly from the south–east toward the north–west.

4. Conclusion

We have presented a new polarization map of stars toward the NGC 6611 cluster. We found that the polarization is not dominated by circumstellar polarization, since there is a very strong alignment in that direction. Therefore, the polarization is mostly interstellar in nature. Also, the polarization is not due to intracluster dust, since polarization does not depend on the membership probability. Finally, some stars also have a circumstellar polarization component, in addition to their strong interstellar polarization. A general scenario that explains all available observations emerges: a cloud of dust with a relatively low density, and with a very uniform magnetic field and well-aligned dust grains lies in front of the NGC 6611 star cluster.

Acknowledgements

This research was supported by the Conseil de Recherche en Sciences Naturelles et Génie du Canada and le Conseil National de Recherche Scientifique de France. We thank the telescope operators of both observatories for all their support during these observations.

References

Arce, H.G., Goodman, A.A., Bastien, P., Manset, N. and Sumner, M.: 1998, *ApJ* **499**, L93.
Belikov, A.N., Kharchenko, N.V., Piskunov, A.E. and Schilbach, E.: 1999, *A&AS* **134**, 525.
de Winter, D., Koulis, C., Thé, P.S., van den Ancker, M.E., Perez, M.R. and Bibo, E.A.: 1997, *A&AS* **121**, 223.
Hillenbrand, L.A., Massey, P., Strom, S.E. and Merill, K.M.: 1993, *AJ* **106**, 1906.
Manset, N. and Bastien, P.: 1995, *PASP* **107**, 483.
Ménard, F., Bastien, P., Montmerle, T. and Bertout, C.: 2004, in preparation.
Myers, P. and Goodman, A.A.: 1991, *ApJ* **373**, 509.
Serkowski, K., Mathewson, D.L. and Ford, V.L.: 1975, *ApJ* **196**, 261.

JETS AND LARGE SCALE OUTFLOWS

JETS AND LARGE SCALE OUTFLOWS: THEORY CONFRONTS THE OBSERVATIONS

RALPH E. PUDRITZ

Department of Physics and Astronomy, McMaster University, Hamilton, Ontario, Canada
E-mail: pudritz@physics.mcmaster.ca

Abstract. This article summarizes recent theoretical, numerical and observational progress on the nature of jets and outflows in regions of star formation. The emphasis is placed on the role of hydromagnetic phenonmena in understanding these new results. The confluence of sophisticated 3D numerical MHD simulations and the recent observations of rotating jets makes it possible, for the first time, to rigorously test jet models.

Keywords: jets, accretion disks, magnetic fields, bipolar outflows

1. Introduction

Astrophysical jets are among the most elegant and beautiful phenomena in astronomy. Their ubiquitousness during star formation ultimately derives from the presence of accretion disks around all young stars. As is well understood from many calculations and numerical simulations, the gravitational collapse of an intially, slightly rotating molecular cloud core produces an accretion disk through which most of a star's material must flow. This implies that the superabundance of angular momentum of gas in molecular cores in comparison with a star's ultimate angular momentum, must be eliminated during this accretion process. If the loss of this angular momentum occurs only because of its outward radial transport in the disk by turbulent stresses, there would be no reason to expect that jets would arise in low mass, slowly rotating, cool T-Tauri stellar systems. The fact that jets are correlated with accretion disks suggests that their energy ultimately derives from accretion power. This implies that jets play an important, if not dominant role in the removal and tranport of angular momentum of the gas that is destined for the central star.

Almost all contemporary theories predict that jets in protostellar systems are disk winds in which the coupling between jets and their underlying disks is mediated by magnetic fields that thread them (e.g. review Königl and Pudritz, 2000). This picture derives from the classic paper of Blandford and Payne (1981) on a centrifugally driven wind model for quasars and radio jets. The X-wind model uses the same physical driver but posits the interaction between an opened dipolar magnetosphere of a modestly rotating star with the inner edge of a surrounding, poorly magnetized disk, as the central engine (e.g. review Shu et al., 2000).

Astrophysics and Space Science **292:** 471–478, 2004.

The use of magnetic fields to tap the gravitational potential energy release of disk accretion in the disk-wind picture, or the spin of a central young star in the X-wind picture, nicely explains why outflows can occur in the absence of strong radiation fields. The efficiency of magnetic torques in removing angular momentum from spinning objects is well appreciated in many astrophysical systems (e.g. magnetic stars, pulsars,) Unfortunately, the hypothesis that magnetic torques play a dominant role in the physics of jets makes detailed testing of these pictures very difficult. How can one reliably measure the geometry and strength of magnetic fields in accretion disks, jets or on the surfaces of TTSs? This conference on *Magnetic Fields and Star Formation: Theory versus Observation* was organized in order to make some progress in answering such questions.

2. Jets and Magnetized Disks: Theory

The basic 2D, analytic theory of jets has a solid foundation that is rooted in the ideas of ideal, axisymmetric MHD. Theorists now face the more difficult challenge of developing a self-consistent 3D, global model for magnetized disks and jets, whose disks also interact with the magnetospheres of their central rotating young stellar objects. While daunting, the availability of parallel computing resources and increasingly sophisticated MHD codes (such as adaptively refined meshes) means this is not insurmountable. The ultimate source for the magnetic field in the disk must be identified and this might involve fields that have either been dragged in from the larger magnetized environment of the molecular cloud core out of which the system collapsed, or that were generated *in situ* by the magneto-rotational instability (Balbus and Hawley, 1991) or related dynamo action. The dynamic behaviour of a magnetosphere that arises from its contact with the surrounding disk is readily seen in the work of Goodson et al (1997). The repeated reconnection events that a stellar magnetosphere undergoes as it is twisted and inflated through the interaction with the surrounding accretion disk is a non-stationary process, which is one means of providing rapid episodic behaviour of jets. This also implies that non-stationary accretion onto the star.

In these proceedings, Ferreira reviews the progress that has been made in 2D self-similar and time-independent models for disk winds which consider the vertical structure of the disk. Earlier disk wind models that employed vertically averaged disk structures have proven to be very useful in predicting the most basic features of jets—such as their ability to self-collimate through the 'hoop stress' that results from a toroidal field produced in a rotating flow; the fact that jets should be rotating which is inherited from the enforced co-rotation of the jet with its underlying disk from the disk surface out to the Alfvén surface; as well as their high efficiency in extracting angular momentum from a disk (e.g. Pelletier and Pudritz, 1992).

As is well known, the torque exerted by an external magnetized jet on the disk gives a fundamental relation between the mass outflow in the wind, and the mass accretion rate through the disk; $\dot{M}_a = (r_A/r_o)^2 \dot{M}_w$. This result simply states that the rate at which the disk loses angular momentum, $\dot{J}_d = \dot{M}_a \Omega_o r_o^2$, is exactly the rate that is carried away by the wind, $\dot{J}_w = \dot{M}_w \Omega_o r_A^2$. The value of the Alfvén lever arm $r_A/r_o \simeq 3$ for most theoretical and numerical models so that one finds $\dot{M}_w/\dot{M}_a \simeq 0.1$ which is in agreement with all of the observations that this reviewer is aware of.

Many of these basic theoretical models assume, for simplicity, that the gas that is launched from the disk is cold, that is, that thermal pressure is not required to launch the outflow. Without a corona to provide material to feed the jet, there must therefore exist a direct link between the physics of magnetized disks, and the origin of outflows—hence the idea of an 'accretion-ejection' structure.

How does mass ejection and wind formation occur in these models? The answer lies in examining the 2D behaviour of material in the accretion disk. Ferreira's review prompts us to imagine what happens to the force balance in a disk as we follow a field line from the midplane of the disk out through the disk surface and then to large distances in a collimated outflow. At the disk mid-plane, the poloidal field must be perpendicular to it ($B_r = 0$). If this were not so, there would be a singular current sheet on the mid-plane that would be unstable and would rapidly heat the disk until the current is dissipated. As one follows the field line from the mid-plane, it curves away from the rotation axis (convex) which is a consequence of the dragging of the field due to the inward accretion flow. For a collimated flow to exist, this field line must begin to curve back towards the rotation axis (concave) of the system somewhere near the surface of the disk. Thus, the field changes its curvature from convex to concave as it traces through the disk. The magnetic tension associated with this curved field line exerts a pinch force down on the disk, compressing it at all heights (denoted by $z_{\rm curv}$) below which the field-line curvature is convex. Above $z_{\rm curv}$, however, the (concave) curvature of the field exerts a tension force upwards, pushing the material slightly away from the disk mid-plane. It is this type of pressure imbalance which feeds gas in the upper layers of a disk into a cold disk wind The formal expression of this physical picture is obtained from the condition of vertical hydrostatic balance (e.g. Ferreira, 1997).

The initiation of the outflow in the upper part of the disk above $z_{\rm curv}$ can be gleaned from the detailed angular momentum balance equation. The toroidal component of the Lorentz force on material rotating in the disk is $F_\phi = J_z B_r - J_r B_z$. One can see that as the radial current J_r falls with height (z) in the disk, this component of the Lorentz force component also changes sign. Below the height $z_{\rm curv}$, this force opposes the gas motion slowing it down slightly below Keplerian value and thereby causing it to fall to a slightly smaller disk radius; i.e. accretion is affected by the magnetic torque. At greater heights in the disk $z > z_{\rm curv}$, however, the force accelerates the gas slightly. It is this magnetic acceleration and upward direction of the force that gets the upper layers of the disk moving, eventually to become a cold, centrifugally driven disk wind.

This physical picture shows that disk winds are very natural. The parameter space needed to find solutions is rather restricted, however. There are two reasons for this. The first is the assumption of self-similarity. The second is that one has assumed 'cold flows'. If disk coronae are present and are the mass sources for the jets, then one may find that the solutions are not so constrained. This possibility has not been yet investigated analytically, although simulations have begun.

Sophisticated numerical simulations of a complete magnetized star-disk-disk wind system have been pursued by several groups The paper by Rekowski et al. (these Proceedings) shows that the MRI instability and stellar magnetosphere both are responsible for generating components of a global outflow. This paper self-consistently simulates the physics of a disk, a central magnetized star, and the dynamo action in a disk that produces a collimate disk-wind or jet along lines suggested by the simulations of Ouyed and Pudritz (1997).

The MRI instability is a weak magnetic field instability and its non-linear evolution in an accretion disk does strengthen the magnetic field until it saturates at a field strength such that the Alfvén speed in the disk is comparable to the disk sound speed; $v_A \simeq c_s$. The process of creating a strong field is not the same as predicted by so-called mean-field theory of dynamo action. Nevertheless, Brandenburg and his associates have shown in a number of papers that this instability does after all have some of the characteristics of dyanmo action. The main conclusion is that a magnetically coupled star and a dynamo-active disk produce a clearly structured flow. These authors find that a hotter, denser and slower stellar wind that is driven by gas pressure occurs towards the centre of this overall outflow. At the same time a cooler, less dense, and faster centrifugally accelerated disk wind is accelerated within a conical shell. These flows are not collimated within a 0.1 AU scale, and are also episodic as the observations demand.

Finally, a class of inflow-outflow solutions has been found that may combine gravitational collapse with accretion-ejection, as discussed in the paper by Combet and Lery (these Proceedings). In these solutions, the region around a protostar is divided into three regions; a jet, an infalling envelope, and a large-scale circulation region in which matter rains down towards the central region and is reflected outwards again from a pressure gradient near the central region. This circulation region is interpreted as being the molecular outflow which is predicted not to arise not from the entrainment of surrounding molecular material by a fast moving jet, but by a separate hydrodynamic circulation flow in its own right. These solutions can also be adjusted so that the ratio of wind outflow to accretion of the correct order; $\dot{M}_a/\dot{M}_w \simeq 0.1$.

3. Jets and Outflows: Observations

The magnetic fields that are predicted in these disk/jet models are very difficult to observe directly. The advent of submillimetre polarimetry on SCUBA at the

James Clerk Maxwell Telescope (JCMT) has brought remarkable advances in this problem because it allows one to map the structure of magnetic fields in relatively dense gas very efficiently (e.g. Matthews and Wilson, 2000). The magnetic fields that thread accretion disks may originate on the larger physical scales characteristic of the molecular cloud cores and filaments. Magnetic fields that are associated with filaments could have a novel helical structure that arises from the twisting of magnetized filaments that occurs in a shearing, turbulent medium (Fiege and Pudritz, 2000a). Such models predict a simple geometric explanation for the 'polarlization holes' (region of low percent polarization) that are often observed towards filaments (Fiege and Pudritz, 2000b).

The paper by Chrysostomou et al. (these Proceedings) presents JCMT submillimetre data for a variety of structures on these larger scales. Their observations find evidence for a number of different magnetic configurations, ranging from hourglass-like configurations to helical fields. The results suggest that magnetic fields are more ordered and therefore less 'passive' in dense regions compared with low density regions, implying that they are relatively lower in energy density on these larger scales. This seems to correspond to the Zeeman results presented by Crutcher (these Proceedings) that suggest that magnetic fields are at levels predicted by the Virial theorem in dense cores, but probably not so in the lower density molecular gas.

Hydromagnetic flows are also accompanied by physical observables that are more directly measurable and that therefore act like diagnostics for magnetic processes. Some examples of these include molecular outflows, shocks and forbidden line emission, as well as the collimation and rotation of jets. This meeting featured novel observational approaches that used several of these diagnostics.

Large scale outflows are an important way of probing the nature of jets. It is now well understood that outflows are probably all driven by underlying jets. Richer's review talk (see also Richer et al., 2001) emphasized that outflows are momentum and not energy driven (the latter leads to the conversion of jet energy to pressure that would inflate bubble-like cavities which are not observed). The collimation of molecular outflows decreases with the increasing age of the outflow which may be a sign that molecular outflows ultimately disrupt the remnants of their natal core and hence undergo less external pressure confinement. There is also evidence that the underlying jets also move around or precess on a time scale that is around 30,000 yr or so for a source like RN043.

One of the most important aspects of outflows for the general theory of star formation is that *they are associated with both low and high mass star formation.* Jet driven flows operate all thc way up to B stars. The paper by Garay et al. (these Proceedings) presents excellent evidence for the existence of a large scale, well collimated, molecular outflow from a forming O-type star. The evidence indicates that this protostar is still undergoing an enormous mass infall rate of $\dot{M}_a \simeq 10^{-2} M_\odot \mathrm{yr}^{-1}$ which is sufficient to quench the development of an ultra-compact HII region. At the same time, one sees direct evidence for the underlying jet which is nicely delineated by shocked H_2 emission that is symmetric about the central source.

The observations of outflows from massive stars are important for star formation theory because they suggest that the accretion paradigm for star formation—namely that individual or binary stars form from the collapse of a dense molecular magnetized cloud core that quickly forms an accretion disk and an outflow—pertains to both low and high mass star formation. This possibility is advocated in the theoretical work of McLaughlin and Pudritz (1997) and McKee and Tan (2003) who show that massive infall rates are expected in the turbulent cores in which massive stars are expected to form.

Bipolar outflows are associated with shocks that are the result of the interaction of the jet with the external magnetized, molecular gas. Shock diagnostics can be used to estimate the field strengths independently of Zeeman measurements. The paper by Cabrit et al. (these Proceedings) shows that shocks can be both of the classical jump (J) type or continuous (C) type depending on the strength of the field and the degree of ionization of the gas. The models presented include time-dependent chemistry such as the finite time for dissociation of H_2 molecules. The spectrum of molecular hydrogen is very different for these two shock regimes, and this turns out to be the useful diagnostic. The results show that shocks in bipolar outflows are generally of C type, and are associated with field transverse field strengths (in μGauss) of the order $B_{\rm trans} \simeq 1 - 10\,(n_H/{\rm cm}^3)^{1/2}$.

Perhaps the most direct test of the nature of the underlying engine comes from examining the dynamics of the jet material itself. High resolution spatial and spectral observations that have recently become available at the CFHT or the HST and these have transformed this field remarkably. One can study the dyamics of the knots that are ubiquitous in jets in great detail (see Pesenti et al.—these Proceedings).

A fundamental prediction of the centrifugally-driven disk wind models for jets is that jets carry off angular momentum from the underlying disks. This angular momentum is shared up between the toroidal component of the field and the rotation of gas in the jet. Gas at each radius r in the jet originated at the footpoint of a field line at a radius r_o on the underlying disk, and transports the angular momentum of the disk that was present at that disk radius.

The observation of the rotation of jets therefore holds great promise for the detailed testing of disk wind models (Bacciotti et al., 2002). The paper by Coffey et al. (these Proceedings) examines the dynamics of jets from Class 1 protostellar systems which are more highly embedded than in the later, less energetic phases of jet evolution. Using near IR [Fe II] forbidden line diagnostics, these authors measure velocity shifts which are then fitted to jet outflow models (Ferreira, 1997) for the outflow in DG Tau. Models that hypothesize 'warm' rather than 'cold' disk winds seem to do a better job in fitting the data. It will be interesting to fit such data to time-dependent numerical simulations of disk winds such as Ouyed and Pudritz (1997) and Krasnopolsky et al. (1999).

4. Next Steps

The confluence of sophisticated numerical simulations with new, high spectral and velocity resolution as well as polimetric observations of jets has opened up a new frontier in star formation studies. Jets and outflows are arguably the best understood and best tested aspects of the entire 'theory' of star formation. The results presented at this meeting suggest that this field has now aquired the sophistication and rigour to actually answer some of the really tough questions.

Perhaps the most important problem to address is on the nature of the engine itself. Are jets disk winds, or X-winds, or neither of the above? Detailed mapping of the velocity field in jets should allow one to determine whether or not the jet material was originally distributed along the surface of a Keplerian accretion disk (as the disk wind model predicts) or whether it came from one particular radius of the disk—its truncation radius with the stellar magnetosphere (as the X-wind picture predicts). It should also be possible to assess what role a stellar component of the overall outflow plays, if any. Answers to these questions are at the very heart of star formation theory because they will help us to chart the evolution of angular momentum flow as protostars are built out of their accreting disks. There is little doubt that ALMA will play a major role in helping to unlock the fine details of jet/disk coupling, so much exciting observational work lies ahead. The future looks bright.

Acknowledgement

I thank Ana Gomez de Castro and the LOC for creating this excellent workshop. My work is supported by a grant from the Natural Sciences and Engineering Research Council of Canada (NSERC).

References

Bacciotti, F., Ray, T.P., Mundt, R., Eilöffel, J. and Solf, J.: 2002, *ApJ* **576**, 222.
Balbus, S.A. and Hawley, J.F.: 1991, *ApJ* **376**, 214.
Blandford, R.D. and Payne, D.G.: 1981, *MNRAS* **199**, 883.
Ferreira, J.: 1997, *A&A* **319**, 340.
Fiege, J.D. and Pudritz, R.E.: 2000a, *MNRAS* **311**, 85.
Fiege, J.D. and Pudritz, R.E.: 2000b, *ApJ* **544**, 830.
Goodson, A.P., Winglee, R.M. and Böhm, K-.H.: 1997, *ApJ* **489**, 199.
Königl, A. and Pudritz, R.E.: 2000, in: V. Mannings, A.P. Boss and S.S. Russell (eds.), *Protostars and Planets IV*, University of Arizona Press, Tucson, p. 759.
Krasnopolsky, R., Li, Z.-Y. and Blandford, R.D.: 1999, *ApJ* **526**, 631.
Matthews, B. and Wilson, C.D.: 2000, *ApJ* **531**, 868.
McKee, C.F. and Tan, J.C.: 2003, *ApJ* **585**, 850.

McLaughlin, D. and Pudritz, R.E.: 1997, *ApJ* **476**, 750.
Ouyed, R. and Pudritz, R.E.: 1997, *ApJ* **482**, 712.
Pelletier, G. and Pudritz, R.E.: 1992, *ApJ* **394**, 117.
Richer, J.S., Shepherd, D.S., Cabrit, S., Bachiller, R. and Churchwell, E.: 2000, in: V. Mannings, A.P. Boss and S.S. Russell (eds.), *Protostars and Planets IV*, University of Arizona Press, Tucson, p. 867.
Shu, F.H., Najita, J., Shang, H. and Li, Z.-Y.: 2000, in: V. Mannings, A.P. Boss and S.S. Russell (eds.), *Protostars and Planets IV*, University of Arizona Press, Tucson, p. 789.

SELF-COLLIMATED STATIONARY DISK WINDS

JONATHAN FERREIRA[1] and FABIEN CASSE[2]
[1]*Laboratoire d'Astrophysique de Grenoble, Grenoble Cedex, France;*
E-mail: jonathan.ferreira@obs.ujf.grenoble.fr
[2]*FOM-Institute for Plasma Physics "Rijnhuizen", Nieuwegein, The Netherlands*

Abstract. In this contribution, we first review the theory of self-collimated jets launched from magnetized accretion disks (disk-winds originating from the first AUs). We show why it is crucial to solve in a self-consistent way the interplay between the resistive accretion disk and the ideal MHD jets. Indeed, this is the only way to get exact values for the disk ejection efficiency ξ (the jet mass load issue). Then, we show self-similar calculations of such accretion-ejection structures: first cold jets, then warm jets obtained in the presence of a hot disk chromosphere. Finally, we present for the first time an accretion-ejection flow crossing all three critical points.

Keywords: accretion, accretion disks, ISM: jets and outflows MHD, stars: pre-main-sequence

1. From Observational Facts to Theoretical Convictions

1.1. Why must jets be self-collimated?

It is now well-known that all young forming stars of low-mass show signatures of both accretion and ejection (Hartigan et al., 1995). Signatures of ejected material are blue-shifted emission in forbidden lines (e.g. OI, SII, NII). The resolved images of some objects like, for example HH30 provided by HST (Ray et al., 1996), are particularly impressive: they show that at distances of about 20 AU from the central star, the ejected material has already reached a typical velocity of about 300 km/s with a very small opening angle. This requires that the underlying physical process accelerating this material into bipolar jets is also responsible for its own confinement. The only process capable of doing it relies on the presence of a large scale magnetic field carried along the jet.

The same observations suggest that these spatially resolved jets are launched from the innermost disk regions, i.e. the central AU or even less. This implies quite short dynamical time-scales (less than 1 yr) while the most energetic ejection events, observed as Herbig-Haro objects, require a much longer variability time scale. As a consequence, steady-state models should be able to catch the main dynamical features of jets from YSOs.

Since magnetic fields are required to simultaneously accelerate and confine jets, one needs to use magnetohydrodynamics (MHD) to describe them. Note that, once jets are launched and have reached their asymptotic state, they may become no longer ionized enough so that an hydrodynamical (HD) description becomes

Astrophysics and Space Science **292:** 479–492, 2004.

acurate. This is not in contradiction with jets being collimated on scales larger than hundreds of AU. Indeed, the opening angle of a HD jet is roughly the inverse of its Mach number, which provides opening angles of order 5° or less, nicely compatible with observations. Anyway, even if they are not anymore relevant, the presence of magnetic fields on those scales could be revealed by the analysis of molecular chemistry in shocks.

1.2. The Crucial Role of Electric Currents

There are three classes of magnetized jet models, depending on the origin of the ejected material and on which energy reservoir is tapped:

Stellar winds assume that field lines are anchored into the rotating star. Both mass and energy are extracted from the star itself (e.g. Weber and Davis, 1967; Sauty et al., 2002).

Disk-winds assume that field lines are anchored into the accretion disk alone. Both mass and energy is therefore extracted from the underlying disk. If jets are launched from a large radial extension in the disk, then one obtains the situation calculated by Blandford and Payne (1982) and extended by Vlahakis et al. (2000). If, on the contrary, the large scale magnetic field is assumed to thread the disk only in a tiny region at the disk inner boundary, then one gets the 'X-wind' picture (Shu et al., 1994; Shang et al., 2002).

Magnetospheric winds are winds that are produced at the interaction between the accretion disk and the protostellar magnetic field. They can carry away mass coming from the disk and tap rotational energy from the star (Ferreira et al., 2000; Matt et al., 2002).

All these models suffer from simplifying assumptions but disk-wind models have been more developed: the capability of T Tauri stars to drive massive jets still needs to be proven (DeCampli, 1981), while magnetospheric winds require numerical simulations that are difficult to control.

Magnetized jets are actually made of nested magnetic surfaces (defined by a constant magnetic flux $a(r, z) = Cst$) that are anchored on a rotating object (hereafter the disc). The axisymmetric poloidal magnetic field writes

$$\vec{B}_p = \frac{1}{r}\nabla a \times \vec{e}_\phi \tag{1}$$

and is usually assumed to be of bipolar topology. The toroidal magnetic field is tremendously important: it produces the magnetic braking on the underlying disc, feeds the jets with energy (MHD Poynting flux) and provides the 'hoop-stress' that confines them. This hoop-stress can be understood as the Laplace force due to presence of a vertical current I, flowing inside a magnetic surface, and the toroidal field. Note however that, because currents must be closed, not all field lines can be self-collimated (Okamoto, 2001). Now, jet acceleration depends also on this current. Indeed, since ejected mass is accelerated by MHD forces (see Eq. (15) in Ferreira, 1997), there is a transfer from the MHD Poynting flux to the kinetic energy flux: $I = 2\pi r B_\phi/\mu_o$ decreases as we go downstream, allowing thereby a current closure. So, from this simple argument, two important issues of MHD jets can be understood: (1) the degree of asymptotic collimation (value of I) is related to the acceleration zone, which is located upstream; (2) any approximation made on the poloidal current has dramatic consequences on jet dynamics.

The source of such an electric current is the unipolar induction effect (Barlow wheel experiment). Any rotating conductor embedded in a magnetic field will produce an electromotive force, which drives a current. So, the current that feeds magnetized jets is driven by the underlying *resistive* accretion disk. This is a very strong constraint, neglected in studies where the disk is a mere boundary condition.

A byproduct is the possibility to launch *asymmetric jets*. Indeed, a symmetric (bipolar) magnetic configuration threading the disk produces one electromotive force. But there are two independent electric circuits, one for each jet. A slight mismatch between the two 'resistances' can produce two different jets with different mass loads and power (Ferreira and Pelletier, 1995).

2. MHD Models of Steady-State Jets

2.1. Invariants and Free Variables

Jets are assumed to be steady, axisymmetric, nonrelativistic and described as a single fluid within the ideal MHD framework. To simplify further, the energy equation is usually replaced by a polytropic description. A magnetized jet is described by eight unknown variables: density ρ, velocity $\vec{u}$, magnetic field $\vec{B}$ (flux function a and toroidal field B_ϕ), pressure P and temperature T. There are eight equations (PDE) allowing to solve the full 2D problem: mass conservation, momentum conservation, induction equation, equation of state (perfect gas) and a polytropic law. Because jets are nondissipative structures in ideal MHD, there are five invariants along each magnetic surface (Jacobi integrals) fixed by boundary conditions. These invariants are

1. Mass flux to magnetic flux ratio $\eta(a)$

$$\vec{u}_p = \frac{\eta(a)}{\mu_o \rho} \vec{B}_p \tag{2}$$

2. Magnetic surface rotation $\Omega_*(a)$

$$\Omega_*(a) = \Omega - \eta \frac{B_\phi}{\mu_o \rho r} \tag{3}$$

3. Total specific angular momentum $L(a) = \Omega_* r_A^2$

$$L(a) = \Omega r^2 - \frac{r B_\phi}{\eta} \tag{4}$$

4. Total specific energy $E(a)$

$$E(a) = \frac{u^2}{2} + H + \phi_G - \Omega_* \frac{r B_\phi}{\eta} - \mathcal{H} \tag{5}$$

5. Specific entropy $K(a)$

$$P = K(a)\rho^\gamma \tag{6}$$

where r_A is the cylindrical radius where the poloidal velocity reaches the poloidal Alfvén velocity, H the enthalpy, $\mathcal{H}$ an heating term (zero if jets are adiabatic) and γ the polytropic index.

To solve the set of MHD equations requires to specify eight boundary conditions. One of them is to assume $\Omega_* = \Omega_K$, i.e. magnetic surfaces rotating at Keplerian speeds (sub-Keplerian speeds for the ejected mass). The second fixes the entropy K: $K = 0$ in Blandford and Payne ('cold' jets), $K \neq 0$ in Contopoulos and Lovelace (1994), Vlahakis et al. (2000) and all numerical simulations ('warm' jets). Three other constraints arise from MHD regularity conditions. Stationarity requires that, as ejected material is accelerated along each magnetic surface, it becomes super slow-magnetosonic (SM), then super-Alfvénic (A) and finally super fast-magnetosonic (FM). This leaves us with *three free and independent* boundary conditions that must be specified at each magnetic surface (anchored at r_0).

Note the crucial fact that, because jet models are in ideal MHD, they do not address the mass loading by the underlying disk: the mass flux is simply parametrized by η.

2.2. An Unsolved Mathematical Problem

Instead of solving eight equations, we can formulate the problem in a much more compact way using the invariants. This leads us to the following (Grad-Shafranov) equation for an adiabatic jet

$$\nabla \cdot (m^2 - 1)\frac{\nabla a}{\mu_o r^2} = \rho \left\{ \frac{dE}{da} - \Omega \frac{d\Omega_* r_A^2}{da} + \left(\Omega r^2 - \Omega_* r_A^2\right)\frac{d\Omega_*}{da} - \frac{C_s^2}{\gamma(\gamma-1)}\frac{d\ln K}{da} \right\} + \frac{B_\phi^2 + m^2 B_p^2}{\mu_o}\frac{d\ln\eta}{da} \tag{7}$$

where $m^2 \equiv u_p^2/V_{Ap}^2$ is the Alfvénic Mach number and $C_s^2 = \gamma k_B T/\mu m_p$ is the jet sound speed. This equation provides the transverse equilibrium (i.e. the degree of collimation) of a magnetic surface. Mathematically, this is a mixed-type PDE providing $a(r, z)$ for a given set of invariants (η, Ω_*, L, E and K). Indeed, the flow is hyperbolic between the cusp (hopefully located at the disk surface) and the SM surface, elliptic between the SM and the FM surface, and again hyperbolic downstream the FM surface. Now, while elliptic zones are fully determined by boundary conditions, hyperbolic domains require to be computed as initial value problems. We have therefore to face two overwhelming difficulties: (1) both shape and localization of the parabolic surfaces (SM and FM) are unknown; (2) the choice of boundary conditions is terribly large.

This is the reason why *there is no consistent solution for 2D steady-state MHD jets* yet. This is an unsolved mathematical problem. Either one solves the time-dependent problem with an MHD code, or one uses some trick: separation of variables (self-similar solutions, eg. Ostriker, 1997), assume the shape of the magnetic surfaces (Lery et al., 1999), solve only in the elliptic domain with ad-hoc boundary conditions (Fendt and Camenzind, 1996; Shang et al., 2002[1]) etc.

2.3. A NOTE ON NUMERICAL MHD SIMULATIONS

Numerical simulations can be cast into four categories, depending on the kind of problem they can really address:

Jet production: These simulations solve both the disk and its atmosphere, in order to compute the interrelations between accretion and ejection. Their main difficulty is to reach a steady-state if the initial solution is too far away from a real equilibrium (Uchida and Shibata, 1985; Kato et al., 2002; Casse and Keppens, 2002).

Jet acceleration and collimation: These simulations allow to study the acceleration and collimation of mass that is continuously injected at the bottom boundary of the computational domain (referred to as the disk surface). There, several fixed boundary conditions are prescribed: a Keplerian profile, the density $\rho(r_0)$, vertical velocity $u_z(r_0)$ and magnetic field $B_z(r_0)$ distributions (Ouyed et al., 2003; Krasnopolsky et al., 1999; Ustyugova et al., 1999).

Jet propagation: These simulations study the formation and shape of bow shocks and internal working surfaces when super-FM jets (i.e. jets that are already collimated) are propagating in an outer medium (Cerqueira et al., 2001; Frank et al., 2000; Stone and Hardee, 2000).

Jet instabilities: Studying jet instabilities, namely the growth rate, non-linear development and mode coupling requires 3D simulations with very high resolution as well as semianalytic calculations (linear codes, MHD spectroscopy) in order to interpret correctly what comes up from the full MHD code (Appl et al., 2000; Lery et al., 2000; Baty and Keppens, 2003).

[1] In this case, boundary conditions were specified at the Alfvén surface itself, not at the FM surface as it should. Some 'connection' has been made afterwards to match the sub-Alfvénic solution to an asymptotic solution.

Complexity of accretion–ejection structures plus the inherent difficulties of using codes (artificial forces due to boundary conditions, numerical resolution and related viscosity, bad initial conditions introducing highly variable transients) forbid to address simultaneously these four physical issues within one code. To our knowledge, only Casse and Keppens (2002) successfully addressed the jet production issue by using the theory described here.

3. MHD Models of Accretion–Ejection Structures

3.1. General concepts

In a standard accretion disk (i.e. no mass loss), the accretion rate is constant with radius. In an accretion–ejection structure, one defines the ejection index as

$$\xi = \frac{d \ln \dot{M}_a}{d \ln r} \tag{8}$$

This parameter (which can vary in the disk) measures the local ejection efficiency. If this efficiency is constant throughout the disk (of inner and outer radii $r_{\rm in}$ and $r_{\rm out}$), then mass conservation provides the ratio of ejection to accretion rates

$$\frac{2\dot{M}_j}{\dot{M}_a} = 1 - \left(\frac{r_{\rm in}}{r_{\rm out}}\right)^{\xi} \simeq \xi \ln \frac{r_{\rm out}}{r_{\rm in}} \tag{9}$$

Observations show that this ratio varies between 0.01 and 0.1, which implies similar values for the ejection efficiency ξ.

A complete theory must provide the allowed values of ξ as a function of the disk properties. One must take into account the full 2D problem and *not* treat the disk as infinitely thin as in a standard disk theory. This requires to treat the physics of both accretion and ejection, from a quasi-Keplerian disk thread by a large scale magnetic field of bipolar topology.

3.1.1. *Accretion*

Steady-state requires the presence of an anomalous magnetic diffusivity allowing the accreting (and rotating) material to cross the magnetic field lines. Ambipolar diffusion could do the job but fully ionized disks (i.e. around compact objects) also display self-collimated jets. So, if one looks for an 'universal' model, one should use another prescription. We assume that an MHD turbulence, triggered and maintained inside the disk, can be described by a local 'effective' diffusivity ν_m. Since rotation is much faster than accretion, there must be a higher dissipation of toroidal field than poloidal one. *A priori*, this implies a possible anisotropy of the magnetic diffusivities associated with these two directions, poloidal ν_m and toroidal ν_m'. Besides, such a turbulence might also provide a radial transport of angular momentum, hence an anomalous viscosity ν_v. To summarize, at least three anomalous transport coefficients are necessary to describe a stationary structure. We will use the following dimensionless parameters defined at the disk midplane:

$$\begin{aligned} \alpha_m &= \frac{\nu_m}{v_A h} \text{ level of turbulence} \\ \chi_m &= \frac{\nu_m}{\nu'_m} \text{ degree of anisotropy} \\ \mathcal{P}_m &= \frac{\nu_v}{\nu_m} \text{ magnetic Prandtl number} \end{aligned} \tag{10}$$

where $v_A = B_o/\sqrt{\mu_o \rho_o}$ is the Alfvén speed. A conservative picture of 3D turbulence would translate into $\alpha_m < 1$, $\chi_m \sim 1$ and $\mathcal{P}_m \sim 1$. But as stated before, the amount of current dissipation may be much higher in the toroidal direction, leading to $\chi_m \ll 1$. Moreover, it is not obvious that α_m must necessarily be much smaller than unity. Indeed, stationarity requires that the time scale for a magnetic perturbation to propagate in the vertical direction, h/v_A, is longer than the dissipation time scale, h^2/ν_m. This roughly translates into $\alpha_m > 1$. Thus, we must be cautious and should freely scan the parameter space defined by these turbulence parameters. The magneto-rotational instability (Balbus and Hawley, 1991) may well be the main source of turbulence in discs. Unfortunately, there is to date no MHD simulation (3D, with a nonzero poloidal flux) providing expected values for the 'viscosity': the shearing sheet approximation brakes down in a few orbital times (Miller and Stone, 2000). No value for the magnetic diffusivity has ever been computed.

One important thing to realize is that the torque due to the large scale field (jet) is always much larger than the 'viscous' (turbulent) torque. Their ratio at the disk midplane writes

$$\Lambda = \frac{\text{magnetic torque}}{\text{viscous torque}} \simeq \frac{B_\phi^+ B_z/\mu_0 h}{\alpha_v P/r} \simeq \frac{r}{h}\frac{B_\phi^+}{B_z}\frac{B_z^2/\mu_0}{\alpha_v P} \sim \frac{r}{h} \tag{11}$$

where $h(r)$ is the local disk vertical scale height (defined with the gas pressure P) and B_ϕ^+ is the toroidal field at the disk surface. Now, such an engine can only be steady with a field close to equipartition, namely $B_z^2/\mu_0 \sim P$. A much smaller field triggers the magneto-rotational instability, while a much larger field simply forbids ejection (huge vertical compression[2]; Ferreira and Pelletier, 1995). Also, unless very special conditions are met (strong tubulence anisotropy; Casse and Ferreira, 2000a), the toroidal field at the disk surface is of the same order than the vertical field.

[2]The magnetic field, heavily compressing the disk, acts against any outward vertical motion inside it. Some authors (eg. Lubow et al., 1994, Cao and Spruit, 2002) did not take this effect into account and found an instability (the more they bend the field, the more mass is ejected). This happens to be wrong as first noted by Königl and Wardle (1996).

3.1.2. *Ejection*
The disk is accreting because the large scale magnetic field is extracting its angular momentum and mass can diffuse through the field. Now, if jets are to exist, then mass that has been loaded onto the field lines (i.e. material located at the disk surface in ideal MHD regime) must be azimuthally accelerated by the magnetic field. This means that the magnetic torque must go from negative (at the disk midplane) to positive (at the disk surface). This necessary condition requires a decrease of the radial current density J_r on a disk scale height (Ferreira and Pelletier, 1995).

Since the magnetic field is compressing the disk (downward pressure gradient due to both bending and shearing), the only force capable of pushing material up is the gas pressure gradient. This has two important consequences: (1) all terms in the disk vertical equilibrium are dynamically important and neglecting one leads to a wrong estimate of the disk ejection capability[3]; (2) the correct range of ξ highly depends on the temperature profile.

3.2. Self-Similar Models of Accretion Discs Driving Jets

Solving the full 2D problem without any approximation (keeping all dynamical terms) can be done using a method of variable separation. Such a separation is made by looking for solutions with the same functional dependency in (r, z) than the dominant force, namely gravity (Ferreira and Pelletier, 1993). The only 'trick' is now to propagate the system of ODE in the resistive MHD regime first, then in ideal MHD. Once in this regime, the five invariants are fixed and three are constrained by the regularity conditions at the SM, A and FM points. General analytic links between disk and usual jet parameters (λ related to $L(a)$ and κ related to $\eta(a)$) can be found in Casse and Ferreira (2000a).

3.2.1. *Cold Jets: Isothermal and Adiabatic Models*
Disks being quasi-Keplerian, the enthalpy carried away by the outflowing mass at the disk surface is negligible (cold jets). The use of isothermal magnetic surfaces (Ferreira, 1997) or adiabatic ones (Casse and Ferreira, 2000a) introduces no significant change. All solutions display Alfvén surfaces located quite far from the disk (at $z_A \sim r_A$), with typical ejection efficiencies ξ of about 0.01. They all recollimate toward the axis because of a dominant 'hoop-stress' (Figure 1) and terminate with a shock at the location of the FM point (which is uncrossable for these solutions). Note however that all solutions are super-FM in the conventional sense (i.e. $u_p > V_{FM}$, see Ferreira, 1997 for more details).

[3]This is the reason why Li (1995) obtained a huge jet parameter space. Wardle and Königl (1993) got a similar parameter space but they simplified the mass conservation equation by using instead $\rho u_z = Cst$: this *always* leads to develop an outwardly directed vertical velocity, whatever the magnetic compression.

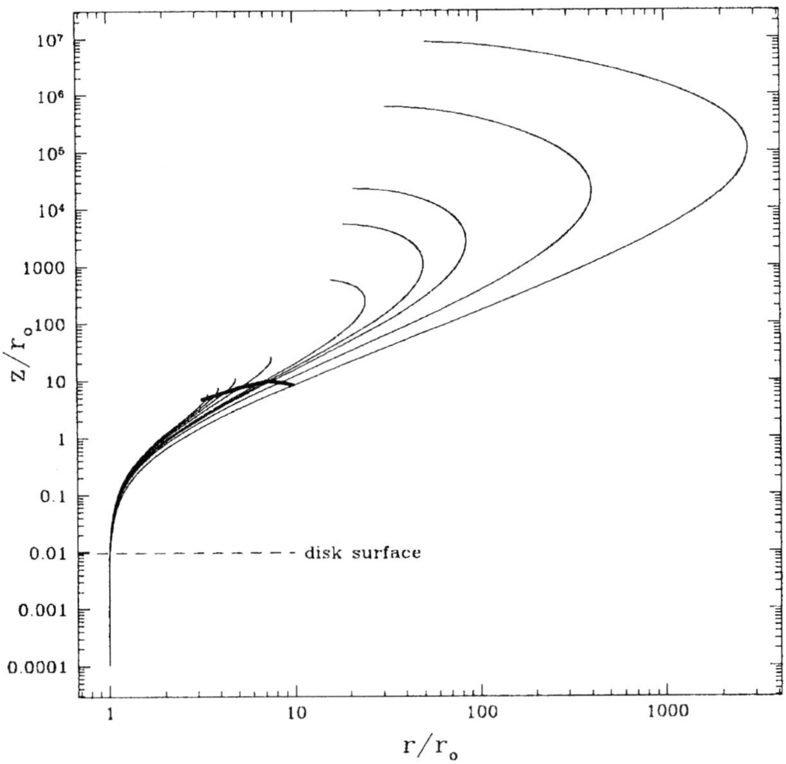

Figure 1. Poloidal magnetic field lines for $\xi = 0.05, 0.04, 0.03, 0.02, 0.012, 0.01, 0.009, 0.007$ and 0.005 (the maximum radius increases with decreasing ejection efficiency). The thick line connects the position of the Alfvén point for each solution (here isothermal, Ferreira, 1997). Note the logarithmic scales: small ξ jets recollimate with angles smaller than 1 degree.

3.2.2. *Warm Jets: Hot Disk Chromosphere*

Warm jets can be produced from quasi-Keplerian disks if the outflowing gas temperature undergoes a sudden rise above the disk. This can be done in mainly two ways:

Illumination: If the central object has a hard surface (not a black hole), then the accretion shock produces hot spots that can illuminate and photoionize (UV or X-rays) the surface layers of the disk. The large X-ray activity of young stars may also contribute to this effect. Above a black hole, pair production and formation of an X-ray zone may also have the same effect on the disk upper atmosphere.

Local dissipation: This may be due to the dissipation of accretion power itself (highly turbulent, magnetized corona) as observations (Kwan, 1997) and numerical simulations (Miller and Stone, 2000) seem to suggest.

Solving the energy equation, Casse and Ferreira (2000b) showed that enhancing the jet temperature has dramatic effects (see Figure 2). For example, the disk vertical

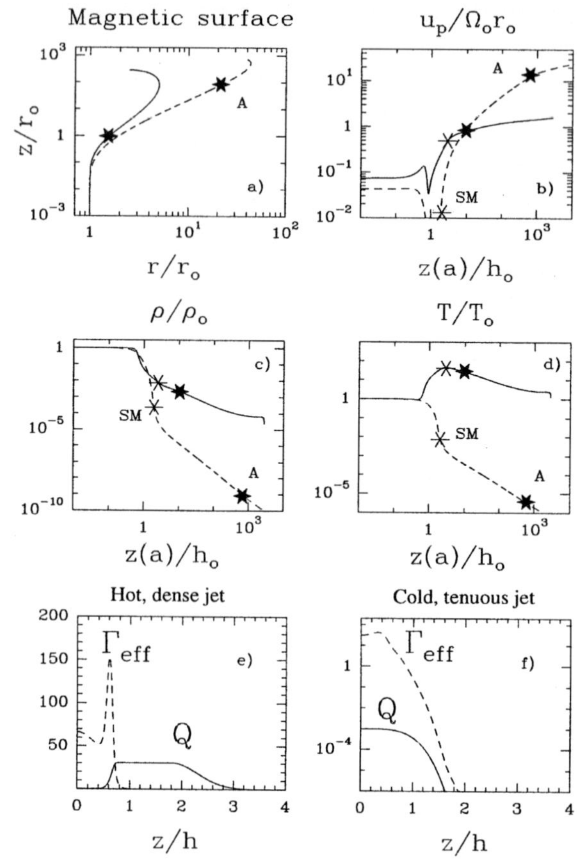

Figure 2. Characteristic plots of two extreme solutions: (a) poloidal cross-section of the magnetic surfaces anchored at a radius r_0; (b) poloidal velocity of the plasma along a magnetic surface in units of the Keplerian speed at its footpoint; (c) and (d), plasma density and temperature along a magnetic surface normalized to their values at the disc midplane. The warm denser jet is drawn in solid line while the cold and more tenuous jet is in dashed line. The cross symbolizes the locus of the slow-magnetosonic (SM) point and the star the locus of the Alfvén (A) point. Pannels (e) and (f) show the effective (viscous and Ohmic) heating term $\Gamma_{\rm eff}$ and the prescribed entropy source Q (Casse and Ferreira, 2000b).

equilibrium can be changed so that a balance can now be achieved with magnetic configurations much more bent. As a result, much smaller values of ξ, down to 0.001, can be obtained. On the other extreme, providing a large enthalpy allows more mass to be loaded onto the field lines: these thermally and magnetically driven jets can accelerate up to $\xi \sim 0.5$ (jet parameter $\kappa \sim 1$).

Now, current numerical MHD simulations of *steady jets* usually have very large mass loads ($\kappa \sim 0.1$ or larger; Ouyed and Pudritz, 1997, 1999). This may be a consequence of code stability, which requires density contrasts not too large in the whole computational domain. Anyway, such large ejection efficiencies can indeed be achieved from geometrically thin disks, but require a hot disk chromosphere. From an astrophysical point of view, this result is very interesting: different objects may have different ejection efficiencies according to the energetics involved.

3.2.3. *Super Fast-Magnetosonic Solutions*

It can be shown that the only way to get super-FM, self-similar solutions is by forcing the Alfvén surface to be close enough to the disk namely, $z_A < r_A$ (Ferreira and Casse, 2003). This is imposed by the self-similar 'geometry' that selects one direction for wave propagation. As a consequence, the critical velocity is not the poloidal velocity (as it would be in general axisymmetric problems), but its projection along that direction (Ferreira and Pelletier, 1995). On the other hand, super-FM jets need to have a large amount of magnetic energy still available at the Alfvén point, so that higher critical speeds can be reached.

The only way to satisfy these two constraints is by building up a large transverse pressure gradient in the sub-Alfvénic region, forcing the magnetic surfaces to open up and pushing thereby the Alfvén surface toward the disk surface. Figure 3 shows such a solution, the first actually that smoothly connects a resistive accretion disk to super-FM jets. Of course, it requires quite large temperatures (up to several

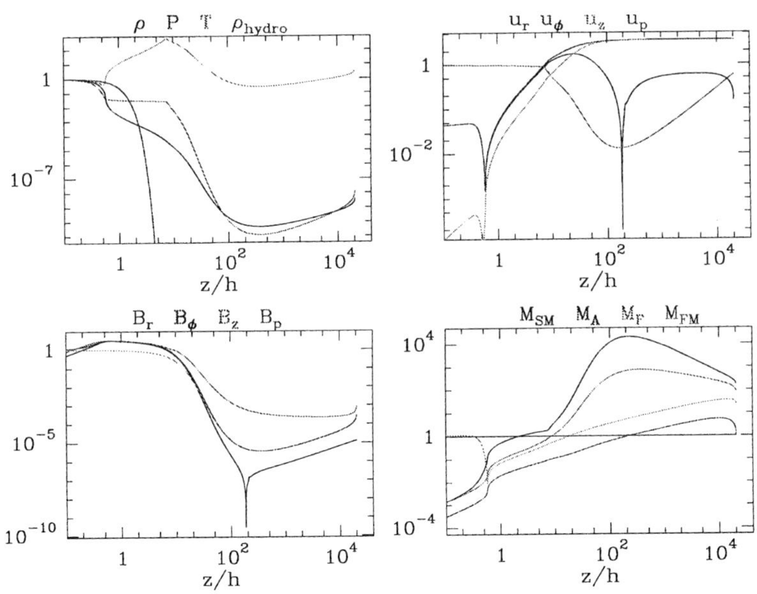

Figure 3. Typical super-FM solution with $\xi = 0.09$ (Ferreira and Casse, 2003). Plots show several quantities along a magnetic surface anchored at a radius r_0. The density (black), pressure (red), temperature (green) and magnetic fields are normalized to their value at the disk midplane, velocities being normalized to the Keplerian velocity at r_0. Note (1) how different a hydrostatic density profile ρ_{hydro} (blue) is from the real profile ρ; (2) that all 3 magnetic field components are comparable up to the Alfvén surface and (3) that jets are conventionaly super-FM (M_F in green).

10^4 K, see also Vlahakis et al., 2000). But this is not in contradiction with current observations of T Tauri stars. Indeed, a hot gas component on scales < 1 AU has already been detected by Gomez de Castro and Verdugo (2001) and Takami et al. (2002) on a few active stars.

4. Concluding Remarks

In this paper, we presented both a theory of *steady jet production* from quasi-Keplerian accretion disks and self-similar models.

The theory describes all relevant physical aspects as well as general analytical relationships between disk and jet physics (for a complete review see Ferreira, 2002). This theory predicts that steady jets require a large magnetic diffusivity, a magnetic field close to equipartition and a radial current density decreasing on a disk scale height. Besides, it has been shown that all dynamical terms are relevant, forbidding toy-models. Such a theory is now firmly settled and has been confirmed by numerical experiments. Indeed, using the VAC code, Casse and Keppens (2002) showed for the first time that, once correct initial conditions (i.e. consistent with the theory) are used, continuous accretion-ejection flows can be computed.

The self-similar solutions showed here are a class of 2D solutions consistent with our theory of steady jets production. The imposed geometry introduces, however, biases in (1) the asymptotic collimation; (2) the FM regularity condition. Self-similar solutions cannot take into account effects of inner and outer pressure (Lery et al., 1999) and all solutions obtained here do recollimate. But recollimation is not imposed by the self-similar ansatz: a class of self-similar solutions of Contopoulous and Lovelace (1994) and Ostriker (1997) does not recollimate (note also that Pelletier and Pudritz (1992) obtained non self-similar but recollimating solutions). The physical cause for recollimation is actually hidden in the transverse gradients of jet MHD invariants (Ferreira, 1997).

We believe that self-similar solutions are a correct description of systems for which $r_{\rm out}/r_{\rm in} \gg 1$ or, alternatively, systems where the Alfvén surface has roughly a conical shape. In the context of YSOs, this raises the question of the amount of magnetic flux trapped in the innermost disk regions. Such a fundamental question remains open. Meanwhile, tests can be made by comparing observational diagnoses with observations of real T Tauri jets (see contributions of O'Brien et al. and Dougados et al., in this volume).

Although the theory of steady ejection from accretion disks is confirmed, several critical questions remain open: (1) origin of the MHD turbulence in presence of an open, large scale magnetic field close to equipartition; (2) theoretical assessment of the chromospheric heating required to produce denser jets; (3) the nature of the stellar-disk interaction.

When the protostellar core builds in from collapsing material, it probably already has some fossil magnetic field trapped in. Such a field would naturally share the same polarity as the field threading the circumstellar disk (at least in the inner

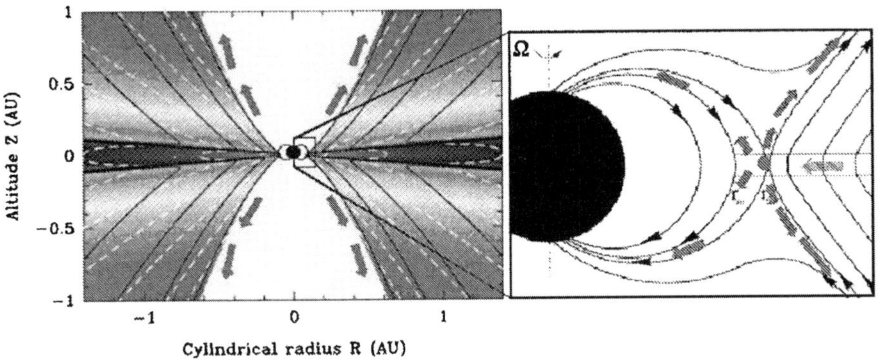

Figure 4. Sketch of a possible magnetospheric interaction between a protostar and a magnetized accretion disk. Two flows are produced in this way: a steady disk wind (solid lines are streamlines, colors: density and dashed white lines are total velocity contours) and a highly variable 'Reconnection X-wind' (arrows). While the former extracts both energy and angular momentum from the disk, thereby allowing accretion, the latter extracts them from the protostar, producing an efficient spin down (Ferreira et al., 2000).

AUs). We can therefore reasonably expect that a magnetic neutral line would form at the equatorial plane, where reconnection of the two fields can take place. Above this reconnection site, a fraction of the accreting plasma can be loaded onto newly opened protostellar field lines (Figure 4). Such Reconnection X-winds arise from a different ejection mechanism: ejected disk material is powered by the protostellar rotational energy. This is the reason why they can brake down very efficiently the protostar, on time scales shorter or comparable to the duration of the embedded phase (Ferreira et al., 2000).

It took almost 30 yr to *prove* that accretion disks can indeed launch jets that carry away a sizeable fraction of their angular momentum. Investigations of the magnetospheric interaction between a protostar and its circumstellar disk are just beginning and the situation is far more complex than for disks.

References

Appl, S., Lery, T. and Baty, H.: 2000, *A&A* **355**, 818.
Balbus, A.S. and Hawley, J.F.: 1991, *ApJ* **376**, 214.
Baty, H. and Keppens, R.: 2003, *A&A* **397**, 913.
Blandford, R.D. and Payne, D.G.: 1982, *MNRAS* **199**, 883.
Casse, F. and Ferreira, J.: 2000a, *A&A* **353**, 1115.
Casse, F. and Ferreira, J.: 2000b, *A&A* **361**, 1178.
Casse, F. and Keppens, R.: 2002, *ApJ* **581**, 988.
Cao, X. and Spruit, H.C.: 2002, *A&A* **385**, 289.
Cerqueira, A.H. and de Gouveia Dal Pino, E.M.: 2001, *ApJ* **560**, 779.
Contopoulos, J. and Lovelace, R.V.E.: 1994, *ApJ* **429**, 139.
DeCampli, W.M.: 1981, *ApJ* **244**, 124.
Fendt, C. and Camenzind, M.: 1996, *A&A* **313**, 591.

Ferreira, J.: 1997, *A&A* **319**, 340.
Ferreira, J.: 2002, in: J. Bouvier and J.-P. Zahn (eds.), *EAS Publ. Series vol. 3* EDP sciences.
Ferreira, J. and Casse, F.: 2004, *ApJ* **601**, L139.
Ferreira, J. and Pelletier, G.: 1993, *A&A* **276**, 625.
Ferreira, J. and Pelletier, G.: 1995, *A&A* **295**, 807.
Ferreira, J., Pelletier, G. and Appl, S.: 2000, *MNRAS* **312**, 387.
Frank, A., Lery, T., Gardiner, T.A., Jones, T.W. and Ryu, D.: 2000, *ApJ* **540**, 342.
Gomez de Castro, A.I. and Verdugo, E.: 2001, *ApJ* **548**, 976.
Hartigan, P., Edwards, S. and Ghandour, L.: 1995, *ApJ* **452**, 736.
Kato, S.X., Kudoh, T. and Shibata, K.: 2002, *ApJ* **565**, 1035.
Konigl, A. and Wardle, M.: 1996, *MNRAS* **279**, L61.
Krasnopolsky, R., Li, Z.-Y. and Blandford, R.: 1999, *ApJ* **526**, 631.
Kwan, J.: 1997, *ApJ* **489**, 284.
Lery, T., Baty, H. and Appl, S.: 2000, *A&A* **355**, 1201.
Lery, T., Heyvaerts, J., Appl, S. and Norman, C.A.: 1999, *A&A* **347**, 1055.
Li, Z.-Y.: 1995, *ApJ* **444**, 848.
Lubow, S.H., Papaloizou, J.C.B. and Pringle, J.E.: 1994, *MNRAS* **268**, 1010.
Matt, S., Goodson, A.P., Winglee, R.M. and Bohm, K.-H.: 2002, *ApJ* **574**, 232.
Miller, K.A. and Stone, J.M.: 2000, *ApJ* **534**, 398.
Okamoto, I.: 2001, *MNRAS* **327**, 55.
Ostriker, E.: 1997, *ApJ* **486**, 291.
Ouyed, R., Clarke, D.A. and Pudritz, R.E.: 2003, *ApJ* **582**, 292.
Ouyed, R. and Pudritz, R.E.: 1997, *ApJ* **482**, 712.
Ouyed, R. and Pudritz, R.E.: 1999, *MNRAS* **309**, 233.
Pelletier, G. and Pudritz, R.E.: 1992, *ApJ* **394**, 117.
Ray, T., Mundt, R., Dyson, J., Falle, S. and Raga, A.: 1996, *ApJ* **468**, L103.
Sauty, C., Trussoni, E. and Tsinganos, K.: 2002, *A&A* **389**, 1068.
Shang, H., Glassgold, A.E., Shu, F.H. and Lizano, S.: 2002, *ApJ* **564**, 853.
Shu, F., Najita, J., Ostriker, E., Wilkin, F., Ruden, S. and Lizano, S.: 1994, *ApJ* **429**, 781.
Stone, J.M. and Hardee, P.E.: 2000, *ApJ* **540**, 192.
Takami, M., Chysostomou, A., Bailey, J., Gledhill, T.M., Tamura, M. and Terada, H.: 2002, *ApJ* **568**, L53.
Uchida, Y., Shibata, K.: 1985, *PASJ* **37**, 515.
Ustyugova, G.V., Koldoba, A.V., Romanova, M.M., Chechetkin, V. M. and Lovelace, R.V.E.: 1999, *ApJ* **516**, 221.
Vlahakis, N., Tsinganos, K., Sauty, C. and Trussoni, E.: 2000, *MNRAS* **318**, 417.
Wardle, M. and Konigl, A.: 1993, *ApJ* **410**, 218.
Weber, E.J. and Davis, L.: 1967, *ApJ* **148**, 217.

OUTFLOWS FROM DYNAMO-ACTIVE PROTOSTELLAR ACCRETION DISCS

BRIGITTA VON REKOWSKI[1], AXEL BRANDENBURG[2], WOLFGANG DOBLER[3] and ANVAR SHUKUROV[4]

[1]*Department of Astronomy and Space Physics, Uppsala University, Uppsala, Sweden; E-mail: Brigitta.vonRekowski@astro.uu.se*
[2]*NORDITA, Copenhagen, Denmark*
[3]*Kiepenheuer-Institut für Sonnenphysik, Freiburg, Germany*
[4]*School of Mathematics and Statistics, University of Newcastle, Newcastle, UK*

Abstract. An axisymmetric model of a cool, dynamo-active accretion disc is applied to protostellar discs. Thermally and magnetically driven outflows develop that are not collimated within 0.1 AU. In the presence of a central magnetic field from the protostar, accretion onto the protostar is highly episodic, which is in agreement with earlier work.

Keywords: protostellar discs, outflows, magnetospheric accretion

1. Introduction

Outflows from protostellar discs have been observed in many star-forming regions. These outflows are generally believed to be driven by magneto-centrifugal acceleration from the accretion disc around the protostar. Yet, most theoretical approaches to understanding stellar outflows have ignored or at least greatly oversimplified both accretion disc physics and magnetic field evolution. Instead, most effort so far has focused around the physics of magnetised outflows and the collimation process assuming a fixed magnetic field. One of the perhaps most convincing simulations of the collimation of outflows have come from Ouyed et al. (1997) and Ouyed and Pudritz (1997) who considered a disc corona with an imposed field aligned with the rotation axis of the disc and modelled the disc physics as a boundary condition.

In recent years several groups have begun to approach the problem from different directions. One approach is to start from accretion disc physics and to extend the simulation domain in the vertical direction, so as to capture parts of the outflow acceleration region. The global simulations of Hawley (2000) are among the most impressive attempts to treat the problem as a whole. Here the magnetic field is generated by the disc dynamo driven by the magneto-rotational instability and some form of outflow is indeed seen in the vicinity of the rotation axis. However, the dynamo still seems to be unable to produce a large scale poloidal field (which seems to be required for driving systematic outflows), and thermodynamics is not included to allow the disc to cool sufficiently to become geometrically thin. Other groups have focused on the dynamo aspect and have used thin disc theory as input to

Astrophysics and Space Science **292:** 493–500, 2004.

their models (Reyes-Ruiz and Stepinski, 1995) or have studied the effect of vertical outflows on the disc dynamo (Bardou et al., 2001).

In the present paper we review the approach of von Rekowski et al. (2003) which is conceptually closest to the dynamo approach mentioned above. The effect of the Lorentz force on the gas motions is taken into account. For a thin accretion disc, it is required that near-Keplerian shear velocity be the result of a balance mainly between centrifugal force and gravity. This is only possible if the disc is cool, and remains cool notwithstanding the adiabatic heating associated with the concentration of gas near the midplane by the vertical component of gravity during the disc formation.

In order to stay conceptually as close as possible to the polytropic setup of Ouyed and Pudritz (1997) we adopt a bi-polytropic model with low entropy in the disc and high entropy in the exterior which we refer to as halo or corona (see the next section).

Since we restrict ourselves to two-dimensional geometry, we model the disc dynamo using mean-field ($\alpha\Omega$ dynamo) theory. The applicability of this theory to protostellar accretion discs, where the turbulence (and the dynamo) is driven by the magneto-rotational (or Balbus-Hawley) instability, will also be discussed in the next section.

2. Modelling a Cool Dynamo-Active Accretion Disc

Like Ouyed and Pudritz (1997) we start with an equilibrium corona, where we assume constant entropy and hydrostatic equilibrium according to $c_p T(r) = GM_*/r$. (Here, T is temperature, G is the gravitational constant, M_* is the mass of the protostar, r is the spherical radius, and c_p is the specific heat at constant gas pressure.) In order to make the disc cooler, we prescribe a geometrical region for the disc (see Figure 1) and choose an entropy contrast between disc and corona such that the initial disc temperature is about 3×10^3 K in the bulk of the disc. Given the degree of simplification of our model, we have refrained from modelling the flaring

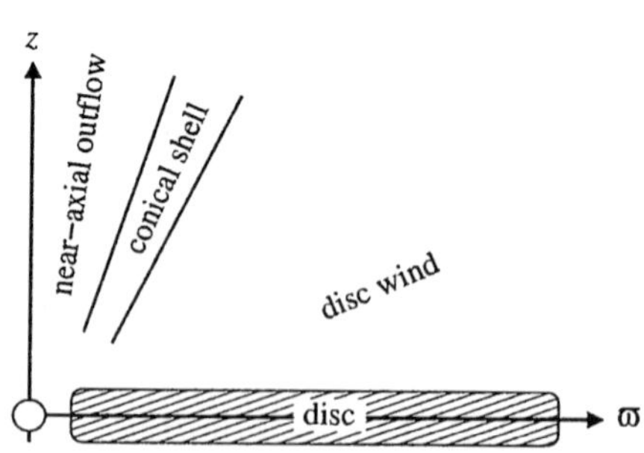

Figure 1. General structure of the outflows typically obtained in our model. The cool, dense disc emits (a) a thermally driven wind (slow, hot, dense, magnetised, rotating), (b) a magneto-centrifugally driven outflow in a conical shell (faster, cooler, less dense, magnetised, quickly rotating), and (c) a thermally driven outflow near the axis (slow, hot, dense, weakly magnetised and weakly rotating).

of the disc. The low disc temperature corresponds to a high disc density of about $10^{-10} \ldots 10^{-9}\,\mathrm{g\,cm^{-3}}$.

Hydrostatic equilibrium requires $(\boldsymbol{u} \cdot \boldsymbol{\nabla})\boldsymbol{u} = -(1/\rho)\boldsymbol{\nabla} p - \boldsymbol{\nabla}\Phi$, where $\boldsymbol{u} = (0, u_\varphi, 0)$ is the velocity field, p is gas pressure, ρ is gas density, and Φ is the gravitational potential. Since specific entropy s is given, it is convenient to express $(1/\rho)\boldsymbol{\nabla} p$ in terms of s and specific enthalpy h, using the relation $\boldsymbol{\nabla} h = (1/\rho)\boldsymbol{\nabla} p + T\boldsymbol{\nabla} s$ (first law of thermodynamics) and $h = c_p T$. We obtain h by integrating the vertical balance equation,

$$-\partial(h + \Phi)/\partial z + h\,\partial(s/c_p)/\partial z = 0, \tag{1}$$

from large z (where $h + \Phi = 0$) down to $z = 0$. The initial rotation velocity, $u_{\varphi 0}$, follows from the radial balance equation,

$$-u_{\varphi 0}^2/\varpi = -\partial(h + \Phi)/\partial\varpi + h\,\partial(s/c_p)/\partial\varpi. \tag{2}$$

In the disc, $h = c_p T$ is small, so $u_{\varphi 0}$ is slightly sub-Keplerian, while the corona does not rotate initially, and is supported by the gas-pressure gradient. Within the disc, we adjust density with a forcing term, so that a similar disc structure is maintained on reasonable time scales.

For the disc dynamo, the most important parameter is the dynamo coefficient α in the mean-field induction equation. The α effect is antisymmetric about the midplane and restricted to the disc. There is evidence that in accretion discs, where turbulence is driven by the magneto-rotational instability, α is negative in the upper disc half (Brandenburg et al., 1995; Ziegler and Rüdiger, 2000). A simple theoretical explanation is given in Brandenburg (1998), and a similar result has been obtained by Rüdiger and Pipin (2000) who considered magnetically driven turbulence; see also Rüdiger et al. (2001).

Is there any evidence that the $\alpha\Omega$ dynamo picture is correct and applicable to accretion discs? One piece of evidence is that the specific dependence of the large scale field geometry on the boundary conditions is the same both in simulations of shearing sheet turbulence driven by the magneto-rotational instability and in $\alpha\Omega$ dynamo models in the same geometry. Since the dynamo-generated magnetic field is of large scale and since it extends to the boundaries, one must expect a detailed dependence of its geometry on the boundary conditions. In discs with negative α in the upper half, with (pseudo-)vacuum boundary conditions one gets an oscillatory large scale field whose horizontal components are symmetric about the midplane. With perfect-conductor or finite-conductivity boundary conditions, one gets a non-oscillatory field whose horizontal components are antisymmetric about the midplane. The *same* is seen both in three-dimensional simulations and in solutions of the $\alpha\Omega$ dynamo problem; see Brandenburg (1998).

Bardou et al. (2001) found that vertical outflows do not change the symmetry or temporal behaviour of the dynamo-generated magnetic field. However, especially in the linear case for negative α in the upper disc half and finite-conductivity in the halo, an outflow tends to switch the disc dynamo off. This happens when the time scale of field stretching by the outflow becomes shorter than the regeneration time scale of the dynamo.

3. Structured Outflow and Acceleration Mechanism

Figure 2, left panel, illustrates that an outflow develops that has a well-pronounced structure. Within a conical shell originating from the inner edge of the disc, the terminal outflow speed exceeds $500\,\mathrm{km\,s^{-1}}$, and temperature and density are lower than elsewhere. The inner cone around the axis is the hottest and densest region where the stellar wind speed reaches about $150\,\mathrm{km\,s^{-1}}$. The wind that develops from the outer parts of the disc has intermediate values of the speed (see Figure 1).

The structured outflow is driven by a combination of different processes. A significant amount of angular momentum is transported outwards from the disc into the wind along magnetic field lines, especially along the strong lines within the conical shell (see Figure 2, right panel). The magnetic field geometry is such that the angle between the rotation axis and the field lines threading the disc exceeds

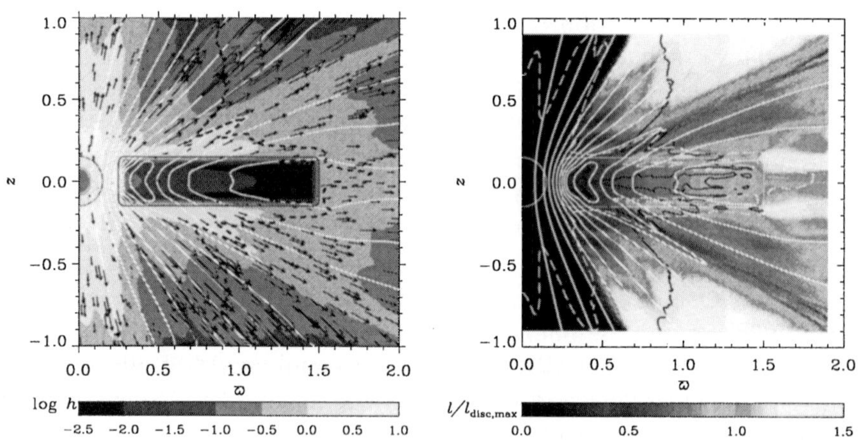

Figure 2. Left: poloidal velocity vectors and poloidal magnetic field lines (white) superimposed on a colour scale representation of $\log_{10} h$. Specific enthalpy h is directly proportional to temperature T, and $\log_{10} h = (-2, -1, 0, 1)$ corresponds to $T \approx (3\times10^3, 3\times10^4, 3\times10^5, 3\times10^6)\,\mathrm{K}$. The black dashed line shows the fast magnetosonic surface. The disc boundary is shown with a thin black line, the stellar surface is marked in red. The dynamo α coefficient is negative in the upper disc half, resulting in roughly dipolar magnetic symmetry. Averaged over times $t \approx 897\ldots906$ days. Right: colour scale representation of the specific angular momentum, normalised by the maximum angular momentum in the disc, with poloidal magnetic field lines superimposed (white). The black solid line shows the Alfvén surface, the white dashed line the sonic surface. Same model as left and averaged over same times.

30° at the disc surface, which is favourable for magneto-centrifugal acceleration (Blandford and Payne, 1982). However, the Alfvén surface is so close to the disc surface at the outer parts of the disc that acceleration there is mainly due to the gas-pressure force. In the conical shell, however, the outflow is highly supersonic but sub-Alfvénic, with the Alfvén radius a few times larger than the radius at the footpoint of the field lines at the disc surface. The lever arm of about 3 is sufficient for magneto-centrifugal acceleration to dominate in the conical shell (cf. Krasnopolsky et al., 1999). The inner cone has very low angular momentum and the magnetic lines are inclined to the axis by less than 30° so that the stellar wind can only be gas-pressure driven. This picture is confirmed by looking at the ratio of magneto-centrifugal to gas-pressure forces, which is much larger than unity in the conical shell, where also the poloidal field lines are strongest. This is also the region where magnetic pressure produced by the toroidal field exceeds gas pressure, indicating that the magneto-centrifugally accelerated outflow in the conical shell is confined by the toroidal field.

4. Magnetic Coupling Between Star and Disc

Observations of T Tauri stars suggest that the stellar spin-down can be connected with a circumstellar accretion disc. Magnetic fields are believed to play an important rôle in the interaction between the star and its surrounding disc, resulting in particular in a spin-down of the star due to magnetic braking. Indeed, magnetic fields as strong as 10^3 Gauss have been detected on T Tauri stars, and there is evidence for hot and cool spots on the stellar surface. Although the structure of the stellar magnetosphere is yet unknown, analytical models have been developed by Hartmann et al. (1994) and Shu et al. (1994) that assume that a stellar dipole threads a surrounding disc. In these static magnetospheric accretion models, disc matter is channelled along magnetospheric field lines to form a funnel-shaped polar accretion flow.

More realistic time dependent numerical simulations by Hirose et al. (1997) and Goodson et al. (1997) show episodic reconnection behaviour between a stellar dipolar magnetosphere and an imposed disc field, resulting in episodic mass transfer from the disc to the star. These authors find inner stellar jets and outer disc winds that are both driven by magneto-centrifugal processes.

Figure 3 shows our model with disc dynamo, where we add a stellar dipole that is anchored within the star, but can evolve with time outside the star (von Rekowski and Brandenburg, 2003). For the strong surface stellar field of about 6 kG considered here, the time dependent accretion is correlated with the magnetic star–disc coupling. When the star is connected to the disc, accretion is along field lines (see Figure 3, left panel). When the star is disconnected from the disc, an enhanced wind carries away excess angular momentum (see Figure 3, right panel). The sum of magnetic, net accretion and viscous torques on the star is positive at radii between star and disc, which means a stellar spin-up. Spin-down torques only

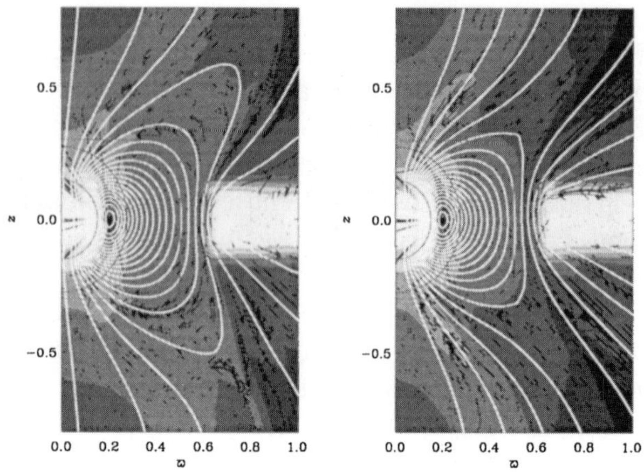

Figure 3. Colour scale representation of the density (bright colours indicate high values; dark colours, low values) with poloidal magnetic field lines superimposed (white) and the azimuthally integrated mass flux density shown with arrows (not shown in the disc). Left: at a time $t \approx 150$ days, when the star is magnetically connected to the disc. Right: at a time $t \approx 156$ days, when the star is disconnected from the disc.

result at latitudes where accretion takes place, because the angular velocity is very small between star and disc.

5. Concluding Remarks

Our simulations show that a system consisting of a magnetically coupled star and dynamo-active accretion disc can maintain a clearly structured outflow. The outflow consists of a (hotter and denser) slow stellar wind and a faster outer disc wind that are both gas-pressure driven, and a (cooler and less dense) fast magneto-centrifugally accelerated inner disc wind within a conical shell (see Figure 4, left panels). In addition, in transition periods, when the magnetic star–disc connectivity changes, there is a gas-pressure driven (small azimuthal velocity and angular momentum, the Alfvén surface close to the disc), hot and dense, but relatively fast outflow between the stellar and disc winds.

Coexisting gas-pressure and magneto-centrifugal driving mechanisms have also been found by Koide et al. (1998) who study the formation of relativistic jets from accretion discs surrounding a black hole by performing numerical simulations using a general relativistic magnetohydrodynamic code. In their model, a global (initially vertical) magnetic field penetrates the accretion disc which then leads to non-steady accretion and ejection processes. According to their simulations, a jet is ejected from the close vicinity of the black hole. The jet has a two-layered shell structure consisting of a fast gas-pressure driven jet in the inner part and a slow magnetically driven jet in the outer part, both of which are collimated by the global poloidal magnetic field penetrating the disc (see Figure 4, right panels). The fast jet is due

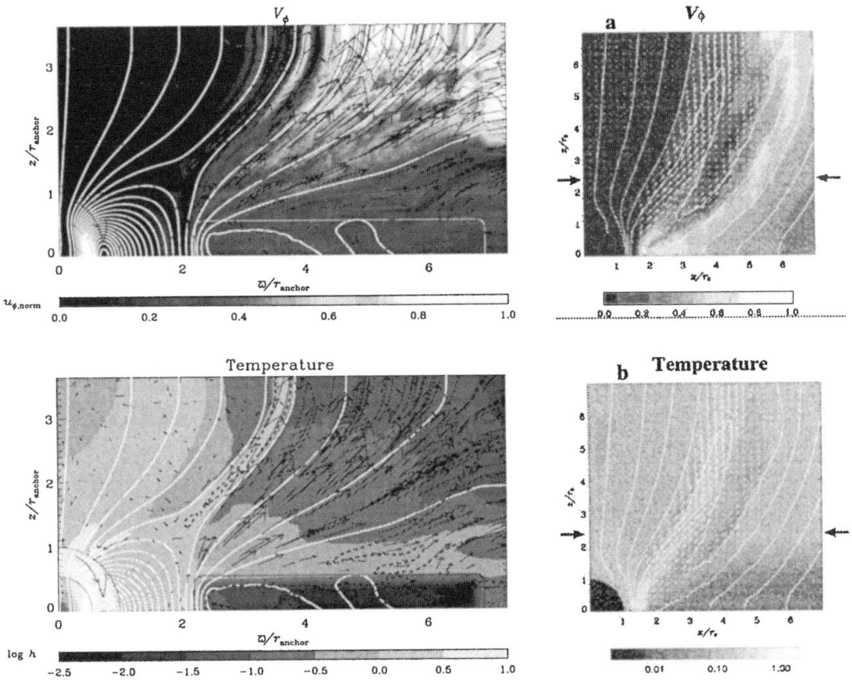

Figure 4. Colour scale representation of the normalised azimuthal velocity (top) and temperature (bottom) with poloidal magnetic field lines (white) and poloidal velocity vectors superimposed. Left: simulations of outflows from protostellar accretion discs by von Rekowski and Brandenburg (2003). Same model as in Figure 3, but at $t \approx 158$ days. Collimation is due to boundary conditions. Right: simulations of outflows from accretion discs around black holes by Koide et al. (1998).

to a strong gas-pressure increase due to shock formation at the inner edge of the disc, caused by a fast accretion flow.

Our relatively fast gas-pressure driven flow might be comparable with the fast jet. However, contrary to the slow jet, our magneto-centrifugally accelerated inner disc wind is even faster.

Acknowledgements

This work was supported by the PPARC Grant PPA/G/S/2000/00528. Use of the supercomputer SGI 3800 in Linköping and of the PPARC supported supercomputers in St Andrews and Leicester is acknowledged. This research was conducted using the resources of High Performance Computing Center North (HPC2N).

References

Bardou, A., von Rekowski, B., Dobler, W., Brandenburg, A. and Shukurov, A.: 2001, *A&A* **370**, 635.
Blandford, R.D. and Payne, D.G.: 1982, *MNRAS* **199**, 883.

Brandenburg, A.: 1998, in: M.A. Abramowicz, G. Björnsson and J.E. Pringle (eds.), *Theory of Black Hole Accretion Discs*, Cambridge University Press, Cambridge, p. 61.
Brandenburg, A., Nordlund, Å., Stein, R.F. and Torkelsson, U.: 1995, *ApJ* **446**, 741.
Goodson, A.P., Winglee, R.M. and Böhm, K.-H.: 1997, *ApJ* **489**, 199.
Hartmann, L., Hewett, R. and Calvet, N.: 1994, *ApJ* **426**, 669.
Hawley, J.F.: 2000, *ApJ* **528**, 462.
Hirose, S., Uchida, Y., Shibata, K. and Matsumoto, R.: 1997, *Publ. Astron. Soc. Japan* **49**, 193.
Koide, S., Shibata, K. and Kudoh, T.: 1998, *ApJ* **495**, L63.
Krasnopolsky, R., Li, Z.-Y. and Blandford, R.: 1999, *ApJ* **526**, 631.
Ouyed, R. and Pudritz, R.E.: 1997, *ApJ* **482**, 712.
Ouyed, R., Pudritz, R.E. and Stone, J.M.: 1997, *Nat* **385**, 409.
Reyes-Ruiz, M. and Stepinski, T.F.: 1995, *ApJ* **438**, 750.
Rüdiger, G. and Pipin, V.V.: 2000, *A&A* **362**, 756.
Rüdiger, G., Pipin, V.V. and Belvedère, G.: 2001, *Sol. Phys.* **198**, 241.
Shu, F., Najita, J., Ostriker, E., Wilkin, F., Ruden, S. and Lizano, S.: 1994, *ApJ* **429**, 781.
von Rekowski, B. and Brandenburg, A.: 2004, *A&A* **420**, 17.
von Rekowski, B., Brandenburg, A., Dobler, W. and Shukurov, A.: 2003, *A&A* **398**, 825.
Ziegler, U. and Rüdiger, G.: 2000, *A&A* **356**, 1141.

H_2 DIAGNOSTICS OF MAGNETIC MOLECULAR SHOCKS IN BIPOLAR OUTFLOWS

S. CABRIT[1], D.R. FLOWER[2], G. PINEAU DES FORÊTS[3,4], J. LE BOURLOT[4] and C. CECCARELLI[5]

[1]*LERMA, Observatoire de Paris, Paris, France; E-mail: sylvie.cabrit@obspm.fr*

[2]*Physics Department, The University, Durham, UK*

[3]*IAS, Université de Paris-Sud, Orsay, France*

[4]*LUTH, Observatoire de Paris, Meudon Cedex, France*

[5]*Laboratoire d'Astrophysique, Observatoire de Grenoble, Grenoble Cedex, France*

Abstract. The shock waves associated with molecular outflows may be of continuous (C) type or jump (J) type, depending on conditions in the preshock gas, notably the magnetic field strength and the degree of ionisation. Intermediate situations also exist, in which a J-discontinuity terminates or is embedded in a C-type flow. We show that proper allowance for the departure of the chemistry from equilibrium (particularly the dissociation/reformation of H_2) and for the inertia of charged dust grains, is crucial for an accurate treatment of the C to J transition. We illustrate the use of H_2 population diagrams and H_2 line profiles, in conjunction with our detailed shock model, to constrain conditions in shocks propagating in molecular outflows. We show that H_2 pure rotational lines yield evidence for C-type precursors in bipolar outflows from young stars, with transverse magnetic field strengths B (μG) $\simeq$ 1–10 $\times \sqrt{n_H/\text{cm}^{-3}}$ similar to those inferred from Zeeman splitting and from the dispersion of dust polarization vectors in dense clouds.

1. Introduction: Magnetic Fields and the Structure of Molecular Shock Waves

The structure of shock waves propagating into shielded molecular media of low ionisation depends crucially on the initial transverse magnetic induction, B_0. Figure 1 compares the thermal and velocity structure of one dimensional, stationary shocks calculated with an initial transverse magnetic induction $B_0 = 0$ (J-type shock) and $B_0 = 100\,\mu$G (Continuous-type shock). In the Jump (J)-type shock ($B_0 = 0$), the gas undergoes a sharp increase in temperature and a sharp drop in velocity (over a few mean free paths), followed by rapid cooling and compression. In the Continuous (C)-type shock wave ($B_0 = 100\,\mu$G), the ionised and neutral fluids decouple dynamically: the ions, attached to the field lines, are slowed down first and then progressively bring the neutrals to rest by collisions (thus also heating the gas). The heated region is thus much broader, and the maximum temperature that is attained lower, than in the J-type shock. The time required to traverse the shock is approximately 250 yr when $B = 0$, but is about 2000 yr when $B_0 = 100\,\mu$G.

In principle, determining which type of shock, J or C, is present in outflows from young stars could thus offer an important means to constrain the transverse

Astrophysics and Space Science **292:** 501–508, 2004.

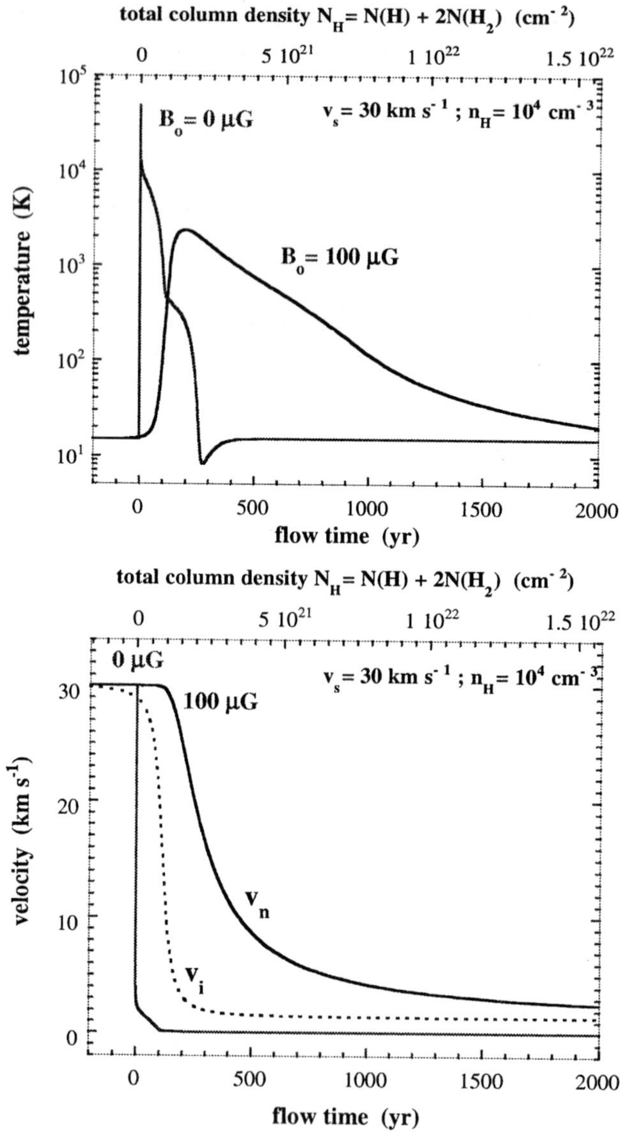

Figure 1. A comparison of the thermal and velocity profiles of a J-type ($B_0 = 0$) and a C-type ($B_0 = 100\,\mu$G) shock, for a shock speed $v_s = 30$ km s^{-1} and a preshock molecular gas of density $n_H = n(H) + 2n(H_2) = 10^4$ cm^{-3}. Note the decoupling between the ion and neutral fluid velocities, v_i and v_n, in the C-type shock. Velocities are expressed in the shock frame.

magnetic field strength (perpendicular to the shock propagation direction) in the preshock gas, independently of other techniques (Zeeman splitting, polarisation measurements, cf. Crutcher, this volume).

In Section 2, we summarize recent calculations of the critical shock speed for the transition from a C- to a J-type shock, as a function of the preshock magnetic field strength, using a detailed 1D shock model including non-equilibrium chemistry and ionisation. In Section 3, we illustrate the use of H_2 and CO infrared lines as diagnostics of C versus J shocks, using Cep A West as an example, and present H_2 line profile predictions.

2. The Transition from C- to J-Type Shocks

For fixed B_0, the transition from a C- to a J-type shock occurs when either (1) the neutral fluid becomes subsonic in the shock frame, due to dissociation of H_2 (the major gas coolant), or (2) the shock speed comes to exceed the magnetosonic speed of the charged fluid, $V_{\text{magneto}} \simeq B/\sqrt{4\pi\rho_c}$.

We have determined the shock speed at which these two phenomena occur using a detailed 1D stationary C-shock model that involves various improvements over previous calculations (e.g. Kaufman and Neufeld, 1996), including: a full non-equilibrium chemical network with 124 species, 895 reactions, allowing to follow the ionisation/recombination of molecules and grains along the shock, and the erosion of refractory dust cores and organic grain mantles by neutral impact (May et al., 2000); the feed-back of the ionisation state of the gas on the shock structure; a level by level non-stationary treatment of H_2 excitation and dissociation by electrons, streaming ions, H, H_2, and grains (see Pineau des Forêts et al., 2001; Le Bourlot et al., 2002).

It was previously believed, based on equilibrium chemistry, that the maximum speed for C-shocks allowed by H_2 dissociation was $\simeq 40$ km s^{-1}(e.g. Draine et al., 1983). Our more detailed calculations have shown that the actual speed limit can be much higher at moderate densities, due to the finite time required for dissociation (Le Bourlot et al., 2002). The maximum C-shock speed set by H_2 dissociation is plotted by a dashed curve in Figure 2 as a function of the preshock magnetic field strength for $n_H = 10^4$ cm^{-3}. It increases from 70 km s^{-1} for $B_0 = 100\,\mu$G to 110 km s^{-1} for $B_0 = 1000\,\mu$G.

The second limit set, by the magnetosonic speed, depends crucially on the abundance of charged dust grains, which dominate the mass density of the charged fluid, ρ_c. The charge of dust grains depends strongly not only on the cosmic ray ionisation rate, but also on the abundance of PAHs, which tend to capture most of the free electrons (Flower and Pineau des Forêts, 2003). The blue and red curves in Figure 2 plot V_{magneto} as a function of B_0 for PAH abundances of 10^{-6} and 10^{-8}, respectively. It can be seen that the transition from C- to J-type shock with increasing magnetic field. Thus, observing a C-shock of speed $\geq$30 km s^{-1} in a medium of preshock density $n_H \simeq 10^4$ cm^{-3} would constrain the magnetic field scaling parameter b to be $\geq$0.5, where b is defined by $(B/\mu G) = b \times \sqrt{n_H/\text{cm}^{-3}}$. It follows that the crucial discriminant is between C- and J-type shocks, and it is to this issue that we now turn.

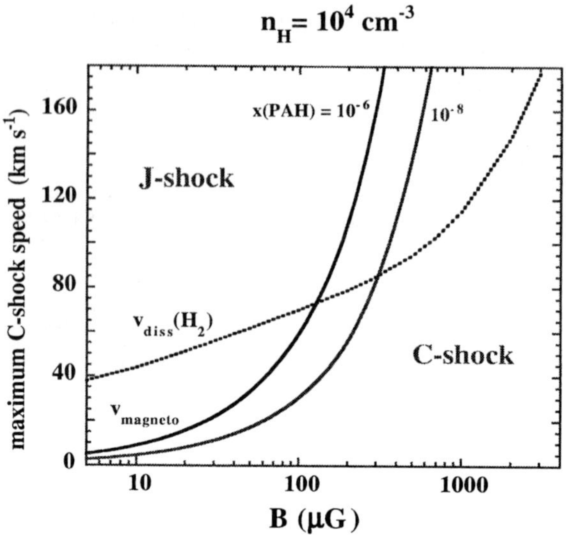

Figure 2. Maximum speed for a C-shock as a function of transverse magnetic field, as determined by H_2 dissociation (dotted curve) and by the magnetosonic speed in the ionised fluid, including the inertia of charged dust grains (blue and red curves).

3. H_2 Excitation As Diagnostics of C- Versus J-Type Shocks

For a given set of shock parameters, the spectrum of molecular hydrogen is very different in a J and a C shock. An important diagnostic tool in this context is the H_2 excitation diagram, which plots the logarithm of the column densities of the H_2 rotational levels, divided by the statistical weights, against their excitation energies. For a Boltzman population, the slope in this diagram is inversely proportional to the excitation temperature.

Examples of such diagrams for a J-type ($b = 0$) and a C-type shock ($b = 1$) of speed $v_s = 25$ km s^{-1} are shown in the top panel of Figure 3. The C-type shock produces a large column of warm gas in the $v = 0$ levels (circular symbols), while the J-type shock produces a large amount of hot gas in levels lying above 10 000 K and little warm H_2, due to fast cooling. These J-shock characteristics do not depend strongly on the shock speed or assumed H_2 reformation state (Flower et al., 2003).

The advantage of H_2 over other tracers of C-shocks such as SiO, CH_3OH, H_2CO, H_2O, or OH (e.g. Schilke et al., 1997; Giannini et al., 2001) is that it is much less dependent on the adopted initial elemental gas phase abundances, grain mantle composition and chemical network. There is a growing body of available observational data on H_2 emission in molecular outflows, for both pure rotational lines and ro-vibrational lines (ISO satellite, near-IR ground-based spectroscopy in the J, H and K bands).

As an example, the bottom panel of Figure 3 shows the population diagram observed in the Cep A West molecular outflow with ISOSWS (Froebrich et al., 2002).

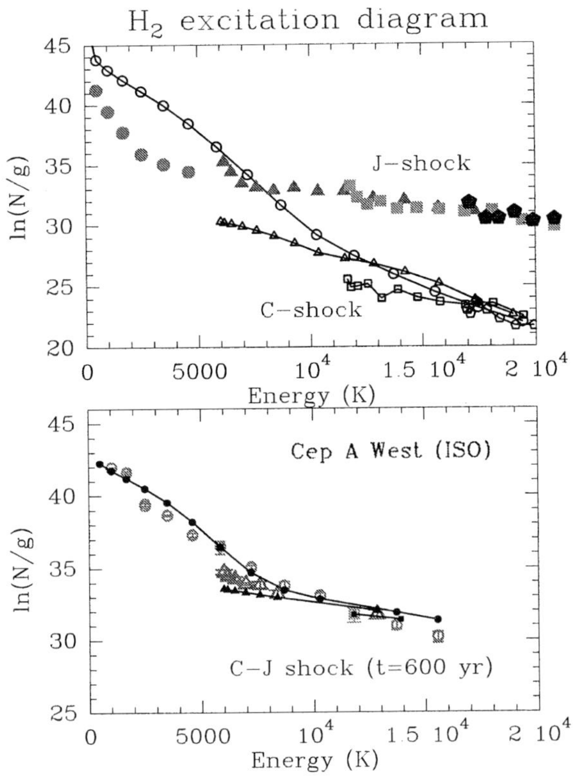

Figure 3. *Top*: Comparison of the H_2 excitation diagrams predicted for a J-type ($b = 0$; color) and a C-type shock ($b = 1$; black). $v = 0$ levels are denoted by circles, $v = 1$ by triangles, and $v = 2$ by squares. *Bottom*: comparison of the predictions for a non-stationary C+J shock with $b = 1$ and filling-factor of 1 (black; see Figure 4) with ISOSWS observations of Cep A West (color; Froebrich et al., 2002)).

Comparing with the top panel of Figure 3, it can be seen that, while levels above 7000 K are compatible with a J-shock, the $v = 0$ levels below 7000 K can only be fitted by a C-shock. Such mixed C and J characteristics are in fact expected in a C-shock wave that has not had time to reach its stationary state, i.e. if the shock age is shorter than the time required to flow through the stationary shock ($\simeq 10^4$ yr at 10^4 cm^{-3}): the C-type precursor is then truncated and followed by a J-type discontinuity (Chièze et al., 1998). An example of the resulting mixed C+J structure for $b = 1$ is shown in Figure 4. The predicted H_2 population diagram is plotted in the bottom panel of Figure 3 and reproduces very well the Cep A observations (Flower et al., 2003). Note that a minimum value of $b \simeq 1$ is required to reproduce the large column density in $v = 0$ levels in Cep A, with a filling-factor of 1. Higher b values with lower filling-factors are also possible.

As an independent test of this model, we compare in Figure 5 the high-J CO lines expected for this C+J shock, calculated with an escape probability approach, with

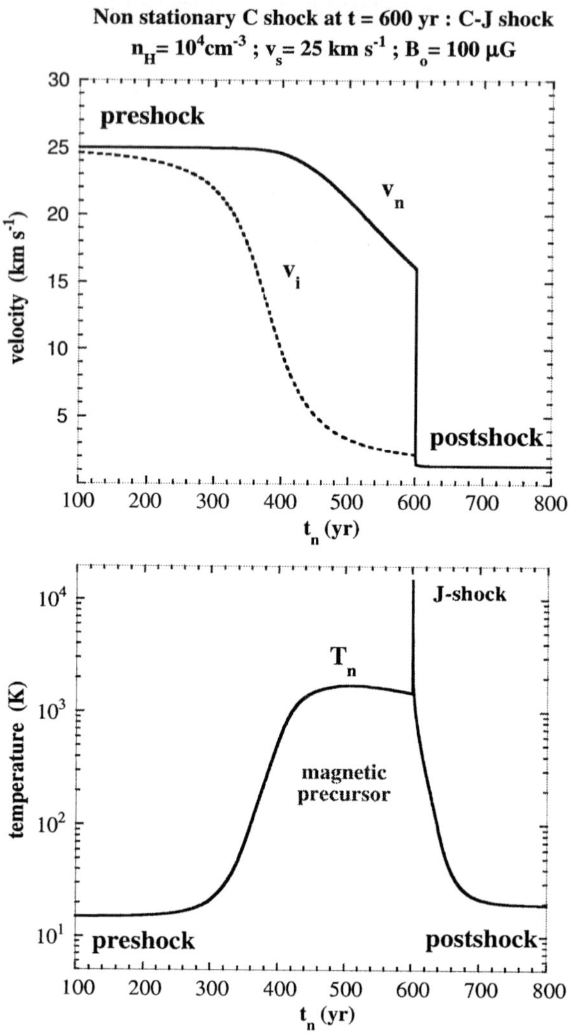

Figure 4. Non-stationary C-shock at an age $t = 600$ yr: The magnetic precursor is truncated by a J-type discontinuity where the neutral and ionic fluids abruptly recouple. Velocities are shown in the shock frame.

the observed line fluxes from ISO LWS (Froebrich et al., 2002). The agreement is quite good. In contrast, the same C-shock in its stationary state predicts too much flux in CO lines, due to the full extent of its magnetic precursor.

4. H_2 Line Profiles

Another diagnostic of J versus C-type shock is the shape of line profiles from hot molecular gas, in particular the $v = 1$–0 S(1) line of H_2. As shown in Figure 6,

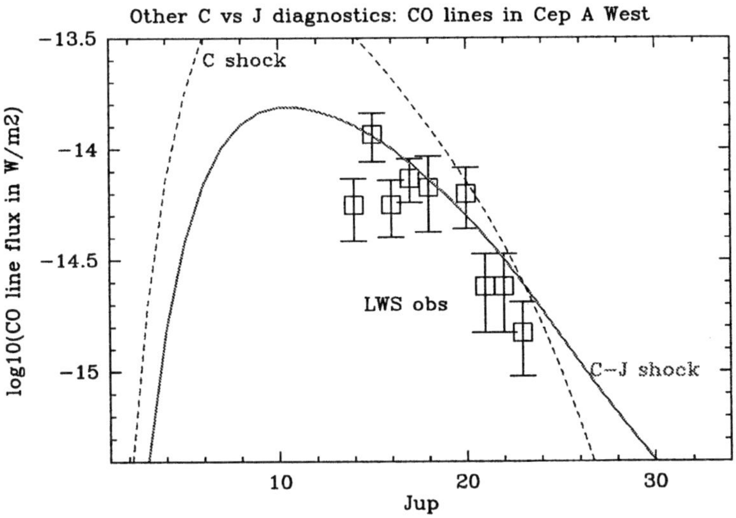

Figure 5. CO line fluxes observed with ISOLWS in Cep A West (squares; Froebrich et al., 2002) compared with predictions for a non-stationary C+J shock model that matches the H_2 population diagram (solid curve) and with the same C-shock in its stationary state (dashed curve). The C+J model shows better agreement.

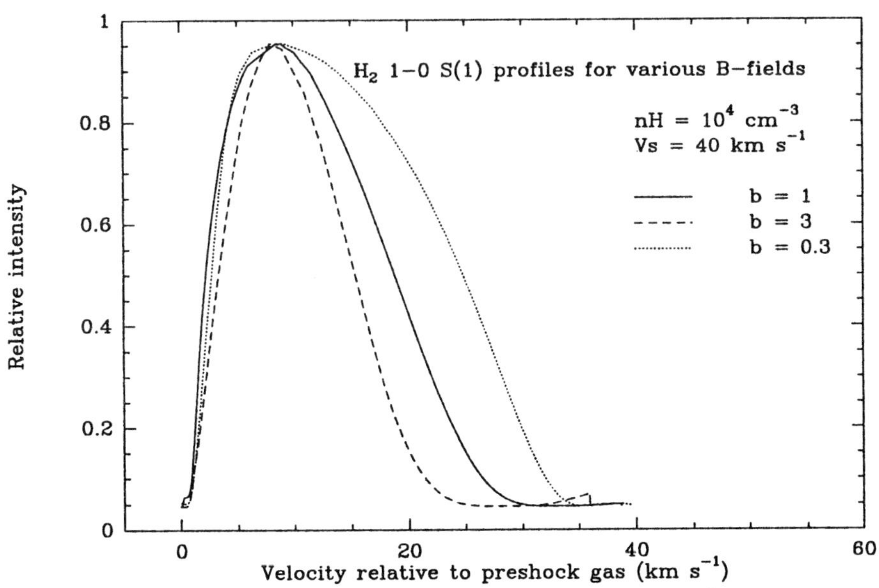

Figure 6. Predicted line profiles in the v = 1–0 S(1) line of H_2 for C-shocks of speed $v_s = 40$ km s^{-1}, a preshock density $n_H = 10^4$ cm^{-3}, and three values of the magnetic field parameter b, defined as $(B/\mu G) = b \times \sqrt{n_H/\mathrm{cm}^{-3}}$. The profile width increases with b but the peak velocity is unchanged. Velocities are in the preshock frame.

most of the emission from this line in a C-shock arises in the region of high ion-neutral drift speed, where the neutrals have not yet been fully accelerated by the shock wave. In the preshock frame, the peak emission is at 8 km s^{-1} for a 40 km s^{-1} shock. The line profile is broad, and its width increases with magnetic field strength. In contrast, acceleration in J-type shocks is almost instantaneous, so that H_2 line profiles would be much narrower and peaked close to the speed of the shock front, i.e. 40 km s^{-1} in this case.

While inclination to the line of sight and non-planar geometries (e.g. a bowshock) could obviously complicate the distinction between these two cases, it could be a powerful discriminant in well-studied regions where inclination and geometry are better constrained. We also note that the very broad line profiles observed in Orion Peak 1 could be reproduced by curved bow C-shocks with a magnetic field parameter $b \simeq 10$ (Chrisostomou et al., 1994), where H_2 can survive up to speeds $\simeq$100 km s^{-1} (Le Bourlot et al. 2000).

References

Chièze, J.P., Pineau des Forets, G. and Flower, D.R.: 1998, *MNRAS* **295**, 672.
Chrisostomou, A., Hough, J.H., Burton, M.G. and Tamura, M.: 1994, *MNRAS* **268**, 325.
Draine, B.T., Roberge, W.G. and Dalgarno, A.: 1983, *ApJ* **264**, 485.
Flower, D.R., Le Bourlot, J., Pineau des Forêts, G. and Cabrit, S.: 2003, *MNRAS* **341**, 70.
Flower, D.R. and Pineau des Forêts, G.: 2003, *MNRAS* **343**, 390.
Froebrich, D., Smith, M.D. and Eisloeffel, J.: 2002, *A&A* **385**, 239.
Giannini T., Nisini, B. and Lorenzetti, D.: 2001, *ApJ* **555**, 40.
Kaufman, M.J. and Neufeld, D.A.: 1996, *ApJ* **456**, 611.
Le Bourlot, J., Pineau des Forêts, G., Flower, D.R. and Cabrit, S.: 2002, *MNRAS* **332**, 985.
May, P.W., Pineau des Forêts, G., Flower, D.R., Field, D., Allan, N.L. and Purton, J.A.: 2000, *MNRAS* **318**, 809.
Pineau des Forêts, G., Flower, D.R., Aguillon, F., Sidis, V. and Sizun, M.: 2001, *MNRAS* **323**, L7.
Schilke, P., Walmsley, C.M., Pineau des Forets, G. and Flower, D.R.: 1997, *A&A* **321**, 293.

THE MAGNETIC ENVIRONMENTS OF YOUNG STELLAR OBJECTS*

ANTONIO CHRYSOSTOMOU, RACHEL CURRAN and DAVID AITKEN

Department of Physical Sciences, University of Hertfordshire, Hatfield, UK;
E-mail: acc@star.herts.ac.uk

Abstract. We report on early results of submillimetre polarimetric measurements towards a sample of young stellar objects. The results allow us to infer the magnetic field structure and show a variety of configurations, providing evidence for axial, helical and pinched (i.e. 'hourglass') magnetic field configurations. We find that in some cases the field is curved over large scales, implying that it is influenced by the gas kinematics in the local environment, and that at these scales at least, the magnetic field plays a passive rôle in the star formation process.

Keywords: star formation, magnetic fields, submillimetre, polarimetry

1. Introduction

Star formation is a field of research striving to understand the processes which convert the mass in a molecular cloud into stars. There has been much work over the last few decades to understand this phenomenon, from a theoretical standpoint supported by observational evidence. We find that our understanding is best improved upon when theory and observation advance together.

Recently, there have been a number of theoretical advances which attempt to understand the rôle that magnetic fields have in the formation of stars (witness, for instance, the number of reviews and contributions in these proceedings). However, the observational evidence in support of these theories has been lacking, and in this contribution we describe (in part) our research programme which attempts to redress the balance and help provide an observational basis to this issue.

1.1. The Importance of Magnetic Fields

The standard theory of star formation (Shu et al., 1987) has the magnetic field playing a fundamental rôle. It can regulate the collapse of material onto the protostar through the action of ambipolar diffusion (Mouschovias and Ciolek, 1999), and is instrumental in driving and collimating molecular outflows.

More recently, magnetic fields have been regarded as important to the evolution of molecular clouds through MHD turbulence. There is a critical ratio of

*The data presented here, apart from Figure 3, are all shown prior to full analysis and are currently being prepared for publication.

Astrophysics and Space Science **292:** 509–515, 2004.

mass to magnetic flux above which the magnetic field can no longer support a cloud from collapse and it is believed that low mass stars form in super-critical cores while high mass stars form in sub-critical cores. However, the observational evidence (see review by R. Crutcher in this volume) suggests that most cores are either at or beyond criticality while Nakano (1998) showed that it is difficult for sub-critical cloud cores to maintain non-thermal velocity widths. Support through turbulence is believed to be important in explaining this apparent lack of sub-critical cloud cores and the sustenance of non-thermal velocity widths.

1.2. Models of Magnetic Fields in Molecular Clouds

There have been a number of recent studies which have looked at the effect of the magnetic field in molecular clouds with particular regard to turbulence (Padoan et al., 2001; Heitsch et al., 2001; Ostriker et al., 2001). Importantly for observers, these simulations all produce maps of the magnetic field structure which can be compared to observations. The results generally show that relatively strong magnetic fields produce uniform field patterns, whereas weak fields produce more random, disorganised patterns. This behaviour is parameterised through the energy density (i.e. strength) of the field measured relative to the energy density in the gas, something which Ostriker et al. (2001) express using the sound speed and the Alfvén velocity, $\beta = c_s^2/v_A^2$.

Other significant models include those of Fiege and Pudritz (2000) and Aitken et al. (2002). Fiege and Pudritz (2000) looked at helical magnetic fields threading through molecular clouds in order to explain the magnetic field structures seen towards the Orion Molecular Clouds (Matthews et al., 2001). As well as providing a working model for filamentary clouds, they found that helical fields provide a simple, geometric explanation for the 'polarization holes' seen in single-dish polarimetry data. Aitken et al. produced working models with the geometry of a protostar (flared disk with a $r^{-1.5}$ radial density profile). By treating the radiative transfer properly they are able to determine the polarization vector pattern for a number of magnetic field configurations at a range of wavelengths between the mid-infrared and the submillimetre, providing the opportunity of constraining the models using a multi-wavelength approach.

1.3. Measuring Magnetic Field Geometries

The classical grain alignment theory has spinning and elongated dust grains which become partially aligned to the local magnetic field direction through paramagnetic relaxation (Davis and Greenstein, 1951). The grains' spin axes precess about the direction of the magnetic field and the radiation emitted by a volume of such grains is polarized. Polarimetry, therefore, allows for a unique determination of the direction and magnitude of the **E**-vector from which the **B**-vector may be inferred.

The observations presented here were all obtained using the polarimeter on the Submillimetre Common User Bolometer Array (SCUBA) on the James Clerk Maxwell Telescope (JCMT) on the summit of Mauna Kea, Hawaii. The millimetre/submillimetre wavelength regime is an ideal probe for star formation studies principally because the radiation is optically thin and naturally traces the densest structures where protostars form. Polarimetry data then provide a column-averaged (and intensity weighted) measurement of the magnetic field direction through the cloud and projected onto the plane of the sky.

2. Magnetic Fields and Outflows

2.1. HH 211 and NGC 1333 IRAS2

In Figure 1 is shown our SCUBA polarimetry of the Class 0 low-mass sources, HH 211 and NGC 1333 IRAS2. The polarization vectors have been rotated by 90° to depict the **B**-vector projected onto the plane of the sky.

The outflow direction for HH 211 lies at a position angle of ~135° (Gueth and Guilloteau, 1999) and the data show that the inferred field pattern is predominantly uniform and axial, i.e. the magnetic field lies along the direction of the outflow axis. However, there is significant departure from a truly axial field with a fair degree of curvature in the projected field pattern through the source. We feel that this may be indicative of a slightly helical field running through the envelope of the young stellar object (YSO).

HH 211 forms part of a campaign at the JCMT to measure the magnetic field geometries towards a number of Class 0 YSOs. The data we have begun to collect seem to be indicating that such a field pattern may be quite common. However, exceptions remain. NGC 1333 IRAS2, for instance, exhibits quite a different scenario. The field pattern appears randomized and the degree of polarization decreases at the source.

Chandler and Richer (2000) studied the radial density profiles of these two sources at 450 and 850 μm. Although not altogether conclusive, they found that they could describe NGC 1333 IRAS2 as an object undergoing free-fall collapse while HH 211 was still enjoying some form of extra hydrostatic support. If this is true, then our results lend evidence to the suggestion that the extra support comes from a relatively strong magnetic field.

2.2. Cepheus A

In Figure 2 we show our polarimetry towards the high-mass YSO, Cepheus A. The polarization vectors have been rotated by 90° to depict the **B**-vector projected onto

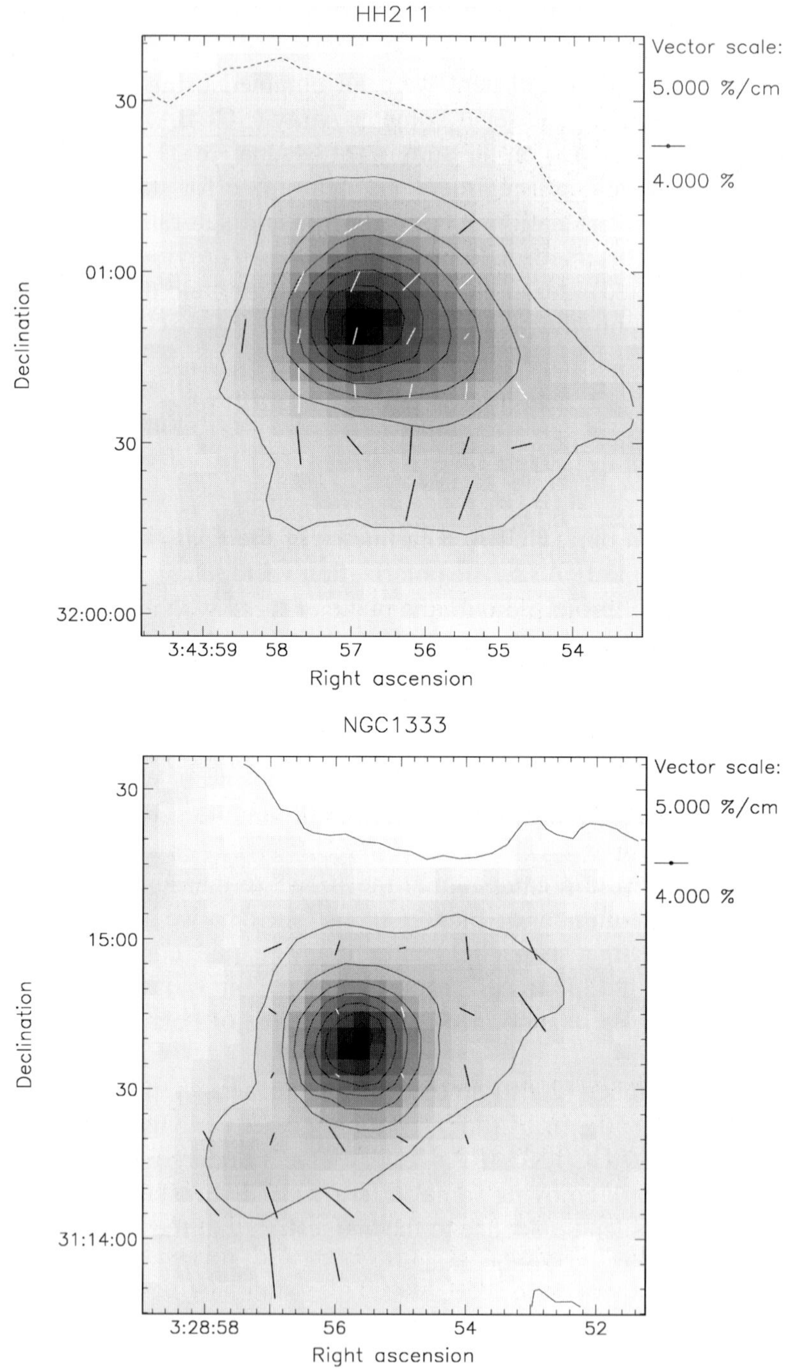

Figure 1. Magnetic field geometry projected onto the plane of the sky for HH 211 and NGC 1333 IRAS2. The greyscale gives the intensity distribution of the 850 μm emission, while the short lines give the degree and direction of the **B**-vector (scale shown on the side). Coordinates are given in J2000.

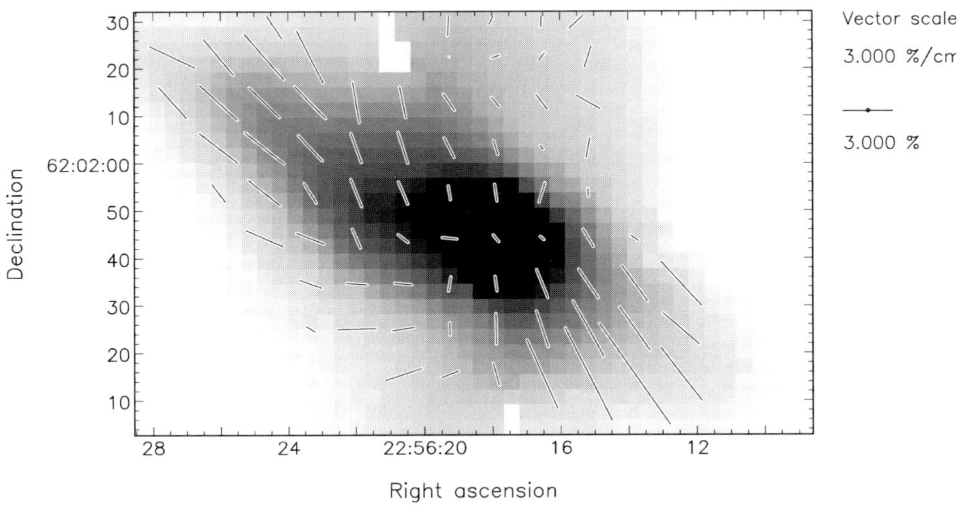

Figure 2. Same as Figure 1 but for Cepheus A.

the plane of the sky. The outflow direction lies at a position angle of ~40°. There are some important features to note here:

- in general, the field runs parallel to the outflow axis,
- the degree of polarization decreases close to the source,
- as they approach the source, the vectors bend in towards it.

This field pattern is strongly reminiscent of pinched, or 'hourglass', field models presented in Aitken et al. (2002)—their model III in particular. Such field patterns have for some time been predicted to be an inevitable consequence of the gravitational collapse of a core into a protostar.

3. Curvature in Magnetic Fields

One of the striking features in our dataset has been the discovery, towards a few of our sources, of significant degrees of curvature in magnetic field patterns. Matthews et al. (2001) find some evidence of this in OMC-3 but it is not clear whether any curvature is due to the dense filamentary structure of the cloud or to the field itself. In Figure 3, we present SCUBA polarimetry of the W 51 region with contours of the 6 cm emission overlaid. Curvature of the field towards this source is seen across the region. A crude comparison to MHD turbulence simulations shows that the vector pattern is *consistent* with a relatively weak field which has $\beta \sim 1$ (Chrysostomou et al., 2002).

There is a quite distinct correspondence between the field pattern and the position and extent of the HII-region, W 51 IRS1. The 6 cm emission shows that the gas from the HII-region is expanding and enveloping itself around the

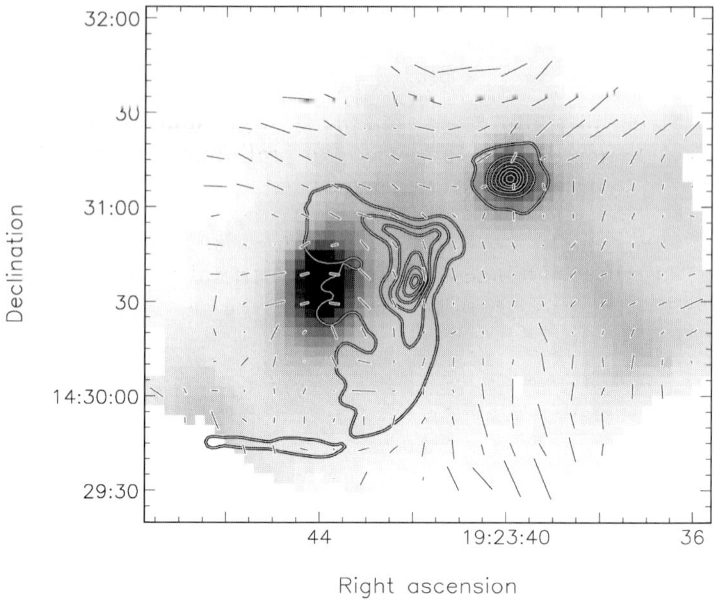

Figure 3. Same as Figure 1 but for W 51. The contours show the 6 cm emission. The core to the north-west is associated with W 51 IRS2. The 6 cm emission in the centre of the field is associated with W 51 IRS1.

dense submillimetre core to the east. The field follows this structure/pattern quite closely[1]

A similar relationship between gas kinematics and the magnetic field structure is seen for NGC 7538 (Momose et al., 2001). Here the correlation is between the field and the molecular outflow, as traced by CO emission. Our observations of S 140 show similar field curvature, although we have as yet not been able to attribute a direct correlation with gas kinematics.

These results imply that, at the scales sampled by the JCMT beam ($\sim$15″), the magnetic field plays a passive rôle, allowing itself to be manipulated by the energy density and pressure gradients in the gas. At closer scales, such as those sampled by interferometers such as the BIMA, this may not be the case. As pointed out in Chrysostomou et al. (2002), interferometric data of the core associated with W 51 IRS1 (Lai et al., 2001) shows polarization vectors consistent with the single-dish data in Figure 3 but show no evidence of a weak field. Instead, we see a uniform field pattern indicative of a relatively strong field ($\beta \ll 1$) and suggesting that the magnetic field may only start to play an active rôle in the star formation process at small scales after cloud collapse and accretion have commenced.

[1]Our measurements have been confirmed through comparison with data taken of W51 at 350 μm using the Hertz polarimeter on the Caltech Submillimetre Observatory. The authors thank and acknowledge John Vaillancourt for allowing this comparison to be made.

4. Conclusions

We have presented preliminary results of a campaign to measure the projected magnetic field structures towards a sample of high- and low-mass young stellar objects. The data are still being analyzed but preliminary results have allowed us to come to the following conclusions.

- We have found evidence to suggest the presence of a variety of magnetic field configurations. These include axial (with the general field orientation parallel to the outflow axis) and helical fields, and even what may be the first direct evidence of an 'hourglass' magnetic field. At the same time, one source in our sample of low-mass objects shows the absence of a definitive magnetic field structure.
- It is important to understand how the magnetic field interacts with the dynamics of the gas in its environment. If correlations with gas dynamics are found, then this opens up the exciting possibility of probing and *mapping* the magnetic field energy density in these systems by simply assuming equipartition between the gas and the field.
- Combining data using single-dish telescopes and interferometers provides us with an understanding of how the field behaves as a function of scale.

References

Aitken, D.K., Efstathiou, A., Hough, J.H. and McCall, A.: 2002, *MNRAS* **329**, 647.
Chandler, C.J. and Richer, J.S.: 2000, *ApJ* **530**, 851.
Chrysostomou, A. et al.: 2002, *A&A* **385**, 1014.
Davis, L. and Greenstein, J.L.: 1951, *ApJ* **114**, 206.
Fiege, J.D. and Pudritz, R.E.: 2000, *ApJ* **544**, 830.
Gueth, F. and Guilloteau, S.: 1999, *A&A* **343**, 571.
Heitsch, F. et al.: 2001, *ApJ* **561**, 800.
Lai, S.P. et al.: 2001, *ApJ* **561**, 864.
Nakano, T.: 1998, *ApJ* **494**, 587.
Matthews, B.C., Wilson, C.D. and Fiege, J.D.: 2001, *ApJ* **562**, 400.
Momose, M. et al.: 2001, *ApJ* **555**, 855.
Mouschovias, T.Ch. and Ciolek, G.E.: 1996, in: C.J. Lada and N.D. Kylafis (eds.), *The Origin of Stars and Planetary Systems*, Kluwer Academic Publishers, Dordrecht, p. 305.
Ostriker, E.C., Stone, J.M. and Gammie, C.F.: 2001, *ApJ* **546**, 980.
Padoan, P. et al.: 2001, *ApJ* **559**, 1005.
Shu, F.H., Adams, F.C. and Lizano, S.: 1987, *ARAA* **25**, 23.

A COLLIMATED STELLAR WIND EMANATING FROM A MASSIVE PROTOSTAR

GUIDO GARAY[1], KATE J. BROOKS[1], DIEGO MARDONES[1] and RAY NORRIS[2]

[1]*Departamento de Astronomía, Universidad de Chile, Santiago, Chile;*
E-mail: guido@das.uchile.cl
[2]*Australia Telescope National Facility, Australia*

Abstract. We report multi-wavelength observations towards IRAS 16547–4247, a luminous infrared source with a bolometric luminosity of $6.2 \times 10^4\ L_\odot$. Dust continuum observations at 1.2-mm indicate that this object is associated with a dust cloud with a size of about 0.4 pc in diameter and a mass of about $1.3 \times 10^3\ M_\odot$. Radio continuum observations show the presence of a triple radio source consisting of a compact central object and two outer lobes, separated by about 0.3 pc, symmetrically located from the central source. Molecular hydrogen line observations show a chain of knots that trace a collimated flow extending over 1.5 pc. We suggest that IRAS 16547–4247 corresponds to a dense massive core which hosts near its central region a high-mass star in an early stage of evolution. This massive YSO is undergoing the ejection of a collimated stellar wind which drives the H_2 flow. The radio emission from the lobes arises in shocks resulting from the interaction of the collimated wind with the surrounding medium. We conclude that the thermal jets found in the formation of low-mass stars are also produced in high-mass stars.

Keywords: IRAS 16547–4247 jets, outflows, radio continuum, stars formation, massive stars

1. Introduction

Whether massive stars (B- and O-type) form via an accretion process similar to that for low-mass stars (Osorio et al., 1999; McKee and Tan, 2003) or instead via collisions with lower-mass stars (Bonnell et al., 1998) is currently under debate (Garay and Lizano, 1999). Highly collimated radio jets (Rodríguez, 1997; Anglada et al., 1998) and Herbig-Haro (HH) flows (Reipurth et al., 2002) are frequently observed towards young low-mass stars. It is now widely accepted that jets are intimately linked to the accretion process and the formation of both bipolar molecular outflows and HH flows (see review by Reipurth and Bally, 2001). The role of collimated jets in massive stars ($M > 8M_\odot$) is less certain. In this higher mass regime observations of jets are difficult, primarily because of the greater distances involved and because the evolutionary time scales of such jets are expected to be much shorter. Of all the reported sources to have collimated radio jets, the three with the brightest luminosity are: HH80-81 ($L \sim 1.7 \times 10^4 L_\odot$; Martí et al., 1998); IRAS 20126 + 4104 ($L \sim 1.3 \times 10^4 L_\odot$; Shepherd et al., 2000); and Cepheus A-HW2 ($L \sim 1 \times 10^4 L_\odot$; Gómez et al., 1999).

Here we present a series of data towards IRAS 16547–4247, a young stellar object with a bolometric luminosity of $6.2 \times 10^4 L_\odot$, equivalent to that of a single

Astrophysics and Space Science **292:** 517–523, 2004.

young O8 star. Our findings show that IRAS 16547–4247 is associated with a radio jet and collimated flow, making it the most luminous young stellar object known to be associated with such phenomena.

2. Observations

IRAS 16547–4247 is one object from a sample of 18 luminous IRAS sources, thought to be associated with massive stars in early stages of evolution, that we are studying in detail. The sources in our sample were chosen from the Galaxy-wide survey of CS(2–1) emission towards IRAS sources with infrared colours typical of compact HII regions (Bronfman et al., 1996). Each source selected for our sample showed line profiles indicative of either inward or outward motions and in some cases broad wings. Their IRAS luminosities were in the range $2 \times 10^4 - 4 \times 10^5 L_\odot$, implying that they all contain at least one embedded massive star ($> 8\ M_\odot$).

Our study of IRAS 16547–4247 incorporates radio continuum data at four frequencies (1.4, 2.5, 4.8 and 8.6 GHz) obtained with the Australia Telescope Compact Array (ATCA), 1.2-mm continuum data using the SIMBA 37-channel bolometer array installed at the Swedish ESO Submillimeter Telescope (SEST), and a series of SEST molecular-line transitions between 85 and 250 GHz (see Garay et al., 2003 for a detailed description). We recently complemented these data with near- and mid-infrared data. Narrow-band images of the $H_2$1–0 S(1) 2.12 μm emission line, together with the adjacent continuum at 2.09 μm, were obtained using ISAAC mounted on the ANTU telescope of the Very Large Telescope (VLT). A mid-infrared image was obtained using TIMMI2 mounted on the 3.6-m telescope in La Silla using the N11.9 μm filter and a pixel scale of 0.3″ pixel^{-1} (see Brooks et al., 2003 for a description of these observations).

3. IRAS 16547–4247: A Dense and Massive Isolated Molecular Core

The 1.2-mm observations show that the dust continuum emission from IRAS 16547–4247 is just resolved, having a diameter of 0.4 pc and a flux density of 16.3 Jy (see Figure 3). Assuming a dust opacity at 1.2-mm of 1 cm^2 g^{-1} (Ossenkopf and Henning, 1994) we derive a mass of $1.3 \times 10^3 M_\odot$. The molecular line observations indicate the presence of a molecular gas core with a diameter of 0.38 pc, a molecular hydrogen column density of 6×10^{23} cm^{-2}, a molecular hydrogen density of 5×10^5 cm $^{-3}$, a mass of $9 \times 10^2 M_\odot$ and a kinetic temperature of 28 K. These properties are similar to those of other massive star-forming cores (e.g. Plume et al., 1997).

The line profiles observed toward IRAS 16547–4247 indicate that the bulk of the molecular gas is undergoing large-scale inward motions, with a characteristic infall speed of 0.75 km s^{-1}. The implied mass infall rate is $\sim 1 \times 10^{-2} M_\odot$ yr^{-1}, large enough to prevent the development of an UC HII region (Walmsley, 1985). The line profiles also show strong wing emission up to radial velocities of ± 18 km s^{-1}

relative to the ambient cloud velocity, indicating outflow activity. We suggest that this massive dense core is in an early stage of collapse, accreting at high rates which prevents the formation of a detectable photoionized UC HII region.

4. Radio Continuum Emission

Figure 1 shows the radio continuum maps at 4.8 and 8.6 GHz towards IRAS 16547–4247. The radio emission arises from three components aligned in a southeast-northwest direction. The outer components are symmetrically located in opposite directions from the central source, with peak positions separated by an angular distance of $\sim 20''$ (0.28 pc at the distance of 2.9 kpc). Using the radio data at all four frequencies the spectral indices for each of the three components were measured. The spectral index for the central object is 0.49 ± 0.12 and is consistent with thermal emission produced by a biconical jet (Reynolds, 1986). The spectral indices of the emission from the outer components are negative (-0.61 ± 0.26 and -0.33 ± 0.04) and are consistent with shock-induced synchrotron emission (Blandford and Eichler, 1987). This non-thermal emission arises at the working surface of the jets, where the collimated winds from the jet interact with the surrounding medium.

The actual thermal radio jet corresponds to the brightest 8.6-GHz emission component in Figure 1b. There is a fainter source offset to the southeast and whose spectral index is not known. It is not certain what role (if any) this source plays. It

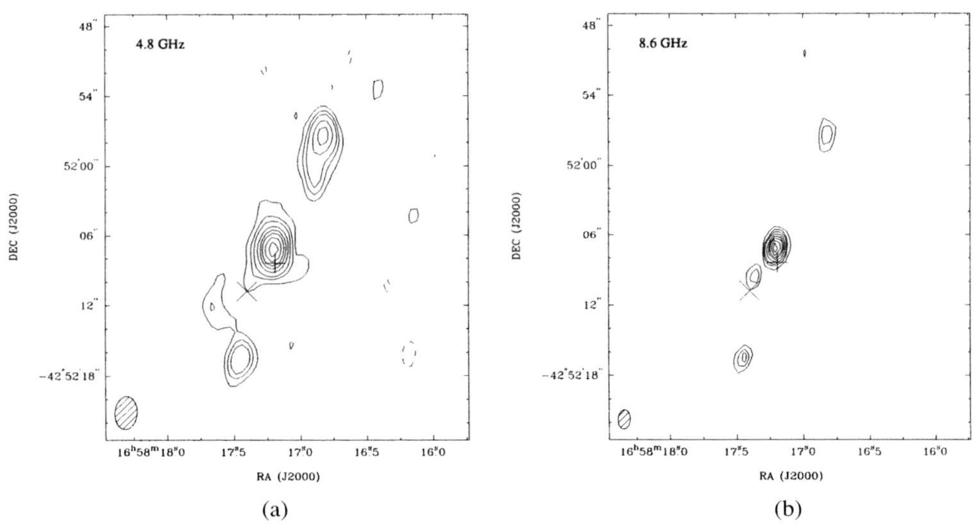

Figure 1. ATCA maps of the radio continuum emission from IRAS 16547−4247. Beams are shown in the lower left corner of each panel. Contour levels are −1, 1, 2, 3, 5, 7, 9, 12 and 15 times 0.30 mJy beam^{-1} ($1\sigma = 0.096$ mJy beam^{-1} at 4.8 GHz and 0.070 mJy beam^{-1} at 8.6 GHz). Also shown are the OH maser position (+) from Caswell (1998) and the H_2O maser position (×) from Forster and Caswell (1989).

is associated with an H_2O maser spot (Forster and Caswell, 1989). An OH maser is located towards the radio jet (Caswell, 1998). It should be noted that there was no detection of any methanol maser emission towards IRAS 16547–4247 (Walsh et al., 1998). The triple radio source is centred on the 1.2-mm continuum emission peak.

5. H_2 2.12 μm Line and 12-μm Continuum Emissions

Figure 2 shows an image of the H_2 2.12 μm emission towards IRAS 16547−4247. There is a complex chain of emission knots extending over 110″ (1.5 pc at the distance of 2.9 kpc). The H_2 emission has the morphological characteristics of HH objects arising from the interaction of a collimated flow with the ambient medium. The outermost emission knots are approximately symmetrically offset from the radio jet and are in the shape of bow-shocks, all pointing away from the direction of the radio jet. Their arrangement is consistent with an elongated outflow cavity.

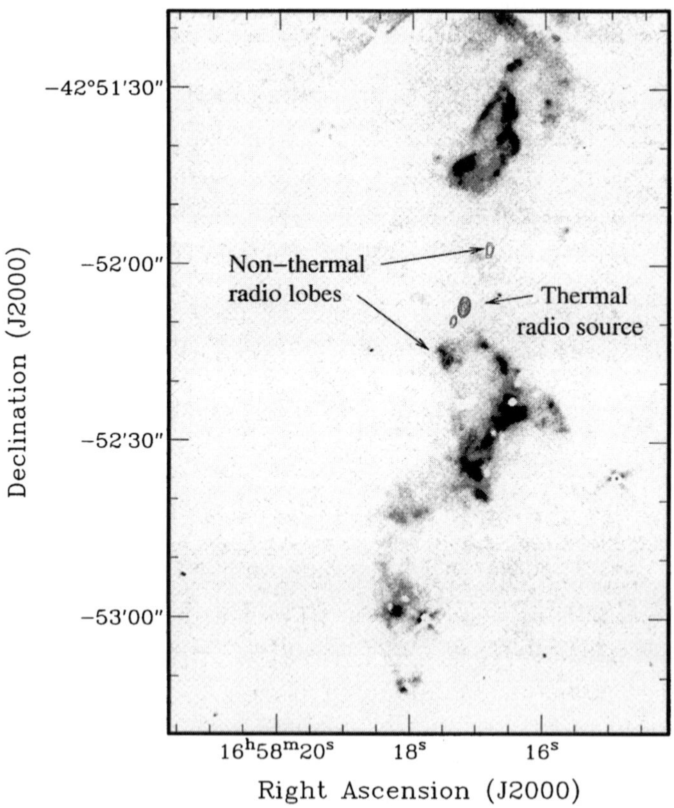

Figure 2. ISAAC H_2 2.12 μm emission image (continuum subtracted) shown in grey scale. Overlaid with contours of the 8.6-GHz continuum emission (see Figure 1).

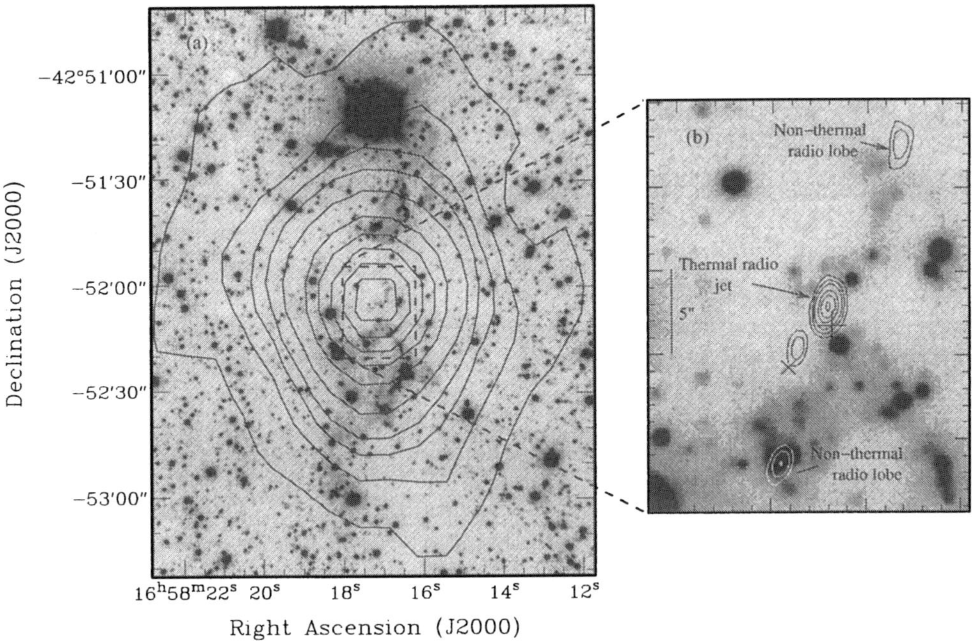

Figure 3. ISAAC H_2 2.12 μm emission image (no continuum subtraction) shown in grey scale. (a) Overlaid with 1.2-mm dust continuum emission. Contour levels are 0.25 (5σ), 0.5, 0.75, 1, 1.5, 2.5, 3.5, 4.5, 5.5, 6.5 Jy beam^{-1}. (b) Overlaid with contours of the 8.6-GHz continuum emission (see Figure 1). Also shown are the OH maser position (+) from Caswell (1998) and the H_2O maser position ($\times$) from Forster and Caswell (1989). There is a positional uncertainty of 1″ in registering the H_2 2.12 μm and 8.6-GHz continuum data.

Figure 3 illustrates the comparison between the H_2 emission (without any continuum subtraction) and (a) the 1.2-mm dust continuum emission and (b) the 8.6-GHz continuum emission. The chain of H_2 emission is oriented in the same direction as the triple radio source and is contained within the molecular core traced by the 1.2-mm continuum emission. One of the H_2 emission knots is associated with the southern non-thermal radio component. The striking alignment of the collimated flow delineated by the H_2 emission and the central location of the radio jet implies that these phenomena are coupled. It is reasonable to assume that the radio jet is the driving force of the collimated flow.

The 11.9-μm image obtained in this study (with a field of view of 96″ × 72″) reveals a single unresolved (<0.6″) emission source with a flux of 0.28 Jy. Within the 5″ positional uncertainty of the mid-infrared data, this source is coincident with the radio jet and is likely to be responsible for the excitation of IRAS 16547–4247. No stellar counterpart to the radio jet was detected in the narrow-band near-infrared images (e.g. see Figure 3b). It is now widely accepted that jets are intimately linked to protostellar accretion disks. Is there an accretion disk in the IRAS 16547–4247 system? Although a definite answer requires observations at infrared and sub-millimeter wavelengths with sub-arcsec angular resolution, we suggest that the 11.9-μm emission arises from such structure.

6. Summary

Dust continuum and molecular line observations show that IRAS 16547–4247 corresponds to a massive dense molecular core with a diameter of $\sim$0.4 pc, a molecular hydrogen density of 5×10^5 cm^{-3} and a mass of $\sim 1 \times 10^3 M_\odot$. The molecular gas is undergoing large-scale inward motions, with a characteristic infall speed of 0.75 km s^{-1}. The implied mass infall rate is $\sim 1 \times 10^{-2} M_\odot$ yr^{-1}.

Radio continuum observations show the presence of a triple radio source consisting of a compact central object and two outer lobes, separated by about 0.3 pc, symmetrically located from the central source. The radio emission from the central object originates in a highly collimated ionized wind, whereas the emission from the lobes results from the interaction of the collimated wind with the surrounding medium. In addition, the H_2 2.12 μm line observations show a chain of knots that trace a collimated flow extending over 1.5 pc. The alignment of the flow and the central location of the radio jet implies that these phenomena are intimately linked. We conclude that IRAS 16547–4247 corresponds to a dense massive core which hosts near its central region a high-mass star in an early stage of evolution. The embedded massive protostar is intensely assembling the bulk of their final mass and undergoing the ejection of a collimated stellar wind which drives the H_2 flow.

IRAS 16547–4247 is the most luminous ($6.2 \times 10^4 L_\odot$) embedded young stellar object known to harbour a thermal radio jet. We conclude that the thermal jets found in the formation of low-mass stars are also produced in high-mass stars. Our finding of a jet and collimated flow towards such a luminous object supports the accretion scenario for the formation of stars across the entire mass spectrum.

Acknowledgements

This work has been partly funded by the Chilean Centro de Astrofísica FONDAP No. 15010003. The ISAAC data were obtained through the ESO Director's Discretionary Time Program.

References

Anglada, G., Villuendas, E., Estalella, R., Beltrán, M.T., Rodríguez, L.F., Torrelles, J.M. and Curiel, S.: 1998, *AJ* **116**, 2953.

Blandford, R. and Eichler, D.: 1987, *Phys. Rep.* **154**, 1.

Bonnell, I.A., Bate, M.R. and Zinnecker, H.: 1998 *MNRAS* **298**, 93.

Bronfman, L., Nyman, L.Å. and May, J.: 1996, *A&AS* **115**, 81.

Brooks, K.J., Garay, G., Mardones, D. and Bronfman, L.: 2003, *ApJ* **594**, L131.

Caswell, J.L.: 1998, *MNRAS* **297**, 215.

Forster, J.R. and Caswell, J.L.: 1989, *A&A* **213**, 339.

Garay, G. and Lizano, S.: 1999, *PASP* **111**, 1049.
Garay, G., Brooks, K.J., Mardones, D. and Norris, R.: 2003, *ApJ* **587**, 739.
Gómez, J.F., Sargent, A.I., Torrelles, J.M., Ho, P.T.P., Rodríguez, L.F., Cantó, J. and Garay, G.: 1999, *ApJ* **514**, 287.
Martí, J., Rodríguez, L.F. and Reipurth, B.: 1998, *ApJ* **502**, 337.
McKee, C.F. and Tan, J.C.: 2003, *ApJ* **585**, 850.
Osorio, M., Lizano, S. and D'Alessio, P.: 1999, *ApJ* **525**, 808.
Ossenkopf, V. and Henning, T.: 1994, *A&A* **291**, 943.
Plume, R., Jaffe, D.T., Evans, N.J., II, Martín-Pintado, J. and Gómez-Gonzalez, J.: 1997, *ApJ* **476**, 730.
Reipurth, B. and Bally, J.: 2001, *A Rev A&A* **2001**, 39.
Reipurth, B., Heathcote, S., Morse, J., Hartigan, P. and Bally, J.: 2002, *AJ* **123**, 362.
Reynolds, S.P.: 1986, *MNRAS* **304**, 713.
Rodríguez, L.F.: 1997, in: B. Reipurth and C. Bertout (eds.) *Herbig-Haro Flows and the Birth of Stars; IAU Symposium No. 182*, Kluwer Academic Publishers, Netherlands p. 83.
Shepherd, D.S., Yu, K.C., Bally, J. and Testi, L.: 2000, *ApJ* **535**, 833.
Walmsley, C.M.: 1985, *RMA&A CS* **1**, 137.
Walsh, A.J., Burton, M.G., Hyland, A.R. and Robinson, G.: 1998, *MNRAS* **301**, 640.

INTERACTION DISK-STELLAR MAGNETOSPHERE

MAGNETIC ACTIVITY AND THE INTERACTION BETWEEN THE STELLAR MAGNETOSPHERE AND THE ACCRETION DISK

A.I. GÓMEZ DE CASTRO

Instituto de Astronomía y Geodesia (CSIC-UCM) Fac. de CC Matemáticas, Universidad Complutense de Madrid, Madrid, Spain; E-mail: aig@mat.ucm.es

Abstract. Analytical and numerical analysis of the interaction between the magnetosphere of the Pre-Main-Sequence (PMS) stars and accretion disks are converging to a robust, fundamental mechanism for jet formation. They provide a reasonably solid base-line to explain the current observations on the formation of solar-like stars. In this summary, the connection between observations and this paradigm are reviewed critically and some observational tests proposed.

Keywords: magnetic fields-stars: formation-circumstellar matter-stars: mass-loss, planetary systems: protoplanetary discs, ISM: jets and outflows

1. Introduction

Magnetic fields are involved in the interaction Pre-Main Sequence (PMS) star-accretion disk. This statement comes from two solid observational facts: the detection of accretion disks (Dutrey, 2004) and the detection of stellar magnetic fields (Guenter et al., 1999; Johns-Krull et al., 1999). Zeeeman measurements have allowed to determine the line-of-sight component of the field which is some few kG and varies in time scales of days (see Valenti and Johns-Krull, 2004).

In the last five years, numerical research on this interaction has gone into an outburst. So far most of the works have analyzed the interaction between a *dipolar* stellar magnetic field and a keplerian accretion disk. The interaction disk-magnetosphere basically transforms angular momentum (differential rotation) into toroidal plasmoids which are ejected from the system. As shown in Uzdensky review (2004), basically all models can be fitted into the basic configuration sketched in Figure 1; there is a current sheet that separates two distinct regions: an inner *stellar outflow* and an external *disk flow*. Magnetic flux dissipation is expected to be produced in the current layer leading to plasmoids ejections, as well as to the injection of high energy particles (cosmic rays) in the environment leading to generation of X-rays and ultraviolet radiation. The phenomenon is nonstationary and controlled by two different temporal scales: the rotation period and the magnetic field diffusion time scale. Stellar rotation is a well know parameter which controls the opening of the field lines towards high latitudes however, plasmoids ejection is controlled by *field diffusion* which is poorly determined (see e.g. Priest and Forbes, 2000).

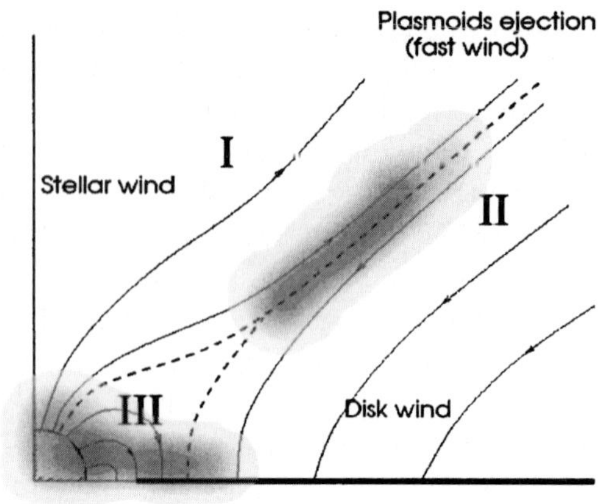

Figure 1. The reference model as described in Uzdensky's contribution and adapted from Lovelace et al., 1995.

Numerical simulations show that the fundamental mechanism for jet formation is robust. The star-disk-outflow system is self-regulating when various initial disk densities, stellar dipolar field strengths and primordial field associated with the disk are tested (Matt et al., 2002) although strong stellar magnetic fields may disrupt the inner parts of the accretion disk temporarily (see Kueker et al., 2004).

All these results define a reasonably solid base-line to explain the current observations on the formation of solar-like stars. In this summary, the connection between observations and this paradigm is reviewed critically and some observational tests are proposed.

2. The Paradigm Versus the Observations

Any model atempting to describe the formation of solar-type stars should explain the observational properties of the outflows, disks, stars and accretion flow. Let us analyze them one by one.

2.1. Outflows

Collimated outflows and/or heavy winds are detected since the very early stages of star formation (see André, 2004).

The radiative output (optical forbidden lines emission) has been computed for the two most widely accepted families of models: centrifugally driven cool MHD disk winds and X-winds (see Pudritz, 2004). However, some recent works point out that cool disk wind models do not fit the observations of the rather evolved classical T Tauri Stars (CTTSs) (Ferro-Fontán and Gómez de Castro, 2003; Dougados et al., 2004). This is not unexpected. At the CTTS phase the accretion rate is moderate

($10^{-7}-10^{-8}$ $M_{\odot}yr^{-1}$), a fair fraction of the inner disk gas and dust will be depleted into planetesimal building-up and the circumstellar X-rays radiation will be able to heat up the inner part of the disk[1] from which disk winds depart. Henceforth, thermal pressure becomes a non-negligible component to load gas into the field lines and warm disk wind solutions seem a reasonable *approach* to the disk outflow in region II (see Figure 1). In the last year, some attempts have been made to model the spectral signature of warm disk winds. Two approaches have been followed: adding turbulent dissipation to the dynamics of the cool disk wind solution (see O'Brien et al., 2004) or making use directly of warm disk wind solutions with pressure (Gómez de Castro and Ferro-Fontán, 2004). In both cases analytic self-similar solutions have been used and the output radiation from optically thin lines has been calculated. These models, rather naive when compared with the complexity of the numerical codes, provide however some important baselines for understanding the observational properties of the unresolved region ($\leq$9AU) where the jet engine is located. For instance, a warm disk wind solution, with correct crossing of all the magnetohydrodynamic surfaces (Vlahakis et al., 2000), allows calculating the profiles of the most important spectral lines. As shown in Figure 2 (from Gómez de Castro and Ferro-Fontán, 2004), the profiles show broad double peaked features or slightly blueshifted asymmetric features depending on the inclination. In edge-on systems, the spectral signature will be very similar to the produced by a rotating envelope however, the double peaked structure is dominated by the axial expansion of the flow. These profiles, when added to the stellar chromospheric emission, will show up as broad and/or asymmetric wings which, indeed, are often observed (Gómez de Castro and Verdugo, 2003). A very interesting property is that warm disk winds have 'chromospheric/transition region' temperatures and could be partially responsible of the optical and UV veiling (see below).

Another important aspect is that a significant fraction of the mass outflow is ejected in a non-stationary manner. Outflows are often resolved into two kinematical components, a slow wind and a high velocity component associated with episodic ejections. It is unclear which is the relative contribution of gas entrainment to the slow wind (see Cabrit, 2004). The time scales for these ejections ranges from few hours (Alencar, 2001; Bouvier et al., 2003; Gómez de Castro and Verdugo, 2003) to some ten years (optical jets observations, see e.g. López-Martín et al., 2003) or some hundreths years (bullets of molecular gas, see e.g. Bachiller, 1996). Recent observations have set-up a *lower limit around 1 h* precluding the association of flares with the few hours time-scale variability in RW Aur; the characteristic decay

[1]The wind becomes lighter as the accretion rate decreases so, the stellar X-ray radiation is able to modify the temperature structure of the wind heating it up to temperatures of some few thousand degrees. At these temperatures the base of a *cool wind* will contribute to the optical veiling and produce significant emission in spectral tracers which are typically considered to be produced in the magnetospheric accretion flow such as the Hα or the NaD lines (Ferro-Fontán and Gómez de Castro, 2003).

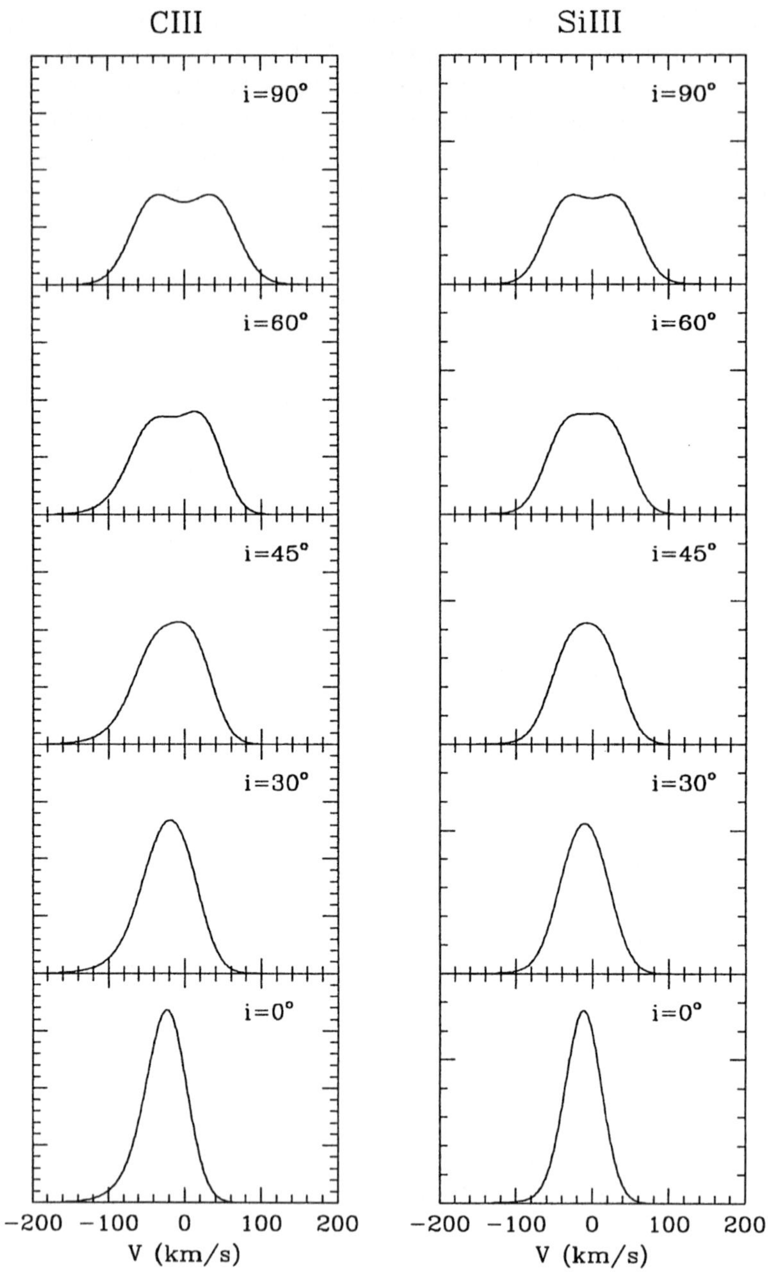

Figure 2. Profiles of the CIII] and SiIII] semiforbidden UV lines produced by the inner 9 AU of a warm MHD disk wind. Notice that there is a slight difference between the SiIII] and the CIII] lines which is caused by the different critical electron densities. The CIII] line is seturated close the source.

time of flares in active stars is some few hundreths of seconds (Gómez de Castro and Verdugo, 2003). Several ejection time-scales co-exist in the same object. For instance, time-scales of ~1 h, ~5.5 d and ~20 yr are observed in RW Aur. The co-existence of several time scales (especially for short time scales) can be readily accommodated into the model due to the uncertainties on the microphysics of field diffusion and the likely development of *dusty* plasma instabilities. This is an open research field that will develop rapidly in the next years. As in the past, the research on cataclysmic variables will provide useful tips.

2.2. Accretion disks

We have not any conclusive observational evidence of the magnetization of protostellars disks although we expect they are, since molecular cores are magnetized (see Chapter 3 contributions). Fields may be fossil or generated through dynamo action; the magneto-rotational instability (MRI) being the source of turbulence and leading to accretion and outflow (see Hawley, 2004). The development of the MRI depends on the effective coupling with the disk, e.g. on the relative density of free charges. The inner region of disks is sufficiently ionized (detection of a rotating ring of ionized gas is reported in these proceedings see Gómez de Castro and Verdugo, 2004), but the outer regions are more problematic. The model sketched in Figure 1 has several regions where high energy events can be produced:

1. The interface between regions I and II. Magnetic field reconnection may be produced in the current layer; e.g. from few R_* to the superalfvénic radius of the wind where strong wind flowing swepts open the newly-reconnected field lines (see Uzdensky, 2004). This region may play an important role in the ionization of the disk since it is above it.
2. Stellar flares.
3. The interaction between the stellar magnetosfere and the circumstellar gas in region III: this includes the possible formation of Corotating Interaction Regions, as the observed in the solar wind.
4. Coronal mass ejections.

The last three types of events are observed in the Solar System and they are known to be a significant source of cosmic rays.

This extra flux of high energy particles has to be added to the calculations of the ionization fraction of the disk based only on galactic cosmic rays (Glassgold et al., 2000).

Numerical simulations of magnetized accretion disks (e.g. without considering the interaction with the stellar magnetosphere) also show that magnetic buoyancy leads to the formation of a disk corona from where the slow wind could be formed. However it is still unclear whether the presence of a large scale ordered field is unavoidable for jet formation (see Hawley, 2004). Such a large scale field will be present in the inner region of the accretion disk because of the interaction between the stellar magnetosphere and the kepplerian disk (see also Matt et al., 2002).

In addition, there is a timing problem. If we trust that T Tauri systems are alike our early solar system, the meteoric evidence set-up an upper limit of some few 10^7 yr for planetesimal differentiation (Wadhwa and Russell, 2000). Accretion and outflow are observed in many T Tauri stars 10^7 yr old. How does MRI co-exist with planetesimal formation? (see Hawley, 2004).

2.3. Star

The photospheric field, e.g. the field on scales $\simeq 1R_*$, is far from dipolar (see Valenti and Johns-Krull contribution or for instance the review by Ménard and Bertout (1999) on the photometric variability of TTSs). Stars with very small accretion rates (Weak Line TTSs or WTTSs) display cold spots with temperatures around 1000 K lower than the photosphere, covering up to a 40% of the stellar surface!. Cold and hot spots co-exist on the surface of the CTTSs which have larger accretion rates. Hot spots are located at lower latitudes than cold spots and cover between ~1% of the stellar surface in CTTSs up to roughly a 10% of the surface in YY Ori which has a large accretion rate: $3 \times 10^{-6}\ M_\odot\ \mathrm{yr}^{-1}$ (Hartigan et al., 1995). The hot and cold spots are variable. Cold spots and, at least, a fraction of the hot spots are associated with the emergence of magnetic flux from the stellar interior.

Most of the insight on the stellar magnetic fields comes from the extrapolation to TTSs of our knowledge of the Sun; models assume that there is a radiative core rotating almost rigidly leading to a shear layer at the bottom of the convection zone. TTSs are assumed to confine just a 5% of their mass in the radiative core (Granzer et al., 2000). Toroidal flux tubes are stored in a layer of overshooting convective motion at the bottom of the convection zone. They are amplified by differential rotation until the Parker instability sets-in. The emergence of magnetic flux from the stellar interior defines the small-scale structure of the field (see Moreno-Insertis, 2004). Transport models show that the Coriollis force leads to strong poleward deflection of the magnetic tubes in rapidly rotating stars, as most of the WTTSs are. Recent slender tube 1D calculations indicate that for rotation rates $\Omega \leq 10\Omega_\odot$, the flux tubes are deflected from the poleward motion towards the equator by the buoyancy force and the emergence latitude does not strongly depend on the rotation rate. For $\Omega \geq 10\Omega_\odot$ the preferred mode is axisymmetric with eruption in the polar regions. These predictions (from Granzer et al., 2000) roughly agree with the photometric properties of the TTSs suggesting that the dynamics of the stellar magnetic flux emergence is rather decoupled from the disk-star interaction in CTTSs and WTTSs. However, slender tube aproximations may not be reliable in a fully convective star. A full 3D study including detailed radiative transfer is required.

Finally, I would like to stress that the reference model in Figure 1 is not well suited to describe the *very early phases* of star formation, e.g. Class 0 sources, when the remnant fossil field may be important and the differentiation disk-protostar is unclear. At this stage convection ought to be important and the physical problem will look more alike the problem of magnetic flux transport in the Sun (see Moreno-Insertis, 2004).

2.4. Accretion Flow

Direct evidence of infalling gas comes from inverse P-Cygni profiles in the Balmer, Paschen and Brackett hydrogen series as well as in some metallic lines (Walker, 1972; Hamann and Persson, 1992; Edwards et al., 1994) and from the detection of redshifted absorption components (Giampapa et al., 1993; Petrov et al., 2001; Bouvier et al., 2003). The observed velocities are of some few hundreths km s^{-1}, e.g. compatible with free-fall from some few stellar radii. Time-lags (see Bouvier et al., 2004) and relative shifts between He I and He II optical lines (Petrov et al., 2001) have been found, opening the possibility to study the dynamical and thermal structure of the accretion flow. The major concern is related with the coexistence of several funnel flows which is required to explain the correlation between magnetospheric line emission and the accretion rate (see e.g. Muzerolle et al., 2001).

Another interesting aspect is that hot-spots are rather estable; if hot spots trace shock-heated gas falling along magnetic funnels from the disk onto the star, magnetic funnels must be estable structures within a very active (flaring) environment where large amounts of magnetic energy are released.

Rotational modulation does not affect all the hot plasma producing the optical veiling and the ultraviolet excess (Simon et al., 1989; Bertout et al., 1996; Gómez de Castro and Fernández, 1996; Gómez de Castro and Franqueira, 1997). This points out that there are other sources of plasma heating in addition to accretion shocks. This non-rotationally modulated excess is correlated with the infrared excess and the strength of the optical forbidden lines produced at the base of the jet. The infrared excess has been often interpreted as direct probe of the accretion rate using α-accretion disk models. However, a re-analysis of the data suggests that the excess (in the CTTS phase) is produced by dust sublimation and therefore is a measure of the surface of the inner accretion disk (Muzerolle et al., 2003). The illumination of the circunstellar environment by X-ray radiation from large scale axisymmetric structures (as the magnetosphere or the current layer between region I and II in Figure 1) could be producing simultaneously optical veiling, UV excess and IR excess.

3. Observational Tests

3.1. Measurements of the Magnetic Field

Direct measurements of the field in protostellar disks and molecular jets are currently unfeasible since high spatial resolution, high sensitivity polarization measurements are required. The new generation of radio interferometers, notably ALMA or the VLA-EVLA update, will be able to measure the expected few percent polarizations and detect non-thermal radiation. Also, the potential detection of paramagnetic molecules such as SO or CN may allow to directly measure the field strength. However, these instruments will be unable to probe the wind engine itself, since it is located within 1 AU ($\simeq$7 milliarcsec at Taurus distance).

3.2. Measurements of Jet Rotation

Jet rotation is searched as an indirect probe: centrifugal drives naturally produce rotating outflows. This is a very hard task since in magnetocentrifugal winds most of the toroidal magnetic flux (and angular momentum) is evacuated radially leading to a dominant poloidal field. As an example, warm MHD disk winds departing from the corona of the inner part of the accretion disk reach rotation velocities below 10 km s^{-1} at a height above some 9 AU over the disk which correspond to 0.06 arcsec at Taurus distance (Gómez de Castro and Ferro-Fontán, 2004). Such small rotation velocities seem to have been marginally detected in optical jets (see Bacciotti, 2004).

3.3. X-ray Radiation

X-ray radiation may be produced either by shocks or by the release of magnetic energy, however it is very difficult to identify the dominant source. Some important constraints are coming from recent observations (see Preibisch, 2004):

1. There is not relation between rotation period and X-ray emission, at least in the thoroughly studied Orion Cluster, however the X-ray luminosity depends on the strength of the Hα and Ca II lines. As mentioned above these lines are produced in warm, thin plasma. This correlation is often interpreted as a correlation between magnetic activity (X-ray flux) and accretion (the Hα and Ca II lines). However, if X-ray radiation is produced in a dense environment it will be rapidly absorbed producing the corresponding cascade and line radiation. In this case, this correlation would just imply that Hα and Ca II radiation comes from photoionized gas.
2. The CTTSs are weaker X-rays emitters than the WTTSs. This may be just indicating that the plasma density is higher in the CTTSs than in the WTTSs (see Preibisch, 2004).
3. The X-ray luminosity is variable; flares are often detected. There is a substantial difference between the understanding of the word 'flare' in solar and space plasma physics and in stellar astrophysics. In the first case, it is well known that this sudden increase of X-rays radiation and high energy particles flux can be produced by, at least, three very different physical processes: flares (e.g. magnetic reconnection events associated with the solar magnetic activity), corotating

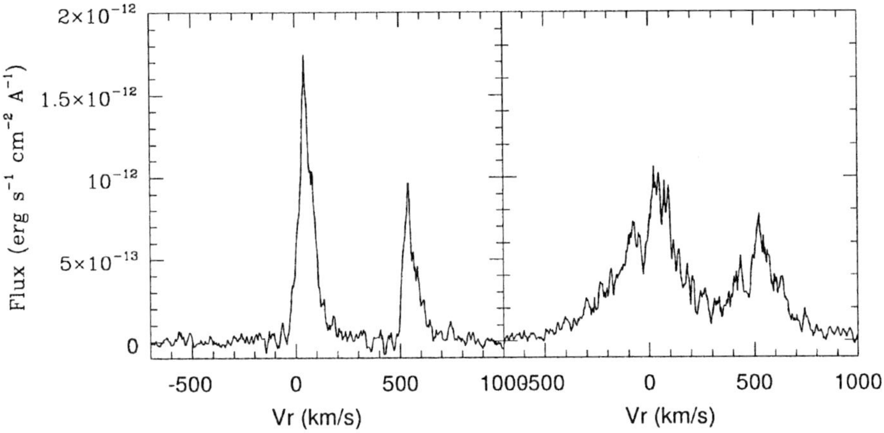

Figure 3. Contribution to the AB Dor C IV[UV1] profile from a normal flare (left) and a transient feature probably associated with a CIR (right). Both events lasted few kiloseconds. The left profile was observed in three events more during the short monitoring time while the last profile was observed only once. Notice the presence of a narrow absorption and the very broad line wings in the right panel profile (see Gómez de Castro, 2002 for more details).

interaction regions or CIRs (shock fronts formed in the interaction between the slow and the fast component of the solar wind) and coronal mass ejections. In our context, we often interpret that flares are only associated with magnetic reconnection events. However, at least in AB Dor, UV spectroscopy has shown to be able to distinguish between bonna-fidae flares and CIRs (see Figure 3).

The key problem with X-ray data is that we lack of spectral resolution to distangle the various possible contributing components by their kinematic signature. In this sense, the high density of the environment may help us to track the kinematics of the emitting gas through the UV radiation coming from nearby photoionized gas. As shown in Figure 4a, emission from UV semiforbidden lines is produced within some 10^9–10^{10} cm from the ionizing source for gas densities between 10^{12}–10^9 cm^{-3} (see Gómez de Castro and Verdugo, 2003).

Another related issue is the interaction of high energy particles with the circumstellar gas (and disk). As shown in Figure 4b (from Antonicci and Gómez de Castro, 2004), relativistic beams of electrons propagating through large gas columns produce 'non-thermal bremstrahlung' radiation in the X-ray range above some 5 keV and prominent Fe Kα emission that has been recently detected indeed (Takagi et al., 2003).

3.4. UV SPECTROSCOPY AND MONITORINGS

The only way to get into the scale of the structures represented in Figure 1 (from a fraction of R_* to some 10 R_*) is by means of high resolution spectroscopy

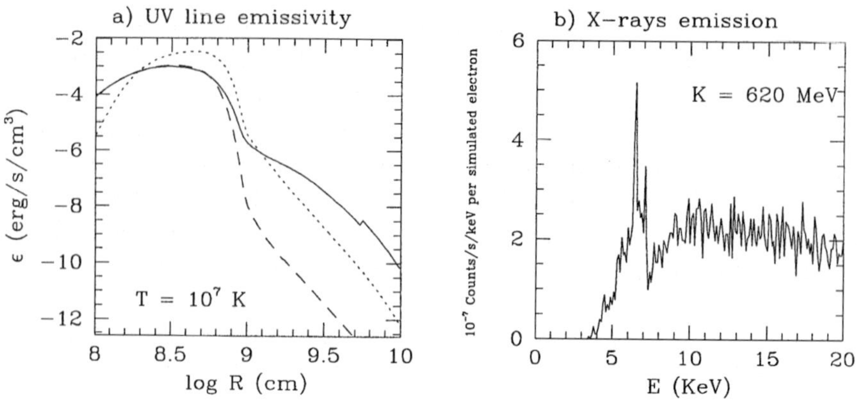

Figure 4. *(a)* Emissivity of the CIII] (solid), SiIII] (dotted) and OIII] (dashed) ultraviolet lines from spherical cloudlets around small reconnecting loops ($R_{loop} = 10^8$ cm^{-3}). The loops radiation is assumed to be thermal bremstrahlung emission from hydrogenic gas at $T_e = 10^7$ K. The cloudlets density is 10^{11} cm^{-3}. *(b)* Output X-rays spectrum from a spherical cloudlet heated by the impact of a beam of relativistic electrons (0.9c). The cloudlet density is 10^{11} cm^{-3} and has a radius of 6×10^{12} cm. Notice the prominent Fe Kα feature and the iron absorption edge at 7.1 KeV.

($R \geq 30\,000$) of warm plasma (from 10 000 to 10 000 K). This allows to sortout the different possible structures-components-regions by its temperature and kinematics. The potential of high resolution UV spectroscopy has been sufficiently outlined in the text. At this point, I would just like to stress the relevance of UV monitorings. From the Archive of the old good *International Ultraviolet Explorer* we know that the fluxes of the UV lines tracing warm-hot plasma are variable (see Gómez de Castro and Franqueira, 1997) and in some cases rotationally modulated (see above). This points out that we have the possibility of distangling hot spots and axisymmetric veiling contributions.

Unfortunately, high spectral resolution UV monitorings have been carried only once and for the very active cool star AB Dor. The monitoring was done with the *Hubble Space Telescope* and the old Goddard High Resolution Spectrograph. The results were most promising as shown in Figure 3; nine events were detected during roughly 10 hours! (Gómez de Castro, 2002). AB Dor is a 30 Myr old star, in many senses fairly similar to some WTTSs as HD283572.

Monitorings may also allow us to understand better the physics of cool active stars in connection with the PMS evolution (see Crespo-Chacón et al., 2004). The existence of large scale magnetic structures, sling-shot prominences, is proposed to explain the detection of material corotating at some few stellar radii. This is observed in many, rapidly rotating, cool stars (see Collier-Cameron, 2001).

4. In Summary

High resolution ultraviolet spectroscopy is instrumental to understand the physics of the warm plasma in the circumstellar environment of the TTSs. This warm plasma has the key to our understanding of the fundamental properties of the engine driving accretion and outflow during the formation of solar like stars. It is also instrumental to understand the emergence of magnetic flux from the interiors of PMS stars.

In addition, spectroscopic UV monitorings open the door to the study of the interaction of the stellar wind with protoplanetary disks in the early phases when planetesimals are differenciated and planets form through the formation of corotating interaction regions. Moreover, UV spectroscopy provides fundamental clues on the radiative field in this epoch and its role in the chemical evolution of disk at the time when planets are built. The understanding of PMS stars will allow us to understand better substellar objects (see Barrado et al., 2004 and Zapatero et al., 2004).

Unfortunately, there are basically no near UV faccilities in the planet Earth (just an instrument in the *Hubble Space Telescope*) and no project is forseen in the near future unless the *World Space Observatory* an international project for a 1.7 m space telescope with imaging and high resolution spectroscopic capabilities which could make this science feasible (URL:wso.vilspa.esa.es).

I could not conclude without some words of joy about the amazing universality of some mechanisms as the role of differential rotation in preventing the development of radial instabilities in plasma flows, so nicely described in Hidalgo, 2004 within the context of research in nuclear fussion by magnetic confinment.

Acknowledgements

This work has been financed by the Ministry of Science and Technology (MCyT) Spain through the research grant: AYA2000-966.

References

André, P.: 2004, *Ap&SS* **292**, 325.
Alencar, S.H.P., Johns-Krull, C.M. and Basri, G.: 2001, *AJ* **122**, 3335.
Antonicci, A. and Gómez de Castro, A.I.: 2004, *A& A* (submitted).
Bacciotti, F., Ray, T.P., Coffey, D., Eislöffel, J. and Woitas, J.: 2004, *Ap&SS* **292**, 651.
Bachiller, R.: 1996, *ARA& A* **34**, 111.
Barrado y Navascués, D., Martín, E.L., Jayawardhana, R. and Mohanty, S.: 2004, *Ap&SS* **292**, 665.
Bertout, C., Harder, S., Malbet, F., Mennessier, C. and Regev, O.: 1996, *AJ* **112**, 2195.
Bouvier, J., Dougados, C., and Alencar, S.H.P.: 2004, *Ap&SS* **292**, 659.
Bouvier, J., Grankin, K.N., Alencar, S.H.P., Dougados, C., Fernández, M., Basri, G., Batalha, C., Guenther, E., Ibrahimov, M.A., Magakian, T.Y., Melnikov, S.Y., Petrov, P.P., Rud, M.V. and Zapatero-Osorio, M.R.: 2003, *A& A* **409**, 169.
Cabrit, S., Flower, D.R., Pineau des Forêts, G., Bourlot, J.L. and Ceccarelli, C.: 2004, *Ap&SS* **292**, 501.
Collier-Cameron, A.: 2001, in: P. Brekke, B. Fleck and J.B. Gurman (eds.), *Recent Insights into the Physics of the Sun and Heliosphere: Highlights from SOHO and Other Space Missions*, IAU Symp. 203, *ASP*, p. 229.
Crespo-Chacón, I., Montes, D., Fernández-Figueroa, M.J., López-Santiago, J., García-Alvarez, D. and Foing, B.H.: 2004, *Ap&SS* **292**, 697.

Dougados, C., Cabrit, S., Ferreira, J., Pesenti, N., Garcia, P. and O'Brien, D.: 2004, *Ap&SS* **292**, 643.
Dutrey, A. and Guilloteau, S.: 2004, *Ap&SS* **292**, 407.
Edwards, S., Hartigan, P., Ghandour, L. and Andrulis, C.: 1994, *AJ* **108**, 1056.
Ferro-Fontán, C. and Gómez de Castro, A.I.: 2003, *MNRAS* **342**, 427.
Giampapa, M.S., Basri, G.S., Johns, C.M. and Imhoff, C.: 1993, *ApJS* **89**, 321.
Glassgold, A.E., Feigelson, E.D. and Montmerle, T.: 2000, in: V. Mannings, A.P. Boss and S.S. Russell (eds.), *Protostars and Planets IV*, University of Arizona Press, p. 429.
Gómez de Castro, A.I.: 2002, *MNRAS* **332**, 409.
Gómez de Castro, A.I. and Fernández, M.: 1996, *MNRAS* **283**, 55.
Gómez de Castro, A.I. and Franqueira, M.: 1997a, *ApJ* **482**, 541.
Gómez de Castro, A.I. and Franqueira, M.: 1997b, *IUE-ULDA Access Guide No. 8: T Tauri Stars*, ESA Scientific Publication ESA-SP 1205.
Gómez de Castro, A.I. and Verdugo, E.: 2003a, *ApJ* **597**, 443.
Gómez de Castro, A.I. and Verdugo, E.: 2003b, *Ap & SS* in press.
Gómez de Castro, A.I. and Ferro-Fontán.: 2004, *MNRAS* submitted.
Gómez de Castro, A.I. and Verdugo, E.: 2004, *Ap&SS* **292**, 691.
Granzer, Th., Schussler, M., Caligari, P. and Strassmeier, K.G.: 2000, *A&A* **355**, 1087.
Guenter, E.W., Lehmann, H., Emerson, J.P. and Staude, J.: 1999, *A& A* **341**, 768.
Hamman, F. and Persson, S.E.: 1992, *ApJS* **82**, 247.
Hartigan, P., Edwards, S. and Ghandour, L.: 1995, *ApJ* **452**, 736.
Hawley, J.F.: 2004, *Ap&SS* **292**, 381.
Hidalgo, C.: 2004, *Ap&SS* **292**, 681
Johns-Krull, C.M., Valenti, J.A., Hatzes, A.P. and Kanaan, A.: 1999, *ApJ* **510**, L41.
Küker, M., Henning, T.H. and Rüdiger, G.: 2004, *Ap&SS* **292**, 599.
López-Martin L., Cabrit, S. and Dougados, C.: 2003, *A&A* **405**, 1L.
Matt, S., Goodson, A.P., Winglee, R.M. and Bohm, K.-H.: 2002, *ApJ* **574**, 232.
Ménard, F. and Bertout, C.: 1999, in: C.J. Lada and N.D. Kylafis (eds.), NATO Science Series *The origin of stars and planetary systems*, Kluwer Academic Publishers, Dordrecht, p. 341.
Moreno-Insertis, F.: 2004, *Ap&SS* **292**, 587.
Muzzerolle, J., Calvet, N. and Hartmann, L.: 2001, *ApJ* **550**, 944.
Muzzerolle, J., Calvet, N., Hartmann, L. and D'Alessio, P.: 2003, *ApJ* **597**, L149.
O'Brien, D.M., Garcia, P.J.V., Ferreira, J., Cabrit, S. and Binette, L.: 2004, Proc. of the workshop "Magnetic fields and star formation: theory versus observations" *Ap&SS* **292**, 609.
Petrov, P.P., Gahm, G.F., Gameiro, J.F., Duemmler, R., Ilyin, V., Laekkonen, T., Lago, M.T.V.T. and Tuominen, I.: 2001, *A&A* **369**, 993.
Preibisch, T.: 2004, *Ap&SS* **292**, 631.
Priest, E. and Forbes, T.: 2000, *Magnetic reconnection : MHD theory and applications*, New York: Cambridge University Press.
Pudritz, R.E.: 2004, *Ap&SS* **292**, 471.
Simon, T., Vrba, F.J. and Herbst, W.: 1990, *AJ* **100**, 1957.
Takagi, S., Murakami, H. and Koyama, K.: 2003, *ApJ* **573**, 275.
Uzdensky, D.A.: 2004, *Ap&SS* **292**, 573.
Vlahakis, N., Tsinganos, K., Sauty, C. and Trussoni, E.: 2000, *MNRAS* **318**, 417.
Valenti, J.A. and Johns-Krull, C.M.: 2004, Proc. of the workshop "Magnetic fields and star formation: theory versus observations" *Ap&SS* **292**, 619.
Wadhwa, M. and Russell, S.S.: 2000, in: V. Mannings, A.P. Boss and S.S. Russell (eds.), *Protostars and Planets IV*, University of Arizona Press, p. 995.
Walker, M.F.: 1972, *ApJ* **175**, 89.
Zapatero-Osorio, M.R., Caballero, J., Martín, E.L., Béjar, V.J.S. and Rebolo, R.: 2004, *Ap&SS* **292**, 673.

MAGNETIC INTERACTION BETWEEN STARS AND ACCRETION DISKS

DMITRI A. UZDENSKY
Kavli Institute for Theoretical Physics, University of California, Santa Barbara, CA 93106; E-mail: uzdensky@kitp.ucsb.edu

Abstract. In this review I consider modern theoretical models of coupled star–disk magnetospheres. I discuss a number of models, both stationary and time-dependent, and examine what physical conditions govern the selection of a preferred model.

Keywords: accretion disks, young stellar objects, magnetic fields

1. Introduction

In this paper, I review recent theoretical progress in our understanding of magnetic interaction between Young Stellar Objects (YSOs), in particular, Classical T-Tauri Stars (CTTSs), and their accretion disks. That such interaction takes place, we have no doubt, as there is now ample observational evidence of strong (of order 10^3 G) magnetic fields in these systems (e.g., Johns-Krull et al., 1999; Guenther et al., 1999).

Most of the theoretical work on magnetically linked star–disk systems, both analytical and numerical, has focussed on examining the structure and role of a large-scale axisymmetric magnetic field with (at least initially) dipole-like topology (see Figure 1). This field presumably arises due to the star's internal magnetic dipole moment. Studying such a large-scale star–disk magnetosphere will also be the focus of this paper. I will thus ignore the effects of any small-scale intermittent loops that may be generated by the turbulent dynamo action in the disk.

Although I am restricting myself to only large *spatial* scales, I will consider a variety of *temporal* scales. The shortest relevant time-scale is the rotation period, typically a few days for CTTSs; the longest time-scale is the accretion disk lifetime, which can be $\sim 10^6$ years or more (Königl et al., 1991; Kenyon and Hartmann, 1995). Among the models developed to date, there exist a dichotomy with respect to the system's behavior on the rotation time-scale. More specifically, in some models a direct magnetic connection between the disk and the star is maintained in a stationary configuration, whereas in other models it is not. In this paper I will review both of these classes of models.

Before I proceed, I would like to list some of the most important questions related to the subject of this review,

Astrophysics and Space Science **292:** 573–585, 2004.

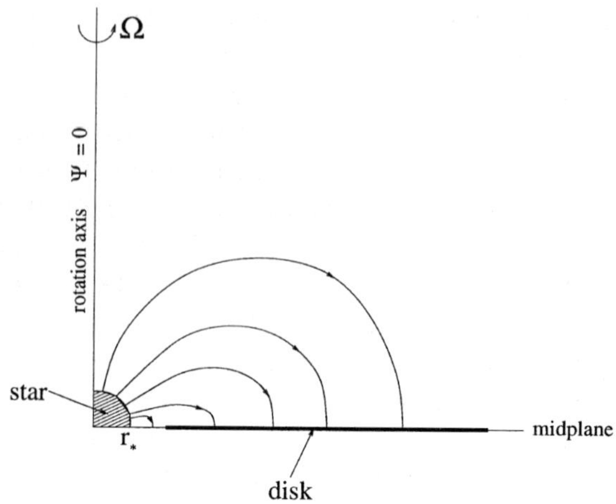

Figure 1. The general geometry of a magnetically-linked star–disk system.

1. What physical parameters determine whether a direct star–disk coupling via a large-scale dipole-like magnetic field can be maintained on the rotation-period time-scale?
2. If a quasi-stationary magnetically-coupled configuration does exist, what is its structure and how does it evolve on the longer (e.g., accretion) time-scale?
3. If the magnetic link is disrupted, then what is the nonsteady process? Are there periodic or quasi-periodic openings and closings of the field (due to magnetic reconnection) or there is a transition to a wind-supported permanent stationary open-field configuration without the link?
4. Is it possible that both scenarios are possible under different physical circumstances?
5. In either scenario, what are the effects of turbulent viscosity and magnetic diffusivity? And what is the role of winds and jets?
6. What are the implications for the time-variability of the accretion flow and for the angular momentum and energy exchange? What are the observational consequences that would allow one to discriminate between the models?

Although I will not be able to answer all of these questions in this review, I will use them as the main guiding themes in my discussion.

2. Nonstationary Models

The study of magnetic interaction between YSOs and their disks has been pioneered by Königl (1991). He has successfully applied the model of Ghosh and Lamb (1978), developed originally for accreting neutron stars, to explain several important observational features of T-Tauri stars, such as their relatively long rotation periods,

UV-excesses, and inverse P-Cygni profiles. In this steady-state model, the stellar magnetic field penetrates the disk over a finite range of radii both inside and outside of the corotation radius $r_{co} \equiv (GM_*/\Omega_*^2)^{1/3}$, where M_* is the mass of the star and Ω_* is its angular velocity. The spin-up magnetic torque due to the field lines connecting to the disk inside r_{co} is balanced by the spin-down torque by the lines connecting to the disk outside r_{co}. Königl has proposed that a magnetic field of 10^3 G at the stellar surface (a value consistent with observations, see Bouvier et al., 2003) can disrupt the disk at a few stellar radii (but well inside the corotation radius) and channel the accretion flow to higher stellar latitudes. He also has estimated the typical time needed to bring the star into the spin equilibrium with the disk to be $\sim 10^5$ years, much shorter than the typical accretion time for CTTSs.

Although this model has been very successful in explaining many spectral and variability features, it has not considered the dynamics of the magnetic field itself. The presence of a strong magnetic field has just been inferred, but no equations governing the magnetosphere structure have been solved. It turns out that there exists a very robust mechanism that leads to the breaking of the magnetic link on the rotation-period time-scale. This presents a serious obstacle for all steady-state models and thus gives us the motivation to consider nonstationary models.

The basic idea can be explained as follows. Both the star and the disk are fairly good conductors and so the magnetic field can generally be considered frozen into them. In addition, they rotate with different angular velocities (except at r_{co}). Therefore, the field lines are twisted by the differential rotation and toroidal magnetic flux is generated out of the poloidal flux. As the toroidal field builds up, the corresponding field pressure tends to push the field lines outward and inflate them. At first, the poloidal field structure changes very little, but, after the relative star-disk twist angle exceeds one radian or so, the field lines start to expand faster and faster (at an angle of $\sim 60^\circ$ with respect to the rotation axis) and tend to open up, thereby destroying the magnetic link between the star and the disk.

This opening process is essentially identical to a similar process that have been studied extensively in solar-corona context (e.g., Aly, 1984, 1995; Low, 1977, 2001; Mikić and Linker, 1994; Uzdensky, 2002b). Indeed, the opening of coronal magnetic arcades, brought about by displacements of the field-line footpoints on the photosphere, is one of the leading mechanisms for Coronal Mass Ejections (Low, 2001). A significant amount of work on this process has also been done in the accretion-disk context. Thus, in the force-free approximation, it has been shown, using both simple analytical and semi-analytical arguments (Aly and Kuijpers, 1990; van Ballegooijen, 1994; Lynden-Bell and Boily, 1994; Uzdensky et al., 2002a; Uzdensky, 2002a; Lynden-Bell, 2003), and via numerical solutions of the force-free Grad–Shafranov equation (Uzdensky et al., 2002a) that such an opening occurs at a finite twist angle (see Uzdensky et al., 2002b for a review). In addition, full numerical 2D MHD simulations (without the force-free assumption) have demonstrated the opening process at work as a part of the overall cycle (Hayashi et al., 1996, 2000; Goodson et al., 1997, 1999; Goodson and Winglee, 1999; Matt et al., 2002).

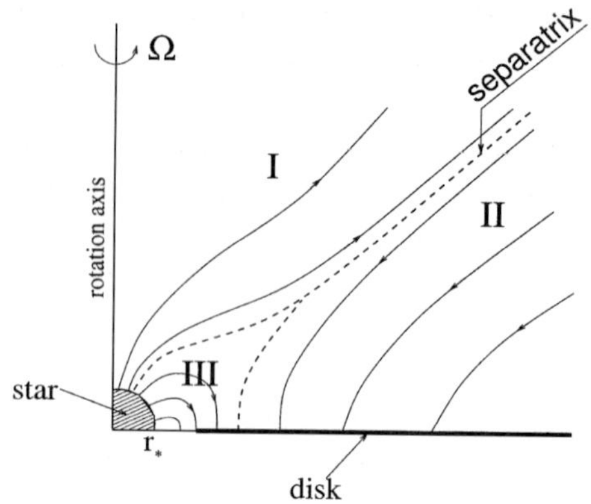

Figure 2. The magnetic configuration of Lovelace et al. (1995).

Thus, at present there is no doubt that, if stellar dipole magnetic field penetrates a conducting disk over a wide range of radii, the twisting of the field lines will open them, thereby breaking the star–disk connection everywhere, with perhaps the exception of the inner disk region where the magnetic field is strong enough to make the matter corotate with the star. This process naturally results in a nonsteady behavior, which has lead to the development of a number of time-dependent models. In addition, however, there exist several alternatives, leading to a small number of distinct stationary models. I shall discuss the nonsteady models first.

After the field lines expand and effectively open up, a natural question to ask is: what happens next? Currently, the situation is not entirely clear and there is no unique answer to this question. There are two drastically-different possibilities that are most often discussed.

In the first scenario (Lovelace et al., 1995), developed in the neutron star context, once the field lines open, they stay open indefinitely. A steady state is then achieved, although it is very different from the original one, as the magnetic link between the disk and the star has been severed on most of the field lines (see Figure 2). One can identify three topologically-distinct regions in the magnetosphere: the stellar wind region (region I), where the field lines extend from the star to infinity, the disk wind region (region II), where the field lines return from infinity to the disk, and the closed field region (region III)—the remnant of the linked magnetosphere, where the field enforces corotation of the disk with the star. Thus, the configuration here is stationary, but the magnetic link extends only over a small part of the disk.

In the second scenario, the situation is really time-dependent, with quasi-periodic cycles of field inflation and opening due to twisting followed by the field closing through reconnection and subsequent contraction back to the initial state. This picture has been suggested by van Ballegooijen (1994) and has subsequently been studied in extensive numerical simulations by a number of authors (Goodson et al., 1997, 1999; Goodson and Winglee, 1999; Hayashi et al., 2000; Matt et al., 2002).

Some recent observational results also seem to favor this point of view (e.g., Bouvier et al., 2003). Let us consider this most interesting scenario in more detail.

First, notice that in the Lovelace et al. (1995) scenario the poloidal magnetic field reverses across the separatrix between regions I and II. This makes the separatrix an obvious prospective site for reconnection. Indeed, the presence of a rather large anomalous or numerical resistivity has routinely lead to reconnection in the numerical simulations by Hayashi et al. (1996, 2000) and by Goodson et al. (1997, 1999), and Matt et al. (2002).[1]

It is also important to realize that most of the toroidal flux, generated in the twisting process, has now been evacuated radially to infinity (since the toroidal flux on an inflated flux tube is concentrated near the tube's apex). Therefore, magnetic field in both open-field regions is essentially poloidal (it is exactly poloidal in the force-free framework, but some toroidal field may be present in the MHD-wind regime where matter inertia is important). This means that if magnetic reconnection does occur somewhere along the separatrix, the inner newly reconnected field lines (connected to both the star and the disk) find themselves out of force-free balance: they have very strong poloidal-field tension that tries to pull them back toward the star but almost no toroidal-field pressure. As a result, in the absence of a powerful outgoing wind (see the discussion below), these inner reconnected field lines contract on the Alfvénic time-scale. If both reconnection (exhibited as a flare) and the subsequent contraction and relaxation occur quickly enough, then the resulting closed field lines have very little residual twist, similar to the original dipole-like state. This sets the stage for a new cycle. The continuing differential rotation gradually twists the lines up again and the whole sequence of events repeats, with the natural period of the order of the rotation period. As for the other, outer, newly reconnected field lines, they, together with the apex regions of the field lines that are still expanding somewhere far away, form a toroidal plasmoid (in a sense, a flying spheromak). The closed magnetic flux surfaces comprising such a plasmoid have the shape of tori nested around a circular line (an O-point in poloidal projection). Each plasmoid is not magnetically connected to either the disk or the star and is out of equilibrium; it then just flies away. If the motion of these plasmoids is collimated toward the axis (i.e., if they are flying mostly vertically), then they can feed the jet, providing an explanation for the observed knotty jet structure (Goodson et al., 1999). The plasmoids will be ejected out with the time intervals equal to the opening/closing period. For CTTSs, however,

[1]We also have to mention that Uzdensky et al. (2002b) were sceptical about the possibility of reconnection, but this is because they had used the van Ballegooijen (1994) self-similar model describing a uniformly-rotating disk. In that model finite-time field-opening occurs without current-sheet formation along the separatrix. In a more realistic case of nonuniform rotation, Uzdensky (2002a) has argued that there will be finite-time partial field opening accompanied by asymptotic thinning of the separatrix current-concentration region, which can be regarded as current-sheet formation. As the current layer becomes thinner and thinner, a reconnection process may be triggered by anomalous resistivity or the Hall effect.

this period is expected to be too short compared with the observed interval between jet knots (Goodson et al., 1999).

Whereas the effective field-opening time is about a fraction of the rotation period, the time between opening and reconnection is not certain. It depends on the intricate details of the reconnection process and, in particular, is intimately related to the so-called reconnection-trigger, or sudden-onset, problem, very well known in studies of flares in solar physics (Priest, 1984). Here is what it means in the context of our problem. As the field lines start to open, one by one, and the current sheet is formed along the separatrix, how long does one wait before reconnection starts? In other words, how much flux is opened before it is reconnected back? For example, one can imagine that reconnection is triggered as soon as the first few field lines have opened; then one will see the ejection of small islands, separated by the time it takes the critical amount of flux to open (much shorter than the rotation period). Or, in the opposite extreme, it may be that a large portion or all of the flux opens and only long after that reconnection somehow starts; then one will see finite-size plasmoids ejected with the time interval equal to the sum of the opening time (days) and the uncertain time delay before reconnection onset [for example, in the simulations by Goodson et al. (1999) the total cycle period was about a month].

Also not clear is how much flux is reconnected in each event before the reconnection process shuts off. This question is important because it determines the size of the ejected plasmoids. Indeed, it may be that reconnection proceeds until a large fraction of the flux is reconnected, in which case there will be large-amplitude oscillations in all of the system's parameters (Hayashi et al., 1996, 2000; Goodson et al., 1999). Alternatively, reconnection may stop very quickly after it has begun, and then one will see very small plasmoids ejected but the larger scale open-field structure will stay intact, as seen in numerical simulations by Fendt (2002). In fact, the model of Lovelace et al. (1995) can be viewed as an extreme manifestation of this latter scenario. Indeed, in this model it has been assumed, without much discussion or argumentation, that there is no reconnection at all. One can actually bring forth some arguments in favor of this point of view.

In general, there is a competition between field-line closing via reconnection and field-line opening by the wind. Reconnection will be stopped if the wind flowing along the open field lines is so strong that any newly reconnected field lines are swept open by it (B.C. Low, private communication). More specifically, if there were no flows along the unreconnected field lines (i.e., no wind), then a newly reconnected closed field line on the inner side of the reconnection region would contract rapidly, with the field-line apex moving out of the reconnection region toward the star with the poloidal Alfvén velocity $V_{A,\mathrm{pol}}$. However, if there is a background outflow such as a wind, then one has to add the velocity of this outflow. If the latter is larger than $V_{A,\mathrm{pol}}$, then the resulting apex motion will be directed outward, i.e., the field line will open again. Thus, I suggest the following physical criterion for determining when the reclosing of the open field lines via reconnection will occur. I propose that if the prospective reconnection site is located outside of the Alfvén radius (along the separatrix field line), so that the wind there is super-Alfvénic with respect to the reconnecting poloidal field, then everything will be swept outward by the wind and hence reconnection will not take place and the magnetic link will not be re-established. One then will get a helmet-streamer configuration like that of Lovelace et al. (1995). In the opposite case, reconnection will occur and will lead to the closing of (a portion of) the field lines, leading to a cyclic behavior.

To summarize, the system's behavior depends on both the reconnection physics (one needs to know where and when reconnection will be triggered) and on the wind physics: one needs to have a model for the wind to determine everything self-consistently.

3. Steady-State Models

I shall now switch gears and discuss the few existing steady-state models in which the magnetic link between the star and the disk remains unbroken. First, to maintain the link, one must find a mechanism that could stop the twisting process. One obvious possibility is the toroidal resistive slippage of the field lines with respect to the plasma in the disk. Let us examine this possibility in more detail and see under what conditions it can work.

The situation depends critically on the disk's effective magnetic diffusivity (which we shall sometimes call the resistivity). Unfortunately, the value of this diffusivity is not very well known (e.g., Bouvier et al., 2003). Nevertheless, one can set a reasonable upper limit by assuming that it is caused by the same turbulence that facilitates angular momentum transport across the disk. Thus, one can set the magnetic diffusivity η to be equal to the Shakura–Sunyaev kinematic viscosity: $\eta \sim \nu_{\text{turb}} = \alpha c_s h$, where c_s is the speed of sound, h is the disk half-thickness, and $\alpha \simeq 0.01 - 0.1$. This range of the α-values is consistent with the results of numerical MHD simulations of the Magneto-Rotational Instability (e.g., Brandenburg et al., 1996; Stone et al., 1996). It is also consistent with the level of MHD turbulence that is necessary for the ejection of disk winds, as follows from the work of Ferreira (1997) combined with the results of Ferro-Fontan and Gomez de Castro (2003).[2] The effective magnetic diffusivity of this kind leads to the toroidal slippage of field lines with respect to the disk with the relative drift velocity $\Delta v_\phi = (\eta/h)|B_\phi/B_z|_d \sim \alpha c_s |B_\phi/B_z|_d$, where the subscript d designates the disk's surface. For a thin disk, $c_s/v_K \sim h/r \ll 1$; thus, the slippage velocity is usually much smaller than the differential rotation velocity $r\Delta\Omega(r) \equiv r[\Omega_K(r) - \Omega_*]$. There are, however, two special circumstances when this is not so. They are very important as they point us toward the ways to get a steady-state configuration. I would like to stress, however, that both of these circumstances are somewhat unusual and hence the resulting steady states are not very natural. The first of the

[2]The conditions for launching MHD winds from accretion disks are found to be close to those necessary for the operation of the Magneto-Rotational Instability (Ferro-Fontan and Gomez de Castro, 2003).

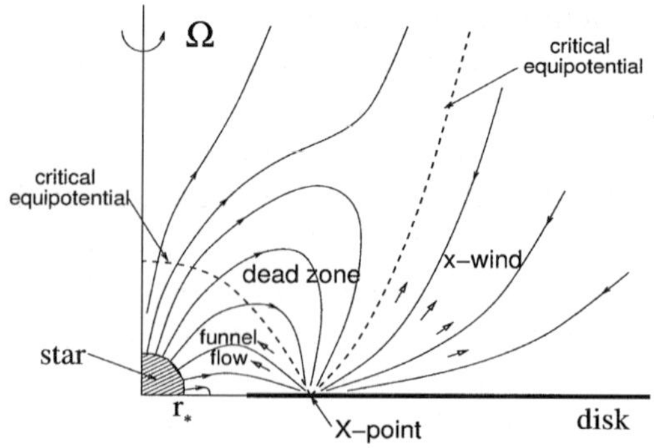

Figure 3. Schematic diagram of the X-wind configuration in the Shu et al. (1994) model.

two schemes is realized when the field lines under consideration are very close to the corotation radius (so that $\Delta\Omega \ll \Omega_K$); it provides the conceptual basis for the model developed by Shu et al. (1994a) (see also Shu et al., 1994b; Najita and Shu, 1994; Ostriker and Shu, 1995). The second scheme requires a very large ratio $|B_\phi/B_z|_d$ and provides the basis for the model developed by Bardou and Heyvaerts (1996) and by Agapitou and Papaloizou (2000).

Let us discuss the first scheme first. Usually, its main idea can be readily dismissed because for a steady state to exist globally, it must exist for all the field lines, and in the majority of models most of the magnetic flux crosses the disk a finite distance away from r_{co}. However, the model developed by Shu et al. (1994a) solves this problem by assuming that almost all of the magnetic flux is "trapped" and concentrated in the so-called X-region, a very close vicinity of the corotation radius (which the authors of the model call the X-point). This model is one of the most promising, well-developed, and sophisticated models and has gained a lot of popularity and observational support (e.g., Johns-Krull and Gafford, 2002) over the last few years. The disk's magnetosphere consists of three parts (see Figure 3). Field lines emanating from the inner part of the X-region connect to the star and form the magnetic funnel that directs the accretion flow. A second portion of the field lines also connects to the star but carries no mass flow; it forms what the authors call the dead zone. Finally, the remaining field lines, those emanating from the outermost part of the X-region, are open and carry the wind that plays a key role in removing the excess angular momentum from the disk. In addition, there are some open stellar field lines that extend from the star to infinity. Thus, the general topology of the poloidal magnetic field is similar to the helmet streamer configuration of Lovelace et al. (1995). An important difference however is that in the Lovelace et al. (1995) model, the poloidal flux is spread smoothly over the entire disk surface and the corotation radius plays no special role, whereas in the Shu model the flux is concentrated close to the X-point, with almost no field at $r > r_{\text{co}}$. Because of that, the differential rotation in the latter model is weak ($\Delta\Omega \ll \Omega_*$), and even a small disk resistivity is sufficient to eliminate the twisting and hence to ensure a steady state. In addition, inside of the corotation radius, the disk density drops rapidly as the plasma is uplifted to form the funnel flow; as a result, the corotation with the star is enforced by the strong magnetic field there.

The inner radius of the disk in the Shu model essentially coincides with the X-point [see, however, Ostriker and Shu (1995), who put it at $r_{\mathrm{in}} = 0.74\, r_{\mathrm{co}}$]. As the accreting matter enters the X-region, some part of it is loaded onto the open field lines and forms the outgoing X-wind. The rest of the plasma gradually diffuses through the magnetic field in the X-region and is loaded onto the funnel-region field lines and then falls onto the star. At the same time, most of the angular momentum of this falling matter is taken away by the magnetic field and is transported back to the inner portion of the disk, while only a small fraction ends up on the star. This provides an effective control mechanism for the star's spin and suggests a plausible explanation of the relatively-long rotation periods of CTTSs (Shu et al., 1994a).

This very interesting model is not free of its own problems and inconsistencies, however. Thus, for example, it is highly unlikely that the poloidal field on the inner side of the X-region will not diffuse toward the star. Indeed, these field lines are very strongly bent so that there is a highly concentrated current, essentially an equatorial current sheet, between the X-point and the inner edge of the disk [i.e., the kink-point of Ostriker and Shu (1995)]. Any small amount of resistivity will then cause the field to slip inward through this current layer. As for the plasma flow, it will not be able to counter this diffusion because it is in the same direction. Thus, the resistive-MHD Ohm's law immediately tells us that this configuration cannot be in a steady state.

The Shu model has also been criticized by Hartmann (1997). He pointed out that their solution requires very fine tuning, so that the inner disk radius (determined by the balance between the stellar magnetic field and the accretion flow) is equal to r_{co} (determined by Ω_*). Hartmann considers this situation unacceptable, citing an example of DR Tau, where accretion rate has been observed to change on the time scale far too short for Ω_* to adjust. He seems to favor the van Ballegooijen (1994) viewpoint and concludes that "the entire magnetosphere might be a complicated, time dependent structure." He also emphasizes the importance of magnetospheric reconnection, noting that it "should lead to substantial heating and flare activity" (Hartmann, 1997).

Now let us consider the second possibility for a steady state. In this scenario, the balance between the twisting due to the differential rotation and the turbulent resistive slippage is made possible by a very large ratio of the toroidal to vertical magnetic field at the disk surface, of the order of $\alpha^{-1}r/h \gg 1$. Such high values are usually considered to be unlikely. Indeed, the angle between the field lines and the disk is determined by the entire solution in the magnetosphere and cannot be arbitrary. The density of matter above the disk is typically so low that magnetic forces completely dominate the dynamics there. In this force-free regime, the toroidal field at the disk surface, $B_{\phi,d}$, increases in proportion to the twist at first, but then reaches a maximum and starts to decline during the rapid-expansion phase; it goes to zero as the field approaches the open state. The maximum value of $B_{\phi,d}$, achievable in a force-free magnetosphere, depends sensitively on the way the poloidal magnetic flux is distributed across the disk, that is on the function $\Psi_d(r)$. Usually, as it turns out, this maximum value is of the same order as the vertical magnetic field and hence the minimum angle between the disk and the projection of the magnetic field vector on the $\theta - \phi$ plane is of order one. In this case, the differential rotation cannot produce the required very large values of the disk toroidal field. The primary physical reason for this is that most of the toroidal magnetic flux, which is being continuously generated by twisting, becomes concentrated near the field-line apex (i.e., the farthest from the star point on a field line). As the field expands, it becomes energetically favorable for the toroidal flux to escape to infinity by opening the poloidal field lines. Coming back to the question of the effects of the disk resistivity, we see that, with the disk toroidal field limited by the opening and flux-escape process, the toroidal resistive slippage, even in a turbulent disk, cannot be fast enough to significantly affect the twisting process.

On the other hand, for a certain class of functions $\Psi_d(r)$, the maximum value $|B_\phi/B_z|_{\rm d,\,max}$ of the ratio of the toroidal to vertical field components at the disk surface, allowed by the force-free solution in the magnetosphere, can be large. In particular, if $\Psi_d(r) \sim r^{-n}$, then $|B_\phi/B_z|_{\rm d,\,max} \sim O(1/n)$ in the limit $n \to 0$ (Lynden-Bell and Boily, 1994; Bardou and Heyvaers, 1996; Agapitou and Papaloizou, 2000; Uzdensky et al., 2002a).

One can then picture the following evolutionary scenario. Let us start with a non-steady cyclic configuration such as that described by Goodson et al. (1997, 1999). During the first part of the cycle, as the field expands and approaches the open state, the field lines at the disk surface are inclined away from the star [i.e., $(B_r/B_z)_d > 0$] and hence diffuse a little bit outward.[3] Then, during the second part of the cycle, as the reconnected field contracts back to the nearly potential state, the field lines may be inclined toward the star at the disk surface [i.e., $(B_r/B_z)_d < 0$]; they will then diffuse inward. (Note also that, for the field lines inside $r_{\rm co}$, such an inclination is conducive to loading of matter onto the field lines; thus, accretion can take place during this phase of the cycle.) Then, one can ask what happens on a much longer time scale, when the radial diffusion of magnetic field has to be included. Here, two possibilities immediately come to mind.

It may be, as suggested by van Ballegooijen (1994), that the little diffusive displacements will, over time, redistribute the disk's magnetic flux so that the net displacement over one cycle will become zero. Then, an averaged steady state will be established, i.e., the cycles of field opening, reconnection, and closing will produce no net secular evolution in the magnetic flux distribution. Since the amount

[3]With the initial velocity of order the rms turbulent velocity in the disk, $v_{\rm turb} \sim \alpha c_s$; hence the characteristic radial footpoint displacement over a rotation period scales as $\Delta r \sim v_{fp}\Delta\Omega^{-1} \sim \alpha c_s/\Delta\Omega \sim \alpha h \Omega_K(r)/\Delta\Omega \ll r$.

of the outward radial displacement depends on the exact moment of reconnection, it follows that the physics of reconnection again plays a crucial role in determining the long-term magnetic flux distribution.

On the other hand, as I have discussed above, if $\Psi_d(r) \sim r^{-n}$, then $|B_\phi/B_z|_{\rm d,\ max} \sim O(1/n)$ in the limit $n \to 0$. Thus, it is in principle possible for the system to achieve exact, not time-averaged, steady state if the disk's flux redistributes in such a way that the corresponding value of $n \equiv -d \ln \Psi_d / d \ln r$ becomes very small. Since the value of $|B_\phi/B_z|_d$, necessary for the balance between the differential rotation and toroidal resistive slippage, is inversely proportional to the disk's effective resistivity, we see that in this case the disk flux distribution $\Psi_d(r)$ is essentially determined by the resistivity. In the case of Shakura–Sunyaev turbulent resistivity, one can obtain the following upper limit: $n < C|B_\phi/B_z|^{-1}_{\rm d,\ max} \sim \eta/rh\Delta\Omega = O(\alpha h/r) \ll 1$, where C is a finite number. Note also that for a steady state to be maintained, one must worry not only about the toroidal direction, but also about the radial direction. This requirement gives not just an upper limit, but in fact determines implicitly the entire function $n(\eta)$, or, more generally, the dependence $\Psi_d(r)[\eta(r)]$ (Bardou and Heyvaerts, 1996; Agapitou and Papaloizou, 2000). Such a stationary field configuration, possible in principle, is very different from the dipole field; in particular, it leads to a dramatic decrease in the torque between the star and the disk (Agapitou and Papaloizou, 2000).

4. Summary

In conclusion, I would like to give the following approximate list of the major theoretical approaches to the problem of magnetically coupled star–disk magnetospheres:

1. Very rich nonstationary scenario (Aly and Kuijpers, 1990; van Balegooijen, 1994; Hayashi et al., 1996, 2000; Goodson et al., 1997, 1999; Goodson and Winglee, 1999; Uzdensky et al., 2002a, 2002b; Matt et al., 2002) with cycles of field inflation, opening, reconnection, contraction, and accretion. Both accretion and outflows occur intermittently, with variability on the differential rotation period (or somewhat longer) time-scale. The amplitude of these oscillations (e.g., how much poloidal flux is opened and then reconnected in each cycle) depends strongly on the physics of reconnection and is not very well constrained. For example, the steady-state model of Lovelace et al. (1995) can be considered a limiting case where no reconnection takes place at all, and thus the oscillation amplitude is zero.
2. The steady-state X-wind model of Shu et al. (1994a). The model of Lovelace et al. (1995) can be considered a bridge model between the Goodson and Shu models.

3. Finally, a steady-state closed magnetosphere with the poloidal vertical magnetic field that threads the disk scaling as $B_z(r) \sim r^{-[2+O(\eta)]}$—models of Bardou and Heyvaerts (1996) and of Agapitou and Papaloizou (2000). These models take into account the field's radial diffusion in the disk over a long (compared with Ω_*^{-1}) time scale.

At present, it is apparently too early to select one of these models as the preferred one based on purely theoretical considerations. More rigorous theoretical work, in conjunction with more sophisticated and thorough numerical simulations and comparison with observations, is needed to sort things out.

Acknowledgement

I am deeply indebted to Arieh Königl and Christof Litwin for many, many insightful and productive discussions, and to Ana Gómez de Castro for her very thoughtful comments and useful suggestions. I am also very grateful to the organizers of the International Workshop on Magnetic Fields and Star Formation (Madrid, April 21–25, 2003) for the invitation to write this review. This research was supported by the National Science Foundation under Grant No. PHY99-07949.

References

Agapitou, V. and Papaloizou, J.C.B.: 2000, *MNRAS* **317**, 273.
Aly, J.J.: 1984, *ApJ* **283**, 349.
Aly, J.J.: 1995, *ApJ* **439**, L63.
Aly, J.J. and Kuijpers, J.: 1990, *A&A* **227**, 473.
Bardou, A. and Heyvaerts, J.: 1996, *A&A* **307**, 1009.
Bouvier, J., Alencar, S.H.P. and Dougados, C.: 2003, preprint (astro-ph/0306553).
Brandenburg, A., Nordlund, A., Stein, R.F. and Torkelsson, U.: 1996, *ApJ* **458**, L45.
Fendt, C.: 2002, preprint (astro-ph/0211601).
Ferreira, J.: 1997, *A&A* **319**, 340.
Ferro-Fontan, C. and Gomez de Castro, A.I.: 2003, *MNRAS* **342**, 427.
Ghosh, P. and Lamb, F.K.: 1978, *ApJ* **223**, L83.
Goodson, A.P., Winglee, R.M. and Böhm, K.-H.: 1997, *ApJ* **489**, 199.
Goodson, A.P., Böhm, K.-H. and Winglee, R.M.: 1999, *ApJ* **524**, 142.
Goodson, A.P. and Winglee, R.M.: 1999, *ApJ* **524**, 159.
Guenther, E.W., Lehmann, H., Emerson, J.P. and Staude, J.: 1999, *A&A* **341**, p.768.
Hartmann, L.: 1997 in: B. Reipurth and C. Bertout (eds.), *Herbig-Haro Flows and the the Birth of Low Mass Stars*, Kluwer, Dordrecht, 391.
Hayashi, M.R., Shibata, K. and Matsumoto, R.: 1996, *ApJ* **468**, L37.
Hayashi, M., Shibata, K. and Matsumoto, R.: 2000, *Adv. Space Res.* **26**, 567.
Johns-Krull, C.M., Valenti, J.A. and Koresko, C.: 1999, *ApJ* **516**, 900.
Johns-Krull, C.M. and Gafford, A.D.: 2002, *ApJ* **573**, 685.
Kenyon, S.J. and Hartmann, L.: 1995, *ApJ Suppl.* **101**, 117.
Königl, A.: 1991, *ApJ* **370**, L39.
Lovelace, R.V.E., Romanova, M.M. and Bisnovatyi-Kogan, G.S.: 1995, *MNRAS* **275**, 244 (LRBK).

Low, B.C.: 1977, *ApJ* **212**, 234.
Low, B.C.: 2001, *Journal Geophys. Res.* **106**, 25–141.
Lynden-Bell, D. and Boily, C.: 1994, *MNRAS* **267**, 146.
Lynden-Bell, D.: 2003, *MNRAS* **341**, 1360.
Matt, S., Goodson, A.P., Winglee, R.M. and Böhm, K.-H.: 2002, *ApJ* **574**, 232.
Mikić, Z. and Linker, J.A.: 1994, *ApJ* **430**, 898.
Najita, J.R. and Shu, F.H.: 1994, *ApJ* **429**, 808.
Ostriker, E.C. and Shu, F.H.: 1995, *ApJ* **447**, 813.
Priest, E.R.: 1984, *Solar Magnetohydrodynamics*, Reidel, Dordrecht.
Shu, F., Najita, J., Ostriker, E., Wilkin, F., Ruden., S. and Lizano, S.: 1994a, *ApJ* **429**, 781.
Shu, F.H., Najita, J., Ruden., S.P. and Lizano, S.: 1994b, *ApJ* **429**, 797.
Stone, J.M., Hawley, J.F., Gammie, C.F. and Balbus, S.A.: 1996, *ApJ* **463**, 656.
Uzdensky, D.A., Königl, A. and Litwin, C.: 2002a, *ApJ* **565**, 1191.
Uzdensky, D.A., Königl, A. and Litwin, C.: 2002b, *ApJ* **565**, 1205.
Uzdensky, D.A.: 2002a, *ApJ* **572**, 432.
Uzdensky, D.A.: 2002b, *ApJ* **574**, 1011.
van Ballegooijen, A.A.: 1994, *Space Sci. Rev.* **68**, 299.

THE EMERGENCE OF MAGNETIC FIELD INTO STELLAR ATMOSPHERES

F. MORENO-INSERTIS

Instituto de Astrofisica de Canarias and Department of Astrophysics, Univ. de La Laguna, 38200 La Laguna, Tenerife, Spain; E-mail: fmi@ll.iac.es

Abstract. The magnetic activity in the atmosphere of the Sun, possibly of most other magnetically active late-type stars, is a consequence of the emergence of magnetic flux from the stellar interior. Research in this field has to consider the rise of magnetized plasma from the depths of the convective envelope through the stellar surface and out across the corona. In the present review, a number of key questions, difficulties and results are discussed.

Keywords: stellar magnetism, stellar atmospheres, dynamo

1. Introduction

Magnetic field is continually emerging from the solar interior into the atmosphere as part of the regular magnetic activity cycle. The emergence occurs sometimes in a rather orderly fashion, with the upcoming plasma just pushing away the matter and magnetic field it encounters in the atmosphere. Quite often, though, in the aftermath of a process of emergence, catastrophic phenomena (at least in the mathematical sense of the word) take place that disrupt the magnetic structure, leading to flares, and, possibly, coronal mass ejections or CMEs (e.g., Low, 2001; Golub and Pasachoff, 1997). The Sun is the only star in which we can observe these processes in detail. However, flux emergence must be a common process in all stars with cyclic magnetic field generation in their interior, and, in particular, in magnetically active late-type stars.

On a larger scale, the process of emergence has to be studied in the framework of the cyclic generation and disposal of magnetic flux in the star. The magnetic field is generated in the stellar interior and transported to the surface through magnetic buoyancy (see Schrijver and Zwaan, 2000; Parker, 1978). The most apparent instance of this process is the birth of an active region, with magnetic flux appearing in large amounts in the photosphere, further rising through the chromosphere and transition region into the corona. This process has been observed in the visible range in the photosphere since the early days of modern astrophysics, including the development of pores, arch filament systems, sunspots, and so on. In recent years, spectacular views of the emergence into the higher atmospheric layers have been obtained through a number of UV and X-Ray detectors on board space missions like Yokhoh, SOHO, TRACE, and so on. They show the rise of bright coronal loops

Astrophysics and Space Science **292:** 587–597, 2004.

rooted in the photosphere, most of them linking the opposite magnetic polarities of an active region, while others establish linkages to neighboring active regions through magnetic reconnection.

Understanding the physics of the magnetic activity in stars requires a thorough investigation of the emergence of flux at all levels where it takes place. However, this is no small undertaking. As just one example, consider the following: in a star like the Sun, there is an extremely large density and pressure contrast between the solar interior and the corona (e.g., some 15 orders of magnitude in density between the bottom of the convection zone and the base of the corona). A magnetized plasma package starting its voyage in the deep convection zone must increase its linear dimensions by some five orders of magnitude (or more, if there are deviations from isotropy) if it is to go all the way up to the corona. Hence, the overwhelming majority of the mass of the initial plasma package must remain in the convection zone. Yet, the tiny fraction that goes all the way up to the higher atmosphere requires detailed study, since it unleashes the magnetic activity observed in the star! The large expansion places severe demands on the resolution power of any computational scheme used to solve the problem numerically. Other complications increase the difficulty of the numerical simulation of this process, like: the timescales of the various processes involved are widely different in the convection zone, photosphere, chromosphere and corona; there may be interaction with convective flows in the interior; radiative transfer is important for the magnetized plasma in the atmosphere; boundary layers and wakes appear as long as the rising magnetized plasma moves relative to the unmagnetized surroundings; and so forth.

In the past 20 years important advances have been made in understanding the physics of flux emergence across the convection zone. In the solar interior a number of crucial simplifications can be applied. First, the plasma is optically thick and always close to thermodynamic equilibrium, so no complicated radiative transfer needs to be applied. Second, consider an individual magnetic tube which later on appears as a bipolar magnetic region at the surface. Given the magnetic flux observed in the latter plus the expected field strengths in the interior, one can estimate the radius of such a tube to be a few thousand kilometers in the deep convection zone (say at a depth of $180,000$ km below the surface). That radius is much smaller than all scale lengths both along the axis of the tube and in the external medium. Hence, one can approximate the rising magnetic tube as a one-dimensional continuum moving in a 3D stratified medium. To that end, one has used the so called *slender flux tube approximation* (Section 2). This fact prompted the development of comparatively simple numerical models in the 1980s and 1990s and led to a good understanding of the physics involved. However, the 1D models cannot be used when the tube reaches the photosphere. In fact, strictly speaking, not even in the final few thousand km before they reach the surface: the radius of the tube in those heights is already large enough (as measured by the local scale heights) that consideration of the transverse structure of the tube is necessary (Section 3).

When studying the rise of flux upward of the photosphere, a full 3D study including detailed radiative transfer is necessary. Additionally, the lower atmosphere is strongly stratified, with typical pressure scaleheight as small as $\mathcal{O}(10^2 \text{ km})$. Finally, the magnetic field develops structures on small scales where opposite magnetic polarities reconnect and the plasma is heated. All those factors together make a full study of the emergence in the atmosphere a big modeling challenge, well beyond the current computational capabilities. So, to achieve progress, one must study the problem in a piecemeal fashion, considering the various basic ingredients separately and hoping to be able to combine them together in a single model in the not too distant future. In this vein, a number of ideal-MHD nonradiative models are appearing right now, which may yield insights into the dynamics and topology of the resulting magnetic loops. These are briefly described in Section 4.

2. Emergence of Flux in the Convection Zone: The 1D Models

As depicted in Figures 1 through 3, these models calculate the evolution of thin magnetic tubes in the convection zone, typically using as initial configuration a toroidal loop in the deep convection zone or in the overshoot region below it (for a summary of the thin tube calculations see the reviews by Moreno-Insertis (1997a, 1997b). In a standard case (e.g., Moreno-Insertis, 1986; Caligari et al., 1995), the tube is unstable to perturbations that bend the field lines upward, the basic mechanism being the classical Parker instability of horizontal magnetic fields in a gravitational potential: if the wavelength of the perturbation is large enough, the buoyancy of the crest of the tube cannot be overcome by the restoring effect of

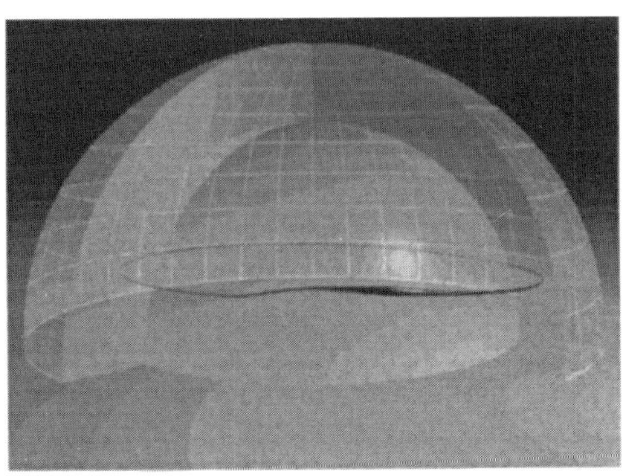

Figure 1. Typical result for the path of a magnetic flux tube obtained in the slender-tube calculations discussed in Section 2. The inner sphere corresponds to the radiative core of the Sun. The magnetic tube depicted is rooted in the deep convection zone, but its upper part (the thicker stretch at the front) reaches close to the photosphere. Note the *tilt* of the crest of the tube with respect to the East-West direction resulting from the evolution. From Caligari et al. (1995).

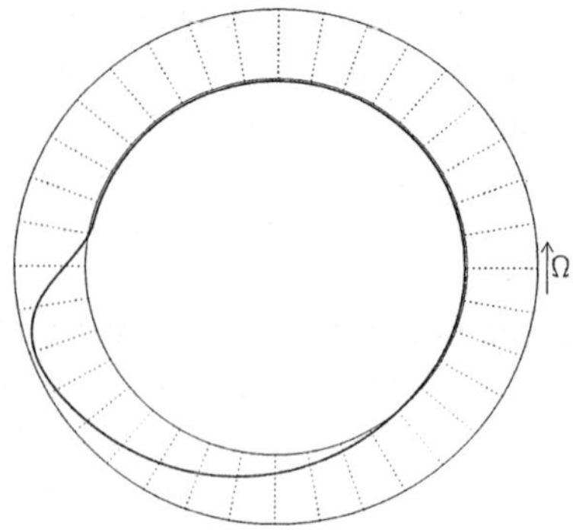

Figure 2. Illustration of the path of a magnetic flux tube rooted in the deep convection zone and close to breaking out at the photosphere. The arrow indicates the direction of rotation of the unperturbed stratification: the angular velocity vector thus points *out* of the plane of the figure toward the reader.)

the magnetic tension so that it travels all the way up to the photosphere (Figure 3). Other authors (e.g., D'Silva and Choudhuri, 1993; Fan et al., 1994) initiate the unstable rise by giving a density distribution (or, equivalently, entropy distribution) along the tube that renders some stretches of the tube underdense, with the ensuing buoyancy being of order of magnitude of the classical magnetic buoyancy of Parker [$(1/\beta)$, with the plasma β being the ratio between gas and magnetic pressure]. In either case, the ensuing evolution is amenable to detailed numerical simulation using the slender flux tube approximation (Defouw, 1976; Roberts and Webb, 1978; Spruit, 1981). The equations used within this model (see, e.g., Caligari et al., 1995) consist of 1D versions of the continuity and magnetic flux conservation equations, the equation of motion, a prescription for the thermal evolution

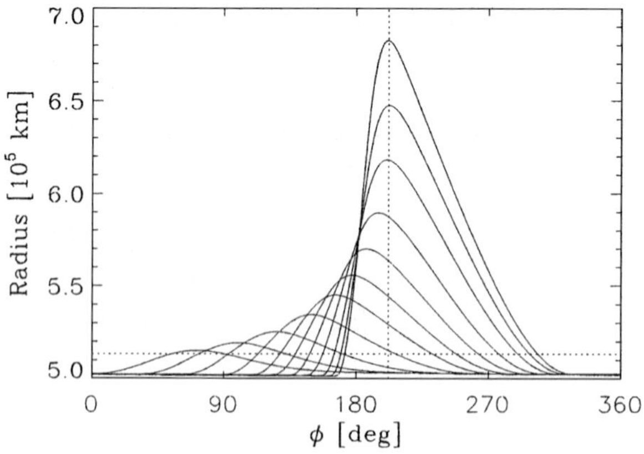

Figure 3. Successive configurations of a magnetic tube rising toward the photosphere (at $R = 6.9\ 10^5$ km) starting from a Parker-unstable toroidal tube in the deep convection zone. Shown are the azimuthal (Φ, abscissa) and the radial coordinate (ordinate) of the plasma elements along the magnetic tube.

of the plasma (typically, a condition of adiabaticity) and the equation of state. The slender-tube calculations typically include a detailed model of the background stratification in spherical geometry. Given that the tube is very thin by assumption, magnetosonic waves can instantly enforce (gas + magnetic) pressure balance across the tube, so that the condition $p_{gas}(\text{external}) = p_{gas} + p_{mag}(\text{internal})$ is fulfilled at all times.

These 1D models have been highly successful in that they provide explanations for kinematic and morphological features observed in young active regions. Seen geometrically, an active region results from the intersection of a rising loop like those shown in Figures 1 through 3 with the photosphere. If the crest of the rising tube is not perfectly aligned with a line of constant latitude (as clearly visible in Figure 1), then the main axis of the resulting active region, i.e., the line joining the centroids of its two polarities, will be seen on the solar surface to be tilted with respect to the East-West direction. In fact, statistical studies of observed active regions show that, in average, their main axis is rotated (counter)clockwise in the (South) North hemisphere with respect to the equator with an angle that is larger the higher the latitude of the active region (see, e.g., Wang and Sheeley, 1989, 1991). There is a simple explanation for the appearance of a tilt angle within the slender-tube models as a consequence of the Coriolis force. As the tube rises it expands sideways as a whole. As in high pressure regions in the Earth's atmosphere, the Coriolis force tends to rotate the expanding matter (counter) clockwise in the (southern) northern hemisphere. When the timescale of rise is of the order of the solar rotation period, the resulting tilt angle matches the observations (D'Silva and Choudhuri, 1993; Schüssler et al., 1994; Fan et al., 1994; Caligari et al., 1995; Fisher et al., 1995; Longcope and Fisher, 1996). Fitting the timescale of rise so that it fulfills said condition yields the order of magnitude of some of the parameters of the tubes in the deep convection zone whose evolution produce active regions in the surface. Their field strength, in particular, must be of order 10^5 G, instead of the substantially lower values that most people had assumed until then. In fact, there are further arguments for this high value of the field strength in the deep convection zone (Moreno-Insertis, 1992; Schüssler et al., 1994). The prediction of the order of magnitude of the field strength for the deep convection zone can be seen as one of the successes of the slender flux tube models.

Another prediction of the thin-tube models basically supported by observations is the asymmetry in the global kinematics of the two opposite polarities of a young active region. The two flanks of the rising loop are seen to have different inclinations with respect to the vertical direction (as is apparent in Figure 3). This is, again, a consequence of the rotational inertial forces acting along the rise: angular momentum conservation tends to push the rising matter elements *backward* with respect to the direction of rotation, so that the leading flank (*leading* as regards the rotation) becomes more inclined than the *follower* one. Now, if the inclination is sufficiently large (as in the case of the figure), when observing the intersection of the rising tube with the photosphere, the leading polarity must be seen to separate more quickly from the neutral line than the follower polarity. This is again borne out by the observation of active regions (Van Driel-Gesztelyi and Petrovay, 1990; Strous et al., 1996). In fact, the numerical models predict a velocity asymmetry close to the observed values (Moreno-Insertis et al., 1994).

In spite of their general success, these models become of limited (or no) use to understand the evolution of the magnetic flux close to the photosphere and at higher atmospheric layers. As explained in the following section, the assumptions that make the 1D approximation valid in the convection zone are no longer applicable even in the uppermost, say, 5% of the convection zone. To understand the physics of the rising magnetic plasma from there upward, a complete three-dimensional simulation is necessary.

3. The Arrival of Flux at the Surface

The plasma in the rising magnetic tubes that reach the final several thousand km below the photosphere is characterized by (1) a high speed of rise with superalfvenic velocities and Mach numbers approaching $\mathcal{O}(10^{-1})$; (2) a typical cross sectional radius in excess of the local pressure scale height and (3) a slight temperature excess with respect to the non magnetic surroundings, as a result of its adiabatic rise in the strongly superadiabatic upper levels of the convection zone.

There are simple reasons for those facts. The key quantity in this is the density difference between a plasma element in the magnetic tube and in the external medium, $\Delta\rho = \rho_i - \rho_e$, where the indices i and e stand for the plasma inside the tube and for the unperturbed surroundings at the same height, respectively. All along the rise in the convection zone, $\Delta\rho/\rho$ is negative and extremely small [e.g., $\mathcal{O}(10^{-4})$] At a given depth z along the rise, elementary thermodynamics coupled with the condition of total pressure equilibrium between the tube and its surroundings yield the following expression:

$$\left(\frac{\Delta\rho}{\rho_e}\right)(z) \approx \frac{S_e(z) - S_{e0}}{c_p} - \frac{1}{\gamma}\left(\frac{1}{\beta} - \frac{c_{p0}}{c_p}\frac{1}{\beta_0}\right), \tag{1}$$

where $S_e(z)$ is the entropy of the external unperturbed stratification and the subindex 0 indicates quantities measured at the initial position of the magnetic tube in the deep convection zone. Quantities with no 0 subindex (e.g., β, ρ_e, c_p) are measured at an arbitrary depth z; c_p is the specific heat at constant pressure. Expression (1) is valid to first order in the small quantities $(\Delta\rho)/\rho$ and $1/\beta$ and for magnetic tubes starting from a condition of mechanical equilibrium, like the Parker-unstable tubes. The expression is also valid for tubes starting from a condition of thermal equilibrium between tube and surroundings by multiplying the $1/\beta_0$ term with $(1 - \gamma)$. The first term in Eq. (1) is the entropy difference in the stratified medium outside the tube and is negative because entropy decreases outward in a convection

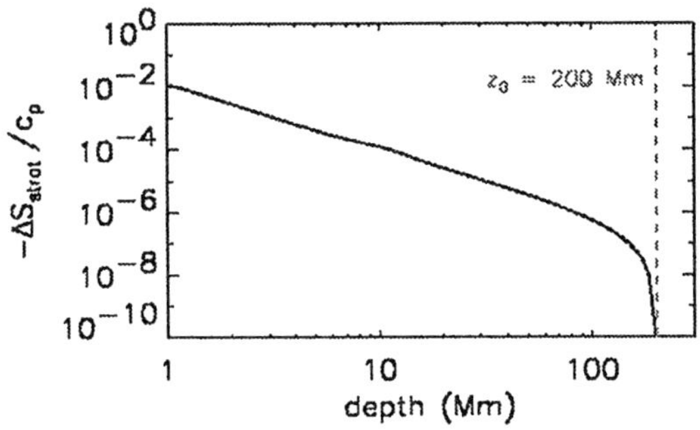

Figure 4. Convection zone stratification: entropy deficit with respect to the bottom of the convection zone, $S_{e0} - S_e(z)$, normalized with the local specific heat $c_p(z)$ in the convection zone. The reference level is situated at the bottom of the convection zone: $z_0 = 2\ 10^5$ km [Calculated from Spruit's (1977) convection zone model.]

zone (Figure 4). In fact, for most of the rise, the entropy term is much smaller than the magnetic term. Except in the initial stages of the rise, the latter is then just given by $1/(\gamma\ \beta)$, since β quickly decreases with respect to the initial value. Therefore, the buoyancy of the tube remains of order the classical Parker buoyancy $(1/\beta)$ for most of the rise. In the uppermost several thousand km, however, the entropy difference grows considerably and may well become the dominant term: depending on the plasma beta of the tube in those heights, the buoyancy of the plasma elements reaching the surface may be as much of a *convective origin* as of a magnetic origin, with the relative density deficit being in any case at least $\mathcal{O}(10^{-2})$. In fact, the thin tube calculations provide no evidence that the plasma beta could go below 10^2 while the tube is still in the convection zone: for that, a large stretching of the plasma along the field lines would be necessary, much larger than seen in those calculations. However, if such stretching were to take place, then the density deficit of the magnetic elements (hence also their buoyancy) could only be larger than the quoted value.

On the other hand, following the steep density decrease as the tube soars in the convection zone, its radius grows by a large factor. The left panel of Figure 5 shows the ratio between the tube radius R and local pressure scale height of the external stratification H_p for the apex of a tube rising toward the surface in a typical slender flux tube calculation with total magnetic flux 10^{21} mx and initial field strength in the deep convection zone $3\ 10^5$ G. As we see, the radius may become comparable to the local pressure scaleheight at a depth of several thousand kilometers. This also signals the boundary where the slender-flux-tube approximation clearly abandons its domain of validity.

Ratios R/H_p no longer small and relative high levels of buoyancy yield large rise speeds. This can be seen as follows: a lower limit for that speed is obtained by prescribing equilibrium between the aerodynamic drag braking the advance of the

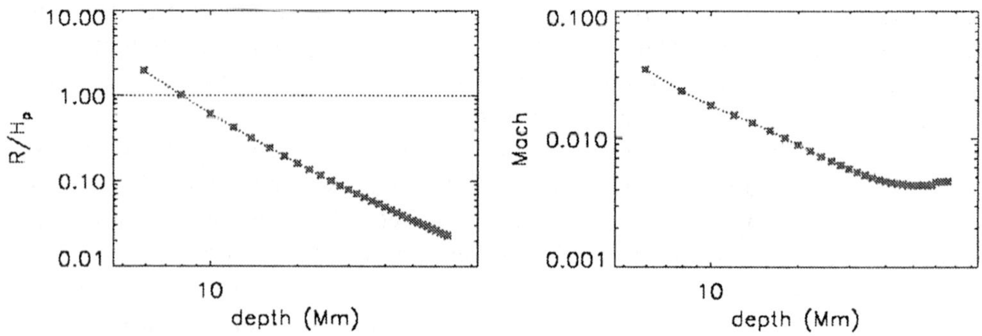

Figure 5. Left panel: Tube radius, R, in units of the local pressure scale height H_p for a magnetic tube rising toward the surface in a typical slender flux tube calculation. As $R/H_p \to 1$, the slender flux tube approximation leaves its domain of validity. *Right panel:* Mach number for the apex of the same tube.

tube and the buoyancy force:

$$\frac{v_{\rm rise}}{c_s} > \mathcal{O}\left(\frac{R}{H_p}\left|\frac{\Delta\rho}{\rho}\right|\right)^{1/2} \tag{2}$$

Using the estimates just discussed, we see that the rise speed in the final few thousand kilometers below the surface may reach values up to a few hundredths the local sound speed, and, also, above the internal Alfven speed. In fact, this fits well with the values obtained in the slender flux tube calculation shown in Figure 5, as can be seen in the right panel of that figure.

Summarizing, the final stages of the rise of a magnetic tube from the deep convection zone shortly before it breaks out at the photosphere must be characterized by a large cross-sectional radius (larger than the local pressure scale height), super-alfvenic speed of rise and large buoyancy, with a relative density difference with respect to the surroundings of close to 1%.

4. The Emergence of Flux Into the Corona

In spite of the impressive increase in observational power of the past decade, including several space missions, there is still no global understanding of the evolution of the magnetic field and plasma encompassing the different layers of the solar atmosphere when fresh magnetic flux bursts out from the interior. Given the wide range of temperature, length and time scales, plasma beta, etc. involved in that process, observations can only provide a patchwork of results that are difficult to fit together. To achieve progress, numerical efforts are called for. However, those same features and the necessity to take into account the radiative transfer (including non-LTE processes, e.g., in the chromosphere) render numerical simulation extremely difficult. Only recently have 3D numerical models begun to appear, even though based on crude assumptions concerning the thermal evolution of the plasma and often using somewhat artificial initial conditions.

So far, two families of models have been proposed. Matsumoto et al. (1998), Fan (2001), Magara and Longcope (2001, 2003) and Archontis et al. (2004) start their calculations by placing a horizontal magnetic flux tube underneath (even though close to) the photosphere. The evolution is initiated either by imposing a density deficit profile along the tube or by forcing a vertical velocity profile in it. In all cases, a rising loop develops and emerges through the photosphere. The initial stratification in these models is a simplified multilayer stratification including a simple convection zone, an isothermal photosphere, a steep temperature rise (mimicking the transition region) and an isothermal corona with temperature of order 100 times the photospheric temperature. The evolution of the plasma was assumed adiabatic (except for numerical diffusion terms). This assumption is patently wrong in the photosphere and chromosphere and its justification is based only on the impossibility, at the present stage, of including detailed radiative transfer, thermal conduction and further cooling/heating processes. They also start with a strongly twisted tube: Fan (2001), imposes a constant pitch such that the field lines turn once around the axis when you advance a distance 2π times the tube radius. This amount of twist helps prevent the splitting of the magnetic tube into vortex fibrils (see Emonet and Moreno-Insertis, 1998), but also force a strongly sheared structure for the magnetic field with height in the atmosphere. On the other hand, Abbett and Fisher (2003) and Fan and Gibson (2003) input magnetic flux into the corona by driving the lower boundary of the integration box with an electric field, albeit in the former case the field is taken from a separate MHD simulation of the evolution of rising magnetic flux in the underlying layers. Abbett and Fisher (2003) also permit the departure from adiabatic behavior by including very simple cooling and heating terms.

In all cases, the magnetic flux quickly fills out a substantial fraction of the available volume in the corona as a result of the strong decline in density with height (as shown in Figure 6, taken from a simulation by Archontis et al., 2004). Fan and Gibson (2003) simulate the emergence of flux up into a corona with a preexisting arcade of potential magnetic field. Some of those models already yield features in the emerging magnetic flux regions which compare favorably with observed properties of active regions. Fan (2001), for instance, provides images of the evolution of the magnetic flux at the photosphere which clearly resemble the observations of incipient active regions by Strous et al. (1996). Matsumoto et al. (1998), Magara and Longcope (2001) and Fan and Gibson (2003) obtain S-shaped structures strongly reminiscent of the *sigmoids* observed in X-rays (Rust and Kumar, 1996; Canfield et al., 1999). Additionally, Magara and Longcope (2003) study the injection of helicity in the corona through the motion of the footpoints of the field lines in the photosphere.

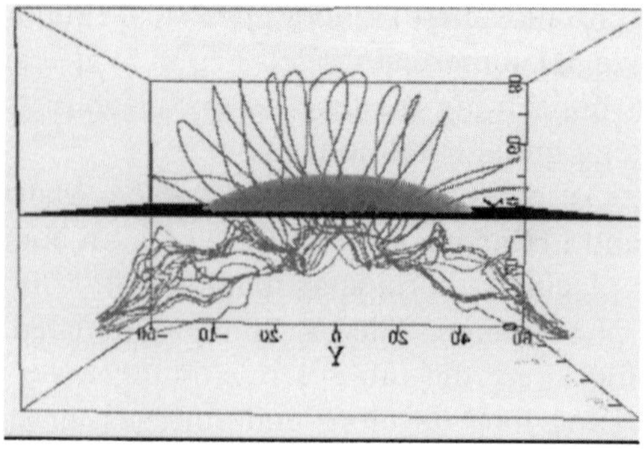

Figure 6. Field lines of a twisted magnetic tube rising out of the convection zone, of which a sheet at the top has gone through the photosphere, further up through the transition region and finally into the corona. The plasma has expanded by a large factor as a result of the strong density gradient in the lower atmosphere. The horizontal surface shown corresponds to the base of the corona. The field lines in the lower half of the box clearly show the initial twist of the magnetic tube. (From Archontis et al. (2004))

Although only a first step toward a more complete numerical modeling, the foregoing simulations represent important progress: not only do they tentatively identify some of their results with observed features in emerging active regions but also, more in general, they provide the basis on which future, more complete numerical studies can be based.

5. Concluding Remarks

The emergence of magnetic flux from the stellar interior through the low atmosphere and into the corona (and further out into space) is one of the fundamental branches of stellar magnetism. Among its open questions there are basic problems like how the stellar dynamo manages to get rid of the flux of one polarity toward the end of each activity half-period, which processes launch sudden events, like flares and CMEs, which have a direct impact onto the Earth's environment, and the connectivity of the various atmospheric levels through the magnetic field. Progress in this area is hampered by the extreme inhomogeneity that all these layers have (not only taken as a whole, but also individually). However, the pace of increase in computational and observational power combined with the theoretical understanding gained in the past decade may well lead to sustantial advances in the coming years.

Acknowledgement

The author is grateful to the organizers of this conference for the opportunity to discuss stellar magnetism in an unusual and stimulating thematic context. Financial support through the PLATON Training Research Network (HPRN-CT-2000-00153) of the European Commission and through the research project AYA2001-1649 of the Spanish Ministry of Science and Technology and the European FEDER Fund is gratefully acknowledged.

References

Abbett, W.P. and Fisher, G.H.: 2003, *ApJ* **582**, 475.
Archontis, V., Moreno-Insertis, F., Galsgaard, K., Hood, A. and O'Shea. E.: 2004, *A&A* in print.
Caligari, P., Moreno-Insertis, F. and Schüssler, M.: 1995, *ApJ* **441**, 886–902.
Canfield, R.C., Hudson, H.S. and McKenzie, D.E.: 1999, *Geophys. Res. Lett.* **26**, 627.
Defouw, R.J.: 1976, *ApJ* **209**, 266.
D'Silva, S. and Choudhuri, A.R.: 1993, *A&A* **272**, 621.
Emonet, T. and Moreno-Insertis, F., 1998, *ApJ* **492**, 804.
Fan, Y.: 2001, *ApJ* **554**, L111.
Fan, Y., Fisher, G.H. and McClymont, A.N.: 1994, *ApJ* **436**, 907–928.
Fan, Y. and Gibson, S.E.: 2003, *ApJ* **589**, L105.
Fisher, G. Fan, Y. and Howard, R.G.: 1995, *ApJ* **438**, 463.
Golub, L. and Pasachoff, J.M.: 1997, *The Solar Corona*, Cambridge: Cambridge Univ Press.
Longcope, D.W. and Fisher, G.H.: 1996, *ApJ* **458**, 380.
Low, B.C.: 2001, *J. Geophys. Res.* **106**, 25141–25164.
Magara, T. and Longcope, D.W.: 2001, *ApJ* **559**, 55.
Magara, T. and Longcope, D.W.: 2003, *ApJ* **586**, 630–649.
Matsumoto, R., Tajima, T., Chou, W., Okubo, A. and Shibata, K.: 1998, *ApJ* **493**, L43.
Moreno-Insertis, F.: 1986, *A&A* **166**, 291–305.
Moreno-Insertis, F., 1992, in: J.H. Thomas and N.O. Weiss (eds.), Sunspots, Theory and Observations, Dordrecht: Kluwer. p. 385.
Moreno-Insertis, F., Caligari, P. and Schüssler, M.: 1994, *Solar Phys.* **153**, 449–452.
Moreno-Insertis, F.: 1997a, in: G. Cauzzi and C. Marmolino (eds.), *The inconstant Sun*, *Mem. Soc. A. It* **68**, 429–447.
Moreno-Insertis, F.: 1997b, in: Viggo H. Hansteen (ed.), *Solar Magnetic Fields*, University of Oslo, p. 3–31.
Parker, E.N.: 1978, Cosmical Magnetic Fields, Oxford Univ Press.
Roberts, B. and Webb, A.R.: 1978, *Solar Phys.* **56**, 5–35.
Rust, D.M. and Kumar, A.: 1996, *ApJ* **464**, L199.
Schrijver, C.J. and Zwaan, C.: 2000, *Solar and Stellar Magnetic Activity*, Cambridge: Cambridge University Press.
Schüssler, M., Caligari, P., Ferriz-Mas, A. and Moreno-Insertis, F.: 1994, *A&A* **281**, L69.
Spruit, H.C., 1977, *PhD Thesis*, Univ Utrecht.
Spruit, H.C., 1981, *A&A* **102**, 129–133.
Strous, L.H., Scharmer, G., Tarbell, T.D., Title, A.M. and Zwaan, C.: 1996, *A&A* **306**, 947.
Van Driel-Gesztelyi, L. and Petrovay, K.: 1990, *Solar Phys.* **126**, 285.
Wang, Y.-M. and Sheeley, N.R.: 1989, *Solar Phys.* **124**, 81.
Wang, Y.-M. and Sheeley, N.R.: 1991, *ApJ* **375**, 761.

MAGNETIC STAR-DISK INTERACTION IN CLASSICAL T TAURI STARS

M. KÜKER[1], TH. HENNING[2] and G. RÜDIGER[1]

[1]*Astrophysikalisches Institut Postdam, An der Sternwarte 16, Potsdam, Germany;*
E-mail: mkueker@aip.de

[2]*Max-Planck-Institut für Astronomie, Königstuhl 17, Heidelberg, Germany*

Abstract. We carry out 2.5D MHD simulations to study the interaction between a dipolar magnetic field of a T Tauri Star, a circumstellar accretion disk, and the halo above the disk. The initial disk is the result of 1D radiation hydrodynamics computations with opacities appropriate for low temperatures. The gas is assumed resistive, and inside the disk accretion is driven by a Shakura–Sunyaev-type eddy viscosity. Magnetocentrifugal forces due to the rotational shear between the star and the Keplerian disk cause the magnetic field to be stretched outwards and part of the field lines are opened. For a solar-mass central star and an accretion rate of 10^{-8} solar masses per year a field strength of 100 G (measured on the surface of the star) launches a substantial outflow from the inner parts of the disk. For a field strength of 1 kG the inner parts of disk is disrupted. The truncation of the disk turns out to be temporary, but the magnetic field structure remains changed after the disk is rebuilt.

Keywords: ISM: jets and outflows, Stars: magnetic fields, Stars: circumstellar matter, Stars: formation

1. Introduction

The rotation rates of Classical T Tauri stars are much smaller on an average than what should be expected for PMS stars contracting toward the main sequence free of any external torques. They must therefore be subject to an efficient braking mechanism, for which angular momentum transfer between star and disk has been proposed (Bouvier et al., 1993; Edwards et al., 1993). Magnetic coupling between a star and a circumstellar accretion disk was discussed for the spin-down of neutron stars by Ghosh and Lamb (1979). The theory has been adopted to the braking of T Tauri stars by Camenzind (1990), Königl (1991), Cameron and Campbell (1993), Shu et al. (1994), Yi (1994), Lovelace et al. (1995) and Ghosh (1995).

So far, most numerical simulations were focused either on the collimation of outflows or the funnel flow down to the stellar surface. Goodson et al. (1997) and Matt et al. (2002) studied the launching of outflows from disks surrounding T Tauri stars assuming the gas in both disk and halo to be resistive. They found the magnetic field to open up and reconnect periodically, launching a pulsed outflow from the center of the system. The formation of a funnel flow from the inner edge of the disk to the polar caps of the star was simulated by Romanova et al. (2002), assuming a turbulence viscosity but no magnetic diffusion. Due to the high density in the halo

Astrophysics and Space Science **292:** 599–607, 2004.

no strong outflows were found, but the field lines originally connecting disk and star opened.

In this work we study the interaction between the stellar magnetic field, the disk, and the surrounding gas. We focus on the innermost parts of the system, especially the vicinity of the corotation radius, to address the problems of magnetic field geometry, disk disruption, and the launching of outflows.

2. The Model

2.1. Set of Equations

The model used here is similar to the one described in Küker et al. (2003). We solve the equations of mass conservation,

$$\frac{\partial \rho}{\partial t} + \nabla(\rho \boldsymbol{u}) = 0, \tag{1}$$

conservation of momentum,

$$\rho \left[\frac{\partial \boldsymbol{u}}{\partial t} + (\boldsymbol{u} \cdot \nabla)\boldsymbol{u} \right] = -\nabla p + \boldsymbol{f} + \nabla \cdot \Pi, \tag{2}$$

conservation of gas energy,

$$c_v \rho \left[\frac{\partial T}{\partial t} + (\boldsymbol{u} \cdot \nabla)T \right] = -p\nabla \cdot \boldsymbol{u} \tag{3}$$

and the induction equation,

$$\frac{\partial \boldsymbol{B}}{\partial t} = \nabla \times [\boldsymbol{u} \times \boldsymbol{B} - \eta_t \nabla \times \boldsymbol{B} + \boldsymbol{u} \times \boldsymbol{B_0}] . \tag{4}$$

in a 2.5 D formulation, i.e. we work with all three components of the magnetic field and the velocity vectors, but assume that none of the variables depends on the azimuthal angle. The force $\boldsymbol{f}$ in Equation 2 reads:

$$\boldsymbol{f} = \rho \nabla \frac{GM_\star}{r} + \frac{1}{4\pi} \nabla \times \boldsymbol{B} \times \boldsymbol{B}. \tag{5}$$

Π denotes the stress tensor.

The code includes the option to solve the equation of radiation energy conservation together with the equations listed above. This is important for finding a realistic disk model, i.e. for the initial setup, but not for the 2D simulations, because the time scale is too short for the radiative energy transport to be of significance. The code used is a finite difference code based on the RHD code by Kley (1989). The magnetic terms are treated as in the ZEUS code of Stone and Norman (1992). The computations are performed in spherical polar coordinates assuming axisymmetry and symmetry with respect to the equatorial plane of the star. The stress tensor Π (with the bulk viscosity set to zero) is described in Klahr et al. (1999).

2.2. Model Setup

The system consists of the central star, which is included through its gravity and the rotation rate at the inner boundary only, the innermost parts of a surrounding accretion disk, and the halo above the disk. The central star is a solar-mass T Tauri star with a radius of three times the solar radius, rotating at 1/10 of the break-up rotation rate, i.e. with a period of 6 days. The gas is a mixture of fully ionized hydrogen and helium with a mean molecular weight $\mu = 0.6$, i.e. we have 10^{24} particles per gram. The disk is defined by its mass accretion rate of $10^{-8} M_{\odot} \mathrm{yr}^{-1}$, an eddy viscosity of the Shakura–Sunyaev-type with a viscosity parameter $\alpha_{\mathrm{SS}} = 0.01$, an opacity law for cool gas, interpolated from the tables of Alexander et al. (1989). The initial disk is interpolated from the results of vertical structure computations using a 1D version of the radiation hydrodynamics code.

With the chosen rotation rate the corotation radius, R_{Ω}, is located at 4.6 stellar radii (from the center of the star). The inner boundary of the computations is located at 1.2 stellar radii, the outer boundary at 20 stellar radii.[1] Outflow boundary conditions are applied at both boundaries. The rotation rate at the inner boundary is fixed to the stellar rotation rate whereas the outer boundary is kept stress-free in the azimuthal component of the equation of motion. The gas density in the halo above the disk is a decreasing function of the radius,

$$\rho = \rho_0 (r/r_0)^{-n}. \tag{6}$$

In this paper, we choose $\rho_0 = 10^{-13}$ g/cm^3 and $n = 4$. The gas in the disk is assumed resistive. Due to the turbulent character of the gas flow in the disk the magnetic diffusion coefficient in Equation 4 is an eddy magnetic diffusivity, and of the same order as the viscosity. We introduce the magnetic Prandtl number, $Pm = \nu_T/\eta_T$, to fix the magnetic diffusion. In this paper we treat the case $Pm = 1$ only.

The initial magnetic field is a pure dipole,

$$B_r = B_\star \cos\theta / r^3, \qquad B_\theta = B_\star \sin\theta / r^3, \tag{7}$$

centered in the star and threads the disk. The field strength parameter $B_\star$, which measures the field strength on the stellar surface close to the poles, is an input

[1]Extending the model down to the stellar surface would have required a higher numerical resolution and a much smaller time step as well as a more sophisticated boundary condition. The outer boundary does not affect the gas flow close to the star and its exact location is therefore somewhat arbitrary.

parameter. Here, we study the cases $B_\star = 100\,\mathrm{G}$ and $B_\star = 1\,\mathrm{kG}$. The initial rotation rate is Keplerian in the disk and zero in the halo except at the inner boundary, which corotates with the star. The poloidal velocity components are zero.

3. Results

Figure 1 shows the initial and final states of a simulation with a field strength $B_\star = 100$ G. In the initial state, the magnetic field is a pure dipole, the disk is undisturbed, and the gas at rest. The right-hand side diagram shows the state at $t = 50$. The time unit is the stellar break-up rotation period, $t_0 = 2\pi\sqrt{R_\star^3/GM_\star}$, where $R_\star$ is the stellar radius and $M_\star$ is the stellar mass. With $R_\star = 3R_\odot$ and $M_\star = M_\odot$ a value of 14.4 h results. There are now three different regions. A region of open-field lines originating at the polar caps of the star, a region of field lines connecting the lower latitudes of the stellar surface to the inner part of the disk, and

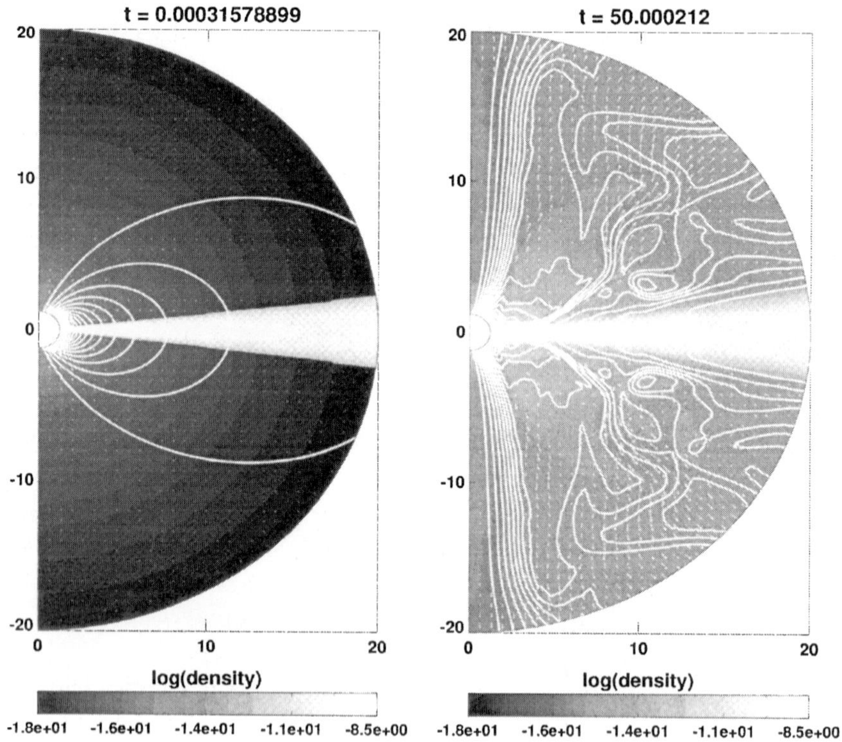

Figure 1. Gas density, poloidal magnetic field lines, and gas velocity for a disk with a mass accretion rate of $10^{-8}\,M_\odot/a$ and a dipole field strength $B_\star = 100$G. The arrows indicate the gas velocity, the solid lines denote (poloidal) magnetic field lines. The unit of the density is g/cm^3 (=10^{24}particles/cm^3). Left: The initial configuration with the stellar dipole. The snapshot was taken a short time *after* the start of the run and therefore already shows the gas in the halo moving slowly inwards. Right: $t = 50$.

a region of open-field lines originating in the outer parts of the disk. The structure of the disk is essentially unchanged, but the halo is filled up with gas from the disk and the transition is less sharp than in the initial state. Far away from the star gas moves outwards at all latitudes, but two regions can be distinguished. At high latitudes the gas density is low and the outflow speed high, whereas at mid-latitudes the gas density is substantially higher and the speed correspondingly lower.

Figure 2 shows a series of snapshots from the $B_\star = 100$ G run. The magnetic field at $t = 0$ is a pure dipole rooted in the star and threads the disk as well as the halo. In the initial phase the geometry of the poloidal field is nearly unchanged, but due to the rotational shear along the poloidal field lines a toroidal field builds up. Because we start with a nonrotating halo and the open boundary condition at the inner boundary does not support a hydrostatic equilibrium a Bondi-type spherical accretion flow develops. At $t = 10$ the field structure has changed significantly. The field has been pushed outwards ad mid-latitudes and the field lines rooted in the polar caps of the star have become almost radial. In the inner parts of the disk, the magnetic field is now bent outwards rather than inwards (as at $t = 0$) above the disk. Gas is moving away from the star and the inner parts of the disk at mid-latitudes. The gas density right above the disk has increased substantially compared to the initial density distribution.

The final field configuration divides the halo into three regions. The innermost parts of the disk are still coupled to the equatorial regions of the star. Field lines with their footpoints on the polar caps of the star have opened up and cones with essentially radial field lines surround the rotation axis. In this region gas still moves inwards onto the polar caps of the star. This is due to the open inner boundary and the lack of a centrifugal force in this region. The field lines with footpoints in the outer parts of the disk are open, too, and the field is mainly radial, like in the region close to the axis. Above the disk there is a layer of about one disk height thick where the gas density is two orders of magnitude higher than in the initial configuration. Inside the corotation radius, the toroidal field is very strong ($\approx$50 G) in this layer, whereas outside R_Ω the field is much weaker and less concentrated. At the boundary between the closed- and open-field line regions gas moves up into the open-field line region, and then outwards along the field lines with speeds up to 200 km/s. Because the magnetic diffusivity is coupled to the density, the gas in the region of enhanced density above the disk and in the outflow is much more diffusive than the region close to rotation axis. Magnetic reconnection occurs frequently in this area and magnetic flux is carried away by the the outflow.

In Figure 3 the result for the same disk model but a magnetic field strength of 1 kG is shown. Except for the greater field strength the initial configuration is identical with that of the 100 G case. At $t = 10$ a cusp-shaped structure has formed in the innermost parts of the disk. The magnetic field is bent outwards in the disk plane and the disk is significantly distorted. Farther outwards the field structure is unchanged. At high latitudes gas is moving inwards along the field lines. At $t = 20$ the inner most part of the disk is completely disrupted and the disk is truncated at a

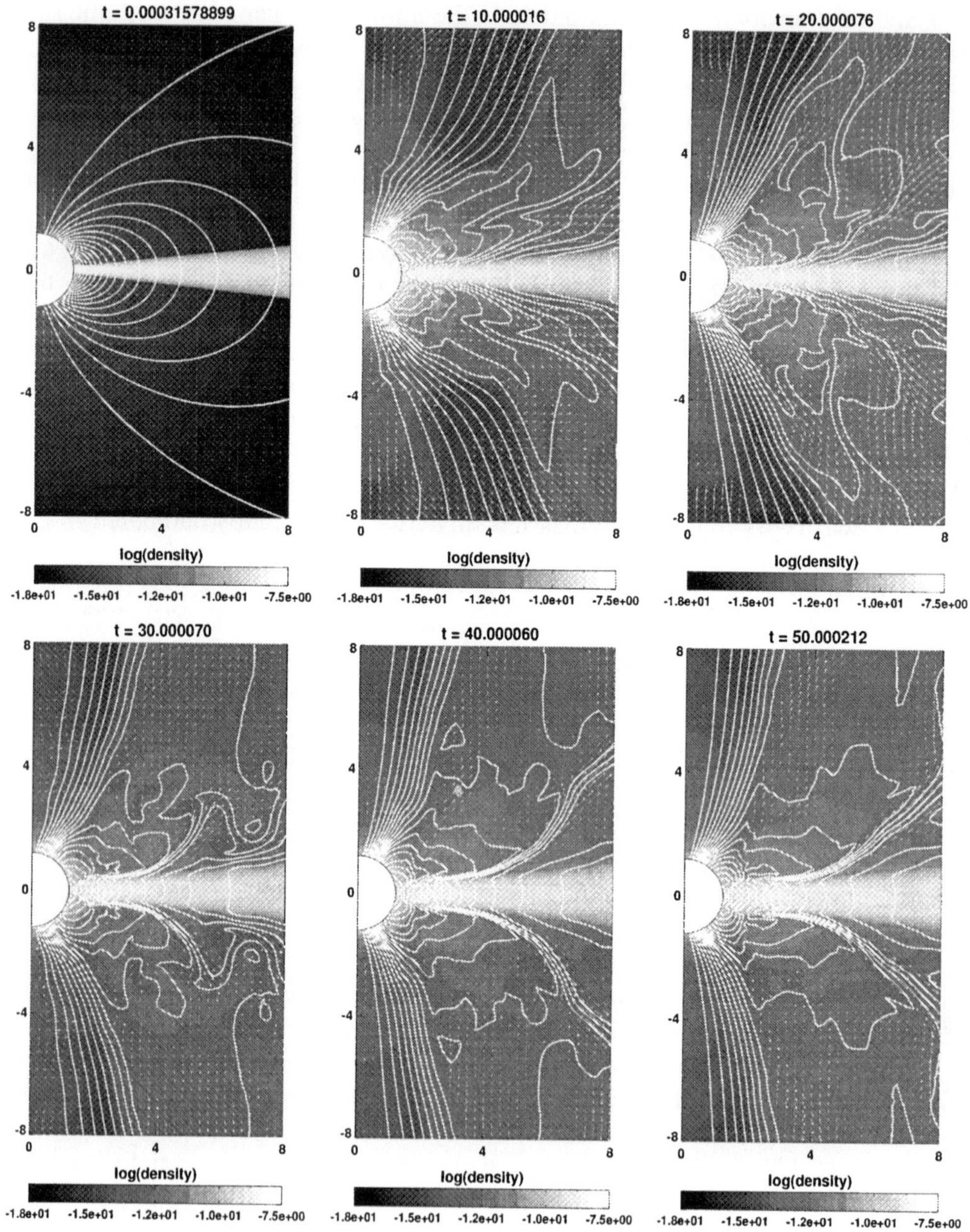

Figure 2. The inner part of the simulation box at $t = 0$, 10, 20, 30, 40, and 50 from the same run as Figure 1. The time unit is stellar break-up rotation period.

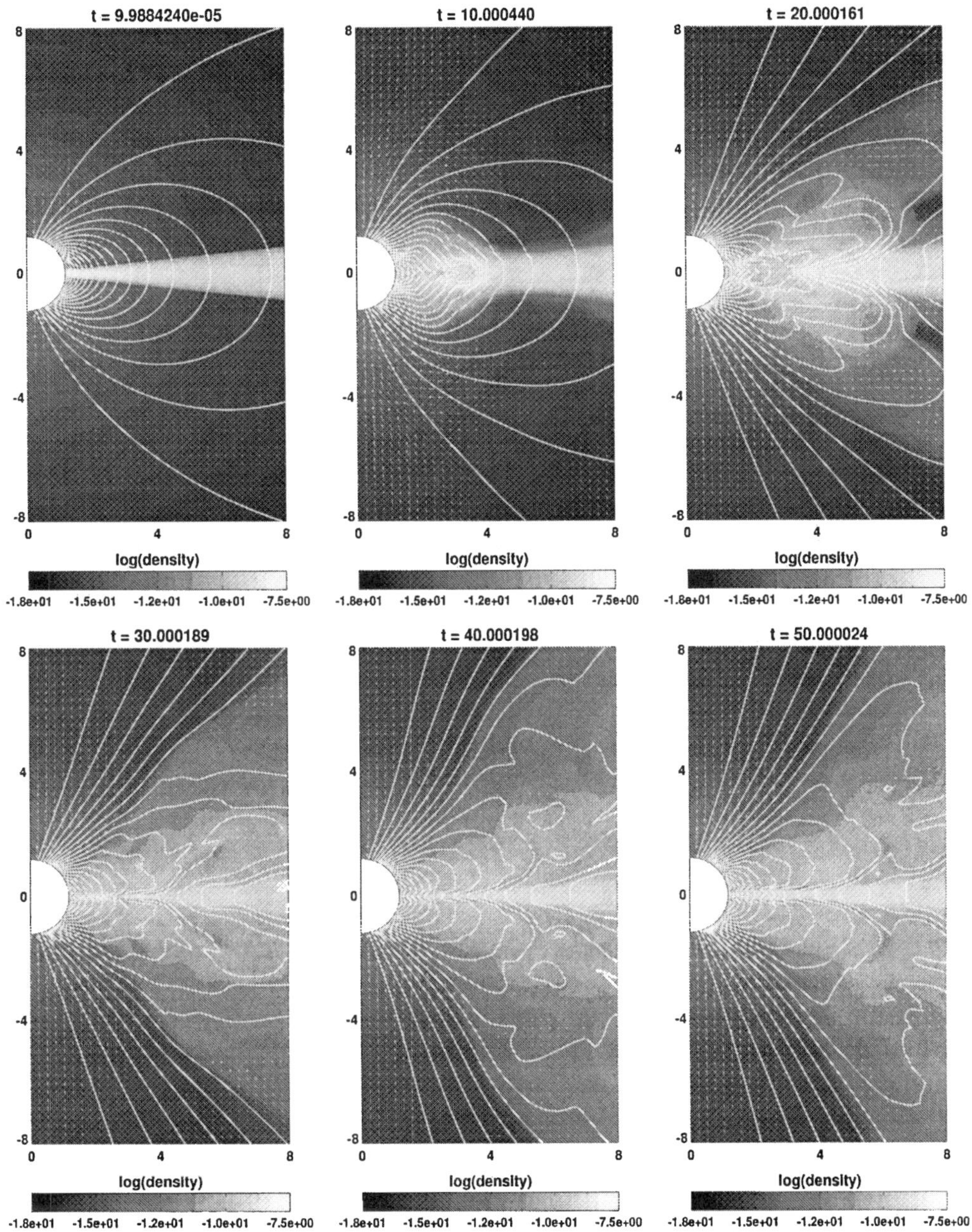

Figure 3. Same as Figure 2 but for a magnetic field strength of 1 kG.

radius of 3.5 stellar radii. At mid-latitudes the magnetic field develops lobes of the same type as in the 100 G case. Although there is still an inflow along the rotation axis the gas now moves outwards at mid-latitudes. At $t = 30$ the inner radius of the disk has moved inwards. The inner parts of the disk have been restored, but it is still very thin. Close to the star the poloidal field structure in the halo is very similar to the original dipole, but the field is bent inwards in the disk, resulting in an almost horizontal field at the upper and lower edges of the disk. At $t = 40$ the thickness of the disk has increased, but it is still substantially thinner than in the initial configuration. This trend has continued at $t = 50$. As in the case of the weaker field, we can now distinguish three regions, namely a region of open-field field lines rooted in the star along the rotation axis, a region of closed-field lines connecting the lower latitudes of the stellar surface to the inner parts of the disk, and another region of open-field lines, rooted in the outer parts of the disk.

Both simulations were run for 50 times the stellar break-up rotation period, a time span long enough for the magnetic field to reach a fairly stationary state but too short to allow any evolution of the disk on the viscous time scale. This shows especially in the $B_\star = 1$ kG case where the disk has not yet been fully rebuilt after its disruption. It is therefore not clear if the disruption of its inner parts occurs only once, because the initial conditions are too far away from equilibrium, or periodically. In both models the transition from closed- to open-field lines occurs at the corotation radius. We do not find a funnel flow, i.e. a truncation of the disk and a gas flow along the magnetic field lines toward the poles of the star. Those parts of the disk lying within the corotation radius are still coupled to the star, and angular momentum is hence transfered from the disk to the star, in contrast to the Ghosh and Lamb (1979) and Shu et al. (1994) scenarios. The main reason is the insufficient strength of the magnetic field, but even for stronger fields the conditions for the formation of a funnel flow seem to be rather special (Miller and Stone, 1997; Romanova et al., 2002).

References

Alexander, D.R., Augason, G.C. and Johnson, H.R.: 1989, *ApJ* **345**, 10149.
Bouvier, J., Cabrit, S., Fernandez, M., Martin, E.L. and Matthews, J.M.: 1993, *A&A* **272**, 176.
Camenzind, M.: 1990, Magnetized Disk-Winds and the Origin of Bipolar Outflows, in: G. Klare, (ed.), *Reviews in Modern Astronomy 3*, Springer-Verlag, Berlin.
Cameron, A.C. and Campbell, C.G.: 1993, *A&A* **274**, 30.
Edwards, S., Strom, S.E., Hartigan, P. et al.: 1993, *AJ* **106**, 372.
Ghosh, P. and Lamb, F.K.: 1979, *ApJ* **232**, 259.
Ghosh, P.: 1995, *MNRAS* **272**, 763.
Goodson, A.P., Winglee, R.M. and Böhm, K.-H.: 1997, *ApJ* **489**, 199.
Klahr, H.H., Henning, Th. and Kley, W.: 1999, *ApJ* **514**, 325.
Kley, W.: 1989, *A&A* **222**, 141.
Königl, A.: 1991, *ApJ* **370**, L39.

Küker, M., Henning, Th. and Rüdiger, G.: 2003, *ApJ* **589**, 397.
Lovelace, R.V.E., Romanova, M.M. and Bisnovatyi-Kogan, G.S.: 1995, *MNRAS* **275**, 244.
Matt, S., Goodson, A.P., Winglee, R.M. and Böhm, K.-H.: 2002, *ApJ* **547**.
Miller, K.A. and Stone, J.M.: 1997, *ApJ* **489**, 890.
Shu, F., Najita, J., Ostriker, E., Wilkin, F., Ruden, S. and Lizano, S.: 1994, *ApJ* **429**, 781.
Stone, J.M. and Norman, M.L.: 1992, *ApJ* **80**, 791.
Romanova, M.M., Ustyugova, G.V., Koldoba, A.V. and Lovelace, R.V.E.: 2002, *ApJ* **578**, 420.
Yi, I.: 1994, *ApJ* **428**, 760.

THE TEMPERATURE AND IONIZATION OF T-TAURI MICRO-JETS

DARREN M. O'BRIEN[1], PAULO J.V. GARCIA[2], JONATHAN FERREIRA[3], SYLVIE CABRIT[4] and LUC BINETTE[5]

[1]*Centro de Astrofísica da Universidade do Porto & Departamento de Matemática Aplicada da Faculdade de Ciências do Porto, Portugal; E-mail: darren@astro.up.pt*

[2]*Centro de Astrofísica da Universidade do Porto & Faculdade de Engenharia da Universidade do Porto, Portugal*

[3]*Laboratoire d'Astrophysique de Grenoble, France*

[4]*LERMA, Observatoire de Paris, France*

[5]*Instituto de Astronomía, Ciudad Universitaria, Mexico*

Abstract. The effects of phenomenological heating functions on the flow thermodynamics of cold T-Tauri disk winds are examined. Turbulent dissipation (mechanical) heating and a warm disk corona are invoked to heat the wind. The temperature and ionization evolution are solved for along the flow. The results allow the construction of synthetic observations; emission maps, forbidden line ratios, line fluxes and line profiles; and successfully reproduce a number of observed trends. Mechanical heating produces line ratios and fluxes that fit very well with observations. Invoking a warm disk corona successfully reproduces forbidden line profile low velocity components.

Keywords: ISM: jets and outflows, stars: pre-main sequence, MHD, line, profiles, accretion disks, coronae

1. Introduction

Collimated mass ejection in young T-Tauri stars (TTS) is observationally found to be correlated with the accretion process (e.g., Hartigan et al., 1995). It is currently believed that magnetic forces are responsible for both high ejection efficiency ($\dot{M}_{\mathrm{jet}}/\dot{M}_{\mathrm{acc}} \sim 0.01 - 0.1$) and the high degree of collimation of these winds (e.g., Konigl and Pudritz, 2000; Shu et al., 2000).

In the regime of cold MHD, ejection is only possible from the disk by magneto-centrifugal launching on field lines sufficiently inclined from the disk axis, as no thermal energy is available to power matter across the gravitational potential barrier. Disk winds from wide range in disk radii (Blandford and Payne, 1982; Ferreira, 1997; Ferreira, these proceedings) have produced sufficiently complete calculations of MHD solutions to allow quantitative comparisons with observations.

Observationally, the information available from jets is in the form of emission lines, therefore to test models, synthetic observations must be generated. This is achieved by modelling the thermal evolution in the jet (Garcia et al., 2001a, hereafter Paper I; Safier, 1993; Shang et al., 2002; Ferro-Fontán and Gómez de Castro, 2003).

This work follows from that of Paper I, which modelled the thermal evolution in T-Tauri jets consistently with the jet dynamical approximation of cold MHD of

Ferreira (1997), including only ambipolar diffusion heating. From the calculated thermal structure, synthetic observations were then generated. These observational predictions successfully reproduced several observed trends, but also encountered a number of problems. The models produce electron densities and LVC (line profile low-velocity component) intensities lower than those observed. The predicted wind terminal velocities and densities were, respectively, higher and lower than the observations.

To overcome some of the above limitations new phenomenological heating terms are studied, namely mechanical heating from the dissipation of energy via weak shocks and heating from a warm disk corona. The effects of these functions are described in the following sections.

2. Mechanical Heating

The first function added to the models of Paper I was a turbulent dissipative heating decribed by a Kolmogorov cascade. A fraction of the flow's kinetic energy is assumed to be in the form of turbulent motion. This situation could arise as a result of weak shocks propagating in the flow. Such a heating term has been applied to the X-wind (Shang et al., 2002) and to AGN photoionized clouds (Bottorff and Ferland, 2002). The heating term is described by

$$\Gamma_{\rm mech} = \alpha\rho v^3/s \tag{1}$$

where ρ, v and s are the density, velocity and distance along the flow respectively and α is a phenomenological constant of proportionality. The parameter space was explored by varying α between 10^{-3} and 10^{-1}.

The temperature of the flow for low values of α (10^{-3}–10^{-2}) was only slightly higher than that calculated in Paper I. However, this slight increase is enough to increase dramatically the number of collisional ionizations. As α was increased further $\Gamma_{\rm mech}$ began to dominate heating and both T and $f_{\rm e}$ began to reach values significantly higher than those of Paper I.

The radiative line cooling (dominated by hydrogen), $\Lambda_{\rm rad}$, which was present but not very significant in Paper I, is increased considerably. This is due in part to collisional ionization of H, and due to a large increase in the electron density as H becomes ionized (Spitzer, 1978). $\Lambda_{\rm rad}$ then becomes the most important cooling mechanism, as opposed to adiabatic expansion, $\Lambda_{\rm adia}$. Although $\Lambda_{\rm rad}$ values approach those of $\Gamma_{\rm mech}$, there is no ionization feedback between these two terms. Hence no temperature plateau is produced, as was in the cases of Paper I and Safier (1993), when the ambipolar diffusion heating and adiabatic expansion cooling were the dominant heating and cooling terms.

Ambipolar diffusion heating, $\Gamma_{\rm drag}$, is greatly reduced in the new calculations. This term arises from a drag force between ions and neutrals, thus it is heavily dependent on the neutral density. With ionization increasing rapidly with distance, it falls away.

The increased ionization yields a wind model whose heating and cooling is dominated by $\Gamma_{\rm mech}$ and $\Lambda_{\rm rad}$, respectively. The main heating and cooling terms are plotted in Figure 1, along with T and $f_{\rm e}$.

A number of important consequences of the addition of $\Gamma_{\rm mech}$ are shown in Figure 2. Shown are the ion abundances with respect to hydrogen for several ions along the flowline anchored in the disk at $\varpi_0 = 0.1$. Ionization in the hottest part of the flow differs greatly from that of Paper I. The high temperature in this region leads some elements to be doubly ionized at higher values of α. It should be noted that the initial ion abundances are set by assuming ionization equilibrium with the incoming radiation field (see Paper I).

Synthetic-emission maps and long-slit spectra along the jet axis were computed in the lines of [O I]λ6300, [S II]λ6731 and [N II]λ6584. No major morphological changes were observed with respect to the work of Paper I.

Comparisons of predicted integrated luminosities in [O I]λ6300 and [S II]λ6731 as functions of $\dot{M}_{\rm acc}$ with observed T Tauri luminosities from Hartigan (1995) are show in Figure (3). Fluxes were integrated out to 200 AU from the star for an edge-on jet, for flowline footpoints anchored between $\varpi_0 = 0.07 - 1$ AU, resulting in a 75 AU radial extent. Predictions made in Garcia et al., 2001b (hereafter Paper II)

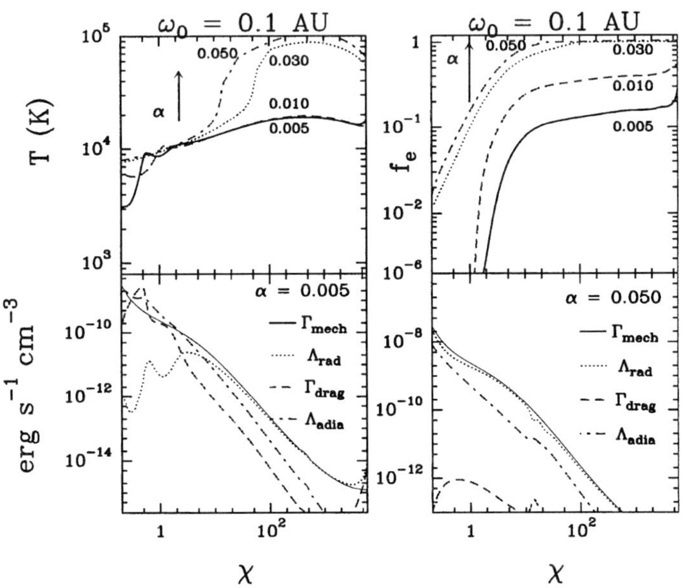

Figure 1. Flow characteristics along the inner flowline with $\varpi_0 = 0.1$ AU and $\dot{M}_{\rm acc} = 10^{-7} M_\odot {\rm yr}^{-1}$. Shown are the evolution with α of T, $f_{\rm e}$, $\Gamma_{\rm mech}$, $\Lambda_{\rm rad}$, $\Gamma_{\rm drag}$ and $\Lambda_{\rm adia}$. The α-values on the T and $f_{\rm e}$ plots are 0.005 (solid), 0.010 (dashed), 0.030 (dotted) and 0.050 (dash dot). $\chi = z/\varpi_0$, denotes the position along a streamline anchored at ϖ_0.

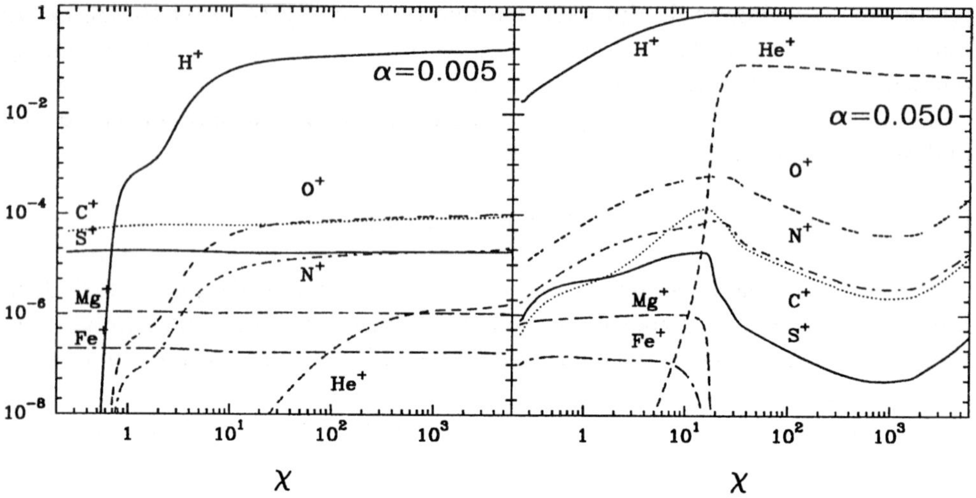

Figure 2. Relative ion abundances with respect to hydrogen along the flowline $\varpi_0 = 0.1$ AU for $\dot{M}_{\rm acc} = 10^{-7} M_\odot {\rm yr}^{-1}$ and for $\alpha = 0.005$ and 0.050.

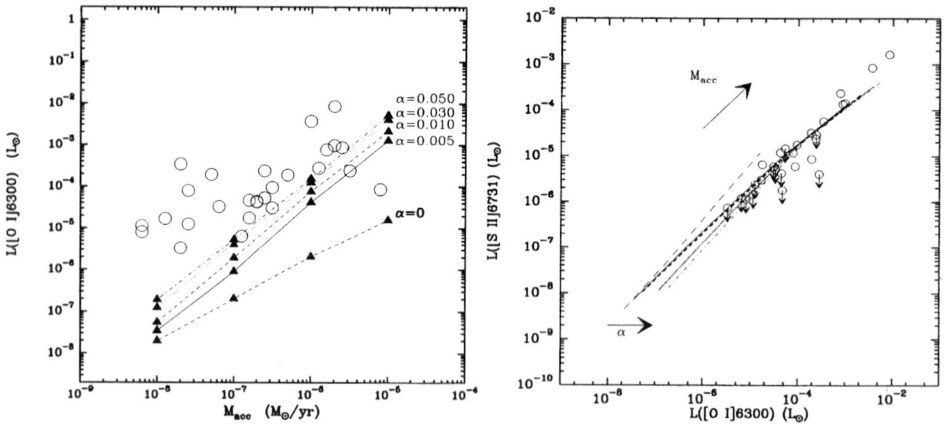

Figure 3. Left hand side Line fluxes of [O I]λ6300 with $\dot{M}_{\rm acc}$ between 10^{-8} and $10^{-5} M_\odot {\rm yr}^{-1}$ using α values 0.005 (solid), 0.010 (dashed), 0.030 (dotted) and 0.050 (dash-dot). Also shown are the predicted fluxes heated predominantly by ambipolar diffusion with no mechanical heating term (Solution A of Paper I, labelled $\alpha = 0$). Observed T Tauri forbidden line luminosities from Hartigan (1995) are shown as open circles. *Right hand side* Ratio of [O I]λ6300 to [S II]λ6731 fluxes for $\dot{M}_{\rm acc}$ between 10^{-8} and $10^{-5} M_\odot {\rm yr}^{-1}$, increasing in the direction of increasing luminosity. α values are 0.005 (solid), 0.010 (dashed), 0.030 (dotted), 0.050 (dashdot) and 0 (long dashed). Observed fluxes are shown as open circles.

are also shown. The mechanical heating increases this line flux above the results of Paper I, and results in a very close correlation with the observational data.

Line ratios([S II]λ6716/[S II]λ6731 vs. [S II]λ6731/[O I]λ6300 and [N II]λ6584/[O I]λ6300 vs [S II]λ6731/[O I]λ6300) were also successfully reproduced; see Figure 4. However, the line ratio [N II]λ6584/[O I]λ6300 does not fit well with the observations at highest value for α, as O becomes doubly ionized at higher temperatures; see Figure 2.

Line profiles are shown in Figure 5. The observed LVC is still not reproduced in these predictions.

3. Coronal Heating

A hot disk corona has been suggested as a means of explaining the LVC of forbidden lines (Kwan, 1997). In Paper II, the relative intensity of LVC versus HVC in [O I]λ6300 was found to be smaller than that observed. To increase the magnitude of the LVC to observed values, a coronal heating function was added. These calculations were made without mechanical heating in order to determine the effect addition of a hot disk corona has on those of Paper I.

The coronal heating was modelled by adding a function to the thermal calculations without modifying the dynamical solution, and is therefore dynamically inconsistent with the cold MHD solution. It is assumed that a fraction of the disk's

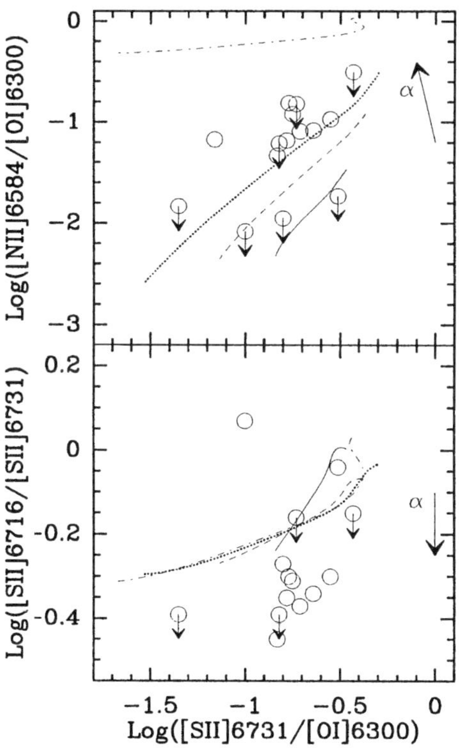

Figure 4. Line ratios [N II]λ6583/[O I]λ6300 and [S II]λ6716/[S II]λ6731 vs. [S II]λ6731/[O I]λ6300 for $\dot{M}_{acc} = 10^{-7}$, $\alpha = 0.005$ (solid), 0.010 (dashed), 0.030 (dotted) and 0.050 (dash dot).

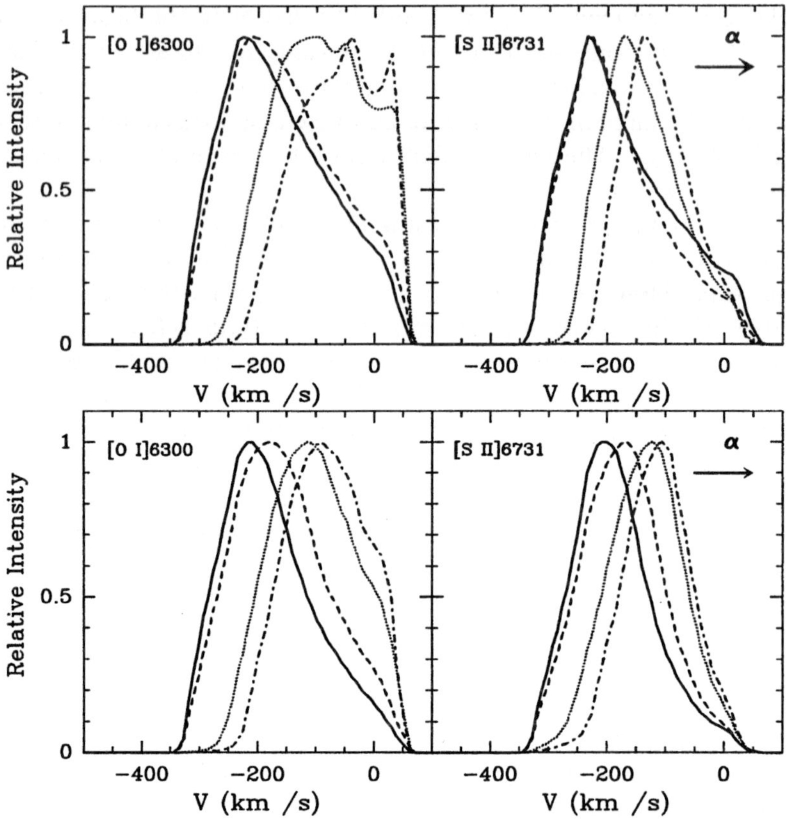

Figure 5. Line profiles of [O I]λ6300 and [S II]λ6731 for $\dot{M}_{\rm acc} = 10^{-7}$ (top) and 10^{-6} (bottom) $M_{\odot}{\rm yr}^{-1}$ shown varying over a range of α; 0.005 (solid), 0.010 (dashed), 0.030 (dotted) and 0.050 (dash dot).

heat generated by accretion flows into the base of the jet as thermal energy (Ferreira and Pelletier, 1993). The function is described by

$$\Gamma_{\rm cor} = \frac{GM_{\star}\dot{M}_{\rm acc}}{8\pi\varpi_0^4}\frac{\Theta}{1+\Lambda}\frac{\exp(-\frac{x-x_{\rm slow}}{\delta_{\rm cor}})}{\delta_{\rm cor}^2} \tag{2}$$

where G is the gravitational constant, $M_{\star}$ is the stellar mass, $\dot{M}_{\rm acc}$ is the accretion rate, ϖ_0 the field line footpoint, $\Theta \ll 1$ a constant representing the fraction of power dissipated in thermal energy, Λ is the ratio of magnetic to viscous torque in the disk, x is the distance along the flow, $x_{\rm slow}$ the position of the slow magnetosonic point (where the calculations start) and the dimensionless $\delta_{\rm cor} = 0.01$ represents the scale height over which the thermal energy is dissipated. Here it is set equal to the disk vertical scale height.

Θ is varied between 0.001 and 0.01 assuming that 0.1–1% of the heat generated by accretion is released into the base of the jet. Even for this small fraction of the available power leads Γ_{cor} to completely dominate the heating of the flow out to ~10AU. For some values of $\dot{M}_{acc}$, T was raised to above 10^4 close to the disk. Figure 6 shows values T and f_e along a streamline anchored in the disk at $\varpi = 0.1$AU. The large increase in T close to the disk leads to high ionization and thus negligible Γ_{drag}. The jet is cooled rapidly by virtue of Λ_{adia}, hence there is initial drop in T and f_e.

The observational predictions constructed yield low fluxes and line ratios uncorrelated with the observations. This is due to the low temperature in the acceleration zone. However, a very strong LVC is reproduced; see Figure 7. This result suggests that there does exist a heating source close to the disk. Recent work studying the role of X-ray radiation in heating the disk and wind (Ferro-Fontán and Gómez de Castro, 2003) reveal that X-ray photoionization is only relevant in the production of such spectral signatures at low accretion rates ($\dot{M}_{acc} = 10^{-9} - 10^{-8} M_\odot \mathrm{yr}^{-1}$). Both these results lend further evidence that warm MHD solutions (Casse and Ferreira, 2000) may more realistically describe micro-jet dynamics (see also the contribution of Dougados in these proceedings).

4. Conclusions

The present results show that by adding new heating functions the limitations of pure ambipolar diffusion heating, i.e. low total fluxes and ionizations, can be overcome. These new heating sources strongly decrease the importance of ambipolar diffusion in the thermal structure. Mechanical heating increases the fluxes and ionizations, while a hot corona produces a strong LVC, thus suggesting that multiple heating sources are at play in these winds.

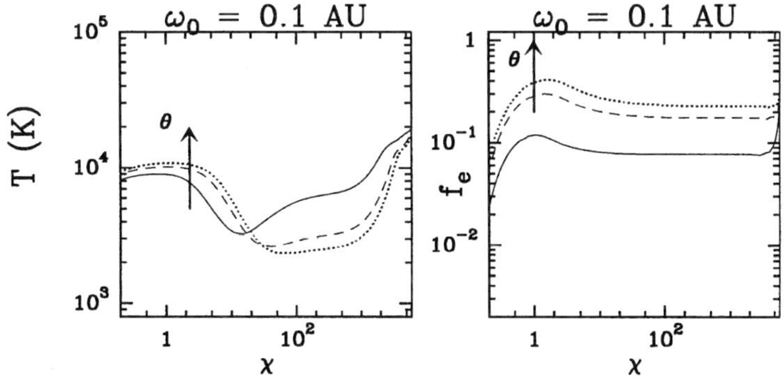

Figure 6. Flow characteristics, T and f_e, along streamlines with $\varpi_0 = 0.1$ AU, $\dot{M}_{acc} = 10^{-6} M_\odot \mathrm{yr}^{-1}$ and $\Theta = 0.001$ (solid), 0.005 (dashed) and 0.010 (dotted).

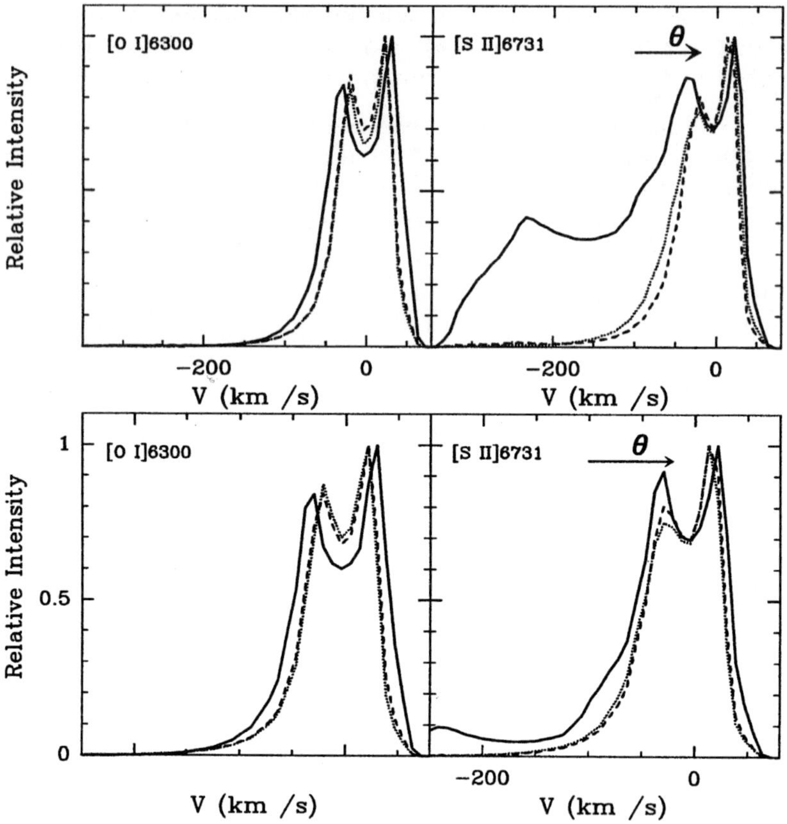

Figure 7. Line profiles of [O I]λ6300 and [S II]λ6731 for $\dot{M}_{acc} = 10^{-7}$ (top) and 10^{-6} (bottom) $M_{\odot}yr^{-1}$ shown over a range of Θ; 0.001 (solid), 0.005 (dashed) and 0.010 (dotted).

Acknowledgements

This work was supported by grant POCTI/1999/FIS/34549 approved by FCT and POCTI, with funds from the European programme FEDER. The authors are grateful to the referee, A. I. Gómez de Castro and to C. Dougados and J. Lima for their helpful comments and discussions.

References

Blandford, R.D. and Payne, D.G.: 1982, *MNRAS* **199**, 883.
Bottorff, M. and Ferland, G.: 2002, *ApJ* **568**, 581.
Casse, F. and Ferreira, J.: 2000, **353**, 1115.
Ferreira, J.: 1997, *A&A* **319**, 340.
Ferreira, J. and Pelletier, G.: 1993, *A&A* **276**, 625.
Ferro-Fontán, C. and Gómez de Castro, A.I.: 2003, *MNRAS* **342**, 427.

Garcia, P.J.V., Ferreira, J., Cabrit, S. and Binette, L.: 2001a, *A&A* **377**, 589.
Garcia, P.J.V., Cabrit, S., Ferreira, J. and Binette, L.: 2001b, *A&A* **37**, 609.
Hartigan, P., Edwards, S. and Ghandour, L.: 1995, *ApJ* **452**, 736.
Konigl, A. and Pudritz, R.E.: 2000, PP IV, p. 759.
Kwan, J.: 1997, *ApJ* **489**, 284.
Safier, P.N.: 1993a, *ApJ* **408**, 115.
Shang, H., Glassgold, A.E., Shu, F.H. and Lizano, S.: 2002, *ApJ* **564**, 835.
Shu, F. H., Laughlin, G., Lizano, S. and Galli, D.: 2000, *ApJ* **535**, 190.
Spitzer, L.: 1978, Wiley-Interscience. New York.

OBSERVATIONS OF MAGNETIC FIELDS ON T TAURI STARS

JEFF A. VALENTI[1] and CHRISTOPHER M. JOHNS-KRULL[2]
[1]*Space Telescope Science Institute, 3700 San Martin Dr., Baltimore, MD, U.S.A.;*
E-mail: valenti@stsci.edu
[2]*Department of Physics and Astronomy, Rice University, Houston, TX, U.S.A.*

Abstract. We present measurements of magnetic field strength and geometry on the surfaces of T Tauri stars (TTS) with and without circumstellar disks. We use these measurements to argue that magnetospheric accretion models should not assume that a fixed fraction of the stellar surface contains magnetic field lines that couple with the disk. We predict the fractional area of accretion footpoints, using magnetospheric accretion models and assuming field strength is roughly constant for all TTS.

Analysis of Zeeman broadened infrared line profiles shows that individual TTS each have a distribution of surface magnetic field strengths extending up to 6 kG. Averaging over this distribution yields mean magnetic field strengths of 1-3 kG for all TTS, regardless of whether the star is surrounded by a disk. These strong magnetic fields suggest that magnetic pressure dominates gas pressure in TTS photospheres, indicating the need for new model atmospheres.

The He I 5876 Å emission line in TTS can be strongly polarized, so that magnetic field lines at the footpoints of accretion have uniform polarity. The circular polarization signal appears to be rotationally modulated, implying that accretion and perhaps the magnetosphere are not axisymmetric. Time series spectropolarimetry is fitted reasonably well by a simple model with one magnetic spot on the surface of a rotating star. On the other hand, spectropolarimetry of photospheric absorption lines rules out a global dipolar field at the stellar surface for at least some TTS.

Keywords: T Tauri stars, magnetospheric accretion, infrared spectroscopy, Zeeman broadening, circular spectropolarimetry

1. Introduction

T Tauri stars (TTS) lie above the main-sequence in the H-R diagram and have one or more other indicators of youth, such as high lithium abundance, a circumstellar disk, or association with a dark cloud. Classical TTS (CTTS) accrete material from a circumstellar disk, which gives rise to excess continuum emission from blue to infrared (IR) wavelengths.

Uchida (1983) and Uchida and Shibata (1984) first suggested that CTTS magnetic fields disrupt the inner accretion disk, lifting material out of the disk plane towards the stellar magnetic poles. This insight inspired many detailed magnetospheric accretion models (e.g., Königl, 1991; Cameron and Campbell, 1993; Shu et al., 1994). These models invoke different forms of magnetic coupling, but all attempt to match the stellar rotation period with the Keplerian orbital period of material near the inner edge of the disk. This underlying assumption leads to

Astrophysics and Space Science **292:** 619–629, 2004.

similar self-consistency relationships between stellar mass (M_*), radius (R_*), rotation period (P_*), dipole magnetic field strength at the equator (B_{eq}), and mass accretion rate in the disk ($\dot{M}$).

Johns–Krull et al. (1999b) used literature values of M_*, R_*, P_*, and $\dot{M}$ for 14 CTTS to calculate the range of B_{eq} values required by each of the three magnetospheric accretion models cited above. The results are quite interesting. For any given model, the range of B_{eq} values spans a factor of 15 to 30. Every model yields a similar ordering of CTTS from lowest to highest B_{eq}, so this ordering might be used to test the rotational equilibrium hypothesis underlying *all* magnetospheric accretion models. However, we will show below that B_{eq} is not necessarily a good measure of the magnetic flux coupling the star and disk, which is the relevant quantity in the models.

2. Zeeman Broadening Measurements

For CTTS with magnetic field measurements described below, the Shu et al. (1994) model predicts values of B_{eq} that range from 0.67 kG for DF Tau to 4.1 kG for BP Tau. When rotational broadening is negligible, Zeeman broadening of Fe I 8468 Å can be detected down to field strengths of about 2 kG. However, TTS typically have $v \sin i$ values of 10 km s^{-1}, so greater sensitivity is required. The splitting of Zeeman components depends quadratically on wavelength, whereas line broadening due to Doppler processes depends linearly on wavelength. For this reason, Zeeman broadening measurements for TTS are best done using the Ti I 2.2 μm lines shown in Figure 1. Robust Zeeman broadening measurements require Zeeman insensitive lines to constrain nonmagnetic broadening mechanisms. Fortunately, CO lines at 2.31 μm have negligible Landé-g factors, making them an ideal null reference.

Figure 2 shows Zeeman sensitive Ti I lines and Zeeman insensitive CO lines in observed and synthetic spectra of two inactive K7 dwarfs. Oscillator strengths affect line depth, but not line width, so Zeeman broadening is detectable even with poor line data. To enable quantitative analysis, however, we determined precise empirical log gf values by fitting these lines in the solar spectrum.

Figure 3 presents an IR spectrum of the CTTS BP Tau, which is also spectral class K7. The Zeeman insensitive CO lines are broader than in Figure 2, but still well fitted by models with enhanced rotational broadening that also accurately reproduce optical line profiles. In contrast, the 2.2 μm Ti I lines cannot be fitted by the nonmagnetic model. Observed BP Tau line profiles are much broader than the corresponding Ti I line profiles in Figure 2, so the excess broadening cannot be due to any process that operates in inactive main-sequence stars of the same spectral type. Zeeman broadening is currently the only known mechanism that can explain the observed line widths. The lines show no evidence of a contribution from the disk, even in spectra of CTTS with much larger mass accretion rates. This new

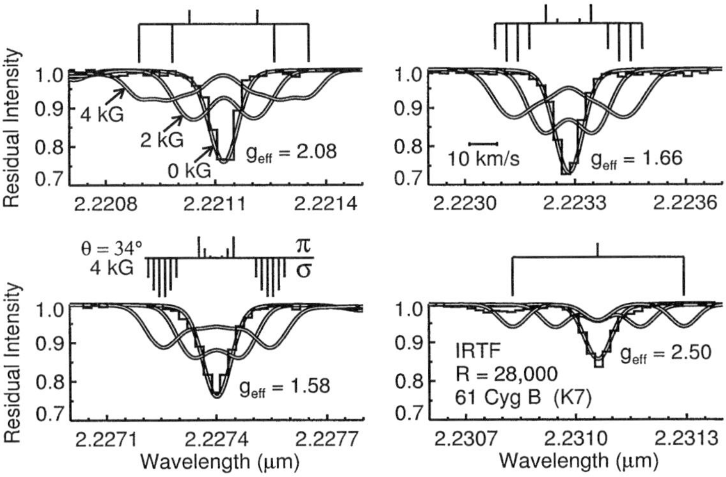

Figure 1. The Zeeman sensitivity of 2.2 μm Ti I lines is illustrated by comparing synthetic line profiles (doubled lines) for 0, 2, and 4 kG radial magnetic fields with an observed spectrum of the inactive K7V dwarf 61 Cyg B (histogram). Above each profile, the splittings and relative intensities of Zeeman components are illustrated for a 4 kG field tilted 34° from the line of sight.

analysis confirms and extends our initial Zeeman broadening detection, based on a single Ti I line (Johns–Krull et al., 1999b).

Optical Zeeman analyses usually assume that the magnetic field strength at any point on the stellar surface is either zero or has a characteristic value. The IR Ti I lines in spectra of TTS cannot be described by such a simple model because the synthetic profiles would have inflections that are not in the observed profiles. To see this, imagine adding the 0 and 4 kG line profiles in Figure 1. Instead, we fit each observed spectrum with a superposition of synthetic spectra for magnetic field strengths of 0, 2, 4, and 6 kG. Minimizing χ^2 yields the relative contributions, f, of each component. The resulting magnetic field distribution is unique because the Zeeman splitting produced by a 2 kG field is comparable to the nonmagnetic width of the Ti I spectral lines. In other words, the Zeeman resolution of the Ti I lines is about 2 kG.

The intensity-weighted mean magnetic field strength, $\langle B \rangle$, over the entire surface of BP Tau is 2.1 kG with field strengths reaching at least 4 kG and probably even 6 kG in some regions. Thus, magnetic fields on BP Tau are stronger than on the Sun, even though the surface gravity on BP Tau is lower by a factor of ten. On the Sun and other main-sequence stars, magnetic field strength seems to be set by an equipartition of gas and magnetic pressure. In contrast, the photospheres of BP Tau and other TTS are apparently dominated by magnetic pressure, rather than gas pressure. This implies that the nonmagnetic model atmospheres used to analyze TTS are missing an important pressure support term. Energy transport by convection should also be reconsidered. Atmospheric models with magnetic fields

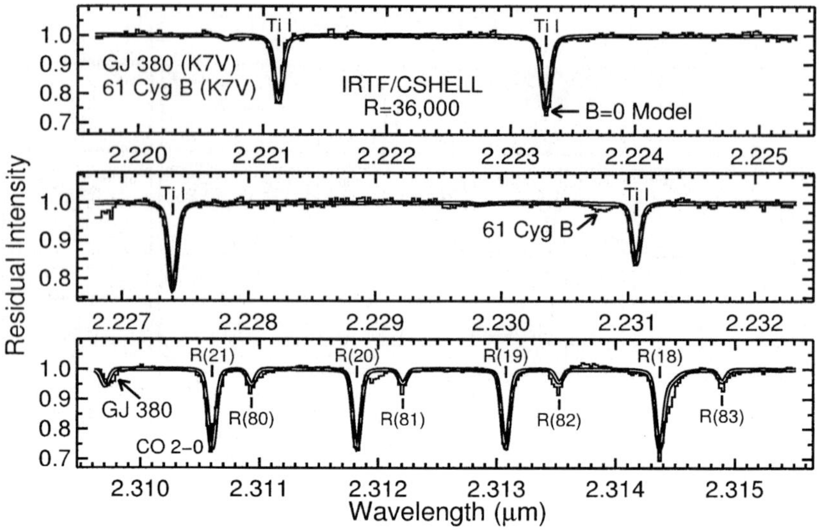

Figure 2. IRTF/CSHELL spectra of two inactive K7 dwarfs (histogram) are well matched by a nonmagnetic synthetic spectrum (doubled curve). In the absence of a significant magnetic field, all the spectral lines have comparable widths. The spectrograph line spread function is well behaved, and blends are minor at this spectral type.

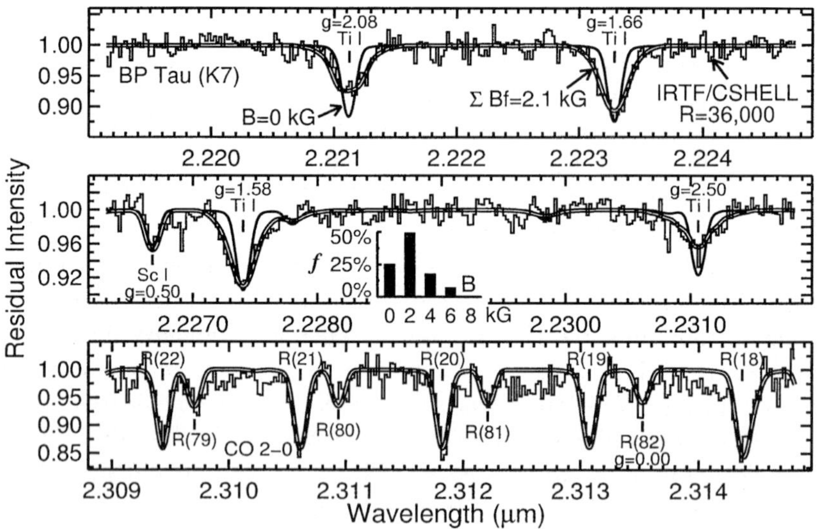

Figure 3. An IRTF/CSHELL spectrum of the K7 CTTS BP Tau (histogram) is compared with synthetic spectra based on magnetic (doubled curve) and nonmagnetic (single curve) models. Zeeman insensitive CO lines are well fit by both models. The Zeeman sensitive Ti I lines are much broader than predicted by the nonmagnetic model. The magnetic model reproduces the observed spectrum, using a distribution of magnetic field strengths (inset histogram) with a mean of 2.1 kG over the entire stellar surface. The effective Landé-*g* factor is given for each atomic line.

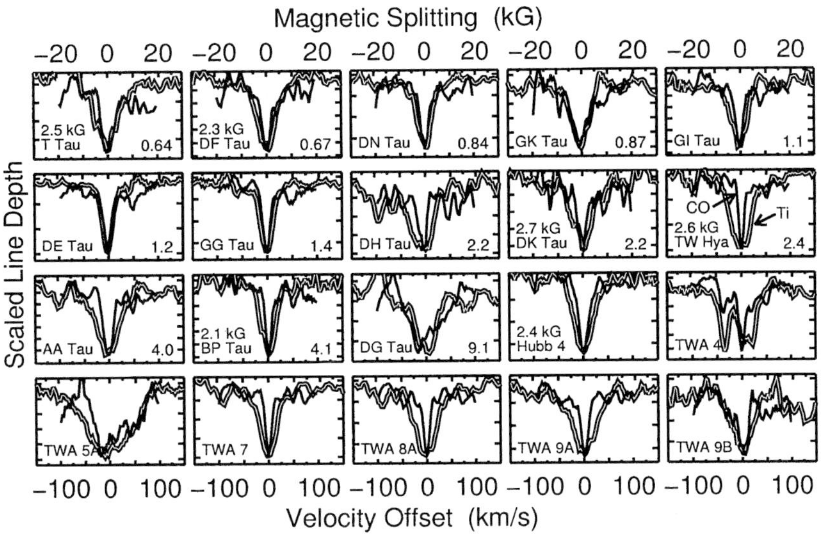

Figure 4. The ubiquity of strong magnetic fields in TTS is demonstrated by overplotting the Zeeman insensitive 2.312 μm CO line (single curve) on top of the Zeeman sensitive 2.223 μm Ti I line (doubled curve). Line profiles have been scaled to a common depth to facilitate visual identification of Zeeman broadening. For accreting TTS, the value of B_{eq} predicted by Shu et al. (1994) is given in the bottom right corner of the panel. TTS without predictions are not accreting. Our measurements of $\langle B \rangle$ appear above the names of certain TTS. All field strengths are in kG.

could conceivably restore equipartition, if the photosphere extends to much deeper layers, where the gas pressure is higher.

Strong magnetic fields are ubiquitous on TTS. By fitting IR spectra, we have determined preliminary magnetic field distributions for BP Tau, DF Tau, Hubble 4, TW Hya, and DK Tau, obtaining $\langle B \rangle$ of 2.1, 2.3, 2.4, 2.6, and 2.7 kG, respectively. Hubble 4 is not accreting, so the broad Ti I lines in this star cannot be due to accretion or the presence of a circumstellar disk. We also see Zeeman broadening in IR spectra of another dozen TTS, including several that are not accreting. Figure 4 shows one Zeeman sensitive Ti I line and one Zeeman insensitive line for every star we have observed. The line profiles have been scaled to a common central depth to facilitate visual identification of Zeeman broadening. In a few cases, the CO line is broadened by rotation or contaminated by a companion. In the remaining cases, Zeeman broadening is apparent. DE Tau has the weakest magnetic field, but $\langle B \rangle$ appears to be about 1–3 kG in the remaining stars.

Recall that the Shu et al. (1994) model predicts B_{eq} of 0.67 kG for DF Tau and 4.1 kG for BP Tau, whereas we obtain $\langle B \rangle$ of 2.3 kG and 2.1 kG, respectively. In fact, examination of the Figure 4 shows that most TTS probably have comparable $\langle B \rangle$, rather than the wide range originally predicted. As we discuss below, magnetospheric accretion models can be fixed by replacing B_{eq} with a more general characterization of the magnetic flux connecting the star and the disk.

3. Spectropolarimetry

Zeeman broadening measurements are sensitive to the distribution of magnetic field strengths, but they have limited sensitivity to magnetic geometry. In contrast, circular polarization measurements for individual spectral lines are sensitive to magnetic geometry, but they provide limited information about field strength. The two techniques complement each other well, as we demonstrate later.

Unfortunately, circular polarization is almost never detected in the spectra of unresolved cool stars. The reason is apparent in any magnetogram of the resolved surface of the Sun. Sunspots and solar active regions tend to come in bipolar pairs, leading to nearly complete cancelation of the net circular polarization signal. The net longitudinal magnetic field (B_z) of the Sun as a star is roughly 1 G, implying a cancelation factor of 10^3. The absence of robust circular polarization detections for unresolved cool stars suggests that similarly complex magnetic geometries are a general feature of cool stars.

However, circular polarization has been detected in radio observations of M dwarf flares and Zeeman Doppler images of very rapidly rotating cool stars. In both cases, observations isolate a small portion of the stellar surface, reducing the impact of polarization cancelation. Johns–Krull et al. (1999a) discovered that this same localization effect occurs in CTTS emission line diagnostics that form predominantly in the accretion shock at the surface of the star. This circular polarization signal is strongest in the narrow component of the He I 5876 Å emission line, but it is also present in the Ca I infrared triplet lines. The peak value of B_z is 2 kG, which is comparable to our measured values of $\langle B \rangle$, implying a cancelation factor of order unity. Magnetic polarity is remarkably uniform in accretion footpoints at the surface of a TTS.

Figure 5 shows our measurements of B_z on six consecutive nights. We obtained these measurements at McDonald Observatory, using the Zeeman analyzer described by Johns–Krull et al. (1999a). More accurate results for many more stars could easily be obtained with a high-resolution spectrograph, fed by fibers from a polarimeter mounted at the Cassegrain focus of a large telescope. Amazingly, no such instrument exists, nor are there serious plans to build one.

Our measured values of B_z vary smoothly on rotational timescales, suggesting that uniformly oriented magnetic field lines in accretion regions sweep out a cone in the sky as the star rotates. Rotational modulation implies a lack of symmetry about the rotation axis in the accretion or the magnetic field or both. For example, the inner edge of the disk could have a concentration of gas that co-rotates with the star, preferentially illuminating one sector of a symmetric magnetosphere. Alternatively, a single large scale magnetic loop could draw material from just one sector of a symmetric disk. Many variants are possible.

Figure 5 shows one interpretation of the He I polarization data. Predicted values of B_z are shown for a simple model consisting of a single magnetic spot at latitude ϕ that rotates with the star. The magnetic field is assumed to be radial with a strength

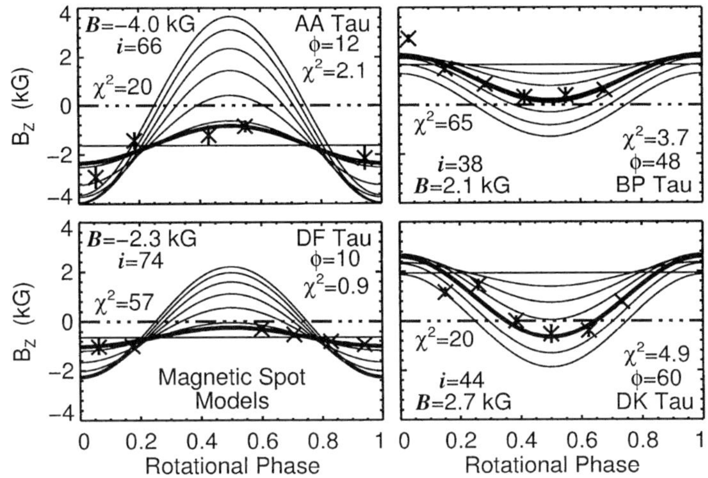

Figure 5. Crosses (×) with vertical error bars indicate the net longitudinal magnetic field (B_z), measured on six consecutive nights using the He I 5876 Å accretion diagnostic. The family of curves show predicted B_z values for a simple model with radial magnetic field lines concentrated in a spot at latitude ϕ. Magnetic field strengths are constrained by independent Zeeman broadening measurements. Reduced χ^2 is 1–5 for the best fitting model, shown as a long dashed curve for each star. Reduced χ^2 is 20–60 for a model that assumes $B_z = 0$.

equal to our measured values of $\langle B \rangle$, except for AA Tau where the unmeasured field strength is a free parameter. Inclination of the rotation axis is constrained by measured $v \sin i$ and rotation period, except that inclination (i) is allowed to float when it exceeds 60° because $v \sin i$ measurements cannot distinguish between these possibilities.

Predicted variations in B_z are plotted for spot latitudes ranging from 0° to 90° in 15° increments. The best fitting model is presented by a heavy curve. The corresponding spot latitude and reduced χ^2 are given on the right side of each panel. Large values of χ^2 on the left side of each panel rule out the hypothesis that no polarization signal is present. In all four cases, this simple magnetic spot model reproduces the observed B_z time series. We have detected, but not yet analyzed, apparent rotational modulation of the He I centroid velocity and equivalent width. Spectropolarimetry yielded robust constraints on magnetic geometry, despite large changes in accretion rate during the observations, offering a possible tool for imposing order on otherwise chaotic observations of CTTS.

As a matter of convenience, existing magnetospheric accretion models assume that intrinsic TTS magnetic fields are dipolar, but this would be unprecedented for cool stars. Higher order multi-polar components do fall off more rapidly with distance, so at the inner edge of the disk a few stellar radii from the surface, the dipolar component of the stellar field probably does dominate. However, at the surface the field is likely to be more complicated. We tested this hypothesis by measuring B_z using Zeeman sensitive Fe I absorption lines that form over the

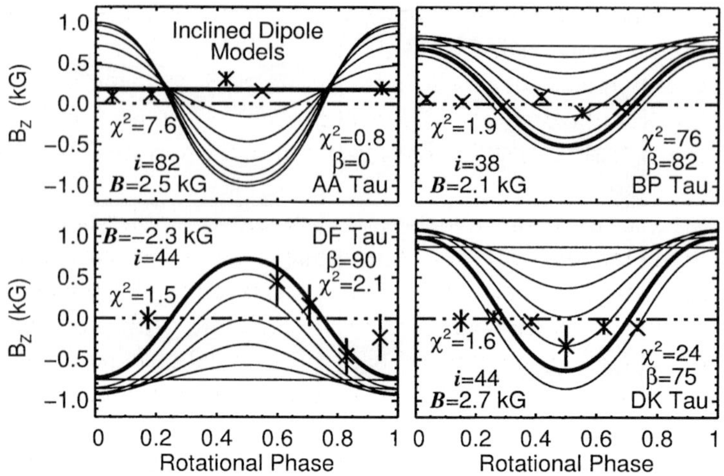

Figure 6. Crosses (×) with vertical error bars indicate the net longitudinal magnetic field (B_z), measured on six consecutive nights using a few Zeeman sensitive absorption lines that form over the entire photosphere. The family of curves show predicted B_z values for a dipole inclined by an angle β with respect to the rotation axis. The inclined dipole model is ruled out for BP Tau and DF Tau, whereas a trivial $B_z = 0$ model works well. The measurements for DF Tau are noisy enough that both models work equally well. AA Tau apparently has a nonzero, constant B_z, which is hard to explain.

entire stellar surface. The same echelle spectra were used to measure both the He I and Fe I line profiles.

Figure 6 shows the B_z time series based on Fe I lines formed over the entire stellar surface. Predicted variations in B_z are shown for dipole models with the magnetic axis tilted by an angle β (in 15° increments) with respect to the rotation axis. The best fitting model is denoted by the thick curve, with the corresponding β and reduced χ^2 given on the right side of each panel. The surface fields on BP Tau and DK Tau are not dipolar. Small values of reduced χ^2 on the left side of most panels indicate that our B_z measurements are generally consistent with no detected circular polarization, despite large values of $\langle B \rangle$, ruling out a dipolar magnetic configuration at the stellar surface. AA Tau is an interesting exception because B_z was slightly positive in all five measurements, which could in principle be caused by a tilted dipole with a finely tuned inclination.

4. Magnetospheric Accretion Models Revisited

Magnetospheric models typically assume an aligned dipolar magnetic field at the surface of a CTTS. More generally, Mahdavi and Kenyon (1998) describe the observational consequences of a tilted magnetic dipole. Nonetheless, time series spectropolarimetry demonstrates that for at least some CTTS, the surface field geometry is not dipolar. Additional data would be useful, but already it is clear that more general models are needed to interpret observations. We hypothesize that the magnetic geometry is complicated at the surface of TTS, as on the Sun and all other cool stars. However, the dipolar component probably dominates at the inner edge of the accretion disk, several stellar radii from the stellar surface. This has practical implications.

Magnetospheric accretion models use an assumed dipolar field geometry to predict which stellar magnetic field lines couple the stellar surface to the disk. Since the actual magnetic geometry is more complicated, the fraction (δ) of the stellar surface with field lines coupled to the disk can differ significantly from predicted values. This means the self-consistency relationships discussed at the beginning of this paper are actually underconstrained, even though we have now measured $\langle B \rangle$.

However, Johns–Krull and Gafford (2002) point out that a more general self-consistency relationship for the Shu et al. (1994) model can be determined by characterizing magnetic coupling in terms of the surface magnetic flux that threads the disk outside corotation, rather than equatorial field strength of an assumed dipole. If we factor this surface magnetic flux into the product $\langle B \rangle \delta$, then Eq. (1) follows from Johns–Krull and Gafford (2002):

$$\delta = 0.15 \frac{\left(\frac{M_*}{M_\odot} \frac{P_*}{1\,\mathrm{dy}} \frac{\dot{M}_*/(1-\varepsilon)}{10^{-7} M_\odot\,\mathrm{yr}^{-1}} \right)^{1/2}}{\left(\frac{R_*}{R_\odot} \right)^2 \frac{B_*}{1\,\mathrm{kG}}} \tag{1}$$

Table I gives predicted values of δ, based on measured stellar and disk accretion parameters, assuming $\varepsilon = 1/3$, where ε is the mass loss rate divided by the disk accretion rate. We use measured values of $\langle B \rangle$ when available, and otherwise assume a typical value of 2.5 kG. Assuming a constant surface field strength yields surface fractions that range from 1–15%, whereas the original models assumed constant δ and predicted a factor of 15 range in surface field strength. Empirically, we believe that δ is more likely to vary than $\langle B \rangle$. Our measurements of $\langle B \rangle$ support this conclusion, and simple models of the observed blue excess in CTTS imply that accretion footpoints cover a similar range of surface fractions (Valenti et al., 1993).

Finally, we note that the original formulation of magnetospheric accretion models predicted very different intrinsic magnetic field strengths for two CTTS with similar stellar properties, but quite different disk accretion rates. For example, DK Tau and DR Tau both have similar stellar parameters, but disk accretion rates that differ by a factor of 20, leading to predicted values of B_{eq} that differ by a factor of 6 (Johns–Krull et al., 1999b). The dependence of intrinsic stellar field strength on circumstellar disk properties seems inconsistent with dynamo theory. Equation (1)

TABLE I
Predicted fractional areas of accretion footpoints

Star	M_* ($M_\odot$)	R_* ($R_\odot$)	P_* (day)	$\log \dot{M}_*$ ($M_\odot$/yr)	$\langle B \rangle$ (kG)	δ (%)
DI Tau	0.31	2.2	7.5	−7.8	2.5	0.9
DN Tau	0.42	2.2	6.0	−7.5	2.5	1.4
T Tau	2.0	3.4	2.8	−7.0	2.5	1.5
GK Tau	0.41	2.2	4.7	−7.2	2.5	1.7
GI Tau	0.30	2.5	7.2	−6.9	2.5	1.9
CY Tau	0.58	1.4	7.5	−8.2	2.5	2.0
DF Tau	0.17	3.9	8.5	−5.9	2.3	2.2
GG Tau	0.29	2.8	10.3	−6.7	2.5	2.3
DE Tau	0.24	2.7	7.6	−6.5	2.5	2.4
DH Tau	0.65	1.9	7.2	−7.5	2.5	2.5
TW Hya	0.70	1.0	2.2	−8.0	2.6	2.8
DK Tau	0.38	2.7	8.4	−6.4	2.7	3.3
GM Aur	0.52	1.6	12.0	−7.6	2.5	3.6
AA Tau	0.38	1.8	8.2	−6.9	2.5	4.5
BP Tau	0.45	1.9	7.6	−6.8	2.1	5.6
DG Tau	0.67	2.3	6.3	−5.7	2.5	12.0
DR Tau	0.38	2.7	9.0	−5.1	2.5	16.6

now allows similar surface field strength for these two stars, as long as δ differs by a factor of 5.

Acknowledgements

We gratefully acknowledge significant contributions to this work by Nikolai Piskunov, Artie Hatzes, and Steve Saar. We thank all of our colleagues for useful discussions that helped clarify some of the implications of these new measurements. We also wish to thank the conference organizers for a truly outstanding event.

References

Cameron, A.C. and Campbell, C.G.: 1993, *A&A* **274**, 309.
Johns–Krull, C.M., Valenti, J.A., Hatzes, A.P. and Kanaan, A.: 1999a, *ApJL* **510**, L41.
Johns–Krull, C.M., Valenti, J.A. and Koresco, C.: 1999b, *ApJ* **516**, 900.
Johns–Krull, C.M. and Gafford, A.D.: 2002, *ApJ* **583**, 685.
Königl, A.: 1991, *ApJL* **370**, L39.
Lynden–Bell, D. and Pringle, J.E.: 1974, *MNRAS* **168**, 603.
Mahdavi, A. and Kenyon, S.J.: 1998, *ApJ* **497**, 342.

Shu, F.H., Najita, J., Ostriker, E., Wilkin, F., Ruden, S. and Lizano, S.: 1994, *ApJ* **429**, 781.
Uchida, Y.: 1983, in: P.B. Byrne and M. Rodono (eds.), *Activity in Red-Dwarf Stars*, Reidel, Dordrecht, The Netherlands, p. 625.
Uchida, Y. and Shibata, K.: 1984, *PASJ* **36**, 105.
Valenti, J.A., Basri, G. and Johns–Krull, C.M.: 1993, *AJ* **106**, 2024.

X-RAY ACTIVITY AND ACCRETION IN YOUNG STELLAR OBJECTS

THOMAS PREIBISCH

Max-Planck-Institut für Radioastronomie, Bonn, Germany; E-mail: preib@mpifr-bonn.mpg.de

Abstract. I discuss recent observational results on the X-ray properties of young stellar objects, based mostly on *Chandra* and *XMM-Newton* observations. The sensitive X-ray data on large, well characterized samples of T Tauri stars (and a number of protostars) allow to study in detail the dependence of magnetic activity on the bulk properties of the young objects and to draw important clues towards the origin of the X-ray emission. The absence of a relation between X-ray activity and rotation for T Tauri stars clearly suggests that their magnetic activity cannot be simply explained by the action of a scaled-up solar-like dynamo. I discuss alternative models for the generation of magnetic fields and also consider the long standing question whether the X-ray properties of the T Tauri stars are related to the presence/absence of circumstellar disks or active accretion.

Keywords: X-rays, young stellar objects, T Tauri stars, accretion

1. Introduction

Young stellar objects (YSOs) in all evolutionary stages from class I protostars to ZAMS stars show highly elevated levels of X-ray activity. Excellent and recent reviews on the X-ray properties of YSOs and on stellar coronal astronomy in general can be found in Feigelson and Montmerle (1999) and Favata and Micela (in press). I therefore keep the first two chapters very brief, with only a minimum of references. For more detailed information and more references the reader is referred to these reviews.

X-ray observations of star forming regions allow to study the high energy processes in these YSOs. Since X-rays with energies above ~1 keV are much less affected by extinction than optical light, the X-ray observations also allow a deep look into dense molecular cloud cores towards the embedded YSOs. The high energy processes in YSOs are of great importance for our understanding of the star formation process. For example, the X-ray emission from a YSO, which still is in its main accretion phase, should photoionize its circumstellar material and thus influence accretion as well as outflow processes, both of which are thought to be based on the interaction of ionized material with magnetic fields. The protostellar X-ray emission might even affect the formation of proto-planets in circumstellar disks.

The primary reference case for the interpretation of the observed X-ray emission from YSOs is, of course, our Sun. The solar corona has an extremely complex and dynamic structure with many different facets. Its high-energy emission can be

Astrophysics and Space Science **292:** 631–641, 2004.

studied in great detail at high spatial, temporal, and spectral resolution. Imaging observatories such as YOHKOH, SOHO and TRACE reveal the details of the different physical structures in the solar corona and their highly dynamical behavior. Stellar X-ray observations, on the other hand, can only investigate the integrated emission, averaged over all the different components of the X-ray emitting plasma; therefore, only some kind of "typical" values can be derived for the plasma properties. Furthermore, high-resolution X-ray spectra, from which information about plasma temperatures and densities could be inferred from the ratios of resolved emission lines, can only be obtained for a small number of very bright sources. For most X-ray sources, only low- or medium-resolution spectra can be obtained, from which only a limited amount of information can be retrieved. This huge difference in information that can be obtained in the case of our Sun on the one hand and other stars on the other hand illustrates, why it is so difficult to figure out what physical processes are responsible for the X-ray emission in YSOs.

2. Basic X-Ray Results on YSOs

The first discoveries of X-ray emission from low-mass YSOs were made with the EINSTEIN X-ray observatory (e.g. Feigelson and DeCampli, 1982). The X-ray luminosities of the detected T Tauri stars (TTS) ranged up to $L_X \sim 10^{31}$ erg/sec, and the plasma temperatures were found to be in the range $T \sim$5–30 MK. Both quantities are much larger than the corresponding values for the solar corona ($L_{X,\odot} \sim 10^{27}$ erg/sec; $T_{X,\odot} \sim$1–2 MK).

The ROSAT observatory increased the number of known X-ray emitting YSOs very considerably. One particularly important result from ROSAT was the discovery of large populations of weak-line T Tauri stars (WTTS) in and around many star forming regions (e.g. Neuhäuser, 1997). Although a significant fraction of these objects were later found to be somewhat older ZAMS stars, ROSAT showed that the stellar populations of star forming regions are considerably larger than suspected by earlier surveys based on classical youth indicators such as Hα emission. Another important result from ROSAT was the detection of X-ray emission from class I protostars (e.g. Grosso et al., 1997).

The ASCA satellite detected X-ray emission from numerous deeply embedded YSOs. Due to its rather poor spatial resolution, however, the proper identification of the X-ray sources was often difficult.

While these X-ray missions provided a lot of very important information about the X-ray properties of YSOs, there were also serious limitations on their results. First, the typical samples of X-ray detected objects in young clusters and star forming regions contained hardly more than approximately 100 objects, too few to allow well founded statistical conclusions to be drawn. Second, a large fraction of the known cluster members (especially low-mass stars) remained undetected in X-rays, and any correlation studies had therefore to deal with large numbers of upper limits. Third, especially in dense clusters, the individual sources could often not be spatially resolved, and so the proper identification of the X-ray sources was difficult or impossible. Finally, the spectral and temporal X-ray properties of only a relatively small number of individual young stars could be studied in detail, and it is not clear whether these stars really are "typical" cases or peculiar objects.

With the advent of *Chandra* and *XMM-Newton*, the situation has improved very dramatically. Due to the large collecting areas of these observatories, their sensitivity is at least an order of magnitude better than that of earlier missions. Due to their wide energy band, extending from ~0.2 keV up to at least 10 keV, they are very well suited to study the strongly extincted and therefore hard X-ray emission from highly obscured and deeply embedded YSOs. Furthermore, *Chandra* has a superb PSF, providing a spatial resolution of better than 1″; this eliminates essentially all resolution/identification problems (see Figure 1).

Deep observations of the following three regions provide good examples of the new quality of X-ray results: A deep *Chandra* observation of the young cluster IC 348 (Preibisch and Zinnecker, 2001, 2002; see Figure 1) revealed 215 X-ray sources, the majority of which could be identified with known and well characterized YSOs. In deep *Chandra* observations of the Orion Nebula Cluster (see Feigelson et al., 2002; Flaccomio et al., 2003; see Figure 2) more than 1000 individual X-ray sources were detected, over 500 of which could be identified with known YSOs. Such observations allow now for the first time to study really large, statistically meaningful samples of YSOs. A deep *Chandra* observation of the ρ Oph star forming region revealed numerous embedded sources; 70% of the known class I protostars were detected in X-rays (Imanishi et al., 2002). A deep *XMM-Newton*

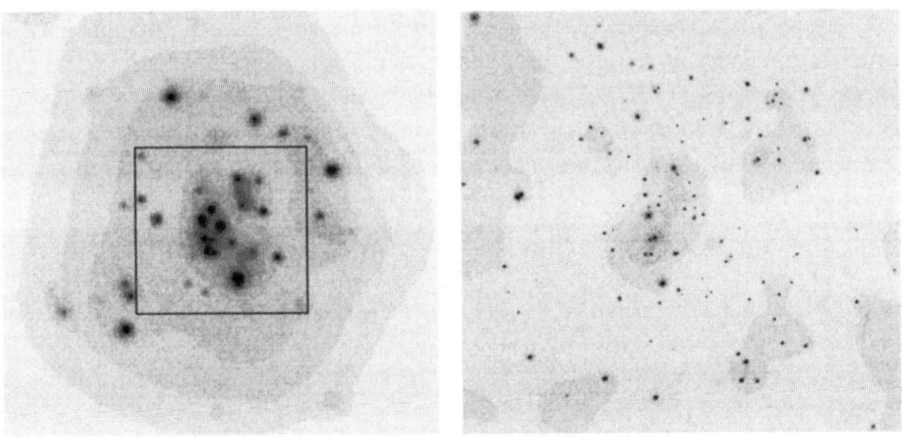

Figure 1. X-ray images of the young stellar cluster IC 348. The left image shows a ~30′ × 30′ region and was obtained with the ROSAT PSPC (Preibisch et al., 1996). The right image shows the central ~12′ × 12′ region (indicated by the box in the ROSAT image) as seen with *Chandra* ACIS (Preibisch and Zinnecker, 2001, 2002).

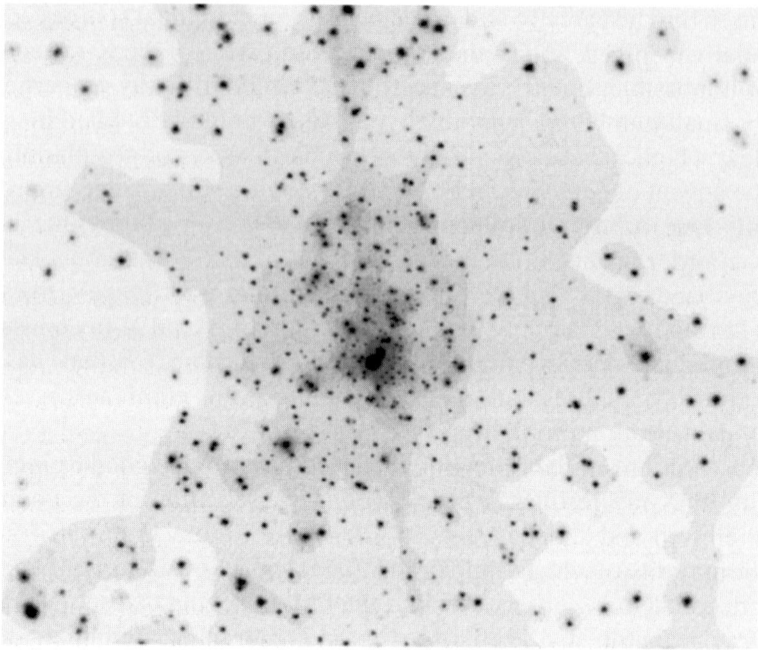

Figure 2. *Chandra* ACIS image of the Orion Trapezium cluster. This $\sim 14' \times 12'$ image was produced from merged archive data with a total exposure time of $\sim$100 ksec ($\sim$28 h) and smoothed with an adaptive filter.

study of the L1551 (Favata et al., 2003) allowed a detailed investigation of the spectral and temporal X-ray properties of numerous TTS.

The main results of these and other studies can be summarized as follows: The X-ray luminosity of YSOs is generally found to be well correlated to their bolometric luminosity ($L_X \propto L_{bol}$). Nearly all YSOs show $L_X/L_{bol} > 10^{-5}$ and are therefore much more X-ray active than the Sun ($L_{X,\odot}/L_{bol,\odot} \sim 10^{-6}$). The most active stars show fractional X-ray luminosities around $L_X/L_{bol} = 10^{-3}$, which is the saturation limit for coronally active stars. There is also a correlation between the X-ray luminosities and the stellar masses of the form $L_X \propto M^{2...3}$. A very important new result concerns the connection between X-ray emission and stellar rotation. In contrast to main-sequence stars, *no relation between rotation period and X-ray activity is found for the YSOs in the Orion Nebula Cluster* (Feigelson et al., 2003; Flaccomio et al., 2003); even slowly rotating YSOs show high levels of X-ray emission.

3. The Origin of the High X-Ray Activity in YSOs

Can the strong X-ray activity of YSOs (X-ray luminosities up to $\sim 10^4$ times, plasma temperatures up to $\sim$30 times higher than in our Sun) be understood as the consequence of enhanced, but in principle solar-like magnetic dynamo processes, or are fundamentally different processes involved? In principle, the X-ray luminosities and plasma temperatures of YSOs are not unlike those of the most coronally active main-sequence stars. For main-sequence stars, the level of X-ray emission is mainly determined by their rotation rate. The very well established activity-rotation relation is given by $L_X/L_{bol} \propto P_{rot}^{-2.6}$ for periods lower than $\sim$ 2–3 days, followed by saturation at $\log(L_X/L_{bol}) \sim -3$ for smaller periods. The plasma temperatures generally increase with the level of X-ray activity, scaling roughly as $T_X \propto (L_X/L_{bol})^{0.5}$. Since most YSOs rotate quite rapidly, neither their X-ray luminosities nor their plasma temperatures *per se* require a non-solar like emission mechanism.

However, the absence of a relation between rotation period and X-ray activity in YSOs strongly puts into question the solar-like dynamo activity scenario for YSOs. Also, the observed correlations of the X-ray luminosity with the stellar bolometric luminosity and the stellar mass are *not* predicted by models assuming solar-like dynamo activity and therefore argue against rotation as the dominant factor in the generation of X-rays. Another argument against the solar-like scenario comes from theoretical considerations: At ages of only a few Myr, the YSOs are still fully convective, and therefore the standard solar-like $\alpha - \Omega$ dynamo, which is anchored at the boundary between the convective envelope and the inner radiative core, cannot work. However, this problem also occurs in the case of fully convective late-type main-sequence stars, some of which are nevertheless strong coronal emitters. There are several alternative dynamo mechanisms that might work in fully convective YSO. One possibility would be an α^2-dynamo, as suggested for example by Küker and Rüdiger (1999); however, the α^2-dynamo would predict a strong dependence on rotation, what is not observed for YSOs. A possibly better alternative might be small-scale dynamo action in a highly turbulent convection zone (cf. Giampapa et al., 1996); such a turbulent dynamo should be only weakly dependent on rotation, in agreement with the observations. A further possibility would be the presence (and amplification) of fossil magnetic fields.

4. X-Ray Emission and Circumstellar Disks

Another possibility is that (some fraction of) the X-ray activity of YSOs originates from magnetic interactions between the YSO and its surrounding circumstellar accretion disk. To investigate this question, many X-ray observations of star-forming regions have been used to search for differences in the X-ray properties of the classical T Tauri stars (CTTS; $W(\mathrm{H}\alpha) \geq 10$ Å), which arc thought to be actively accreting via circumstellar disks, as compared to the weak-line T Tauri stars (WTTS; $W(\mathrm{H}\alpha) < 10$ Å), which seem to lack disks and active accretion. The results were quite confusing: in some regions, no significant differences could be found between the X-ray luminosity functions of CTTS and WTTS, while in other regions the X-ray luminosity functions seem to be clearly different, with the WTTS being the stronger X-ray emitters. A detailed investigation of this question was performed by Preibisch and Zinnecker (2002), based on the *Chandra* observation of IC 348. Figure 3 shows the X-ray luminosities of the stars as a function of $W(\mathrm{H}\alpha)$. Considering only the X-ray detected TTS, there is no significant difference in the X-ray properties of the WTTS as compared to the CTTS. However, many CTTS were not detected as X-ray sources; including these non-detections in the analysis, one finds a significant difference in the X-ray luminosity function of the WTTS as compared to the CTTS.

However, the use of Hα emission as a tracer to find TTS can introduce very serious selections effects, when it comes to WTTS. CTTS are quite easy to recognize by their prominent Hα emission even if they are very faint, whereas WTTS of similar brightness are much harder to identify; in fact, the majority of WTTS in many star forming regions have been found in X-ray observations (cf. Neuhäuser, 1997). Therefore, if we wish to obtain a complete, brightness-limited sample of TTS (both CTTS and WTTS) in a given region, it is necessary to use other selection criteria. A better way to discern between YSOs with and without circumstellar material is to look for near-infrared excess emission. Preibisch and Zinnecker (2002) therefore used $JHKL$ data to discern between excess sources and non-excess sources in IC 348. As can be seen in Figure 3, no systematic differences in the X-ray luminosities of excess sources versus non-excess sources are found, if a complete, luminosity limited, sample (based on the J-band flux) is used. A detailed statistical analysis confirms that the distributions of X-ray luminosities are equal for both samples. In conclusion, *we found no evidence that the presence of circumstellar*

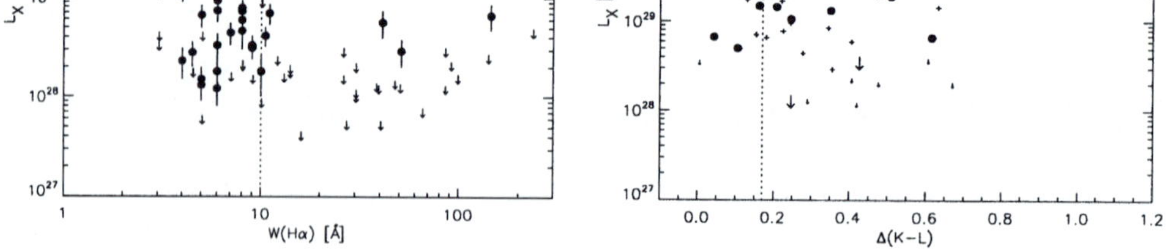

Figure 3. *Left*: X-ray luminosities for all optically selected TTS in IC 348 versus their Hα emission line equivalent widths. Upper limits are marked by arrows. *Right*: X-ray luminosity of the TTS in IC 348 versus their near-infrared excess. Objects with $\Delta(K-L) \geq 0.17$ show infrared excesses. The TTS in the complete sample are shown by the big dots, the fainter objects, for which the sample is incomplete, by the small crosses.

material makes any difference to the X-ray properties of the TTS in IC 348. Similar results have been derived for other star forming regions (e.g. Grosso et al., 2000; Getman et al., 2002). The apparent difference in the X-ray luminosity functions of WTTS versus CTTS can, at least in part, be explained by strong selection effects, favoring the optical detection of faint CTTS (which are faint in X-ray too) with strong Hα emission, as opposed to faint WTTS. The lack of faint WTTS in the sample of young stars tends to shift the median X-ray luminosity of WTTS to higher values, while the inclusion of faint CTTS decreases the median X-ray luminosity of CTTS.

The picture changes, however, if one looks for the presence/absence of accretion signatures rather than disks. Flaccomio et al. (2003) used the strength of the 8542 Å Ca II line to discern between stars with high accretion rates and stars with no accretion (or low accretion rates) in the Orion Nebula Cluster. They found that the X-ray luminosity functions of these two samples of TTS are significantly different, with the accreting stars having 2–3 times lower X-ray luminosities.

The conclusion from these results is that *the X-ray luminosities of TTS do not depend on the presence or absence of circumstellar material, but seem to depend on the presence or absence of disk accretion*. What is the reason for this difference?

Possibility I: Magnetically braked CTTS have weaker dynamos

It has been suggested that CTTS are weaker X-ray emitter because their rotation velocity is braked by magnetic disk locking, what should lead to weaker dynamo action and therefore less X-ray emission. While recent studies of the rotation rates of TTS seem to confirm the magnetic braking scenario (the mean rotation period for CTTS is ~8 days versus ~2 days for WTTS, see e.g. Herbst et al., 2002), the non-existence of a rotation-activity relation for TTS means that this difference in rotation should not cause any difference in the level of X-ray emission. *The magnetic braking scenario for CTTS can therefore not explain their lower X-ray luminosities*.

Possibility II: Changes in the magnetic field structure

The pressure of the accreting material can distort the large-scale stellar magnetic field considerably (e.g. Miller and Stone, 1997) and the magnetospheric transfer of material to the star can give rise to instabilities of the magnetic fields around the inner disk edge and cause reconnection events. The presence of accreting material also leads to higher densities in (parts of) the magnetosphere. In contrast to the WTTS, which probably have loops with moderate, solar-like plasma densities, some fraction of the magnetic field lines in CTTS would be mass loaded and therefore have much (~100×) higher densities. If now a magnetic reconnection event liberates a given amount of energy, this can heat the plasma in the low-density loops of WTTS to X-ray emitting temperatures ($\gtrsim$10 MK), while the plasma in the mass loaded loops of the CTTS would be only heated to much lower temperatures, and remain too cool to emit X-rays. This might account for the lower X-ray luminosities of the CTTS as compared to WTTS.

Possibility III: Accretion changes the stellar rotation pattern
The accretion process probably changes the internal stellar structure, the rotation pattern, and thereby might influence the magnetic field generation process. In our Sun, the rotational shear due to differential rotation is the main source of magnetic field energy. Any magnetospheric star-disk coupling will affect the differential rotation at the stellar surface. If the coupling happens closer to the stellar equator than to the poles, the magnetospheric braking effect thought to be at work in CTTS will also reduce the amount of differential rotation; this might lead to weaker dynamo action, less magnetic field generation, and finally weaker X-ray emission.

5. X-Ray Emission From Accretion Shocks?

It has sometimes been argued that X-rays might originate from hot accretion shocks on the surfaces of CTTS. The power of the mass accretion process in CTTS can reach fractions up to $\sim$10% of the stellar luminosity, and according to the magnetospheric accretion scenario, the accreted matter crashes onto the stellar surface with velocities of up to several 100 km/sec, causing hot shocks that might be capable to emit soft X-rays.

Two recent studies have claimed to see accretion-induced X-rays in CTTS. Kastner et al. (2002) analyzed high-resolution *Chandra* X-ray spectra of the CTTS TW Hya and derived quite high plasma densities of the order of $n \sim 10^{12}$–10^{13} cm^{-3} in combination with a plasma temperature of $T \sim 3$ MK. They argue that in coronally active stars, similarly high densities are usually found only in conjunction with much hotter plasma, and that their results are therefore suggestive of a model in which the bulk of the X-ray emission is generated via mass accretion from a circumstellar disk. However, we note that the plasma temperature and density of TW Hya are not so terribly different from those found for coronally active stars (e.g., densities of $n \sim 10^{12}$ cm^{-3} are found for plasma temperatures of $T \sim 5$ MK on RS CVn stars; see Osten et al., 2003). Also, the observed X-ray luminosity would require an accretion rate much higher than the rate deduced from Hα.

Favata et al. (2003) analyzed *XMM-Newton* data on the CTTS XZ Tau and found large-amplitude variations on time scales of several hours with large changes in the intervening absorption. They interprete this as evidence that the X-ray emitting region is eclipsed by an accretion stream and argue that the X-ray emission might originate from the accretion shock.

However, a closer look at magnetospheric accretion shock models (e.g. the investigation of the shock structure and emission by Calvet and Gullbring, 1998) does not suggest the accretion shock to be a significant source of X-ray emission from TTS. Directly in the shock zone, temperatures of up to a few MK can actually be reached. However, the soft X-rays (mostly at energies $\lesssim$0.3 keV) emitted from the

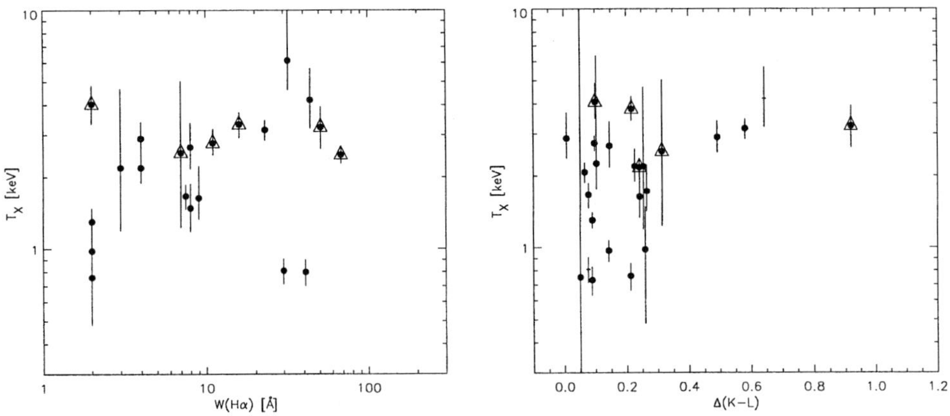

Figure 4. Plasma temperatures of the TTS in IC 348 derived from their *Chandra* spectra are plotted versus the Hα equivalent width (left) and versus the infrared excess (right).

hot shock plasma are nearly completely absorbed and thermalized within or close to the shock, because the typical column density above the shock is $\gtrsim 10^{23}$ cm^{-3}. The *observable* radiation from the shock, coming from an optical depth $\tau = 2/3$, has typical effective temperatures of ~8000–10 0000 K and causes the well known (blue) optical and UV excesses of CTTS, but no high-energy emission is expected to be observed.

Exceptions might perhaps be possible under special conditions, e.g. if the accretion stream is seen almost exactly perpendicular to the stream direction, so that the column density along the line-of-sight towards the shock is considerably lower. However, it is important to note that even if accretion would actually contribute to the X-ray emission of CTTS, this would still not explain the observation that the CTTS seem to be systematically *weaker* X-ray emitters than the WTTS; one would rather expect the opposite trend, as there would be an additional X-ray emission mechanism in the CTTS and these should therefore be stronger X-ray sources. Also, the plasma temperatures of accreting TTS should then be systematically lower (shock temperatures of $\lesssim 1$ MK) than those of the non-accreting TTS (coronal emission with $T \gtrsim 10$ MK). No such trend has been observed. Figure 4 shows the temperatures derived from the *Chandra* spectra for TTS in IC 348 and indicates no relation between plasma temperatures and Hα emission or infrared excess.

In conclusion, no really convincing observational signs have yet been found which would suggest that X-ray emission from accretion shocks is a significant component in the X-ray activity of TTS.

6. X-Ray Emission From Class I Protostars

In contrast to the TTS, the X-ray emission of the much younger class I protostars may well be dominated by non-coronal emission processes. While TTS usually show only low-level residual accretion at rather low rates, the class I protostars are still in their main accretion phase and actually derive most of their energy from the accretion process. The high accretion rates can be expected to strongly affect the magnetospheric coupling between the protostar and its disk. For example, the dipolar stellar magnetic field lines anchored to the inner part of the accretion disk should be twisted around because of the differential rotation between the star of the disk. Theoretical work suggests that the twisted field lines periodically reconnect and release magnetic energy as hot, X-ray emitting plasma (Hayashi et al., 1996; Montmerle et al., 2000; Uzdenski, 2003 [these proceedings]). In this way, the X-ray emission may be connected to the accretion process. Observational support for this scenario comes, for example, from observations of quasi-periodic X-ray flaring in the class I protostar YLW 15 (Montmerle et al., 2000). The magnetospheric interaction models also provide an interesting basis for a unified explanation of not only the accretion process and the X-ray emission, but also the launching of strong winds and jets (e.g. Goodson et al., 1999).

7. Further Progress: The *Chandra* Orion Ultradeep Project

The best way to address the open questions is to obtain very sensitive X-ray observations of large, homogenous, and unbiased samples of YSOs with well known stellar parameters. A huge step forward can be expected from the "*Chandra* Orion Ultradeep Project" (COUP; PI: Eric Feigelson), in which an 850 ksec (i.e. 10 full days !) exposure of the Orion Nebula Cluster was recently obtained with the ACIS camera, by far the deepest and most sensitive X-ray observation ever obtained for a young cluster. This unique data set will allow us to study the X-ray spectra and the temporal variability of hundreds of YSOs in an unprecedented way. It will also provide a well-founded basis for the statistical analysis of the relations between X-ray activity and stellar properties with really large sample of YSOs, and should lead to a much better understanding of the origin of their strong X-ray activity.

Acknowledgement

I would like to thank the organizers of this very interesting conference for inviting me to give this review talk.

References

Calvet, N. and Gullbring, E.: 1998, *ApJ* **509**, 802.
Favata, F. and Micela, G.: 2003, *Space Sci. Rev.*, **108**, 577.
Favata, F., Giardino, G., Micela, G., Sciortino, S. and Damiani, F.: 2003, *A&A* **403**, 187.
Feigelson, E.D. and DeCampli, W.M.: 1981, *ApJ* **243**, L89.
Feigelson, E.D. and Montmerle, T.: 1999, *ARA&A* **37**, 363.
Feigelson, E.D., Broos, P., Gaffney, J.A. III, Garmire, G., Hillenbrand, L.A., Pravdo, S.H., Townsley, L. and Tsuboi, Y.: 2002, *ApJ* **574**, 258.

Feigelson, E.D., Gaffney, J.A. III, Gramire, G. and Hillenbrand, L.A., Townsley, L.: 2003, *ApJ* **584**, 911.
Flaccomio, E., Damiani, F., Micela, G. et al.: 2003, *ApJ* **582**, 398.
Getman, K.V., Feigelson, E.D., Townsley, L., Bally, J., Lada, C.J. and Reipurth, B.: 2002, *ApJ* **575**, 354.
Giampapa, M.S., Rosner, R., Kashyap, V., Fleming, T.A., Schmitt, J.H.M.M. and Bookbinder, J.A.: 1996, *ApJ* **463**, 707.
Goodson, A.P., Böhm, K.-H. and Winglee, R.M.: 1999, *ApJ* **524**, 142.
Grosso, N., Montmerle, T., Feigelson, E.D. et al.: 1997, *Nat* **387**, 56.
Grosso, N., Montmerle, T., Bontemps, S., Andre, P. and Feigelson, E.D.: 2000, *A&A* **359**, 113.
Hayashi, M.R., Shibata, K. and Matsumoto, R.: 1996, *ApJ* **468**, L37.
Herbst, W., Bailer-Jones, C.A.L., Mundt, R., Meisenheimer, K. and Wackermann, R.: 2002, *A&A* **396**, 513.
Imanishi, K., Tsujimoto, M. and Koyama, K.: 2002, *ApJ* **572**, 300.
Kastner, J.H., Huenemoerder, D.P., Schulz, N.S., Canizares, C.R. and Weintraub, D.A.: 2002, *ApJ* **567**, 434.
Küker, M. and Rüdiger, G.: 1999, *A&A* **346**, 922.
Miller, K.A. and Stone, J.M.: 1997, *ApJ* **489**, 890.
Montmerle, T., Grosso, N., Tsuboi, Y. and Koyama, K.: 2000, *ApJ* **352**, 1097.
Neuhäuser, R.:1997, *Science* **267**, 1363.
Osten, R.A., Ayres, T.R., Brown, A., Linsky, J.L. and Krishnamurthi, A.: 2003, *ApJ* **582**, 1073.
Preibisch, Th.: 1997, *A&A* **320**, 525.
Preibisch, Th. and Zinnecker, H.: 2001, *AJ* **122**, 866.
Preibisch, Th. and Zinnecker, H.: 2002, *AJ* **123**, 1613.
Preibisch, Th., Zinnecker, H. and Herbig, G.H.:1996, *A&A* **310**, 456.

THE ORIGIN OF JETS FROM YOUNG STARS: MHD DISK WIND MODELS CONFRONTED TO OBSERVATIONS

C. DOUGADOS[1], S. CABRIT[2], J. FERREIRA[1], N. PESENTI[1], P. GARCIA[3] and D. O'BRIEN[3]

[1]*Laboratoire d'Astrophysique de Grenoble, France;*
E-mail: Catherine.Dougados@obs.ujf-grenoble.fr
[2]*Observatoire de Paris, France*
[3]*University of Porto, France*

Abstract. We discuss in this contribution constraints on the origin of mass-loss from young stars brought by recent observations at high angular resolution ($0.1'' = 14$ AU) of the inner regions of winds from T Tauri stars. Jet widths and collimation scales, the large extent of the velocity profile as well as the detection of rotation signatures agree with predictions from magneto-centrifugal disk wind ejection models. However dynamically cold disk wind solutions predict too large terminal velocities and too low jet densities and ionisation fractions, suggesting that thermal gradients (originating in an accretion heated disk corona for example) may play an important role in accelerating the flow.

Keywords: T Tauri stars, high-angular resolution, jets, magneto-hydrodynamic

1. Introduction

The physical mechanism by which mass is ejected from young stars and collimated into jets remains a fundamental open issue in star formation theory. The strong correlation between ejection and accretion found in PMS stars, with a mass flux ratio as high as 0.1 (Cabrit et al., 1990; Hartigan et al., 1995) has favored accretion-driven magneto-hydrodynamic (MHD) wind models. However, it is not yet established whether the jet originates from the stellar surface (Sauty and Tsinganos, 1994) the magnetosphere/disk interface (Shu et al., 1995), or a wide range in disk radii ("disk winds": Ferreira, 1997, hereafter F97); and whether it is launched mostly by magneto-centrifugal forces or by a strong thermal pressure gradient in an accretion-heated corona. Distinguishing between these various scenarii is crucial in particular for models of exoplanet formation, as they have distinct implications on the internal structure, angular momentum transfer, and heating/irradiation processes in the inner regions of protoplanetary disks.

The small scale jets detected in the vicinity of the Classical T Tauri stars (see Figure 1) offer a unique opportunity to test current ejection theories: they give access to the innermost regions of the wind ($\leq$100 AU) where models predict that most of the collimation and acceleration processes occur; in addition, the stellar and accretion disk properties are well characterized in these systems. We summarize in this contribution constraints brought by the detailed comparison of high-angular

Astrophysics and Space Science **292:** 643–650, 2004.

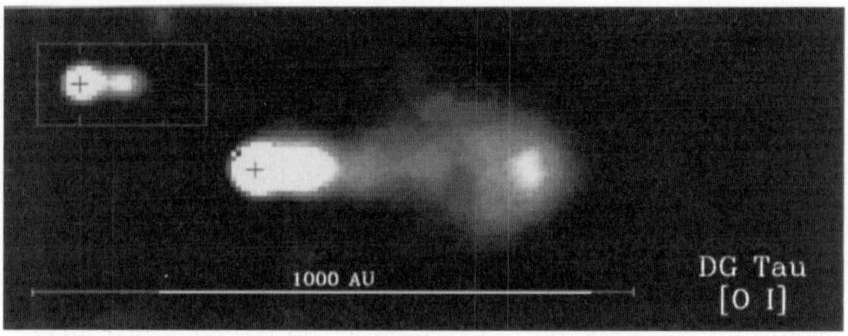

Figure 1. Deconvolved [O I] 6300 Å + continuum narrow-band image of the DG Tau microjet obtained with the adaptive optics system PUEO at CFHT. Achieved spatial resolution is 0.1″ = 14 AU at the Taurus distance. Adapted from Dougados et al. (2002).

resolution studies of a few prototypical T Tauri jets with predictions from stationary self-similar disk wind models.

2. Disk Wind Model Predictions

We concentrate our study on two classes of self-similar disk wind solutions, which extend the work of Blandford and Payne (1982) with a self-consistent treatment of the accretion flow in the underlying resistive keplerian disk. F97 investigated dynamically *cold* solutions where jet enthalpy is assumed to be negligible for accelerating the flow. Casse and Ferreira (2000) generalized this study to *warm* solutions, including the effects of entropy generation at the wind base, powered by an accretion-heated disk corona for example.

Both wind solutions are characterized by three free parameters (see Ferreira these proceedings): $\epsilon = h/r_0$, where h is the disk scale height at the cylindrical radius r_0, $\alpha_m = \nu_m/V_A h$, related to the magnetic diffusivity parameter ν_m and describing the level of turbulence, and ξ which controls the mass loading onto the field lines: $\dot{M}_{\rm acc}(r) \propto r^{\xi}$. ξ is related to the total ejection to accretion rate ratio by $2\dot{M}_{\rm jet}/\dot{M}_{\rm acc} = \xi \times \ln(r_{\rm out}/r_{\rm in})$, where $r_{\rm out}$ and $r_{\rm in}$ are the outer and inner radii of the disc involved in the ejection process. We consider given models with $\epsilon = 0.1$ (consistent with observations of HH 30 by Burrows et al., 1996) and $\alpha_{\rm m} = 1$ (high magnetic torque). Solutions that propagate far from the Alfvén surface are then found for ξ in the range 0.005 to 0.01 for the cold disc winds, ξ up to 0.5 for the *warm* disc winds. Thermal gradients have a strong influence on the streamlines mass-loading, allowing to reach larger ejection efficiencies in the warm case.

Detailed comparison with observations requires solving for the thermal and ionization structure of the flow. We concentrate here on two main heating mechanisms: ambipolar diffusion (Garcia et al., 2001a), and extra mechanical energy deposition due to the dissipation of shocks (O'Brien et al., 2003). The full computation of the thermal structure has been performed for the *cold* solution only. For the *warm* solution, we show preliminary predictions assuming uniform temperature (10^4 K) and the same prescription for hydrogen ionisation fraction as in Cabrit et al. (1999). We consider streamlines anchored in the disk at radii ranging from $r_{\rm in} = 0.07$ AU, the typical disk corotation radius for a T Tauri star, to $r_{\rm out} = 1$ AU, where molecules should start to form. In the following, a central stellar mass of $0.5M_\odot$ and accretion rate $\dot{M}_{\rm acc} = 10^{-6}M_\odot$ yr^{-1}, typical of the sources investigated here, will be used. We further take into account beam dilution and projection effets. The computed emissivity grids are projected onto the plane of the sky, to construct emission maps and long slit spectra. Synthetic maps are then convolved by a two-dimensional gaussian beam, representative of the instrumental spatial and spectral resolutions.

3. Jet Widths and Collimation Scales

Shown in Figure 2 are transverse FWHM of the optical forbidden lines emission for the five microjets for which we currently have constraints on scales $\leq$200 AU from the central driving source (HL Tau, HH 30: Ray et al., 1996; DG Tau, CW Tau, RW Aur: Dougados et al., 2000, Woitas et al., 2002). At the highest spatial resolution achievable today ($0.1'' = 14$ AU), the inner collimation region is still

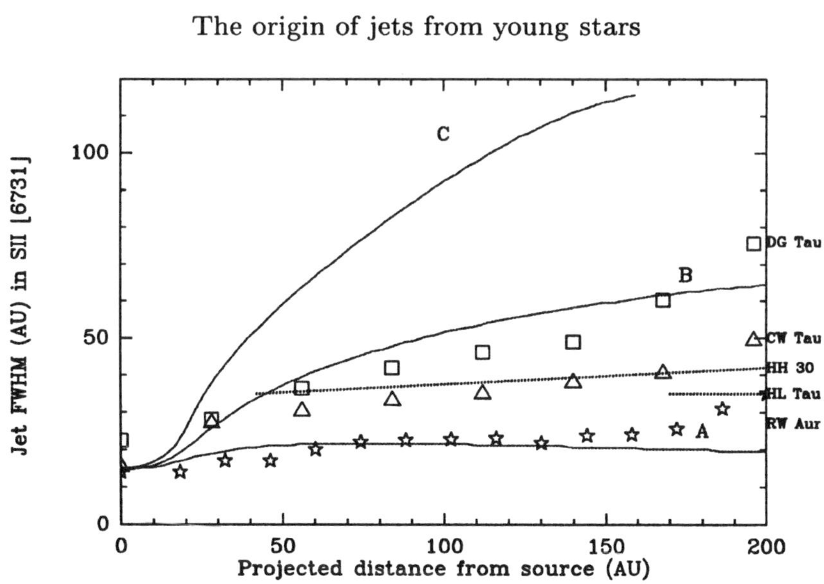

Figure 2. Jet widths and collimation scales: jet FWHM (symbols and dotted lines) derived from CFHT/PUEO and HST/STIS observations (see text for more details) are compared with predictions from cold disk wind models heated by ambipolar diffusion (solid lines) with varying ξ values: model A ($\xi = 0.01$), B ($\xi = 0.007$), C ($\xi = 0.005$). Models are convolved with a 14 AU (FWHM) gaussian beam. Adapted from Garcia et al. (2001b).

unresolved. Jet widths are resolved beyond 30 AU with typical FWHM of 20–40 AU at $z_{proj} = 100$ AU then slowly increase with opening angles $\leq 5°$.

The *cold* disk wind solutions are characterized by a wide opening of the field lines beyond the Alfvén surface (located at z/r $\simeq 1$). The expansion factor r_∞/r_0 increases from 50 to 2000 when ξ decreases from 0.01 to 0.005 (F97). The strong variation of the collimation properties of the model with the ejection efficiency parameter is illustrated in Figure 2. Both observed collimation scales and jet widths favor solutions with moderate to high ejection efficiency: $\xi \geq 0.007$, *i.e.* $\dot{M}_{ej}/\dot{M}_{acc} \geq 2 \times 10^{-2}$, in agreement with observations (Hartigan et al., 1995).

4. Kinematics

We illustrate in Figure 3 the comparison between observed and predicted position-velocity diagrams along the inner regions ($\leq$200 AU) of the DG Tau microjet. The disk wind models reproduce the large velocity distribution close to the central souce. In the framework of self-similar disc winds, terminal poloidal velocities for a given streamline are proportional to the keplerian velocity at the launching radius r_0 (see Eq. (1)). A large range in initial disk launching radii therefore naturally induces a corresponding extended range in asymptotic poloidal velocities. The observed decrease of line width with distance is also well reproduced by the models. It results from the gradual collimation of streamlines (Figure 3). Due to the rapid cylindrical density stratification (see Cabrit et al., 1999), inner streamlines quickly dominate the emission and produce a high-velocity, collimated jet component (the so-called jet optical illusion, previously pointed out for the X-wind model by Shang et al., 1998). However, two major discrepancies occur between model and observations. Firstly, the *cold* models predict too little emission at intermediate velocities

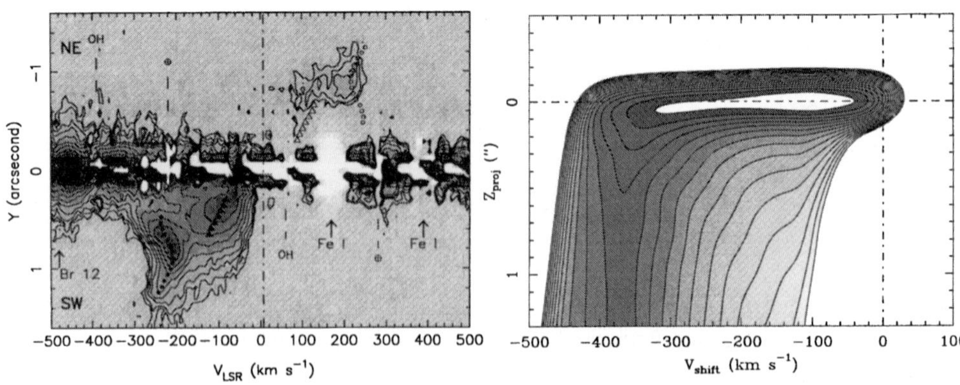

Figure 3. Position-velocity diagrams along the DG Tau jet in the [Fe II] 1.257 μm line: Left Observations by Pyo et al. (2003) of the central 200 AU of the microjet, Right Cold disk wind model predictions ($i = 45°$, $\xi = 0.01$, $\dot{M}_{acc} = 10^{-6} M_\odot$ yr^{-1}) on the same spatial scale convolved with a 22 AU $\times$ 30 km s^{-1} beam. Adapted from Pesenti et al. (2003).

($\simeq -100$ km s^{-1}). A more efficient heating mechanism seems required in the outer field lines at the wind base. Secondly, predicted terminal poloidal velocities appear to exceed observed ones by a factor 1.5 to 3 (Figure 3, Dougados et al., 2003). Under the assumption that all available energy is converted into jet kinetic power, the asymptotic value of the poloidal velocity, independent of the ejection model, can be written as (F97):

$$V_{\rm pol} = V_{\rm kep}(r_0) \times \sqrt{2\lambda - 3}, \tag{1}$$

where $\lambda \simeq (r_A/r_0)^2$ is the magnetic lever arm ($r_{\rm A}$ is the cylindrical radius at the Alfvén surface) and $V_{\rm kep}(r_0)$ is the keplerian velocity at the disk launching cylindrical radius r_0. Decreasing the terminal poloidal velocities could be achieved by either increasing r_0 or decreasing the magnetic lever arm λ.

Bacciotti et al. (2002) recently reported the detection of rotation signatures in the DG Tau microjet from STIS/HST observations. Velocity shifts of 6 to 15 km s^{-1} are detected at a few tens of AU from the jet axis and between 20 and 90 AU above the disk plane, in the component with poloidal velocity $\simeq$$-80$ km s^{-1}. We compare in Figure 4 these observations with predictions from the disk wind solutions. We note that the observed trend of increasing velocity shifts with radial distance from the axis is due to projection and beam convolution effects and does not reflect the true underlying azimuthal velocity profile (Pesenti et al., 2004). In fact, simulations demonstrate that, in the case of the DG Tau observations, distances from the axis $\geq$30 AU are required for a proper estimate of V_ϕ. The cold models (with $\lambda \simeq 50$) predict velocity shifts that exceed the observed ones by a factor 2 to 3. In contrast, preliminary predictions for a warm disk solution with lower magnetic lever arm ($\lambda = 8$) appear in very good agreement with the DG Tau observations. These results independently confirm the simplified asymptotic derivation of Bacciotti et al. (2002) and Anderson et al. (2003) who find that $\lambda \simeq 6$–10 and $r_0 \simeq 2$–3 AU are required to account for the rotation signatures in DG Tau.

5. Excitation Conditions

Constraints on the excitation conditions (n_e, T_e, $x_e = n_e/n_H$), independently of the heating process, can be derived from the analysis of ratios involving the prominent optical forbidden emission lines, using the inversion method developped by Bacciotti and Eislöffel (1999). This method has been successfully applied to the study of the inner regions of the DG Tau and RW Aur microjets (Lavalley et al., 2000; Bacciotti, 2002; Dougados et al., 2002). In both cases, strong excitation gradients are observed in the central regions: $n_{\rm e}$ decrease from $>2 \times 10^4$ cm^{-3} to a few 10^3 cm^{-3} and $x_{\rm e}$ increase from a few 10^{-2} to a few 10^{-1} along the central 200 AU. Higher electronic densities and ionisation fractions are observed at larger flow velocity (Lavalley-Fouquet et al., 2000; Bacciotti, 2002).

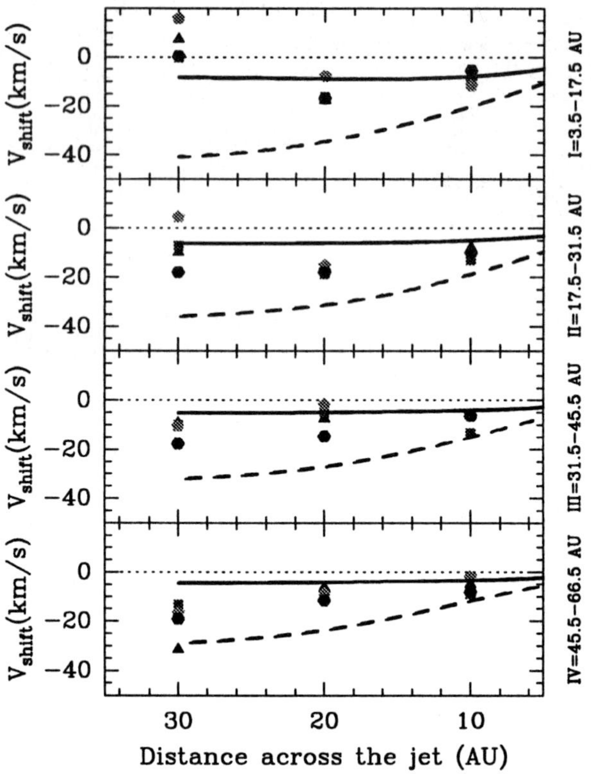

Figure 4. Rotation signatures in the DG Tau microjet. Symbols show measured velocity shifts with transverse distance from the jet axis at different projected altitudes above the disk. The different symbols correspond to measurements in the strongest optical emission lines (see Bacciotti et al., 2002 for more details). Also shown are predictions from a *cold* ($\xi = 0.01, \lambda = 50$; dashed curve) and a *warm* ($\xi = 0.07, \lambda = 8$; full curve) disk wind solution, after convolution by a 50 km s^{-1} $\times$14 AU beam. In both cases $i = 45°$ and $\dot{M}_{acc} = 10^{-6} M_{\odot}$ yr^{-1}. Adapted from Pesenti et al. (2004).

This density and ionisation stratification with flow velocity is indeed expected in disk wind models where the faster streamlines originate from the inner, denser regions of the disk. However, cold disk wind solutions heated by ambipolar diffusion alone fail to account for the observed ionisation fractions (Garcia et al., 2001b; Dougados et al., 2003). Line ratios in the inner regions of the DG Tau and RW Aur jets appear best reproduced by planar shock models with moderate shock velocities (Lavalley-Fouquet et al., 2000; Dougados et al., 2002). Internal shocks due to time variability in the ejection process therefore seem to play a dominant role in the excitation process.

Derived total jet densities range from a few 10^5 to $\geq 10^6$ cm^{-3} at projected distances $\leq$100 AU. These values exceed cold disk wind model predictions for a typical streamline launched at $r_0 = 0.15$ AU by one order of magnitude (proper comparison awaits taking into account projection and beam convolution effects).

Cold disk wind solutions also fail to reproduce integrated line fluxes in a large sample of cTTs, unless large amount of mechanical heating is included, implying very high ionisation fractions inconsistent with observations (O'Brien et al., 2003 these proceedings). The higher density warm disk wind solutions may help solve this discrepancy. Detailed comparison to the observations awaits proper derivation of their thermal structure.

6. Concluding Remarks

High spatial resolution studies of the inner regions of the wind in a few prototypical T Tauri jets, compared with detailed predictions from stationary MHD wind models, brought fundamental new constraints to ejection models. Observed jet widths, as well as derived upper limits for the collimation scale ($\leq$14 AU), are consistent with predictions from disk wind solutions with moderate to high ejection efficiency (ξ between 0.01 and 0.1). The onion-like structure of the flow (with faster, denser material concentrated along the axis), the distribution of the velocity profile close to the central source as well as the detection of rotation in the DG Tau microjet also strongly support magneto-centrifugal disk ejection scenarii. However, cold disk wind solutions, with high magnetic lever arm ($\lambda \simeq 50$), predict too large poloidal and azimuthal velocities (by a factor 1.5 to 3) and too low emission line fluxes on scales >30 AU.

An additional heating source at the base of the wind (e.g. an accretion-heated disk corona) would help enhance the mass-loading efficiency on the field lines and reduce the magnetic lever arm, allowing both to decrease terminal wind velocities and to increase jet densities. These *warm* disk wind solutions, taking into account the effects of thermal gradients, have been investigated by Casse and Ferreira (2000). Computation of their thermal structure, allowing detailed comparison with observations, is currently under way. Preliminary results show promising agreeement with the rotation signatures detected in the DG Tau microjet.

Obviously, the detailed analysis of a larger sample of microjets and comparison with predictions from other classes of wind models, including stellar winds (Sauty et al., 2002) and X-winds (Shang et al., 2002), are critically required before deriving general conclusions on the origin of mass-loss in young stars. We additionally note the limitations of 1-D self-similar models to adequately represent the physics traced by the observations. More sophisticated ejection models, including in particular the effect of time variability, are clearly needed.

References

Anderson, J.M., Li, Z.-Y., Krasnopolsky, R. and Blandford, R.D.: 2003, *ApJ* **590**, L107.
Bacciotti, F.: 2002, *RMxAC* **13**, 8.
Bacciotti, F. and Eislöffel, J.: 1999, *A&A* **342**, 717.
Bacciotti, F., Mundt, R., Ray, T.P., Eislöffel, J., Solf, J. and Camenzind, M.: 2000, *ApJ* **537**, L49.

Bacciotti, F., Ray, T., Mundt, R., Eislöffel, J. and Solf, J.: 2002, *ApJ* **576**, 222.
Blandford, R.D. and Payne, D.G.: 1982, *MNRAS* **199**, 883.
Burrows, C.J. et al.: 1996, *ApJ* **473**, 437.
Cabrit, S., Edwards, S., Strom, S.E. and Strom, K.M.: 1990, *ApJ* **354**, 687.
Cabrit, S., Ferreira, J., Raga, A. 1999 A&A 343, L61.
Casse, F. and Ferreira, J.: 2000, *A&A* **361**, 1178.
Dougados, C., Cabrit, S., Lavalley, C. and Ménard, F.: 2000, *A&A* **357**, L61.
Dougados, C., Cabrit, S. and Lavalley-Fouquet, C.: 2002, *RMxAC* **13**, 43.
Dougados, C., Cabrit, S., Lopez-Martin, L., Garcia, P. and O'Brien, D.: 2003, *Ap&SS* **287**, 135.
Ferreira, J.: 1997, *A&A* **319**, 340 (F97).
Ferreira, J. and Pelletier, G.: 1995, *A&A* 295, 807.
Garcia, P.J.V., Ferreira, J., Cabrit, S. and Binette, L.: 2001a, *A&A* **377**, 589.
Garcia, P.J.V., Cabrit, S., Ferreira, J., and Binette, L.: 2001b, *A&A* **377**, 609.
Hartigan, P., Edwards, S. and Ghandour, L.: 1995, *ApJ* **452**, 736.
Lavalley-Fouquet, C., Cabrit, S. and Dougados, C.: 2000, *A&A* **356**, L41.
O'Brien, D., Garcia, P., Ferreira, J., Cabrit, S. and Binette, L.: 2003, *Ap&SS* **287**, 129.
Pesenti, N., Dougados, C., Cabrit, S., O'Brien, D., Garcia, P. and Ferreira, J.: 2003, *A&A* **410**, 155.
Pesenti, N., Dougados, C., Cabrit, S., O'Brien, D., Garcia, P. and Ferreira, J.: 2004, *A&A* **416**, L9.
Pyo, T.-S., Kobayashi, N., Hayashi, M. and Terada, H., et al.: 2003, *ApJ* **590**, 340.
Ray, T., Mundt, R., Dyson, J., Falle, S. and Raga, A.: 1996, *ApJ* **468**, L103.
Sauty, C. and Tsinganos, K.: 1994, *A&A* **287**, 893.
Sauty, C., Trussoni, E. and Tsinganos, K.: 2002, *A&A* **389**, 1068.
Shang, H., Shu, F. and Glassgold, A.: 1998, *ApJ* **493**, L91.
Shang, H., Glassgold, A.E., Shu, F.H. and Lizano, S.: 2002, *ApJ* **564**, 853.
Shu, F., Najita, J., Ostriker, E.C. and Shang, H.: 1995, *ApJ* **455**, L155.
Woitas, J., Ray, T.P., Bacciotti, F., Davis, C.J. and Eislöffel, J.: 2002, *ApJ* **580**, 336.

TESTING THE MODELS FOR JET GENERATION WITH HUBBLE SPACE TELESCOPE OBSERVATIONS

F. BACCIOTTI[1], T.P. RAY[2], D. COFFEY[2], J. EISLÖFFEL[3] and J. WOITAS[3]

[1]*I.N.A.F. - Osservatorio di Arcetri, Firenze, Italy; E-mail: fran@arcetri.astro.it*

[2]*Dublin Institute for Advanced Studies, Dublin, Ireland*

[3]*Thüringer Landessternwarte Tautenburg, Tautenburg, Germany*

Abstract. Many magneto-hydrodynamic (MHD) models have been developed to describe the acceleration and collimation of stellar jets, in the framework of an infall/outflow process. Thanks to high angular resolution instrumentation, such as the one on-board the Hubble Space Telescope (HST), we are finally able to test observationally the proposed ideas. We present the results obtained by us from the first 0".1 resolution spectra of the initial portion (within 100–200 AU from the source) of the outflows from visible T Tauri stars, taken with the Space Telescope Imaging Spectrograph (STIS). We obtain the jet morphology, kinematics and excitation in different velocity intervals, and we derive the jet mass and momentum fluxes. These results confirm the predictions of magneto-centrifugal models for the jet launch. Recently we have also found indications for rotation in the peripheral regions of several flows. The derived rotational motions appear to be in agreement with the expected extraction of angular momentum from the star/disk system caused by the jet, which in turn allows the star to accrete up to its final mass. Improvements to resolution are expected from observations with STIS in the ultraviolet, and with the forthcoming AMBER spectrometer to be mounted at the VLTI.

Keywords: T Tauri stars, YSO jets, high resolution, angular momentum

1. Introduction

Herbig-Haro jets powered by the central 'engine' of young stellar objects (YSOs) are believed to be a key element of the star formation process, as they might be responsible, for example, for the removal of excess angular momentum from the disk and for the final dispersal of the infalling material (Reipurth and Bally, 2001). Since most of the properties of the models are set up in the first few AUs from the source, to validate them observationally we need very high angular resolution. Despite many sources of 'classical' jets being heavily embedded, the less powerful flows from more evolved T Tauri stars (TTSs) provide us with a 'window' on the central engine, as they can be traced back to their origin in optical and near-infrared (NIR) forbidden and permitted lines. To date, the highest resolution data in the optical ($\sim 0''.1$, i.e. about 14 AU in Taurus) have been obtained with the instruments on board the *Hubble Space Telescope*. In the NIR, adaptive optics facilities and interferometry are becoming available, or they will be in a near future. In this contribution we will describe the current status of the observations for the initial portion of stellar jets, and their interpretation in the framework of the proposed models.

Astrophysics and Space Science **292:** 651–658, 2004.

2. Kinematic and Excitation Properties

According to the most popular theoretical scenario, magnetic and centrifugal forces act together to launch the jets along magnetic field lines in a 'bead-on-a-wire' fashion, either from the vicinity of the star ($\leq$0.1 AU for the *X-wind* (Shu et al., 2000)), or from the disk (up to a few AUs from the source for the *disk wind* (e.g. Königl and Pudritz, 2000 and contributons by Ferreira and Pudritz, this volume)). Despite the existing differences, all these models predict similar properties for the jet in the initial 100–200 AU from the source. In particular, the flow should have an onion-like overall structure, in which the gas distributed along the flow 'layers' is progressively faster, denser and more excited as the axis is approached.

Testing these general predictions has finally become possible using spectra taken with the Space Telescope Imaging Spectrograph (STIS). Thanks to the combination of STIS with HST resolution, we have literally been flooded with new and important information. We are now able to analyse the physical properties of the flow not only close to the launching region but also across the flow width, which is usually not larger than $0''.5$.

For example, we have observed the well-known small-scale jets from the TTSs DG Tau and RW Aur, with slits parallel to the symmetry axis, stepping the slit position across the flow every $0''.07$. (For an illustration of the slit set-up, see Figure 2, left panel.) In this way we obtained spectral 'data-cubes' from which 2-D images of the jets in different velocity intervals could be reconstructed. Examples of such 'channel maps' are shown in Figure 1 for the blue-shifted flow from DG Tau. This jet shows an onion-like kinematic structure (Bacciotti et al., 2000), being more collimated at higher velocities and excitation (traced by the [NII] lines).

The bi-polar jet from RW Aur has also been imaged with this technique (Woitas et al., 2002), and despite being much more collimated than the DG Tau flow, it shows similar properties in the examined velocity intervals. Also, the known velocity asymmetry between the red and blue lobes (the former flows at twice the speed than the latter) persists down to $0''.2$ from the star, and it is probably related to the ejection mechanism.

For both flows we estimated the diameter as the Full Width Half Maximum of the reconstructed images, finding that it varies from a few tens of AU at the jet base (i.e. within 200–300 AU from the source) to about 200 AU further out along the beam. Collimation appears to be achieved very early, within 15–20 AU from the source: here the observed opening angle is at most a few degrees (see Woitas et al., 2002, and Dougados et al., 2004).

The application of standard and original spectroscopic diagnostic techniques has then allowed us to derive from the observed line ratios important information about the gas physics. The *electron density* n_e is easily found from the [SII] doublet,

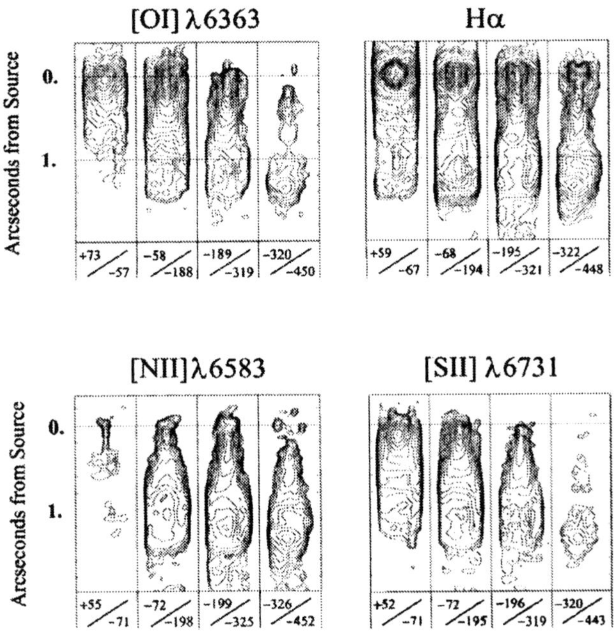

Figure 1. Contours of the velocity channel maps obtained in various lines for the first arcseconds of the jet from DG Tauri. The numbers indicate limiting velocities (in km s^{-1}) in each interval.

and in the initial portion of the jet it can increase from 10^3 to more than 2×10^4 cm^{-3} approaching the source (up to 10^6 from [OI] lines). Our multiple spectra of the DG Tau jet have provided 2-D maps of n_e within $1''.5$ from the star and in different velocity intervals. These maps confirm that n_e is higher closer to the star, the axis, and at higher velocity (Bacciotti, 2002; Bacciotti et al., 2003). The hydrogen ionization fraction x_e and the electron temperature T_e in the same region of the flow have then been determined applying the method described in Bacciotti (2002). This technique uses the ratios of the fluxes of [OI]λ6300, [NII]λ6583, and [SII]λλ6716,6731, without making assumptions about the gas heating mechanism. In the jet, one typically finds $0.02 < x_e < 0.5$ and $8 \times 10^3 < T_e < 2 \times 10^4$ K. At this point one can determine the total hydrogen density, $n_H = n_e/x_e$. Using the values derived for n_H and the jet diameter, we estimated the mass loss rate in the flow, which turns out to be $\sim 10^{-7} M_\odot$ yr^{-1}, and is ~1/25 of the mass accretion rate through the disk, as prescribed by MHD models. In all cases one finds that in the initial portion of the jet the ionization fraction rapidly rises with distance from the star, reaches a plateau at about 100–200 AU from the source, and slowly decreases along the jet following a recombination curve. At the same time, T_e falls by a factor of 2–3 within the first 50–100 AU from the source, while in the same region n_H falls from 10^6 to 10^4 cm^{-3}. For a theoretical interpretation of these findings (see Garcia et al., 2001 and Shang et al. 2002).

3. Jet Rotation

Very recently, a number of studies have been conducted to determine if stellar jets rotate around their symmetry axis. Rotation is an important property for these objects since, following the magneto-centrifugal scenario, the jet may be fully responsible for the extraction of the excess angular momentum from the disk, thus allowing the accretion of matter onto the star.

The first hints of rotation were found in the HH 212 flow at large distances (2×10^3 to 10^4 AU) from the source (Davis et al., 2000), in the form of velocity shifts of a few km s^{-1} in the profile of the H_2 line emitted from opposite edges of the flow. Independently, our group found evidence of rotation in the first 100 AU of the flow from DG Tau, with a set of seven HST/STIS 'parallel' spectra described previously. Systematic differences in the radial velocity of about 6 to 20 km s^{-1} were found for each pair of slits displaced symmetrically with respect to the axis (Bacciotti et al., 2002). If interpreted as rotation of the flow, these values lead to typical toroidal velocities of 5 to 15 km s^{-1} at a few tens of AU from the axis, and between 20 and 90 AU above the disk plane. Such velocities, and the derived angular momentum extracted by the jet, are in the range predicted by the different disk-wind models (Bacciotti et al., 2002; Anderson et al., 2003) and Dougados et al., this volume). Moreover, the sense of rotation is in agreement with that of the underlying disk (Testi et al., 2002). To confirm these encouraging findings, we examined also our multiple 'parallel' STIS spectra of the RW Aur jet (see Figure 2). The analysis transpired to be much more difficult than for DG Tau, as this jet is more collimated, and the core of the jet is not spectrally/spatially resolved in the innermost slits. Nevertheless, we have found clear signatures of rotation at the edges of the jet in the form of velocity shifts of about 15–20 km s^{-1} in all emission lines (Woitas et al., 2004).

We then started, with HST/STIS, a systematic search for rotation signatures in jets from eight objects, but this time placing the slit in the direction perpendicular to the jet axis, at $0''.2$–$0''.3$ from the source (see Figure 3). This set-up gives a much higher efficiency, since less spacecraft orbits are required and the data analysis is extremely simplified. Spectra already acquired include the jets from THA15-28 (in Lupus3, at 170 pc), RW Aur, and LkHa321 (in Cygnus, at 550 pc). An example of a position–velocity contour plot is shown in Figure 3, top-right panel, for the red-shifted lobe of the TH 28 jet. The skew in lower order contours is consistent with the presence of rotation in the outer jet channel, while the on-axis higher velocity component remains unresolved. We then compare the emission profiles along horizontal cuts at mirrored positions with respect to the jet axis (dotted and dashed lines in Figure 3). The resulting displacement in wavelength gives the velocity differences across the flow: for the TH 28 jet these are summarized, for various emission lines and in both the blue- and the red-shifted lobe, in Figure 3 bottom panels.

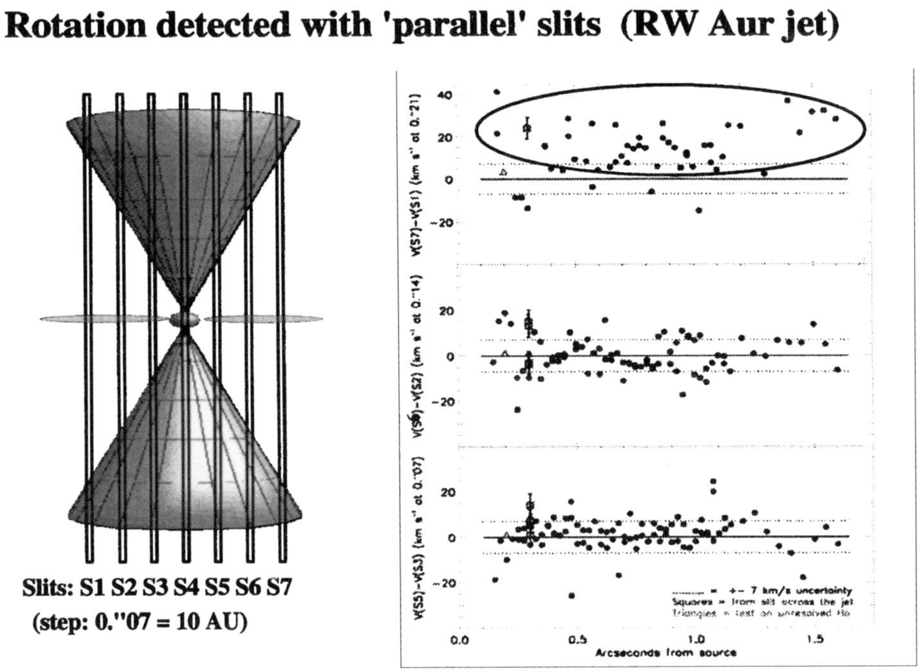

Figure 2. Left: Configuration of the 'parallel' slits for the HST/STIS observations of the RW Aur and DG Tau jets. Right: Velocity shifts in the line profiles observed in the red lobe of the RW Aur jet at mirrored positions with respect to the axis (S7-S1, S6-S2, S5-S3). The data were corrected for uneven slit illumination effects. Evidence of rotation can be seen in the uppermost panel, for the points encircled by the ellipse. The velocity of the unresolved stellar Hα emission, the triangle, acts as a control. The "perpendicular-slit" data (see Figure 3) are overlaid as squares.

For all the cases examined we were able to detect systematic velocity differences of 10–25 km s^{-1} (accuracy $\pm$ 5 km s^{-1}), determined with both gaussian fitting and cross-correlation techniques (Coffey et al., 2004). For the bi-polar jets, the sign of the velocity difference turns out to be the same in both lobes, as should be expected from evidence of rotation. Also the data for RW Aur compare well with the measurements derived from the 'parallel' slit configuration (see Figure 2). In all cases the derived values for the toroidal velocity from these jets are similar to those found for the DG Tau jet, and are in agreement with theoretical predictions, which also justify the apparent increase of the observed shift with distance from the axis (Pesenti et al., 2004). More details can be found in the contribution by Coffey et al. (this volume).

Other observational programs have commenced, or are planned, to determine conclusively whether stellar jets rotate. These include observations with the ESO Very Large Telescope (VLT) of flows from Class I and Class II sources emitting in NIR lines, and new approved observations of jets from TTSs with HST/STIS, for lines emitted in the near-ultraviolet (NUV). In the latter case, the spatial resolution

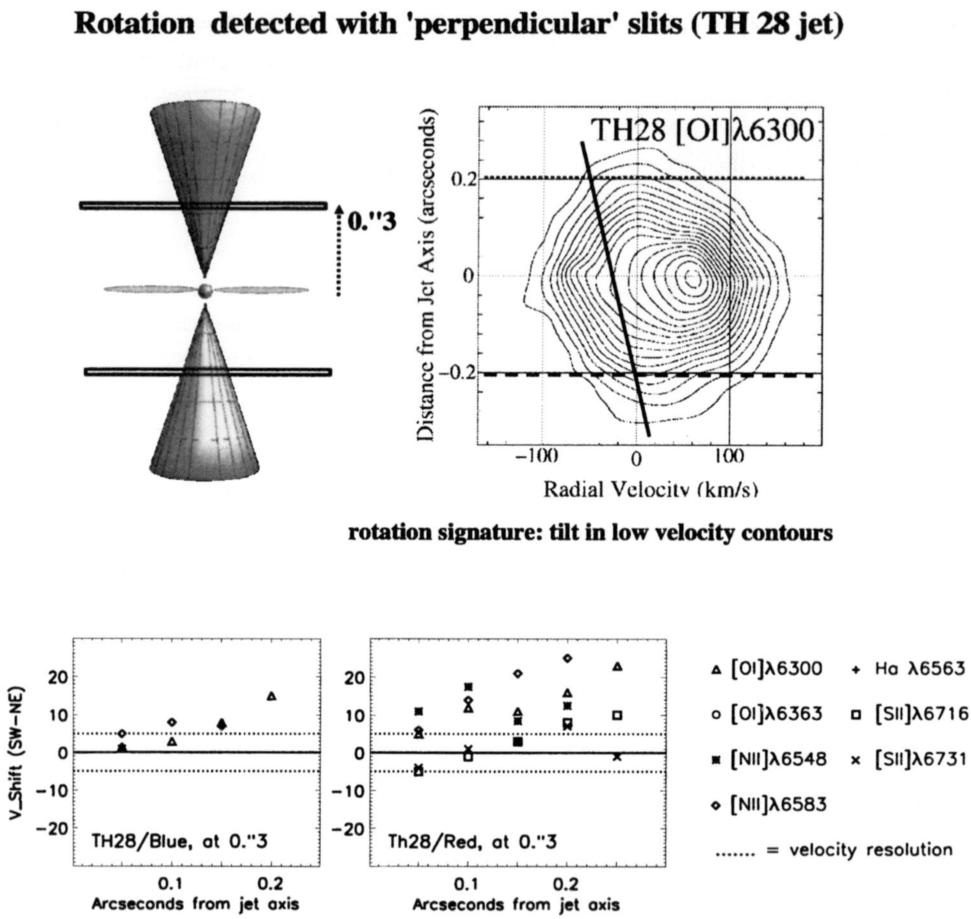

Figure 3. Top left: Configuration of the 'perpendicular' slits for the HST/STIS observations of the jets from Th 28, RW Aur and LK Hα321. Top right: Position–velocity contour plot of the [OI]λ6300 line from the red-shifted lobe of the TH 28 jet. The skew in lower order contours (oblique line), is indicative of rotation. The velocity shift is measured comparing the profiles along horizontal cuts at mirrored positions with respect to the axis (dotted and dashed lines). Bottom: Summary of the measured velocity shifts in both lobes of the Th 28 jet, in various emission lines.

is doubled with respect to the optical data, while the spectral resolution provided in the NUV by the STIS echelle gratings is five times higher. We therefore expect a factor 10 improvement in our measurements. This will hopefully allow us to resolve both spatially and spectrally the high velocity jet core, which rotates faster, and therefore should present even more evident signatures of rotation. The ultimate confirmation of the validity of the magneto-centrifugal mechanism, however, will probably come with the planned observations at the VLT Interferometer (VLTI). A joint project has been started by several European institutes to examine, using the VLTI spectrograph AMBER, the morphology and kinematics of the jets from two bright southern TTSs: namely CS Cha and RU Lupi (Bacciotti et al., 2003). With the VLTI resolution of 1 mas we will be able to access the innermost region of the flow down to 0.1 AU from the source (and hence below the so-called Alfvén surface), where most of the acceleration and collimation take place.

4. Conclusions

High angular resolution observations are required to test the mechanism proposed for the acceleation of YSO jets. In this respect the results recently obtained with the Space Telescope Imaging Spectrograph (STIS/HST) proved to make the difference in our ability to constrain the models. With the HST we resolve the flow at distances of 10–15 AU from the sources in systems closeby. This allowed us to reach the edge of the acceleration and collimation zone and to resolve the flow transversely, thus giving important information on the kinematic and excitation properties in the initial jet channel. Recently we concentrated on the search for signatures of rotation in the first 100 AUs of the flow, as rotation is a fundamental property of the magneto-centrifugal launching mechanism. We have now collected the detected evidences of rotation for all five systems examined so far. The overall set of derived results seem to confirm the validity of popular magneto-hydrodynamic models. As final confirmation of this statement, however, we need to observe the core of the central engine at even higher spectral and spatial resolution. In this respect, the facilities being prepared for optical/NIR interferometry will be a key tool for our investigations.

References

Anderson, J.M., Li, Z.-Y., Krasnopolsky, R. and Blandford, R.: 2003, *ApJ* **590**, L107.

Bacciotti, F., Mundt, R., Ray, T.P., Eislöffel, J., Solf, J. and Camenzind, M.: 2000, *ApJL* **537**, 49.

Bacciotti, F.: 2002, in: W.J. Henney, W. Steffen, A.C. Raga and L. Binette (eds.), *Emission Lines from Jet Flows, Rev. Mex. Astron. Astroph.* Conf. Series, vol. **13**, p. 8.

Bacciotti, F., Ray, T.P., Mundt, R., Eislöffel, J. and Solf, J.: 2002, *ApJ* **576**, 222.

Bacciotti, F., Testi, L., Marconi, A., Garcia, P.J.V., Ray, T.P., Eislóffel, J. and Dougados, C.: *Ap&SS* **286**, 157.

Bacciotti, F., Eislöffel, J., Ray, T.P., Mundt, R. and Solf, J.: in preparation, *A&A*.

Coffey, D., Bacciotti, F., Woitas, J., Ray, T.P. and Eislöffel, J.: 2004, *ApJ* **604**, 758.

Davis, C.J., Berndsen, A., Smith M. D., Chrysostomou, A. and Hobson, J.: 2000, *MNRAS* **314**, 241.

Garcia, P., Ferreira, J., Cabrit, S. and Binette, L.: 2001, *A&A* **377**, 589.

Königl, A. and Pudritz, R.: 2000, in: V. Mannings, A.P. Boss, and S.S. Russell (eds.), *Protostars and Planets IV*, University of Arizona Press, Tuscon, p. 759.

Pesenti, N., Dougados, C., Cabrit, S., Ferreira, J., Casse, F., Garcia, P. and O'Brien, D.: 2004, *A&A* **416**, L9.

Reipurth, B. and Bally, J.: 2001, *ARA&A* **39**, 403.

Shang, H., Glassgold, A.E., Shu, F.H. and Lizano, S.: 2002, *ApJ* **564**, 853.
Shu, F.H., Najita, J.R., Shang, H. and Li, Z.-Y.: 2000, in: V. Mannings, A.P. Boss and S.S. Russell (eds.), *Protostars and Planets IV*, University of Arizona Press, Tuscon, p. 789.
Testi, L., Bacciotti, F., Sargent, A.I., Ray, T.P. and Eislöffel, J.: 2002, *A&AL.* **394**, 31.
Woitas, J., Ray, T.P., Bacciotti, F., Davis, C.J. and Eislöffel, J.: 2002, *ApJ* **580**, 336.
Woitas, J., Bacciotti, F., Ray, T.P., Eislöffel, J., Marconi, A. and Coffey, D.: 2004, *A&A* submitted.

TIME DEPENDENT MAGNETOSPHERIC ACCRETION IN T TAURI STARS

J. BOUVIER[1], C. DOUGADOS[1] and S.H.P. ALENCAR[2]

[1]*Laboratoire d'Astrophysique, Observatoire de Grenoble, France;*
E-mail: jbouvier@obs.ujf-grenoble.fr
[2]*Departamento de Física, ICEx, UFMG Belo Horizonte, Brazil*

Abstract. We review recent observational results which suggest that magnetically channeled accretion in T Tauri stars is a highly time dependent process on timescales ranging from hours to months.

Keywords: T Tauri stars, magnetospheric accretion

1. Introduction

T Tauri stars are low-mass stars with an age of a few million years, still contracting down their Hayashi tracks towards the main sequence. Many of them, the so-called classical T Tauri stars, show signs of accretion from a circumstellar disk (see, e.g., Ménard and Bertout, 1999 for a review). Surface magnetic fields of order of 1–3 kG have recently been derived from Zeeman broadening measurements (Johns Krull et al., 1999, 2001; Guenther et al., 1999) as well as from the detection of electron cyclotron maser emission (Smith et al., 2003). These strong stellar magnetic fields are believed to alter significantly the accretion flow in the circumstellar disk close to the central star. Based on models originally developed for magnetized compact objects in cataclysmic binaries (Ghosh and Lamb, 1979) and *assuming* that T Tauri magnetospheres are predominantly dipolar on the large scale, Camenzind (1990) and Königl (1991) showed that the inner accretion disk is expected to be truncated by the magnetosphere at a distance of a few stellar radii above the stellar surface for typical mass accretion rates of 10^{-9} to $10^{-7} M_{\odot}\mathrm{yr}^{-1}$ in the disk (Basri and Bertout, 1989; Hartigan et al., 1995; Gullbring et al., 1998). Disk material is then channeled from the disk inner edge onto the star along the magnetic field lines, thus giving rise to magnetospheric accretion columns. As the free falling material in the funnel flow eventually hits the stellar surface, accretion shocks develop near the magnetic poles. The successes and limits of static magnetospheric accretion models (MAMs) in accounting for the observed properties of classical T Tauri systems (CTTS) have been reviewed by Bouvier, Alencar and Dougados (2003). We discuss here recent results which suggest that the interaction between the star's magnetosphere and the inner disk is a highly dynamical and time dependent process.

Astrophysics and Space Science **292:** 659–664, 2004.

2. Magnetospheric Accretion : A Dynamical Process

Evidence for magnetically mediated accretion mostly comes from snapshot studies and it is only recently that the stability of the phenomenon has started to be thoroughly investigated. In this section, we briefly review model predictions regarding the temporal evolution of the process (a more detailed account is given by Uzdensky, this volume) and recent observational results which indicate that it is highly time dependent indeed.

2.1. Time dependent models

Most MAMs assume that the stellar magnetosphere truncates the disk close to the corotation radius, i.e., the radius in the disk where the Keplerian angular velocity of the disk material equals that of the star. Field lines threading the disk at this radius thus corotate with the central object. However, due to the finite radial distance over which the stellar magnetosphere interacts with the inner disk, the footpoints of most field lines rotate differentially, one being anchored into the star, the other into the Keplerian disk.

Recent numerical simulations indicate that magnetic field lines are thus substantially distorted by differential rotation on a timescale of only a few Keplerian periods at the inner disk. One class of models predict that differential rotation leads to the expansion of the field lines, their opening and eventually their reconnection which restores the initial (assumed dipolar) magnetospheric configuration (e.g. Aly and Kuijpers, 1990; Goodson et al., 1997; Goodson and Winglee, 1999; Uzdensky et al., 2002). Magnetospheric *inflation cycles* are thus expected to develop and to be accompanied by violent episodic outflows as field lines open and reconnect as well as time dependent accretion rate onto the star (Hayashi et al., 1996; Romanova et al., 2002). The most recent 3D MHD simulations of disk accretion onto an inclined stellar magnetosphere are presented in Romanova et al. (2003) and illustrate well the extreme complexity of the process.

Other models, however, suggest that the field lines respond to differential rotation by drifting radially outwards in the disk, leading to magnetic flux expulsion (Bardou and Heyvaerts, 1996). The response of the magnetic configuration to differential rotation mainly depends upon the magnitude of magnetic diffusivity in the disk, a parameter of the models which is poorly constrained from first principles.

2.2. Observational evidence for time dependent magnetospheric accretion

Spectrophotometric monitoring studies of CTTS have brought clues to a time dependent interaction between the inner disk and the stellar magnetosphere. Thus, episodic high velocity outbursts, possibly connected with magnetospheric reconnection events, have been reported for a few systems based on the slowly varying velocity of blueshifted absorption components in emission line profiles on a timescale of hours to days (Alencar et al., 2001; Ardila et al., 2002). Possible evidence for magnetic field lines being twisted by differential rotation has also been reported for SU Aur by Oliveira et al. (2000). These authors measured a time delay of a few hours between the appearance of high velocity redshifted absorption components in line profiles formed at different altitudes in the accretion column. They interpreted this result as the crossing of an azimuthally twisted accretion column on the line of sight. Another possible evidence for magnetic field lines being twisted by differential rotation and leading to quasi-periodic reconnection processes has been reported for the embedded protostellar source YLW 15 based on the observations of quasi-periodic x-ray flaring (Montmerle et al., 2000).

Synoptic studies lasting for several weeks were performed for a handful of CTTS and provide new insight into the magnetospheric accretion process and its temporal evolution. Two of these have targeted the CTTS AA, Tau, which proved to be ideally suited to probe the inner few 0.01 AU of the disk-magnetosphere interaction region. Due to its high inclination ($i \simeq 75^\circ$, see Bouvier et al., 1999), the line of sight to the star intersects the region where the stellar magnetosphere threads the inner disk. The peculiar orientation of this otherwise typical CTTS maximizes the variability induced by the modulation of the magnetospheric structure and thus provides the strongest constraints on the inner disk and the magnetospheric cavity.

A first monitoring campaign (Bouvier et al., 1999), led to the discovery of recurrent eclipses of the central object with a period of 8.2 days. The eclipses were attributed to a nonaxisymmetric warp of the inner disk edge, located at about 8 $R_\star$, which periodically obscures the central star as it orbits it at Keplerian speed. Such an inner disk warp is expected to develop as the disk encounters an *inclined* magnetosphere (Terquem and Papaloizou, 2000; Lai, 1999; Romanova et al., 2003).

A second campaign combined photometry and high resolution spectroscopy (Bouvier et al., 2003). Spectroscopic variability yielded evidence for accretion columns and accretion shocks with their respective signatures (redshifted absorptions, continuum excesses) being modulated on a rotation timescale. Furthermore, a time delay of about one day was found between the flux variations of lines forming at different altitudes in the accretion column, from H_α near the disk inner edge to HeI close to the accretion shock at the stellar surface. The measured time delay is consistent with nonstationary accretion propagating downwards the accretion columns at free fall velocity, starting from a distance of about 8 $R_\star$, the radius at which the stellar magnetosphere disrupts the inner disk.

Over longer timescales, of order of one month, significant variations were also observed in the line and continuum excess flux, indicative of a smoothly varying mass accretion rate onto the star. Simultaneously, a tight correlation was found between the radial velocity of the blueshifted and redshifted absorption components in the H_α emission line profile (Figure 1). Since the former is a wind diagnostics while the latter forms in the accretion flow, this correlation provides additional evidence for a physical connection between (time dependent) inflow and outflow in

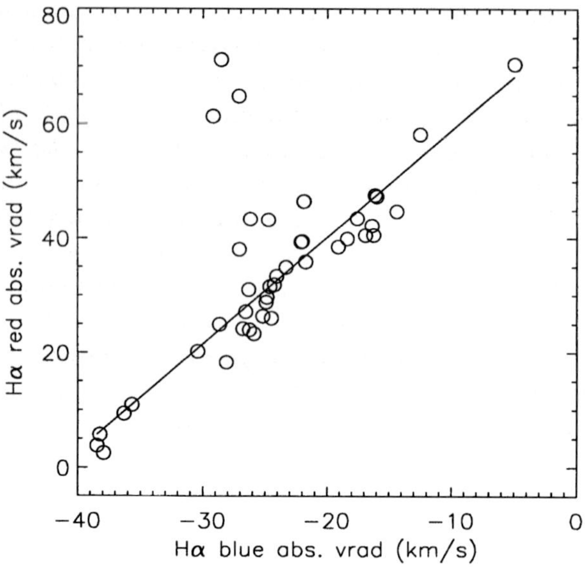

Figure 1. The correlation between the radial velocity of the blueshifted and redshifted absorption components in the H_α emission line profile of AA Tau (from Bouvier et al., 2003).

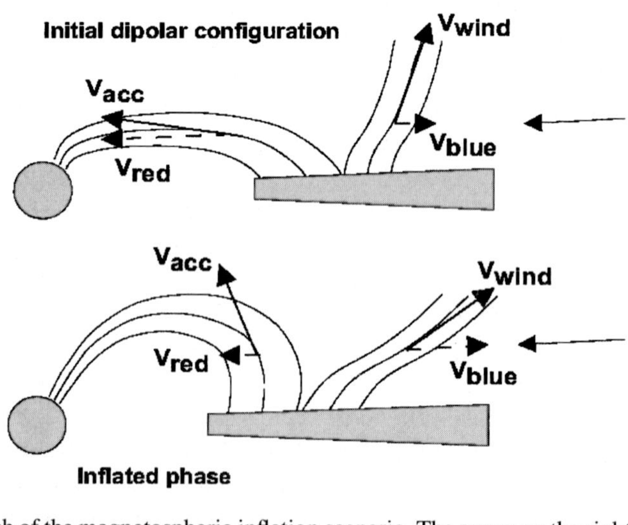

Figure 2. A sketch of the magnetospheric inflation scenario. The arrow on the right side indicates the line of sight to the AA Tau system (from Bouvier et al., 2003).

CTTS. Bouvier et al. (2003) argued that these flux and radial velocity variations can be consistently interpreted in the framework of *magnetospheric inflation cycles*, as predicted by recent numerical simulations of the star-disk interaction.

This is schematically illustrated in Figure 2. As magnetic field lines expand due to differential rotation between the star and the inner disk, the radial velocity of the accretion (resp. wind) flow decreases (resp. increases) due to projection effects on the line of sight, thus resulting in the observed correlation between the velocities of redshifted and blueshifted H_α absorptions (Figure 1) as the magnetosphere inflates. At the same time, the loading of disk material onto inflated field lines becomes increasingly difficult owing to the large angle field lines make relative to the disk plane. This results in a reduced accretion rate onto the star, as deduced from the depressed line and continuum fluxes observed at this phase (see Bouvier et al., 2003).

The last synoptic campaign on AA Tau thus yields the first evidence for global instabilities developping on a timescale of a month in the large scale magnetosphere, a result which yields support to the predictions of time dependent models of magnetospheric accretion. Whether these magnetospheric instabilities are truly cyclic, being driven by differential rotation between the star and the inner disk, as predicted by numerical models, will require additional monitoring lasting for several months.

3. Conclusion

Observational evidence for magnetospheric accretion being instrumental in classical T Tauri stars has accumulated in recent years. Several key properties of these young stars are naturally accounted for by assuming that the stellar magnetic field governs the accretion flow close to the star. The strong variability of CTTS on all timescales, from hours to months (and possibly years, Bertout, 2000), further suggests that the magnetically mediated interaction between the accretion disk and the central object is a highly dynamical and time dependent process.

The implications of the nonsteadiness of magnetospheric accretion in CTTS are plentiful and remain to be fully explored. They range from the evolution of their angular momentum (Agapitou and Papaloizou, 2000), the origin of inflow/ouflow short term variability (Woitas et al., 2002, Lopez-Martin et al., 2003), the modeling of the near infrared excess of CTTS and of its variations both of which will be affected by a nonstandard and time dependent inner disk structure (Carpenter et al., 2001; Eiroa et al., 2002; Johns-Krull and Valenti, 2003), the origin of CTTS variability which is expected to be a complex combination of modulation by hot and cold spots and variable circumstellar extinction (e.g. DeWarf et al., 2003), and possibly the halting of planet migration close to the star (Lin et al., 1996).

References

Agapitou, V. and Papaloizou, J.C.B.: 2000, *MNRAS* **317**, 273.
Alencar, S.H.P., Johns-Krull, C.M. and Basri, G.: 2001, *AJ* **122**, 3335.
Aly, J.J. & Kuijpers, J.: 1990, *A&A* **227**, 473.

Ardila, D.R., Basri, G., Walter, F.M., Valenti, J.A. and Johns-Krull, C.M.: 2002, *ApJ* **566**, 1100.
Bardou, A. & Heyvaerts, J.: 1996, *A&A* **307**, 1009.
Basri, G. and Bertout, C.: 1989, *ApJ* **341**, 340.
Bertout, C.: 2000, *A&A* **363**, 984.
Bouvier, J., Chelli, A., Allain, S. et al.: 1999, *A&A* **349**, 619.
Bouvier, J., Grankin, K., Alencar, S. et al.: 2003, *A&A* **409**, 169.
Bouvier, J., Alencar, S., Dougados, C.: 2003, in: J.Lépine and J.Gregorio-Hetem (eds.), Open Issues in Local Star Formation, *Astrophys. Space Science* 299.
Camenzind, M.: 1990, *Reviews of Modern Astronomy* **3**, 234.
Carpenter, J.M., Hillenbrand, L.A. and Skrutskie, M.F.: 2001, *AJ* **121**, 3160.
DeWarf, L.E., Sepinsky, J.F., Guinan, E.F., Ribas, I. and Nadalin, I.: 2003, *ApJ* **590**, 357.
Eiroa, C. et al.: 2002, *A&A* **384**, 1038.
Ghosh, P. and Lamb, F.K.: 1979, *ApJ* **234**, 296.
Goodson, A.P., Winglee, R.M. and Boehm, K.: 1997, *ApJ* **489**, 199.
Goodson, A.P. and Winglee, R.M.: 1999, *ApJ* **524**, 159.
Guenther, E.W., Lehmann, H., Emerson, J.P. and Staude, J.: 1999, *A&A* **341**, 768.
Gullbring, E., Hartmann, L., Briceno, C. and Calvet, N.: 1998, *ApJ* **492**, 323.
Hartigan, P., Edwards, S. and Ghandour, L.: 1995, *ApJ* **452**, 736.
Hayashi, M.R., Shibata, K. and Matsumoto, R.: 1996, *ApJ* **468**, L37.
Johns-Krull, C.M. and Valenti, J.A.: 2004, *ApJ* in press.
Johns-Krull, C.M., Valenti, J.A. and Koresko, C.: 1999, *ApJ* **516**, 900.
Johns-Krull, C.M., Valenti, J.A., Saar, S.H. and Hatzes, A.P.: 2001, *ASP Conf. Ser.* 223: 11th Cambridge Workshop on Cool Stars, Stellar Systems and the Sun, **11**, 521.
Königl, A.: 1991, *ApJl* **370**, L39.
Lai, D.: 1999, *ApJ* **524**, 1030.
Lin, D.N.C., Bodenheimer, P. and Richardson, D.C.: 1996, *Nature* **380**, 606.
Lopez-Martin, L., Cabrit, S. and Dougados, C.: 2003, *A&A* **405**, L1.
Ménard, F. and Bertout, C.: 1999, *NATO ASIC Proc. 540*: The Origin of Stars and Planetary Systems, 341.
Montmerle, T., Grosso, N., Tsuboi, Y. and Koyama, K.: 2000, *ApJ* **532**, 1097.
Oliveira, J.M., Foing, B.H., van Loon, J.T. and Unruh, Y.C.: 2000, *A&A* **362**, 615.
Romanova, M.M., Ustyugova, G.V., Koldoba, A.V. and Lovelace, R.V.E.: 2002, *ApJ* **578**, 420.
Romanova, M.M., Ustyugova, G.V., Koldoba, A.V., Wick J.V. and Lovelace, R.V.E.: 2003, *ApJ* **595**, 1009.
Smith, K., Pestalozzi, M.,Güdel, M., Conway, J. and Benz, A.O.: 2003, *A&A* **406**, 957.
Terquem, C. & Papaloizou, J.C.B.: 2000, *A&A* **360**, 1031.
Uzdensky, D.A., Königl, A. and Litwin, C.: 2002, *ApJ* **565**, 1191.
Woitas, J., Ray, T.P., Bacciotti, F., Davis, C.J. and Eislöffel, J.: 2002, *ApJ* **580**, 336.

CLASSICAL T TAURI STARS AND SUBSTELLAR ANALOGS

Classification Based on Empirical Criteria

D. BARRADO Y NAVASCUÉS[1], E.L. MARTÍN[2], R. JAYAWARDHANA[3] and S. MOHANTY[4]

[1]*Laboratorio de Astrofísica Espacial y Física Fundamental (LAEFF-INTA), Madrid, Spain; E-mail: barrado@laeff.esa.es*

[2]*Institute for Astronomy, University of Hawaii, U.S.A.*

[3]*Department of Astronomy, University of Michigan, U.S.A.*

[4]*Harvard-Smithsonian Center for Astrophysics, Cambridge, U.S.A.*

Abstract. We propose a spectroscopic criterion based on Hα equivalent width and spectral type to classify classical T Tauri stars and substellar analogs. We argue that accreting objects can be identified from low-resolution optical spectroscopy, when their Hα flux is stronger than the saturation limit at Log {Lum(Hα)/Lum(bol)} = −3.3. Additional criteria, such as the relation between HeI5876 or HeI6678 and Hα, or the ratios between the components of the CaII infrared triplet, are also discussed. We have tested the reliability of these criteria by applying them to several objects with masses in the range 0.11–0.025 $M_\odot$, which belong to nearby star forming regions and the TW Hya association.

Keywords: classical T Tauri, brown dwarfs

1. Introduction

In the past, different criteria have been used to classify classical and weak-line T Tauri stars (CTTS and WTTS). The Hα equivalent width—W(Hα)—has been profusely used in the literature, both as a fixed value —W(Hα) = 5, 10 or 20 Å—or as depending on the spectral type (Martín, 1997; White and Basri, 2003). With the wealth of information available nowadays, we have redefined this criterion, and we have extended it into the brown dwarf or substellar regime (i.e., objects with masses below ~0.072 $M_\odot$, unable to fuse hydrogen in the core). We have defined substellar classical and weak-line T Tauri analogs (SCTTA and SWTTA) as very young substellar-mass objects (age <10 Myr, mass <0.072 $M_\odot$) that show W(Hα) stronger than the saturation limit—for SCTTA—indicating accretion activity, or weaker than that value in the case of the SWTTA. Our criteria can be used for statistical studies of the accretion activity in large populations of low-mass stars and substellar-mass objects, only requiring low-resolution optical spectroscopy.

2. Defining Classical and Weak-Line T Tauri Phenomenology

Classical T Tauri stars present in their spectrum clear sign-post of accretion such as narrow forbidden lines, wide and asymmetric permitted lines, ultraviolet, blue and

Astrophysics and Space Science **292:** 665–672, 2004.

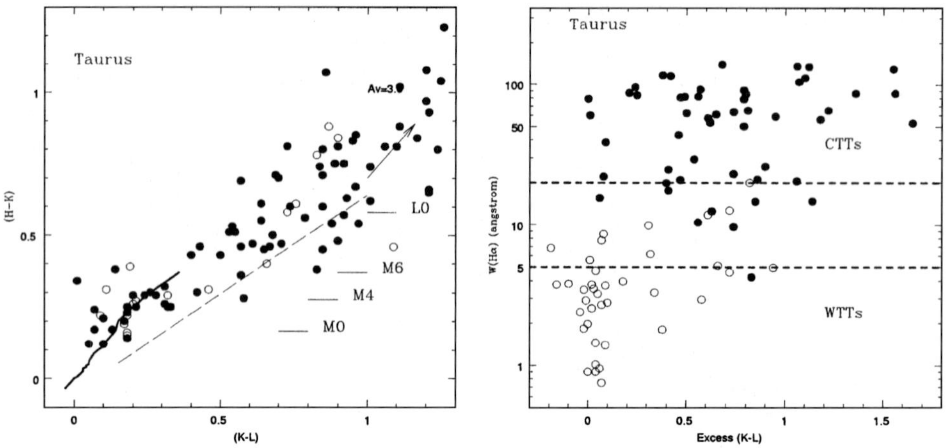

Figure 1. Left.—Infrared color-color diagram for Taurus members. Data from Kenyon and Hartmann (1995). The solid and the dashed lines correspond to the loci for MS field and CTT stars (Leggett, 1992 and Meyer et al., 1997). Right.—Infrared excesses versus Hα equivalent widths. The dashed segments represent previous proposed CTT/WTT criteria. In both panels, CTT and WTT stars are represented with solid and empty circles, respectively.

infrared excesses (see the reviews by Appenzeller and Mundt, 1989 and Bertout, 1989). Although only the study of the spectral features allows to perform an accurate classification, it is possible, from the statistical point of view, to use their photometric properties to differentiate them from weak-line and post-T Tauri stars (WTTS, PTTS). Figure 1 (left) displays an infrared color-color diagram for CTTS and WTTS belonging to Taurus (after Kenyon and Hartmann, 1995). The figure shows that photometric excesses are not enough to separate both groups. The comparison between the infrared excess with the Hα equivalent width does a better job (Figure 1, right). However, there is a region where it is possible to find both CTTS and WTTS stars (i.e., between 5 and 20 Å for this particular dataset).

Strong Hα emission is an adequate accretion indicator which is thought to be produced by magnetospheric infall even in very low-mass objects (Muzerolle et al., 2000). Martín (1997) established that the actual borderline depends on the spectral type. This criterion has been improved by White and Basri (2003). Nevertheless, there are other phenomenologies which do produce large W(Hα), as discussed below. White and Basri (2003) also provide a robust criterion based on the Hα width at 10% of the maximum intensity, at least for objects with veiling. However, this demands high resolution spectroscopy. Moreover, Jayawardhana, Mohanty and Basri (2003) have found that objects with small accretion rates have a Hα width smaller that the critical value (270 km/s) proposed by White and Basri (2003).

Figure 2 (left) compares the Hα equivalent width versus the spectral types for three young open clusters. The maximum chromospheric activity—long-dashed line—coincides with the saturation limit—solid line—at Log $\{$Lum(Hα)/

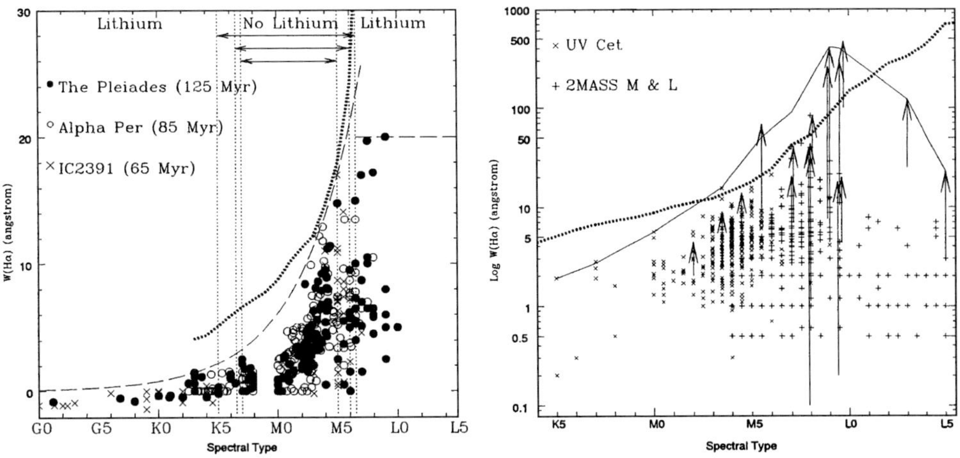

Figure 2. Left.—Hα equivalent width versus the spectral type. The dashed and dotted curves correspond to the young cluster and saturation criteria, whereas the vertical dotted segments mark the lithium gap in each cluster (after Barrado y Navascués and Stauffer 2003). Right.—As left panel, except than in this case field objects are displayed. The maximum activity due to flares are included as a thin-solid line.

Lum(bol)} = −3.3 for M2-M4. The vertical dotted segments and the horizontal arrows discriminate the spectral type ranges where lithium, a youth indicator and therefore helpful to establish the evolutionary status, has been already depleted (from top to bottom, the Pleiades, Alpha Per and IC2391). In fact, even flare stars such as those observed in UV Cet type stars and in late-M and L-type dwarfs do not surpass the saturation limit (they normally have W(Hα) smaller than the upper envelope for the open clusters). They are shown as crosses (UV Cet type stars from Gershberg et al., 1999) and plus symbols (late M and L spectral type stars and brown dwarfs discovered by the 2MASS survey. Data from Kirkpatrick et al., 2000; Gizis et al., 2000; Reid et al., 2002; and Mohanty et al., 2003) in Figure 2 (right). This only happens for a few, very active and exceptional objects, (represented as arrows, going from minimum to maximum values ever recorded in the literature), and only for short time intervals. Moreover, when this saturation limit is compared with previous CTT and WTT classifications (Figure 3, left-top), it is evident that it is an excellent, although very restrictive, indicator of accretion. Essentially, all accretors are above the line, and few of them are below (and this might be due to the veiling, which diminishes the measured equivalent width). Therefore, we propose the saturation limit as a criterion to classify CTT and WTT stars and substellar analogs. Additional details can be found in Barrado y Navascués and Martín (2003).

In addition to this criterion, other information can be very helpful. Figure 3 depicts W(HeI5876) and W(HeI6678) versus W(Hα), and the ratio between the lines of the CaII infrared triplet. In these diagrams, the location of the CTT stars and substellar analogs is clearly detached from other types.

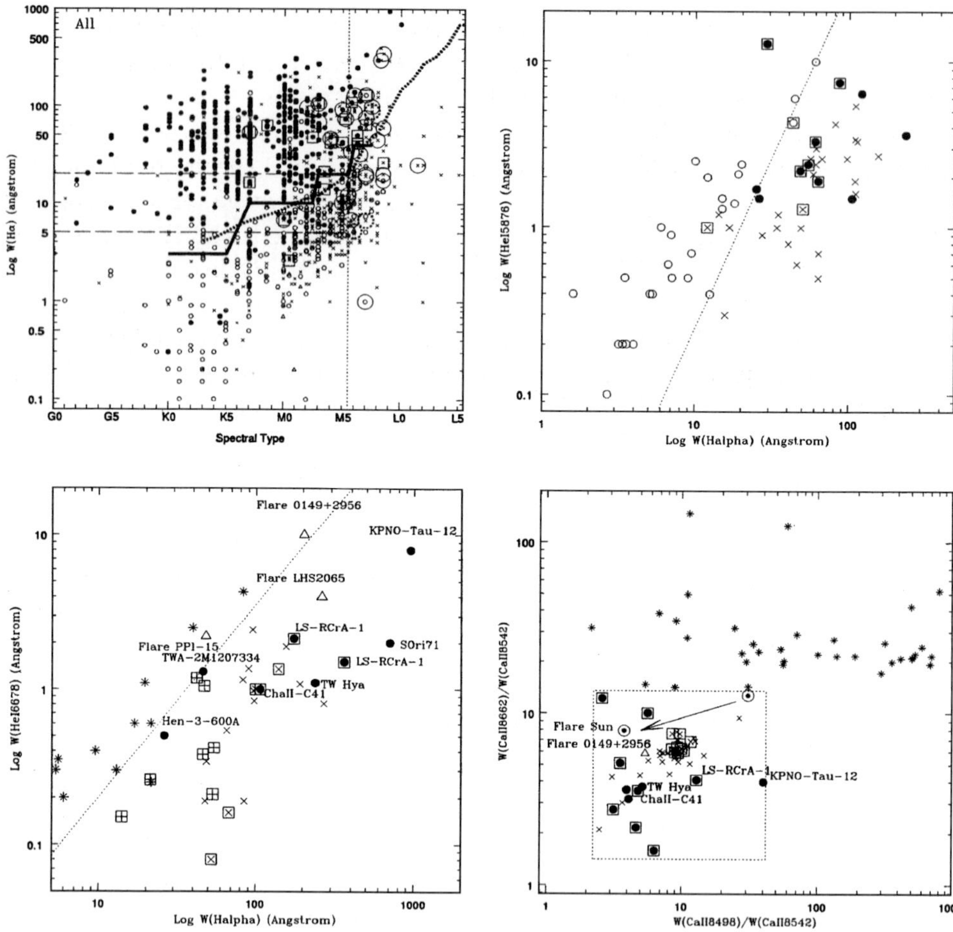

Figure 3. Left-top.—The proposed CTT/WTT criterion and its comparison with data from several star forming regions (Orion, Taurus, IC348, ChaI, TWA, σ Ori, Sco-Cen complex). Right-top and left-bottom.—Comparison between the equivalent width of HeI5876 or HeI6678 and Hα. Right-bottom.—Ratios between the components of the CaII infrared triplet. Objects with IR excess or forbidden lines are indicated with large circles or squares. CTT and WTT are shown as solid or open circles, whereas flares appear as open triangles. Crosses correspond to CTTS from Taurus (Briceño et al., 1998 and Alencar and Basri, 2000) or other T Tauri stars from Hamann and Perssons (1992). Plus symbols indicate the location of very low mass stars and brown dwarfs in σ Orionis (Zapatero Osorio et al., 2002). Other non-accretors are shown with asterisks (Hamann and Perssons, 1992; Montes et al., 2001).

3. Examples at the Substellar Limit and Beyond

We have applied these criteria to several objects with masses around the substellar limit, namely ChaII-C41 (Vuong et al., 2001), LS-RCrA 1 (Fernández and Comerón, 2001)—we estimate their masses as 0.07 $M_\odot$ and 0.055 $M_\odot$, respectively—, and two M8 field objects discovered by Gizis (2002), which are candidate members of the TW Hya association (TWA). Using a 10 Myr isochrone from Baraffe et al. (1998), their mass is estimated as $\sim$0.030 $M_{\odot}$ in both cases.

In March 2003, we collected low and medium resolution (R = 600 and 2500) spectra at the Magellan I 6.5 m telescope, using the B&C spectrograph and the 300 and 1200 l/mm gratings. Based on these data, we derived spectral types by comparison with spectral templates observed with the same setup. The spectra of these objects are displayed in Figures 4 and 5 (left), respectively. In the diagrams, we have labeled several important features, such as Hα, forbidden lines from oxygen, nitrogen and sulfur, and the calcium II infrared triplet. The insets contain the range around Hα and LiI6708Å (all of them show lithium in their medium resolution spectrum). Figure 5 (right) indicates that ChaII-C41 (M5.5) has a strong K-band excess, whereas LS-RCrA 1 (M6.5) only shows a small interstellar reddening. They do show forbidden lines in emission, as well as HeI6679 Å, together with intense Hα. The profiles might be asymmetric too, but the medium resolution spectra are not good enough to establish this fact. Clearly, both have a disk and are accreting from it. The two TWA objects, now confirmed as members, do not have any photometric measurement in the I filter. Nevertheless, they do not have K excess in a (J-H,H-K) diagram. The profile of Hα is wider in the case of the brown dwarfs with the large emission, indicating accretion. It also has HeI6678 Å in emission. In fact, the location of these objects in the diagram shown in Figure 3 satisfies the proposed CTT/WTT empirical criteria (i.e., all but TWA-2M1139511 are CTT or substellar analogs). Moreover, Mohanty, Jayawardhana and Barrado y Navascués (2003) have confirmed that TWA-2M1207334 belongs to the association and it is

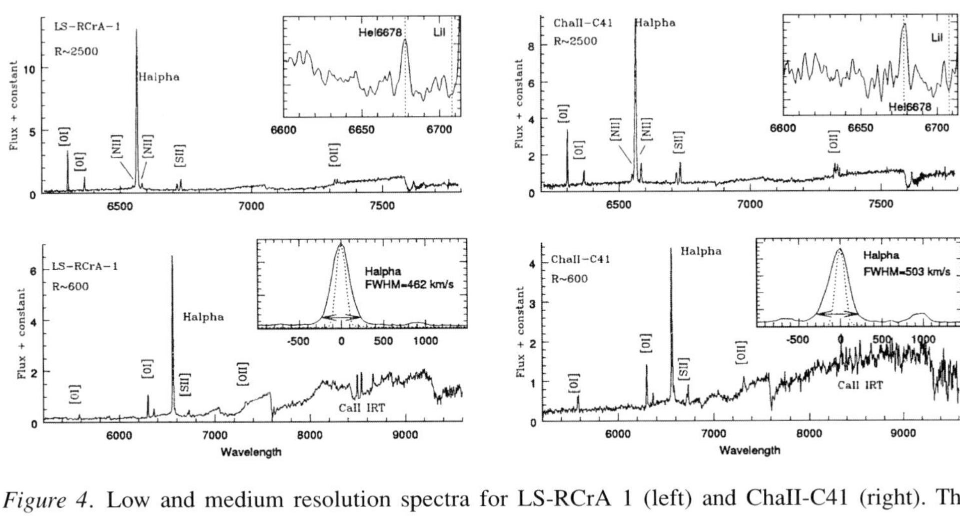

Figure 4. Low and medium resolution spectra for LS-RCrA 1 (left) and ChaII-C41 (right). The zooms show the spectral range around HeI6687/LiI6708 and Hα (in this case, the instrumental profile is included as a dotted line).

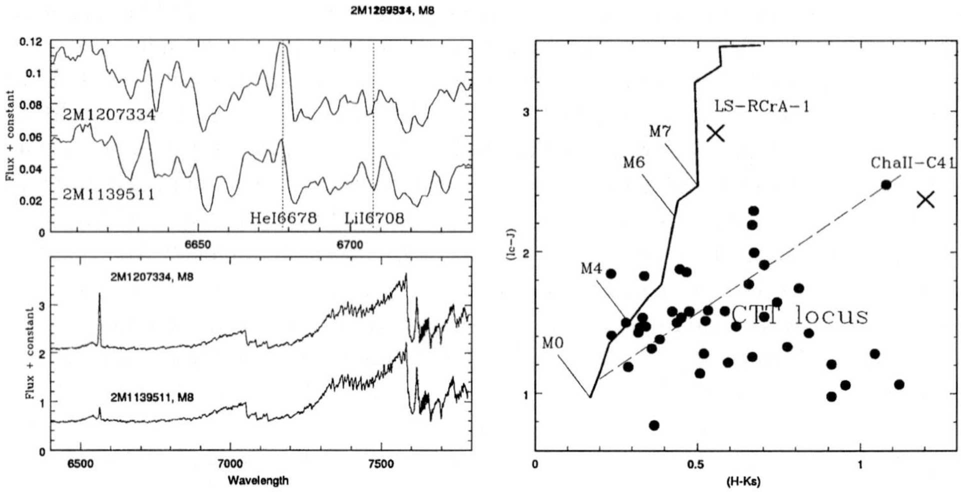

Figure 5. Medium resolution spectra for two proposed M8 members of TWA (left). Right.—Optical-infrared color-color diagram. The solid circles represent CTTS in Orion (Herbig and Bell, 1988).

accreting, based on high resolution echelle spectroscopy, the broad, asymmetric Hα profile and their accurate measurement of its radial velocity. Similar conclusion, membership to the R CrA association and bona fide accretion, has been achieved by Barrado y Navascués, Mohanty and Jayawardhana (2003) regarding LS-RCrA 1.

These four objects are overplotted in Figure 6 (left), where we have defined four areas, based on the saturation limit and the maximum chromospheric activity

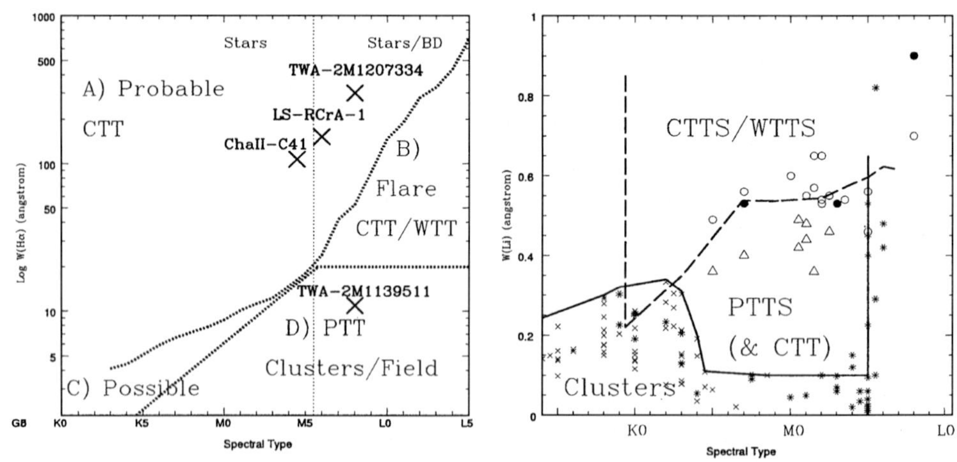

Figure 6. Left.—A schematic diagram with our empirical Hα criterion. The saturation limit and the maximum limit found in open clusters discriminate four regimes. Right.—Lithium equivalent widths versus the spectral type for TWA members (solid and open circles, and open triangles represent CTT, WTT and PTT). The figure also includes data for IC2391 and the Pleiades (asterisks and crosses).

as defined by young open clusters. Region A is populated by CTTSs and their substellar analogs. They have W(Hα) above the line defined by the saturation (i.e., the origin cannot be chromospheric). In region (B) classical, weak-line T Tauri and flaring objects can be found. Most of the objects in this area should be brown dwarfs. In particular, all WTTS and CTTS with spectral type cooler than M5.5-M6. This area is delimited by the saturation criterion (up) and the maximum chromospheric activity measured in upper clusters (down). In region C there are classical, weak-line and post-T Tauri stars. The limits are the same as for region B, but in this case, this region is located at the warmer side of the diagram. Region D is populated by cluster and field stars and brown dwarfs. This area also includes weak-line and post-T Tauri objects and is delimited by the upper envelope of the activity measured in young open clusters. The ambiguity in the B and C areas can be solved using the lithium content as measured from the 6708 Å doublet. Figure 6 (right) displays the lithium equivalent width versus the spectral type for TWA members ($\sim$10 Myr), which contains classical, weak-line and post T Tauri stars.

This is an example of how our empirical criteria, based on low resolution optical spectroscopy, can be used to classify CTTS, WTTS, SCTTA and SWTTA. We find that our criteria are generally consistent with other criteria based on high-resolution spectroscopy and/or near-infrared colors.

References

Alencar, S.H.P. and Basri, G.: 2000, *AJ* **119**, 1881.

Appenzeller, I. and Mundt, R.: 1989, *A&A Rev.* **1**, 291.

Baraffe, I., Chabrier, G. and Allard, F. et al., 1998, *A&A* **337**, 403.

Barrado y Navascués, D. and Martín, E.L.: 2003, *AJ* **126**, 2997.

Barrado y Navascués, D. and Stauffer, J.R.: 2003, in: E.L. Martín (eds.), *Brown Dwarfs, IAU Symposium 211*, p. 155.

Barrado y Navascués, D., Mohanty, S. and Jayawardhana, R.: 2004, *ApJ* **604**, 284.

Bertout, C.: 1989, *Ann. Rev. Astron. Astroph.* **27**, 351.

Briceño, C., Hartmann, L., Stauffer, J.R. and Martín, E.L.: 1998, *AJ* **115**, 2074 .

Fernández, M. and Comerón, F.: 2001, *A&A* **380**, 264.

Gershberg, R.E., Katsova, M.M., Lovkaya, M.N., Terebizh, A.V. and Shakhovskaya, N.I.: 1999, *A&A Suppl. Series* **139**, 555.

Gizis, J.E., Monet, D.G., Reid, I.N., Kirkpatrick, J.D., Liebert, J. and William, R.J.: 2000, *AJ* **120**, 1085.

Gizis, J.E.: 2002, *ApJ* **575**, 484.

Hamann, F. and Perssons, S.E.: 1992, *ApJ Suppl. Series* **82**, 247.

Herbig, G.H. and Bell, K.R.: 1988, *Lick Observatory Bulletin*, Lick Observatory.

Kenyon, S.J. and Hartmann, L.: 1995, *ApJ Suppl. Series* **101**, 117.

Kirkpatrick, D., Reid, I.N., Liebert, J., Gizis, J.E., Burgasser, A.J., Monet, D.G., Dahn, C.C., Nelson, B. and Williams, R.J.: 2000, *AJ* **120**, 447.

Leggett, S.K.: 1992, *ApJ Suppl. Series* **82**, 351.

Martín, E.L.: 1997, *A&A* **321**, 492.

Meyer, M.R., Calvet, N. and Hillenbrand, L.A.: 1997, *AJ* **114**, 288.

Mohanty, S., Basri, G. and Shu, F. et al.: 2002, *ApJ* **571**, 469.

Mohanty, S. and Basri, G.: 2003, *ApJ* **583**, 451.
Mohanty, S., Jayawardhana, R. and Barrado, y Navascués, D.: 2003, *ApJ Letters* **593**, 109.
Montes, D., López-Santiago, J., Fernández-Figueroa, M.J. and Gálvez, M.C.: 2001, *A&A* **379**, 976.
Muzerolle, J., Calvet, N., Bricño C., Hartmann L. and Hillenbrand, L.: 2000, *ApJ Letters* **535**, 47.
Muzerolle, J. et al.: 2003, *ApJ* **592**, 266.
Reid, I.N., kirkpatrick, J.D., Liebert, J., Gizis, J.E., Dahn, C.C. and Monet, D.G.: 2002, *AJ* **124**, 519.
Vuong, M.H., Cambrésy, L. and Epchtein, N.: 2001, *A&A* **379**, 208.
White, R.J. and Basri, G.: 2003, *ApJ* **582**, 2109.
Zapatero-Osorio, M.R., Béjar, V.J.S., Martín, E.L., Barrado y Navascués, D. and Rebolo, R.: 2002, *ApJ Letters* **569**, 99.

CLUES TO SUBSTELLAR FORMATION: ROTATION AND THE LOW-MASS END OF THE INITIAL MASS FUNCTION

MARÍA ROSA ZAPATERO OSORIO[1], JOSÉ CABALLERO[2], EDUARDO L. MARTÍN[3], VÍCTOR J.S. BÉJAR[2] and RAFAEL REBOLO[2]

[1]*LAEFF-INTA, Madrid, Spain; E-mail: mosorio@laeff.esa.es*
[2]*Instituto de Astrofísica de Canarias, La Laguna, Tenerife, Spain*
[3]*IfA, University of Hawaii, Honolulu, U.S.A.*

Abstract. We have monitored S Ori 45, a young, low-mass (20 $M_{\rm jup}$) brown dwarf of the σ Orionis cluster (~3 Myr, 352 pc), using optical and near-infrared filters. S Ori 45 (spectral type M8.5) is found to be multi-periodic with a dominant modulation at 2.5–3.5 h, and a short modulation at about 46 min. We ascribe the longer of these modulations to a rotation period. After comparing these results with observations of more massive cluster brown dwarfs and field brown dwarfs, we conclude that substellar objects present rotational and angular momentum evolution.

We have also obtained intermediate-resolution near-infrared spectroscopy of S Ori 70, which is a T-class, free-floating planetary candidate member in the σ Orionis cluster. Its observed spectrum has been compared to data of field brown dwarfs of similar types and to theoretical spectra computed for different surface temperatures and gravities. We conclude that S Ori 70 has a significantly cool, low-gravity atmosphere. This supports the young age of this object and its membership in the cluster. From state-of-the-art evolutionary models, the mass of S Ori 70 is estimated at 3 times the Jovian mass ($^{+5}_{-2}$ $M_{\rm jup}$), challenging current stellar/substellar formation models. S Ori 70 remains the lowest mass object so far identified in any open cluster.

Keywords: young star clusters, brown dwarfs, planetary-mass objects, angular momentum evolution

1. Introduction

Rotation and the initial mass function (IMF) are major key parameters to understand the mechanisms that lead to the formation of stars, brown dwarfs (BDs) and free-floating planetary-mass objects. Substellar objects (BDs and planets) are unable to attain high interior temperatures to fuse hydrogen stably because of their very small mass[1] ($M \leq 75\ M_{\rm jup}$ for solar metallicity). The deuterium-burning mass limit at 13 $M_{\rm jup}$ (Saumon et al., 1996) has recently been suggested to separate BDs (75–13 $M_{\rm jup}$) from planetary-mass objects or planets ($M \leq 13\ M_{\rm jup}$). We will use this criterion to refer to either BDs or free-floating planetary-mass objects, throughout this paper.

The presence of circumstellar disks from which young, central nascent objects accrete material, plays an important role in the early rotational evolution of low-mass objects. Accreting low-mass stars are usually slow rotators. This is commonly

[1] 1 $M_\odot \sim 1000\ M_{\rm jup}$.

Astrophysics and Space Science **292:** 673–679, 2004.

believed to result from the magnetic interaction between the rotating star and the accretion disk, known as the "disk-locking" mechanism for regulating the angular momentum evolution during the pre-main sequence phase (Herbst et al., 2000, and references therein). If the formation of stellar and substellar objects is alike (as the most recent observations suggest), BDs and stars should undergo similar rotational evolution. Here, we present our photometric monitoring of S Ori 45 (Béjar et al., 1999), a 20-$M_{\rm jup}$ BD of the young ($\sim$3 Myr) σ Orionis cluster. Further details are given in Zapatero Osorio et al. (2003).

The BD mass distribution of the σ Orionis cluster is a rising function (in linear mass units), which can be described by a power law equation with an idex $\alpha = 0.8 \pm 0.4$ (Béjar et al., 2001). These authors studied the cluster IMF in the BD domain, and found evidences for an extension toward the planetary regime (Zapatero Osorio et al., 2000). The determination of "the lower end" of the IMF would extraordinarily constrain stellar and substellar formation models. At the present time, the least massive σ Orionis candidate member is the mid-T class dwarf S Ori 70 (Zapatero Osorio et al., 2002a). If its youth is confirmed, we would be able to estimate its planetary mass. To achieve this goal, we have obtained intermediate-resolution near-infrared spectroscopy. Further details are given in Martín and Zapatero Osorio (2003).

2. Variability of the Young, Low-Mass Brown Dwaf S Ori 45

S Ori 45 is an M8.5-type dwarf that shows strong Hα emission over a period of three years of spectroscopic observations (Béjar et al., 1999), (Zapatero Osorio et al., 2002b), (Barrado y Navascués et al., 2003). Béjar et al. (1999) published the photometry and optical spectroscopy of this object, concluding that it is a bona-fide BD member of the star cluster σ Orionis. Thus, its age and mass are estimated at around 3^{+5}_{-2} Myr and 20^{+20}_{-5} $M_{\rm Jup}$, respectively. Despite its strong Hα emission, S Ori 45 does not exhibit any near-infrared excess at 2.2 μm.

We monitored S Ori 45 in the I-band and J-band filters using various telescopes at the Calar Alto Observatory (CA), the Teide Observatory (TCS), and the Roque de los Muchachos Observatory (INT and NOT). Five different observing campaings took place between December 1998 and October 2002. The log of the observations (UT dates, filters, individual exposure times, and total time coverage per night) is provided in Table I. Seeing measurements ranged from 0.9$''$ to 2.3$''$, with the mean value at 1.4$''$ in the I-band and 1.9$''$ in the J-band. Images were collected at airmasses between 1.2 and 2.3. Raw data were reduced using standard techniques of optical and near-infrared wavelengths.

To obtain the light curves of S Ori 45, we compared the target apparent brightness with the brightness of two surrounding comparison stars. Differential photometry cancels the effects of apparent brightness changes due to variable extinction, instrument performance or exposure time. Aperture photometry was performed using the

TABLE I
Log of observations of S Ori 45

Obs. Date (starting UT)	Telescope	Fil.	Exposure (s)	Coverage time (h)	No. of data points
1998 Dec 5	2.2 m CA	J	$9 \times 5 \times 20$	5.2	12
1998 Dec 6	2.2 m CA	J	$9 \times 5 \times 20$	6.8	19
1998 Dec 7	2.2 m CA	J	$9 \times 5 \times 20$	6.7	13
2000 Dec 30	2.5 m INT	I	1500	1.9	5
2000 Dec 31	2.5 m INT	I	1500	4.7	10
2001 Jan 1	2.5 m INT	I	1500	3.3	6
2001 Oct 8	1.5 m TCS	J	$10 \times 6 \times 20$	3.7	9
2001 Dec 8	1.5 m TCS	J	$10 \times 6 \times 10$	3.1	15
2002 Oct 25	2.5 m NOT	I	300	1.7	17
2002 Oct 26	2.5 m NOT	I	300	1.7	17

IRAF script VAPHOT (Deeg et al., 1998). Changes in the seeing conditions and in the telescope focus throughout a night implied variations in the size of the point-spread function (PSF) of the sources from frame to frame. To take this into account, we derived an average PSF of the comparison stars by fitting a circular Gaussian on each frame. Apertures were chosen as the FWHM of the average PSF. Relative magnitudes of S Ori 45 with respect to the sum of the flux of the reference stars were derived. Hence, an increase in the differential magnitude represents a dimming of the target. We checked that our differential photometry of S Ori 45 does not correlate with either seeing or airmass.

S Ori 45 does not display large photometric variability in any of the filters, being the peak-to-peak variation of the order of 0.2–0.3 mag. We searched for periodic signals by calculating the power spectrum (or periodogram) of the 1998, 2000 and 2001 data separately. We used the Lomb–Scargle method (Lomb, 1976; Scargle, 1982). A cluster of peaks were found between 2.09 and 3.6 h with low false-alarm probabilities, which may indicate the presence of a significant periodic signal. We also attempted to remove the spectral window function from the data using the CLEAN algorithm (Roberts et al., 1987). In principle, window functions can create an ambiguity in the determination of the real period. We found that S Ori 45 is a multi-periodic object with a dominant period at 2.5–3.5 h as derived from the light curves obtained between 1998 and 2001. There is a shorter probable period of 46.4 ± 1.5 min as derived from the most recent set of observations taken in 2002. Some light curves folded on their respective periods are illustrated in Figure 1. The amplitude of the light curves is similar in both filters, ranging from 34 up to 81 mmag.

The origin of the 46.4 min period is unknown to us. It might be related to the presence of a surrounding accretion disk. S Ori 45 shows a persistent, strong Hα

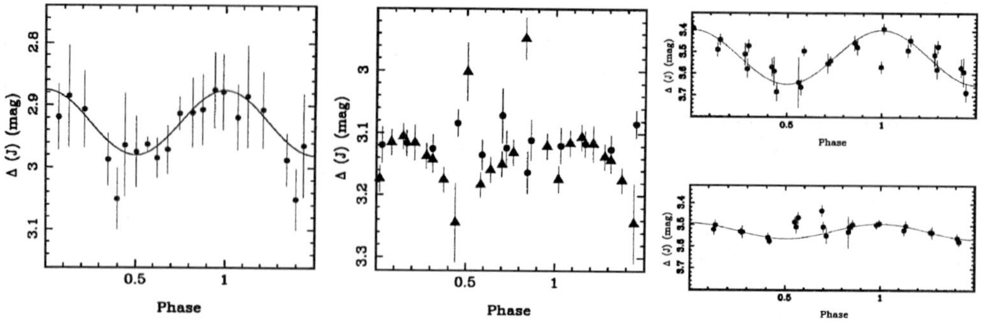

Figure 1. Phased light curves of S Ori 45. From left to right, periods are: 2.6 h, 3.6 h, and 46.4 min. In the middle panel, different symbols stand for different observing nights. To the right, the upper and bottom panels display the phased light curves of two consecutive nights; note the different amplitudes of the photometric modulation.

emission with pseudo-equivalent widths between 20 and 60 Å. On the other hand, about 6–20% of the σ Orionis members display infrared excesses, which suggests the presence of dusty disks (Oliveira et al., 2002), (Barrado y Navascués et al., 2003).

We have investigated different possible mechanisms (rotation, binarity, and presence of circumstellar material) to explain the 2.5–3.5 h period, and concluded that it may likely be related to a rotation period. The younger brown dwarfs of Chamaeleon I rotate significantly slower (Joergens et al., 2003). Thus, the rapid rotation of S Ori 45 can be understood as spin-up due to gravitational contraction following disk unlocking. The observed photometric modulation could be due to surface magnetic dark spots and/or dust clouds that typically form in very cool atmospheres. This is supported by the long-term change of the light curves. Nevertheless, our data do not explicitily exclude other possible scenarios.

The short rotational period of S Ori 45 (1–8 Myr) suggests that young BDs are rapid rotators, as are older field BDs of similar and cooler spectral types (Clarke et al., 2002; Bailer-Jones and Mundt 1999, 2001). However, the radius of S Ori 45 is about three times larger according to state-of-the-art evolutionary models (Chabrier et al., 2000; Burrows et al., 1997). There seems to be some rotational and angular momentum evolution in BDs.

3. Low Gravity of the Planetary Cluster Member S Ori 70

S Ori 70 is the coolest (T6-class) and faintest object so far detected in any young cluster or association (Zapatero Osorio et al., 2002a). The determination of its true nature is relevant to constrain the substellar IMF of the σ Orionis cluster. We observed S Ori 70 at J wavelengths with the Keck II near-infrared spectrometer NIRSPEC on 2002 November 24. The observing technique and data reduction were standard in the near-infrared wavelengths. For comparison purposes, we used the

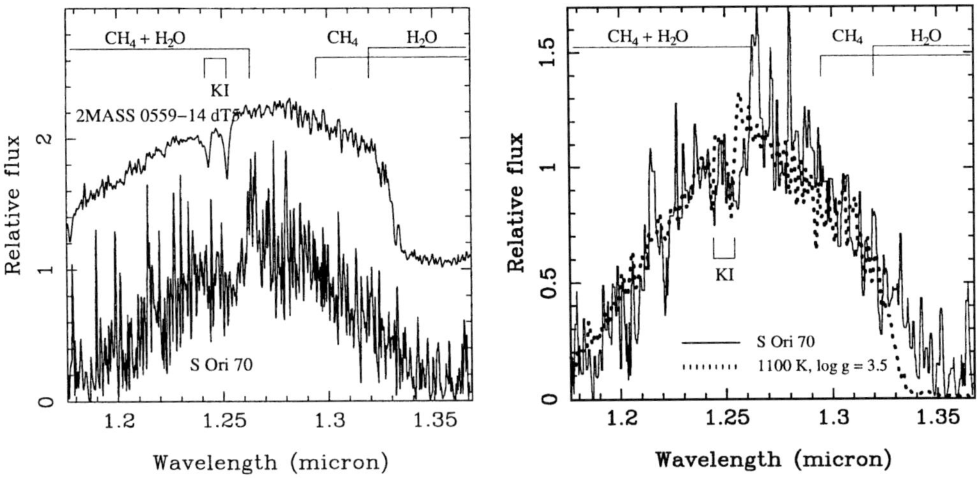

Figure 2. (Left panel) NIRSPEC J-band spectra of S Ori 70 (~T6) and the T5 field brown dwarf 2MASS J0559191–140449. Data are normalized to unity at 1.234–1.24 μm, and have been shifted for clarity. Note the strong water vapor and methane absoptions present in the spectrum of S Ori 70. Major molecular and atomic features are indicated. (Right panel) The spectrum of S Ori 70 (which has been median filtered) is compared to the COND synthetic spectrum with T_{eff}=1100 K and log g=3.5 cm s^{-2}.

spectrum of the relatively bright T5 field dwarf 2MASS J0559191–140449 kindly provided by Adam Burgasser (Burgasser et al., 2002).

The resulting spectrum (resolving power of $R = 1500$) of S Ori 70 and the similar-resolution spectrum of 2MASS J0559191–140449 are plotted in Figure 2 (left panel). S Ori 70 displays stronger molecular absorptions, which may indicate different physical conditions in the photospere. Lucas et al. (2001) showed that triangular shapes in near-infrared spectra of L dwarfs could be explained by low gravity atmospheres. We will show that the triangular shape observed in the J-band spectrum of S Ori 70 is caused by a surface gravity lower than that of the 2MASS dwarf, consistent with having a larger radius as expected for a younger object.

Our spectroscopic data have been compared with a grid of theoretical atmospheric spectra based on the COND models by Allard et al. (2001), which assume that dust grains lie below the photosphere. These models include the latest AMES list of H_2O lines (Partridge and Schwenke, 1997). Low-gravity ultra-cool atmospheres show enhanced molecular absorptions in the near-infrared, particularly water vapor, which provides low-gravity spectra with triangular shapes. The strength of the K I doublet at 1.25 μm is also greater for lower gravities. Using a least square minimization technique, we have obtained that the best fit to our data of S Ori 70 is given by the model with $T_{eff} = 1100$ K and log $g = 3.5$ cm s^{-2}. This is shown in Figure 2 (right panel). The uncertainties of our fitting technique is $^{+200}_{-100}$ K and ± 0.5 cm s^{-2}. These results are consistent within the error bars, with those estimated in Zapatero Osorio et al. (2002a).

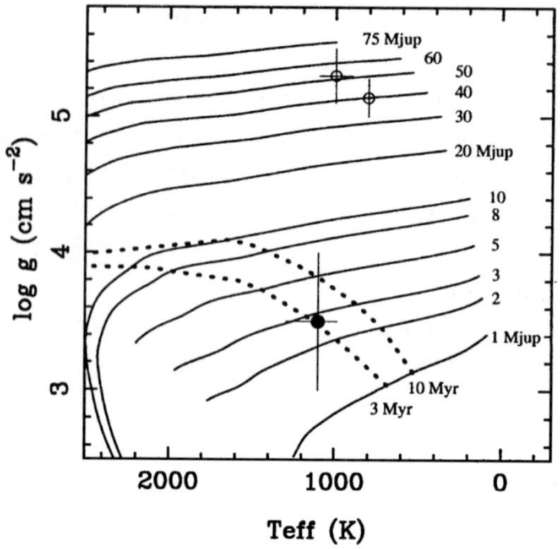

Figure 3. The surface gravity and temperature of S Ori 70 are rather low. From the comparison with evolutionary models by Burrows et al. (1997), we infer that S Ori 70 is younger than 10 Myr, with a likely age of 3 Myr, and has a planetary mass of 3^{+5}_{-2} times the Jovian mass. Open circles correspond to the data of two T-type field brown dwarfs.

The spectroscopically derived gravity and surface temperature of S Ori 70 allow us to place this object in an evolutionary diagram like the one shown in Figure 3. This diagram is independent of the distance to S Ori 70. This object lies on the 3-Myr isochrone for a mass of three times the mass of Jupiter, according to the tracks by Burrows et al. (1997). We obtain a similar result by using the evolutionary models of Chabrier et al. (2000). The low gravity of S Ori 70 implies a very young age that is fully consistent with the age of the σ Orionis cluster (1–8 Myr).

S Ori 70 does not appear to be bound to any other member of the cluster, either stellar or substellar. The nearest identified σ Orionis star is located at a projected separation of ~62500 AU for a distance of 352 pc, and the closest cluster BD is at a projected separation of 26700 AU. Béjar (private communication) has estimated that the typical projected separation among cluster members is of the order of 50000 AU. S Ori 70 lies at a projected distance of 182700 AU from the multiple star system σ Orionis (HR 1931, total mass of 60 $M_\odot$). S Ori 70 is very likely bound to the cluster, but is not particularly bound to any cluster star or BD.

The presence of objects like S Ori 70, which are free-floating in young star clusters, challenges any stellar/substellar formation model (e.g., see Bate et al., these proceedings). Because of the very young age of S Ori 70, it seems that isolated planetary-mass objects form rapidly, in time scales of a few Myr. The finding of this object in a rather small survey (Zapatero Osorio et al., 2002a) could indicate that the population of planetary-mass objects in the σ Orionis cluster is numerous (see Béjar et al., these proceedings). Actually, its discovery is consistent with a smooth extrapolation of the cluster BD IMF (Béjar et al., 2001) into the planet domain.

Acknowledgements

We thank France Allard, Adam Burrows and Adam Burgasser for sending us their data in electronic version. We also thank D. Barrado y Navascués, A. Burgasser, B. Montesinos and E. Solano for useful discussions. The data presented here have been collected with the following telescopes: the KeckII telescope of the Keck Observatory (Hawaii), the Isaac Newton Telescope (INT) and the Nordic Optical telescope (NOT) of the Roque de los Muchachos Observatory (La Palma), the Carlos Sánchez telescope (TCS) of the Teide Observatory (Tenerife), and the 2.2 m telescope of the Calar Alto Observatory (Almería). This work has been financed by the Spanish "Programa Ramón y Cajal" and the National Science Foundation grant no. 1339-CO2.

References

Allard, F. et al.: 2001, *ApJ* **556**, 357.
Bailer-Jones, C.A.L. and Mundt, R.: 1999, *A&A* **348**, 800.
Bailer-Jones, C.A.L. and Mundt, R.: 2001, *A&A* **367**, 218.
Barrado y Navascués, D. et al.: 2003, *A&A* **404**, 171.
Béjar, V.J.S. et al.: 1999, *ApJ* **521**, 671.
Béjar, V.J.S. et al.: 2001, *ApJ* **556**, 830.
Burgasser, A. et al.: 2002, *ApJ* **564**, 421.
Burrows, A. et al.: 1997, *ApJ* **491**, 856.
Chabrier, G. et al.: 2000, *ApJ* **542**, 464.
Clarke, F.J., Oppenheimer, B. and Tinney, C.: 2002, *MNRAS* **332**, 361.
Deeg, H.J. et al.: 1998, *A&A* **338**, 479.
Herbst, W. et al.: 2000, *AJ* **119**, 261.
Joergens, V. et al.: 2003, *ApJ* **594**, 971.
Lomb, N.R.: 1976, *Ap&SS* **39**, 447.
Lucas, P. et al.: 2001, *MNRAS* **326**, 695.
Martín, E.L. and Zapatero Osorio, M.R.: 2003, *ApJ* **593**, L113.
Oliveira, J.M. et al.: 2002, *A&A* **382**, L22.
Partridge, P. and Schwenke.: 1997, *J. Chem. Phys.* **106**, 4618.
Saumon, D. et al.: 1996, *ApJ* **460**, 993.
Roberts, D.H., Lehár, J. and Dreher, J.W.: 1987, *AJ* **93**, 968.
Saumon, D. et al.: 1996, *ApJ* **460**, 993.
Scargle, J.D.: 1982, *ApJ* **263**, 835.
Zapatero Osorio, M.R. et al.: 2000, *Science* **290**, 103.
Zapatero Osorio, M.R. et al.: 2002a, *ApJ* **578**, 536.
Zapatero Osorio, M.R. et al.: 2002b, *A&A* **384**, 937.
Zapatero Osorio, M.R. et al.: 2003, *A&A* **408**, 663.

ON THE NATURE OF TRANSPORT IN FUSION PLASMAS

C. HIDALGO

Asociación EURATOM-CIEMAT, Avda. Complutense 22, 28040 Madrid, Spain;
E-mail: carlos.hidalgo@ciemat.es

Abstract. A view of the mechanisms underlying particle and energy transport in fusion plasmas devices (tokamak and stellarators) is given. The heat and energy transport in fusion plasmas is generally due to turbulent processes. Several driving (plasma gradients) and damping (electric field) mechanisms of turbulence have been investigated. The development of methods to modify and control plasma turbulence by electric field effects has opened a new era on fusion confinement research. Recently, a new approach based on multi-field probability density function statistical analysis has been proposed to describe transport in non-equilibrium systems and to unravel the overall picture connecting transport, gradients and flows in fusion plasmas.

Keywords: plasmas, fusion, turbulence

1. Introduction

Nuclear fusion energy is released when the nuclei of light elements join together to form heavier ones. The easiest reaction to achieve is between the two isotopes of hydrogen (deuterium and tritium) (Wesson, 1997; Fusion Special Issue, 1998):

$$\mathrm{D}^2 + \mathrm{T}^3 \longrightarrow \mathrm{He}^4\ (3.5\ \mathrm{MeV}) + n^1(14.1\ \mathrm{MeV})$$

Most of the energy released in the fusion reaction is carried out by high speed neutrons. The remaining energy goes to He^4 nuclei (alpha particles). In a fusion reactor, a jacket around the reactor region would slow down the neutrons, converting their energy into heat.

As deuterium is a common component of water, there is virtually an inexhaustible supply in the oceans. On the contrary, tritium does not occur naturally in any significant quantities and must be manufactured. This can be achieved by using reactions that occur between neutrons formed in the fusion reactions and the light metal lithium:

$$\mathrm{Li}^6 + n^1 \longrightarrow \mathrm{T}^3 + \mathrm{He}^4.$$

It is possible to use neutrons released in the deuterium-tritium (D-T) fusion reaction to breed tritium from lithium, of which there are sufficient reserves to enable world electricity generation using fusion reactors for several hundreds of years.

Fusion reactions can only take place if the nuclei are brought close to one another. By heating the D-T fuel to very high temperatures, sufficient energy can be given to the nuclei that the repulsive (coulomb) force can be overcome and they fuse together. The fusion brought in this way is called thermonuclear fusion. In the D-T reaction temperatures of 100 million degrees (i.e. 10 keV) are required. At such temperatures the fuel is fully ionized and its state is given the name of plasma. In a fusion reactor it would be necessary to confine the energy of a sufficiently dense plasmas for a (confinement) time which allows an adequate fraction of the fuel to react. In a fusion reactor the values of ion temperature (T_i), density (n_i) and energy confinement time must be such that their product is higher than 5×10^{21} s keV m^{-3}. For density around 10^{20} m^{-3} and ion temperatures in the range 10–20 keV, the confinement time is of the order of 1 sec.

Because a plasma is a mixture of charged (positive and negative) particles it is influenced by magnetic fields. With a suitable magnetic field configuration ("magnetic bottle"), it should be possible to confine the plasma in the required density and temperature range and a sufficiently long time (confinement time) to obtain a net energy gain. So far, the magnetic configuration that has advanced furthest to achieve fusion reactor conditions is the tokamak (Wesson, 1997). Besides the tokamak, the most mature line is the stellarator which offers attractive properties like inherent steady state operation. In the stellarator concept, the magnetic field configuration is fully produced by a set of coils (toroidal, poloidal, helical). Figure 1 shows a view of the TJ-II stellarator in operation at the CIEMAT site in Madrid (Alejaldre, 1990). The TJ-II stellarator is the cornerstone of the Spanish magnetic confinement fusion program. The goals of the TJ-II physics programe will be focused on three main areas of investigation: (a) confinement studies in low collisionality plasmas, (b) operational limits at high beta (ratio of plasma pressure to magnetic field pressure) regimes and (c) plasma confinement optimization (role of electric fields).

2. Turbulent Transport in Fusion Plasmas

A confined plasma is necessarily out of equilibrium. The most obvious relaxation mechanism is by coulomb collisions; in toroidal devices (such as tokamaks and stellarators) the resulting particle and energy flows are termed neoclassical transport. However, the transport which actually occurs in tokamaks and stellarators does not agree with the calculated (neoclassical) values. In particular the energy transport by electrons can be up to two orders of magnitude higher than the one predicted by neoclassical theory; transport is anomalous.

It is now fully accepted that anomalous transport is due to plasma turbulence (Connor and Wilson, 1994; Carreras, 1997). Understanding the underlying physics of anomalous transport in fusion devices remains as one of the key issues in magnetic fusion research.

Broadband electrostatic and magnetic fluctuations have been observed in the core and edge regions of magnetically confined plasmas (Wootton et al., 1990; Hidalgo,

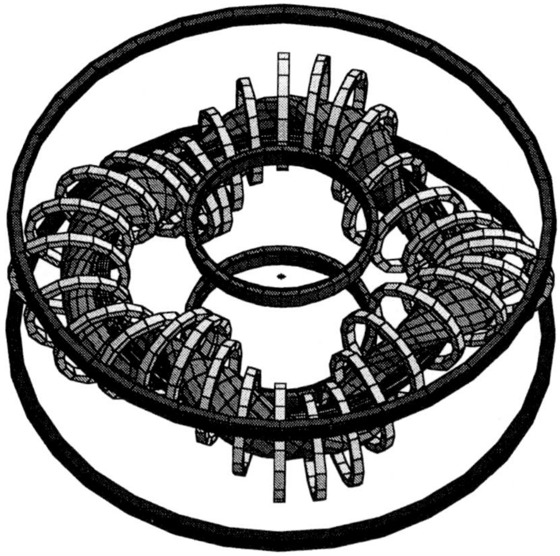

Figure 1. View of the TJ-II stellarator. The major radius is 1.5 m and the magnetic field is 1 Tesla. The magnetic configuration is generated by a system of toroidal, central, poloidal and vertical coils.

1995). Edge fluctuations show well developed turbulence features with fluctuations dominated by frequencies below 500 kHz. Density fluctuation levels are in the range $\delta n/n \approx (0.1 - 0.3)$ and edge magnetic fluctuation levels of $\delta B_r/B \approx 10^{-4}$ to 10^{-5} have been found in tokamaks.

Characterization of fluctuation driven particle and energy fluxes requires experimental techniques for measuring the variations in parameters such as density, temperature and magnetic and electric fields with good temporal and spatial resolution. With the present state of the art in plasma diagnostics this kind of measurement is mostly limited to the plasma edge where material probes can be used, giving a good opportunity to identify the driving forces of the turbulence. Electrostatic fluctuation induced particle transport is large enough to account for the total particle transport in the plasma boundary region. The measurements of the fluctuation induced flux at the plasma edge indicate that the particle loss is intermittent and the flux is bursty as shown in Figure 2, that is to say, a significant fraction can be attributed to the presence of large and sporadic transport events. As a consequence, in order to go deeper into the mechanisms underlying turbulent transport it is important not only to compute the average value of the fluctuation induced transport, but also the statistical properties of the time resolved turbulent flux. There is a strong similarity between the statistical properties of the turbulent fluxes in different devices (tokamaks and stellarator), supporting the hypothesis of the universal character of plasma turbulence in magnetic confinement devices.

Although our understanding of the plasma turbulence has improved during the last few years, the dominant free energy source responsible for the turbulence in the fusion plasmas has not been fully identified yet.

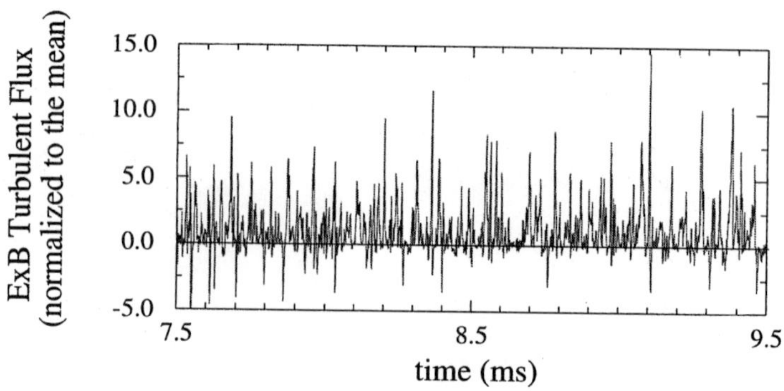

Figure 2. Time evolution of electrostatic ExB fluctuation induced particle transport in the plasma boundary region of fusion devices.

In the plasma core region, instabilities driven by the ion temperature gradient poses a significant threat to confinement in high temperature fusion plasmas. This is because the heat that is generated from the charged particles produced by fusion reactions (alpha particles) will produce very substantial gradients in the ion temperature, whereas the density gradient is expected to be small in the main part of the plasma.

A critical parameter determining reactor efficiency is the maximum attainable ratio of plasma pressure to magnetic field pressure (β); the thermonuclear power (P_{th}) is given by, $P_{th} \approx \beta^2 B^4$, where B is the toroidal magnetic field. Pressure gradient instabilities ("interchange instability") seem to play an important role on maximum β value which can be obtained. Figure 3 shows the way in which this instability can be developed in the plasma edge region. We assume sharp boundary between the plasma and the containing magnetic field which upwards to the page and decreasing as we go away from the plasma (Figure 3a). In Figure 3b has developed a slight pump in which the ions and electrons can drift in opposite directions under the influence of the inhomogeneous magnetic field, resulting an electric field (Figure 3c). Under the combined influence of this field and the magnetic field the slight bump develops into an increasingly large one, which is a typically unstable situation.

More recently, a new approach to plasma confinement and turbulence based on the idea of self organized criticality (SOC) has been suggested (Diamond and Hamh, 1995). From the experimental point of view the question to answer is the following: Is there any evidence of large-scale transport events in fusion plasmas? In the framework of SOC models these large-scale transport events could provide a mechanism to connect remote regions of the plasma. Because self-organized criticality can explain some of the properties of transport in magnetically confined plasmas (such as the bursty behavior of transport), techniques for quantifying probability density functions of fluctuations and transport are needed (Hidalgo et al., 2002; Carreras et al., 1999).

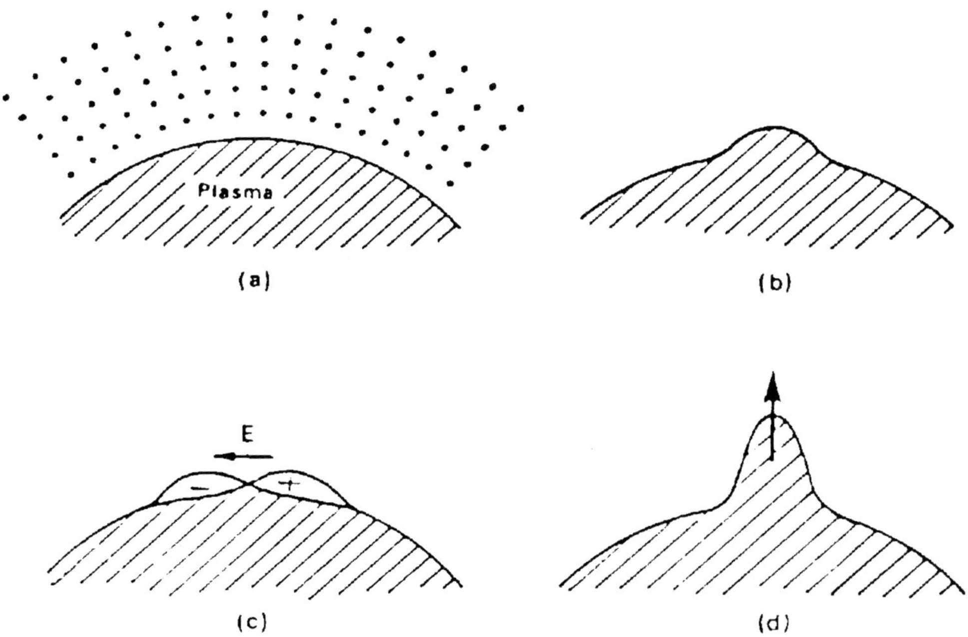

Figure 3. Development of the interchange instability in a toroidally confined plasma. For simplicity we asumme a sharp boundary between the plasma and the magnetic field which we take as upwards through the page and decreasing as we go away from the plasma.

3. Transport and Radial Electric Fields

The radial electric field plays a key role to explain transport in fusion plasmas (Burrell, 1997; Carreras, 1997; Terry, 2000). Sheared flows can influence the turbulence by shear decorrelation mechanisms and, as a consequence, modify the turbulent induced transport. A strong radial gradient in the radial electric field (E_r) produces strong shear in the particle ($E_r \times B$) drift velocity (B being the toroidal magnetic field). This tears apart the coherent cellular pattern of the unstable modes, as shown in Figure 4. This effect can be expressed by a decorrelation time which is proportional to the inverse of the radial gradient of the radial electric field, $\tau_c \approx (dE_r/dr)^{-1}$.

On the other hand, the decorrelation time due to the background diffusion can be expressed by, $\tau_b \approx l^2/D$ where l is the radial scale length of fluctuations and D is the diffusion coefficient.

When τ_c is smaller than τ_b then the sheared poloidal flow reduces the radial scale (correlation) length of fluctuations, and as a consequence the level of anomalous transport.

Toroidal particle simulation including the influence of poloidal shear flow on plasma turbulence are shown in Figure 5. In Figure 5 (a) no external poloidal flow is applied and radially coherent structures connecting the core and edge

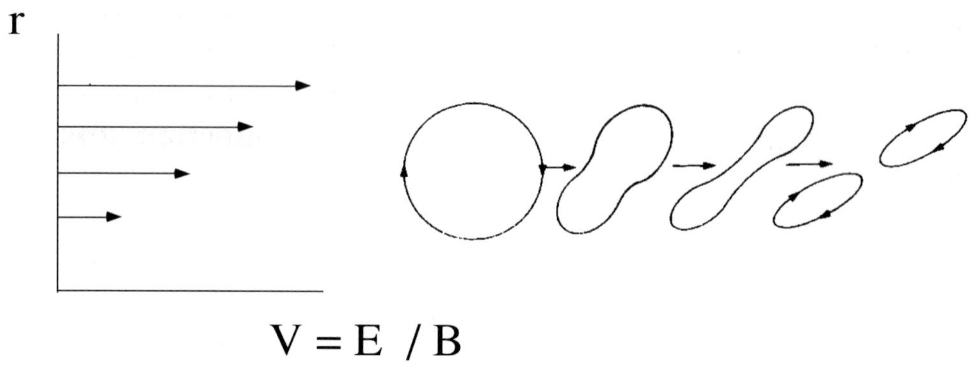

Figure 4. Effect of $E_r \times B$ sheared flow on a turbulent convective cell.

Figure 5. Nonlinear simulation of plasma turbulence (Horton et al., 1996).

plasma regions are evident. In Figure 5 (b) the external sheared poloidal flow reduces the radial correlation length and the level of fluctuations (Horton et al., 1996).

Understanding the mechanisms which drive ExB sheared flows is a key issue to understand the transition to improved confinement regimes. There are different sources of the radial electric fields in plasmas (Terry, 2000). From the radial force balance it follows that the radial electric field can be generated via poloidal and toroidal flows and by pressure gradients. Concerning the physics of poloidal flows several driving mechanisms have been proposed: ion temperature gradients, ion orbit losses (Shaing et al., 1989) and turbulence via Reynolds stress (Kim, 1991). Understanding the physics underlying the generation of flows is a key issue in the plasma physics fusion community.

Figure 6 shows how instabilities driven by different free energy sources (gradients) can produce turbulence (by non-linear saturation mechanisms). This

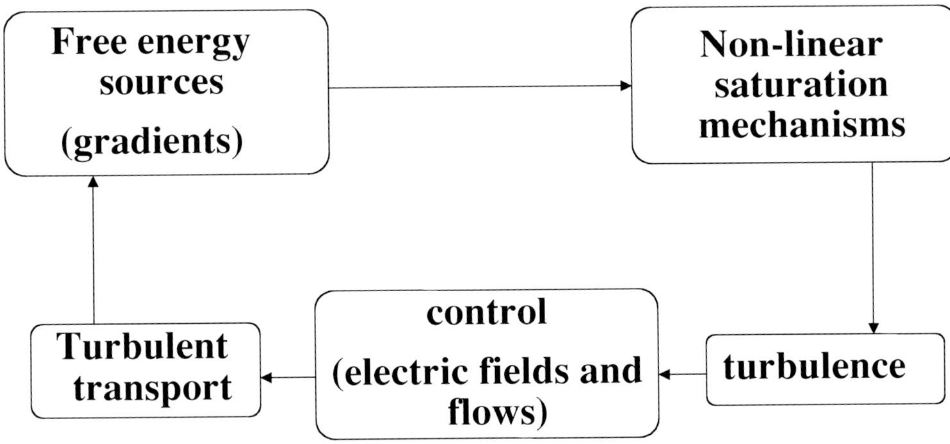

Figure 6. The connection between the plasma free energy sources, turbulence, sheared flows and turbulent transport processes.

turbulence can be controlled by sheared poloidal flows (electric fields) and the resulting transport can partly determine the gradients (free energy source) themselves.

The stabilizing effects of sheared flows on plasma turbulence has opened a new era on nuclear fusion research. The best fusion performance in the existing fusion devices has been recently obtained in plasma conditions where turbulent transport reduction by shear decorrelation is taking place. For the first time, it has been possible to reduce the rate of ion thermal transport to the minimum level set by particle collisions (neoclassical transport) over the whole plasma (Lazarus et al., 1996). These results represents a revolutionary step forward in the control of plasma turbulence and transport (Terry, 2000).

4. Statistical Description of Transport

Several distinct mechanisms have been invoked to explain the connection between turbulent fluxes and gradients, that provide a free energy source to drive instabilities in magnetically confined plasmas. Non-linear relations between heat fluxes and gradients, in which heat fluxes increases non-linearly as gradient becomes steeper, can explain the confinement degradation with the heating power reported in magnetically confined fusion plasmas. Instabilities governed by a threshold may lead to a self-organized critical (SOC) system by producing transport events at all scales, called avalanches. In the context of these models, the functional dependence between heat and particle transport is expected to show sharp jumps as the system crosses instability thresholds. The transition to improved regimes implies non-monotonic relations (bifurcations) between gradients and transport (Itoh et al., 1999).

It has been recently argued that in order to go deeper into the underlying physics of those mechanisms, a multi-field statistical approach is needed to characterize transport (Hidalgo et al., 2003; Gonzalves et al., 2002). This new way of thinking emphasizes the importance of the statistical description of transport processes in fusion plasmas as an alternative approach to the traditional way to characterize transport based on the computation of effective transport coefficients. Recent improvement in edge plasma diagnostics has allowed to compute the joint probability function of transport and gradient fluctuations.

Figure 7 shows the probability density function (PDF) for fluctuations in density gradients, and the expected value of the ExB turbulent flux for a given density

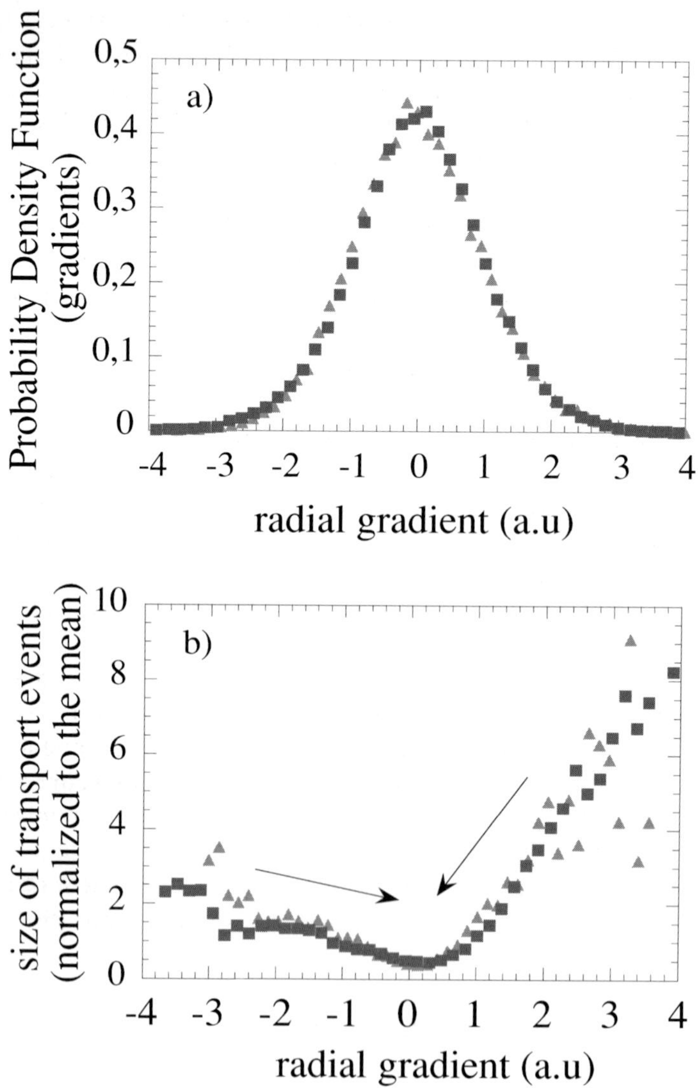

Figure 7. Probability Density Function of fluctuations in gradients (a) and dynamical coupling between transport events and fluctuations in gradients (b).

gradient in the plasmas boundary region of fusion plasmas. The results show that most of the time the plasma is at its average gradient and the size of transport events has minimum amplitude. Large amplitude transport events take place when the plasma displaces from the most probable gradient average value. Experimental results show that the bursty and strongly non-gaussian behavior of turbulent transport is strongly coupled with fluctuations in gradients. As the density gradient increases above the most probable gradient the ExB turbulent driven transport increases and the system perform a relaxation which tends to drive the plasma back to the marginal stable situation which minimized the size of transport events.

Recent experiments have shown that the dynamical relation between fluctuations in gradients and transport is strongly affected by the presence of sheared flows. Furthermore, the dynamical coupling between turbulent transport and parallel flows has been investigated in the plasma boundary region of fusion plasmas, showing that there is a dynamical relationship between transport and parallel flows. As the size of transport events increases parallel flows also increase. These results show that turbulent transport can drive parallel flows in the plasma boundary of fusion plasmas. This new type of measurement is an important element to unravel the overall picture connecting radial transport and flows in fusion plasmas.

5. Conclusions

Understanding the underlying physics of transport in fusion plasmas remains as one of the key issues currently confronting nuclear fusion research. It is now accepted that particle and energy losses are predominately due to plasma turbulence. However, in spite of the recent progress in this area, the dominant free energy sources of plasma turbulence have not been fully identified. Sheared flows can explain the turbulence reduction and confinement improvement observed in toroidal magnetically confined fusion plasmas.

A new approach, based on multi-field statistical approach to describe transport in non-equilibrium systems, has been proposed to unravel the overall picture connecting transport, gradients and flows in fusion plasmas. It emphasizes the importance of the statistical description of transport processes in fusion plasmas as an alternative approach to the traditional way to characterize transport based on the computation of effective transport coefficients.

With an eye on the future, the understanding and development of methods for controlling plasma turbulence have opened a new path in plasma physics research and contributed to design a nuclear fusion plant.

References

Alejaldre, C. et al.: 1990, *Fusion Technol.* **17**, 131.
Burrell, K.H.: 1997, *Phys. Plasmas* **4**, 1499.
Carreras, B.A.: 1997, *IEEE Trans. Plasma Sci.* **25**, 1281.

Carreras, B.A. et al.: 1999, *Phys. Rev. Lett.* **83**, 3653.
Connor, J.W. and Wilson, H.R.: 1994, *Plasma Phys. Control. Fusion* **36**, 719.
Diamond, P.H. and Hamh: 1995, *Phys. Plasmas* **2**, 3640.
Fusion Special Issue: 1998, *Europhysics News* **29**, 6.
Gonzalves, B. et al.: 2002, *Nucl. Fusion* **42**, 1205.
Hidalgo, C.: 1995, *Plasma Phys. Control. Fusion* **37**, A53.
Hidalgo, C. et al.: 2002, *Plasma Phys. Control. Fusion* **44**, 1557.
Hidalgo, C. et al.: 2003, *J. Nucl. Materials* **313**, 863. *Phys. Rev. Lett.* **91**, 065001.
Horton, W. and Tajima, T. et al.: 1996, *Comments on Plasma Physics and Controlled Fusion* **17**, 205.
Itoh, K. et al.: 1999, *Tokamaks and Structural Plasmas*, IOP Publishing.
Kim, Y.B. and Diamond, P.H.: 1991, *Phys. Fluids B* **3**, 1626.
Lazarus, E.A. et al.: 1996, *Phys. Rev. Lett.* **77**, 2714.
Shaing, K.C. and Grume, E.C.: 1989, *Phys. Rev. Lett.* **63**, 2369.
Terry, P.W.: 2000, *Rev. Mod. Phys.* **72**, 109.
Wesson, J.: 1997, *Tokamaks*, Oxford University Press, Oxford.
Wootton, A.J. et al.: 1990, *Phys. Fluids B* **12**, 2879.